Studies in Economic Theory

Editors

Charalambos D. Aliprantis
Department of Mathematical Sciences
IUPUI
402 N. Blackford Street
Indianapolis, IN 46202-3216
USA

Nicholas C. Yannelis
Department of Economics
University of Illinois
Champaign, IL 61820
USA

Springer-Verlag Berlin Heidelberg GmbH

Titles in the Series

M. Ali Khan and Nicholas C. Yannelis (Eds.)
Equilibrium Theory in Infinite Dimensional Spaces

Charalambos D. Aliprantis, Kim C. Border and
Wilhelmus A. J. Luxemburg (Eds.)
Positive Operators, Riesz Spaces, and Economics

Donald G. Saari
Geometry of Voting

Charalambos D. Aliprantis and Kim C. Border
Infinite Dimensional Analysis

Jean-Pierre Aubin

Dynamic Economic Theory

A Viability Approach

With 30 Figures

Springer

Jean-Pierre Aubin
Université de Paris-Dauphine
Viabilité, Jeux, Contrôle
F-75775 Paris cx (16), France

ISBN 978-3-642-64542-6 ISBN 978-3-642-60756-1 (eBook)
DOI 10.1007/978-3-642-60756-1

Cataloging-in-Publication Data applied for

Die Deutsche Bibliothek - CIP-Einheitsaufnahme
Aubin, Jean-Pierre: Dynamic economic theory: a viability approach / Jean-Pierre Aubin. - Berlin; Heidelberg; New York; London; Paris; Tokyo; Hong Kong; Barcelona; Budapest: Springer, 1997
(Studies in economic theory; 5)

Originally published by Springer-Verlag Berlin Heidelberg New York in 1997
Softcover reprint of the hardcover 1st edition 1997

SPIN 10487288 42/2202-5 4 3 2 1 0 – Printed on acid-free paper

A Viability Approach

"Il pourrait au contraire sembler à quelques uns qu'étant donnée cette complexité des phénomènes économiques, l'économie mathématique est justement beaucoup trop simple. Ceux-là n'auraient qu'à se rappeler que les premières propositions de la géométrie sont, elles aussi, très simples, sans que cette simplicité interdise en rien les complications ultérieures.

Enfin, si quelques-uns, tout en reconnaissant que la méthode mathématique n'est pas superflue pour constituer la science de l'économie publique, et tout en reconnaissant sans doute que cette science n'atteint pas toute la complexité du réel vivant, mais que ces résultats cependant sont toujours sous-jacents, pour ainsi dire, à ce réel, se défiaient de cette science elle-même, et de son importance, et craignaient qu'on ne s'y enfermât un peu complaisamment, ceux-là n'auraient pas été frappés des admirables paroles, et décisives, où le savant fait lui-même la part de la science : la réforme sociale doit procéder à la fois du sentiment socialiste et de la science économique.".

CHARLES PÉGUY

Un économiste socialiste : Léon Walras
La Revue Socialiste, 1897

This book is dedicated
to Jean Cartelier, Richard Day and Donald Saari

Acknowledgments

I would like to thank IIASA and the Scuola Normale di Pisa, which provided me for several years shelters and intellectual environments which allowed me to escape the pressures bearing upon university professors, even when they are lucky enough to be part of the faculty of Université de Paris-Dauphine.

At IIASA, I met and discussed during many summers with a great number of colleagues who shared with me the importance of evolution under uncertainty of "macrosystems" involved in economics, ecology and environmental issues: A. Andersson, A. Cellina, G. Dosi, D. Foray, A. Kirman, N. Keyfitz, A. Nagy, R. T. Rockafellar, D. Saari, K. Sigmund, R. Wets and S. Winter to quote a few.

At Scuola Normale di Pisa, I owe G. da Prato a warm friendship and ... the adaptation of viability theory to stochastic differential inclusions.

N. Bonneuil, J. Cartelier and J. Weber deserve special thanks for all the sound counsels and challenging discussions concerning economic theory and demographics for which this book owes much.

At the Université de Paris-Dauphine, besides the closest of mathematical collaborations with Hélène Frankowska who developed many of the results presented in this book, I am much indebted to Daniel Gabay for long discussions and to the very active, stimulating team of young scientists participating to the *Groupe de Recherche Viabilité, Jeux, Contrôle* who made us the pleasure to trust us: P. Cardaliaguet, F. Chiaramonte, L. Doyen, K. Müllers, M. Quincampoix, P. Saint-Pierre, who collaborated to parts of the manuscript, as well as N. Caroff, H. Clément-Pitiot, O. Dordan, R. Duluc, A. Gorre, P. Lacoude, N. Maderner, J. Mattioli, L. Najman, N. Seube and C. Vigneron whose contributions are scattered in several parts of this monograph. This is also the opportunity to recall that without P.-M. Larnac, F. Roure and above all, P. Tabatoni, mathematics and mathematical economics would not have existed at Université de Paris-Dauphine. Twenty five years later, this book is a reward for the help they provided when I founded the Department and the Centre de Recherches de Mathématiques de la Décision (CEREMADE) for developing mathematics motivated by the sciences of organizations in the broad sense. I am grateful to D. Blondel for welcoming me to IRIS-TS when after 25 years these motivations were no longer the main objectives of CEREMADE. I owe R.H. Day my initiation to economics when he invited me to first at University of Wisconsin and then, for several monthly visits, at University of Southern California. I do not forget A. Bensoussan, I. Ekeland and the late C. Kipnis who influenced my ideas on the articulation of management, economics, environment and mathematics.

I am finally very grateful to Professor C. Aliprantis and Springer-Verlag for having accepted to publish this book in the *Studies in Economic Theory* series.

Preface

This book is intended to provide theoretical economists with mathematical tools necessary to handle the concepts of **evolution** and **adaptation** arising in economics, pursuing the Arrow-Debreu-Hahn legacy.

It may be now the time to answer the wish J. von Neumann and O. Morgenstern expressed in 1944 at the end of the first chapter of their monograph *"Theory of Games and Economic Behavior"*:

"Our theory is thoroughly static. A dynamic theory would unquestionably be more complete and therefore, preferable..."
"Our static theory specifies equilibria ... A dynamic theory, when one is found — will probably describe the changes in terms of simpler concepts."

Economies in the real world are normally "out of equilibrium". Actually, this does not signify much, because the concept of equilibrium is polysemous. The mathematical one, which we adopt here, expresses stationary — time independent — evolution, that is, no evolution. This is the static concept specified by von Neumann and Morgenstern. The other meaning is connoted by the **viability constraints** to which the state of the system is subjected, and that it is not allowed to violate during its evolution. Namely, the articulation between dynamics and constraints means here that from any initial state starts at least one evolution satisfying the constraints at each instant (such an evolution is called viable). An equilibrium may be viable or not viable, these are two independent issues.

The fact that many scarcity constraints are presented in terms of "balance", such as the balance of payments, may be at the root of the misunderstanding. Indeed, the image of the balance conveys both the concept of equalization of opposite forces, hence of constraints, and the resulting stationarity — often called the "stability", again, a dangerous word connoting too many different meanings. Therefore, speaking of disequilibrium does not necessarily mean putting at stakes the viability of a system. Rather, "disequilibrium" situations — nonstationary viable situations — appear to induce more or less efficient reactions of economic agents to adapt to viability constraints, the condition *sine qua non* on the economic evolution (Schumpeter has put particular emphasis on the role of entrepreneurs in this process).

It becomes then natural to specify the "minimal" conditions under which an economy can work and to specify classes — as large as possible — of reasonable economies whose evolution does not violate these viability conditions (as well as other specifications). This is not always possible, and a preliminary task is to find — and compute — the vi-

ability kernel, which is the set of states from which starts at least one viable solution.

The next task is to derive from the confrontation of the dynamics and the constraints the concealed regulation mechanisms governing viable evolutions, and to select some of them according to some further principle. This allows us to derive "adjustment laws" instead of founding the modelization process based on such a law. This goes against the tradition of theoretical economy, where the adjustment of some variables for reaching an equilibrium plays a basic and prominent role. The so-called "law of supply and demand" states that prices react in a determined direction in response to a difference between supply and demand in the market: the price of a particular commodity is assumed to vary according to the sign of the excess demand of this commodity. Instead of reasoning with a law of adjustment *a priori* given, and which does not produce viable evolutions, we shall build "dedicated" laws of supply and demand which shall provide viable solutions. In some way, this reverse approach allows us to "explain" *a posteriori* the role of such an adjustment law instead of scrutinizing the consequences of an *a priori* given law for possible justifications.

As a consequence, whether in a time-independent environment an evolution — even viable — converges to an equilibrium is not the major issue that I shall address in this book. Instead, *the main purpose of the viability approach to dynamical economics is to explain the evolution of a system, governed by given nondeterministic dynamics and viability constraints, to reveal the concealed regulation laws which allow the system to be regulated and provide selection mechanisms for implementing them.*

It assumes implicitly an "opportunistic" and "conservative" behavior of the system: a behavior which enables the system to keep viable solutions as long as its potential for exploration (or its lack of determinism) — described by the availability of several evolutions — makes possible its regulation.

The reader may agree or not with the proposed mathematical metaphors for explaining the evolution of a decentralized dynamic economic system, comparable to but different from the standard tâtonnement and nontâtonnement models. Relevant or not[1], the point is to illustrate how these new tools can be used to journey from a static analysis built on

[1]I naturally hope that the approach I suggest is so. If it is unfortunately a fallacy, I wish that it meets at least this Vilfredo Pareto criterion : *"Give me a fruitful error anytime, full of seeds, bursting with its own corrections. You can keep your sterile truth for yourself."* To which Johann von Goethe added however that *"if it becomes a kind of credo admitting no doubt or scrutiny, that is the real evil, one which has endured through the centuries.* I may borrow to Denis Diderot the lines with which he begins his *Pensées philosophiques* : *"Je compte sur peu de lecteurs, et n'aspire qu'à quelques suffrages. Si ces pensées ne plaisent*

concepts of optima, equilibria and attractors to a contingent dynamic framework.

Naturally, many attempts had already been made to introduce time and evolution in economic models, in the framework of market dynamics (one of the main motivations of this book), as well as sustainable development (regarded here as viable development), economic growth theory[2] or the theory of business cycles, which are mentioned in the bibliography. Let me just mention here the recent achievements of "evolutionary economics"[3]. Dynamical models have also been used in the framework of game theory: let us quote for instance the theory of repeated games on the discrete side[4], differential and dynamical games on the continuous side[5], evolutionary game theory[6] for which there are already several available books and an extensive literature.

However, I have attempted to comply to the request of J. von Neumann and O. Morgenstern by proposing other (and sometimes new) tools to tackle these issues which elude a stationary context and which supplement the theory of nonlinear dynamical systems[7].

In the extent where the latter is confined to the study of stability or chaotic behavior around attractors, time and evolution are actually pushed aside, since one ignores the **transient** states before they converge to an attractor. If this makes sense in some physical problems where the transient stage is very short by human standards, economic evolutions remain transient for our time scale. What happens when time approaches infinity may not be much more relevant than looking for stationary solutions — equilibria. "In the long term, we are all dead", Keynes reminded us. Local stability analysis gives poor intuitions about the effective evolutions and asymptotic properties are not sufficient for economists interested in the evolution *hic et nunc* of the economy. Stability properties of equilibria, which are too special evolutions, although important, are not the end of the story. In contrast to traditional dynamical economical analysis where local stability and asymptotic properties of models are the heart of the matter, the viability approach is concerned with contingent evolutions over time. This

à personne, elles ne pourront qu'être mauvaises ; mais je les tiens pour détestables si elles plaisent à tout le monde".

[2]dealing with growth of outputs, since the pioneer studies of Harrod, Domar and Solow after World War II.

[3]after the publication of the book [276, Nelson & Winter].

[4]See for instance [214, Fudenberg & Tirole].

[5]See for instance [105, Cardaliaguet, Quincampoix & Saint-Pierre].

[6]We refer for instance to [232, Hofbauer & Sigmund] and the popular book [332, Sigmund]

[7]whose economic applications were initiated by R.H. Day and which are exposed in his recent books [147,149, Day].

is one point.

Another one deals with the fact that until now the field of nonlinear systems and chaotic behavior deals with **deterministic systems** which are quite **unstable** and sensitive to initial conditions on some invariant domains (the attractor, for instance). In economics, I believe we face a completely orthogonal situation, with **nondeterministic systems** (caused by deeply ignorant and/or non rational economic agents, and the immense variety of possible decisions to be taken at each instant by economic agents). Nondeterministic (contingent) dynamical economic systems have actually to produce evolutions as **regular or stable** (in a very loose sense) as possible for the sake of adaptation and viability.

Contingent uncertainty does not mean stochastic uncertainty. Expectations of economic agents do not result from a cold probabilistic calculation. Following Keynes, *"the outstanding fact is the extreme precariousness of the basis of knowledge"* on which expectations are made. In order to capture the kind of uncertainty relevant for economics, it seems better to rely upon non–deterministic non–stochastic models.

I also advocate in this book another distinctive approach by not appealing to intertemporal criteria, *which involve some knowledge of the future*[8] (even of a stochastic nature.) This is of particular importance when experimentation is not possible[9] or when the phenomenon under study is not periodic. For example, in biological evolution as well as in economics and in the other systems we shall investigate, *the dynamics of the system disappear and cannot be recreated.*

Hence, **forecasting or prediction of the future are not the issues which we shall address here**. *La prévision est un rêve duquel l'événement nous tire,* wrote Paul Valéry.

However, the conclusions of the theorems allow us to reduce the choice of possible evolutions, or to single out impossible future events, or to provide explanation of some behaviors which do not fit any reasonable optimality criterion.

The complexity[10] of dynamic economic systems stems from such a non

[8]Most systems we investigate do involve myopic behavior; while they cannot take into account the future, they are certainly constrained by the past.

[9]Experimentation, by assuming that the evolution of the state of the system starting from a given initial state for a same period of time will be the same whatever the initial time, allows one to translate the time interval back and forth, and, thus, to "know" the future evolution of the system.

[10]This word is used here with its usual meaning. We shall however investigate one aspect of complexity, ***connectionist complexity***, as one mechanism to adapt to more and more viability constraints, which implies the emergence of links between the components of a system and their evolution, together with — or instead of — using decentralizing "messages". We shall compare it with other dynamical decentralization mechanisms, such as price decentralization. The emergence of the cyberspace may give a new life to connectionist mechanisms, since

teleological collective evolution of the set of agents, even though many individual economic agents think of themselves as being pursuing definitive and rational aims, instead of **adapting permanently** to the many viability constraints (among which scarcity constraints) they face under uncertainty, either stochastic or contingent. This theme has been introduced and studied in economic theory under the name of "bounded rationality". Indeed economic agents are humans, not computers, are seldom rational, obey inertia principles, are poor forecasters, actually very myopic. They base many decisions not on rational grounds, but on faith, beliefs and bets, rules of thumb, moods, rumors, and the like. They are more inductive than deductive in their learning processes. Actually, we can adopt Peirce's terminology and look at them as "abductive", i.e., as making conjectures rather than predictions[11]. They are most often more conservative than innovative, afraid of changes when they are mot perceived as improving their situation. They may prefer to adopt a herd behavior instead of choosing dissident ways opening new avenues.

Even though we are far away of being able to capture mathematically the indomitable psychological, sociological, political and cultural[12] behavior of human beings which warns their purely economic behavior[13], some recent concepts where diffidently introduced during the last decade for analyzing these features, some of them simple despite their novelty. **Most of them were discovered or redirected for answering questions of economic theory in an evolutionary perspective.** Among these, let us mention a sample: differential inclusions, differential inclusions with memory, viability theorems for differential inclusions and stochastic differential inclusions governing the evolution under constraints, mutational equations for studying the evolution of sets (such as production sets), differential calculus of set-valued maps and set-valued analysis for many reasons, nonsmooth analysis for studying stability problems by the Lyapunov method or for revealing concealed utility functions, etc. Some of these theories are exposed in scattered journals and books, most often for a mathematical audience. We shall survey them in the third part of this book — the Mathematical Tool-Box — without reproducing the proofs when they appeared elsewhere and can be easily retrieved.

Naturally, getting apprised with these mathematical tools requires from the nonprofessional mathematician a huge investment in time and efforts, since most concepts and results are too recent to be familiar and

---

any two agents would soon be able to communicate at each instant without the mediation of a price or other messages providing the necessary information.

[11]Non mathematical accounts of such questions can be found in [36, Aubin].

[12]in which the science and technology components of culture are included.

[13]restricted to production, distribution and consumption of commodities and services.

accepted by a clear consensus. However, I share the conviction of some of us that these tools — and others to be invented — will be mandatory for tackling evolutionary issues in economics (and social sciences). If we accept that physics studies much simpler phenomena than the ones investigated by social and biological sciences, and that for this very purpose, they motivated and used a more and more complex mathematical apparatus, we have to accept also that social sciences require a new and dedicated mathematical arsenal which goes beyond what is presently available. It has been said that physics could be defined as the part of the environment which can be understood through mathematical metaphors. By dropping the adjective "political" which characterized so well their domain of investigation and by looking for mathematical metaphors, some economists wished to transform their science into a physical — hard ? — one. We would rather hope that in return, some mathematicians would add this qualifying adjective to their domain of investigation to encourage them to direct their quest to the analysis of human and social behavior. Paradoxically, the very fact that the mathematical tools useful for social sciences are and have to be quite sophisticated impairs their acceptance by many social scientists and economists, and the gap menaces to widen.

Jean-Pierre Aubin Poillé sur Vègre, February 19, 1997

Outline of the Book

I open this book with a plea[14] for decentralized "ecodynamics", or rather, "eco^2-dynamics, in order to "house"[15] both economical and ecological dynamical analysis under a same roof. It raises the issues in the Arrow-Debreu framework of Walrasian economics, but in an evolutionary perspective, comparing the approach when the Walras equilibrium is derived from an *a priori* law of supply and demand and the viability approach where "viable" laws of supply and demand are derived from a dynamical description of the behavior of consumers faced to scarcity constraints.

The second chapter presents a list of low dimensional examples of several explicit models where the actual computation — analytical or obtained by computer programs —of the viability kernels is provided, illustrating how the concepts of viability kernels can bring different informations in the dynamical analysis of such systems than the study of stability properties of equilibria and their asymptotic behavior.

The second part of this book entitled "ecodynamics" is then devoted to the introduction of the theorems dealing with evolution under non-stochastic (also called contingent) uncertainty and viability issues in the dynamic version of the basic problem of resource allocations and more general dynamical economies.

Chapter 3 states the main viability theorems in the case of allocation of scarce resources. The idea is to "differentiate" so to speak the set of allocations of available commodities among the consumers, so that we can obtain the "envelope" of all dynamical systems compatibles with the viability constraints imposed by the scarcity of the available commodities. Among them, we introduce a class of dynamical systems

[14]written in collaboration with Jean Cartelier.

[15]The Greek root "oiko" means house. The word "economy" appeared in French in 1615 in the title of the treatise *Traité d'économie politique* by Antoine de Montchrestien, in opposition to the word " mesnagerie" meaning at the time private economy. The word "ecology" has been coined by Ernst Haeckel in 1866, defined as : *"...the global science of relations between organisms and their environment, in which we include the living conditions"*. The choice of the term "eco-viability" has been suggested by Luc Doyen and Philippe Cury.

parametrized by messages such as prices. We look for prices such as the associated dynamics belongs to this envelope: This provides the "regulation law" associating with any allocation the set of prices which "regulate" the dynamical allocation of scarce resources. The dual version of the regulation law reveals "budgetary rules" allowing the consumers to share the available commodities at each instant. Among them, one find the dynamical version of the Walras law stating that it is forbidden to spend more monetary units than received during the transactions. We then show how to integrate time both in the dynamics and the constraints, how to "govern" the evolution of the sets of resources through "mutational equations" and to how to incorporate the past both in the dynamics and the constraints.

Chapter 4 then shows how to regard the problem of dynamical resource allocation as a particular case of "dynamical economies" governing viable solutions, for which "planning" mechanisms are devised. Among them, we single out "slow evolutions" which are the evolutions regulated by prices with minimal norms. It is shown that in some cases, they are the same than evolutions governed by the "least complex" connection matrices, illustrating mathematically that in this case, the decentralization role of prices corresponds to minimal complexity of interactions between agents.

Chapter 5 uses the dual characterization of the regulation law to characterize prices as "viability multipliers" associated with the constraints, as in optimization theory under constraints. This allows us to design dynamical economies for which the viability is guaranteed.

These three chapters motivate a further step forward in the details of viability theory. Hence, in Chapter 6, we shall in particular answer two basic and natural questions: When a constrained subset is not viable,

1. what is the set of initial states from which starts at least one viable solution? We shall call this set the **viability kernel**.
2. what is the set of initial states from which starts a least one solution reaching the constrained set in finite time? We shall call this set the **capture basin**.

These concepts are central for the viability analysis of many problems.

Some of them are needed for implementing mathematically in Chapter 7 the "inertia principle" and the concepts of "viability niches", "punctuated equilibria" and "heavy solutions", as well as to derive the "adjustment laws" such as the law of supply and demand from the confrontation of a dynamical economy with constraints. They are obtained by "differentiating" the regulation law, so to speak. When several regulees are involved in a regulation mechanism, we advocate the selection of evolutions in the context of "hierarchical dynamical games" in which the different regulees obey a cascade of different hierarchical inertia.

Chapter 8 is devoted to "dynamical inequalities" involving inequalities that must be always satisfied, which are viability constraints of a special type in disguise. They involve the study and the construction of "utility functions" which increase along at least one evolution, the "Lyapunov functions" so useful for studying asymptotic stability, the "exit time function" associating with any initial state the largest time interval in which one solution obeys the constraints, the "crisis function" measuring the time spent violating the constraints and the "value functions" of intertemporal optimal control problems. Being viability constraints in disguise, these dynamical inequalities are amenable to viability theorems.

The ecodynamics part of the book is concluded in Chapter 9 by a very short introduction to a viability approach of the difficult topic of dynamical games, in which the role of two players acting on two different regulees is either agonistic, as in comprehensive economies, or antagonistic, as in dynamical games against nature.

The third part of the book presents a "Mathematical Tool Box". It surveys the available tools that I believe to be relevant and useful for investigating the issues which make the topic of this monograph. It is more "mathematically technical" and can be used mostly as an enriched "mathematical glossary".

Chapter 10 is devoted to the concepts of pseudo inverses of linear operators which are needed for the characterization of specific regulation laws such as slow or heavy solutions and tensor products of linear operators necessary to deal with the evolution of connection and technological matrices.

Chapter 11 gives an account of the definitions of set-valued analysis and Chapter 12 surveys the main results of viability, invariance and equilibrium theorems in the framework of differential inclusions, of time-dependent differential inclusions, of differential inclusions with memory and mutational equations.

Chapter 13 and 14 summarizes classical results on convex optimization and equilibrium theory, with a double objective: Give to the reader not familiar static mathematical economics the basic instruments of convex and nonlinear analysis on one hand, provide a short account of some results of general equilibrium theory on the other.

Chapter 15 is a gate to future investigations, because it presents the stochastic version of some of the basic viability theorems, where both the drift and the diffusion are set-valued. These are tools needed in some domains of finance and actuarial analysis, where the present consensus admits that uncertainty is governed by stochastic processes satisfying almost surely given constraints at each instant. It also attempts to reconcile two mathematical representations of uncertainty through stochastic differential inclusions: a "stochastic uncertainty", driven by a Wiener process, and a "contingent uncertainty", driven by a set-valued map[16]. But the price to pay in mathematical technicality may become prohibitive.

Usually, there are few disputes among "modelers" when they are listing the constraints looming in an economic problem. Serious disagreements may begin when behavioral assumptions have to be made.

Using a dynamical approach such as viability theory instead of a static one such as theory optimization or equilibrium theory is a first element of controversy. Utilizing dynamical games against nature or stochastic differential equation is another element of contention.

In order to weaken such controversies, or to prorogue the ultimate choice of a behavioral description of the economic agents, I have tried to offer in this book an array of options for modelers, knowing too well that this is too short a list and that too ineffective are its items.

[16]Not to mention another approach dealing with "fuzzy viability" which is not considered in this book.

Contents

I

Evolutionary Perspective

Chapter 1

Decentralized Ecodynamics: a Manifesto

1.1 The Issues

We shall devote the beginning of these lectures to the simplest economic problem we can think of:

how to allocate scarce resources among consumers

by complying to the basic economic scarcity constraint

It is impossible to consume more physical goods than available

In other words, let us introduce the set of allocations of these scarce resources among the consumers. If M denotes the set of scarce resources available to the consumers, then the set of allocations of scarce commodities to n consumers[1] is the set K of $x = (x_1, \ldots, x_n)$ satisfying $y = \sum_{i=1}^{n} x_i \in M$

This means that each consumer receives a commodity the sum of which is viable in the sense that the total consumptions is an available resource.

This problem looks at first glance somewhat silly and simple minded, since it amounts to pick up an element in this allocation set (i.e., to choose an allocation) in the case of static models, or to evolve in this set, regarded as a viability set, in the case dynamical systems. However, it elucidates the basic difficulties characteristic of economic theory, which has to explain the viability of allocation mechanisms, from "dictatorial" mechanisms where a planning bureau computes and imposes an allocation to more sophisticated mechanisms ("social rules" imposed by institutions) involving decentralization

[1] "Exchange economies" are the particular case when $M := \{0\}$ is reduced to 0. In this case, x denotes the net exchange. For such exchange economies, viability theorems are trivial.

of decisions through prices, shortages, taxes, etc. which allows consumers to choose freely and independently their commodities in such a way that the scarcity constraints are satisfied.

In the framework of an allocation model decentralized by prices (and only by prices), our ambition is to present a mathematical metaphor of a decentralized dynamical mechanism. We shall characterize all — if any — the "supply and demand laws" regulating the evolution of prices that are **viable in the sense that they allow continuous transactions of consumers in a decentralized way**. The question whether or not in a stationary environment such a mechanism converges to an equilibrium (stationary allocation) is an independent issue which is not the one which is primarily addressed, even though it became a predominant criterion of economic relevance.

Indeed, static models assign one or several elements in the allocation set. We study here some mechanisms which govern the **evolution** of allocations of scarce resources[2]. In these dynamical models, the laws which govern the evolution of allocations are most often represented by differential equations (or differential inclusions) with or without memory (functional differential inclusions).

Static models are particular cases of (time-independent) dynamical models yielding "constant evolutions", which are also called "equilibria". By the way, the concept of equilibrium often covers two different meanings in economics:

1. The first one, the meaning we use in these lectures, is derived from mechanics, where an equilibrium is a constant evolution, or a "rest point" of an underlying dynamical system. In this case, equilibrium means **stationarity**.

   When a solution of a dynamical system converges asymptotically to a limit, this limit is an equilibrium. The question arises conversely to obtain an equilibrium which is **asymptotically stable**, i.e., an equilibrium which is obtained as a limit of the solutions to the dynamical process starting nearby.

2. The second meaning is covered here by what we call the **viability constraints**, such as the total consumption must be less than or equal to the total supply, etc.

Stationarity — the first meaning of equilibrium — cannot be an issue whenever the behavior of consumers evolves with time and whenever the set of scare resources can be depleted by consumption or enlarged by production,

[2]By the way, in dynamical models, we can and shall assume that the subset of allocations evolves with time, and may also depend upon the history of the evolution. In this chapter, for the sake of simplicity of exposition, the set M of resources is assumed to remain constant, i.e., resources are not depleted by consumption.

technological advances and so on. These are reasons which led some of us to abandon such stationarity requirements.

However, in a time independent environment, stationarity may describe an aspect of satisfaction: when satisfied, no one has any reason to change the situation. But even *homo oeconomicus* may be frustrated most of the time[3].

But viability constraints — even fiduciary ones — cannot be violated, and we have to devise evolutionary models which provide consumption paths which respect them at each instant.

This is then the first issue which is dealt with by viability theory: find evolutions which obey at each instant the scarcity constraints. A second issue is, in a time independent environment, to prove whether among these solutions one can find equilibria. A third issue is then to know whether some of these equilibria are asymptotically stable.

It is known that the classical Walrasian tâtonnement is neither viable nor necessarily asymptotically stable: For some of them, there may exist an equilibrium, even unique, which is not asymptotically stable[4]. This failure can be overcome : But for finding a way to get out of this impasse, we have to agree to perform some modifications of the mathematical translation of the basic ideas and to change the questions, as we shall explain now.

Instead of starting with a supply and demand law such as the Walrasian tâtonnement, even though it aggregates sensible individual behaviors, even though it is so seducing — prices goes up when demand exceeds supply and go down when supply exceeds demand — to be "believed" in, we propose to start with the actual behavior of consumers among which scarce resources have to be allocated and **derive** what are the supply and demand laws which allow an allocation decentralized by the prices. The supply and demand law is no longer a **primitive** of the model, but a **conclusion**, in a sense made precise later: **the supply and demand law shall emerge from the confrontation of the diverse wishes of the consumers and the scarcity of available commodities.**

The next problem is then to select among such supply and demand laws compatible with the scarcity of constraints the ones which satisfy supplementary requirements. We shall suggest to chose the ones which satisfy the **inertia principle**, in the sense that prices evolve only when scarcity is at stakes.

---

[3]If *homo oeconomicus* is not born at equilibrium — and thus, eternally happy — he only reaches blissfulness when he dies at very old — infinite — age!

[4]See the famous Scarf's counterexample in [321, Scarf]. However, there are instances where Walrasian tâtonnement does converge, thanks to the theorem of [136, Crandall & Pazy] on differential inclusions for maximal monotone set-valued maps.

1.2 Decentralization

1.2.1 Centralized and Decentralized Mechanisms

We begin by distinguishing between centralized and decentralized models. In the first category of models, consumers delegate their decision power to another "agent" who, knowing the behaviors of the consumers and the set of scarce resources, solves the problem at the global level.

For instance, consumers must agree to describe their behavior by a collective utility function

$$x := (x_1, \ldots, x_n) \mapsto U(x) = U(x_1, \ldots, x_n) \in \mathbf{R}$$

Then, this agent (planning bureau, big computers or big brothers, ...) knowing U and the subset M, decides to maximize U over the allocation set K: Find an allocation $\bar{x} = (\bar{x}_1, \ldots, \bar{x}_n)$ such that $\sum_{i=1}^{n} \bar{x}_i \in M$ and

$$U(\bar{x}_1, \ldots, \bar{x}_n) = \max_{\sum_{i=1}^{n} x_i \in M} U(x_1, \ldots, x_n)$$

The problem is then transferred to the question of choosing the collective utility function U.

Or, in the dynamical version, they agree to represent their behavior by, say, a system of differential equations

$$x_i'(t) \;=\; h_i(x_1(t), \ldots, x_n(t)) \quad (i = 1, \ldots, n)$$

where the variations of the consumption of each agent depend upon the knowledge of the choices of every other agents. This representation of the dynamical behavior of the agents must also take into account the whole set of scarce resources. Nagumo's Theorem — the first viability theorem for ordinary differential equations published in German in 1943, and thus, forgotten and rediscovered at least fourteen times — provides necessary and sufficient conditions for the dynamics h_i to provide solutions to this system of differential equations satisfying the viability condition

$$\forall\, t \geq 0, \;\; \sum_{i=1}^{n} x_i(t) \;\in\; M$$

Such necessary and sufficient conditions can be regarded as "social" rules imposed on the consumers for respecting the scarcity conditions. Actually, there are many dynamical systems satisfying such rules[5].

In both cases, we deal with centralized models.

[5] which, for exchange economies, take the form

$$\sum_{i=1}^{n} h_i(x_1, \ldots, x_n) \;=\; 0$$

so that there are as many mechanisms than ways to allocate 0 in the form of velocities of consumers.

1.2.2 Connection Mechanisms

We denote by $Y := \mathbf{R}^l$ the commodity space. Assume for instance that, if there were no scarcity constraints to comply with, the dynamical behavior of each consumer would be fully decentralized, and governed by differential equations of the form

$$\forall\, i = 1, \ldots, n, \;\; x_i'(t) \;=\; g_i(x_i(t))$$

There are no reason why scarcity constraints

$$\forall\, t \geq 0, \;\; \sum_{i=1}^{n} x_i(t) \;\in\; M$$

of a collective nature would be satisfied by such independent behaviors.

We set $X := Y^n$. One first solution which comes up to the mind is to "connect" these dynamics through a connection matrix $W := (w_i^j) \in \mathcal{L}(X, X)$ and to govern the evolution of consumptions according to an evolution law of the form

$$\forall\, i = 1, \ldots, n, \;\; x_i'(t) \;=\; \sum_{j=1}^{n} w_i^j(t) g_j(x_j(t))$$

where $w_i^j \in \mathcal{L}(Y, Y)$. Naturally, we obtain a fully decentralized economy when the connection matrix is the identity matrix.

We propose mathematical metaphors designing evolution laws governing the evolution of connection matrices linking the autonomous dynamics of each agent in order to sustain the viability constraints prescribed by the environment of the isolated system.

Our point is the following: Starting with a disconnected system — for which the nondiagonal elements of the connection matrix are all equal to zero — the viability can be maintained whenever each agent of the system is linked with other through an evolving connection matrix, **dictating cooperation among the agents**. In the mathematical version of the metaphor, a class of evolution laws governing the evolution of these connection matrices can be designed from the disconnected dynamical system and the set of constraints.

We shall single out for instance the problem of **maximal decentralization** which requires to find connecting matrices $W(t)$ as close as possible to the identity matrix $\mathbf{1}$ — **slow evolution** — or which evolve as slowly as possible while sustaining the viability of the allocation of scarce resources — **heavy evolution** — in such a way that the connection matrix remains constant as long as the viability of the system is not at stakes. We advocate the latter mode, which has the property of locking in the connection matrices which can regulate forever a nonempty subset of states, called the **viability niche of the connection matrix**.

This attempt to sustain the viability of the system by connecting the dynamics of its agents may be a general feature of "complex systems"[6]. Is not complexity meaning in the day-to-day language the labyrinth of connections between the components of a living organism or organization or system ? Is not the purpose of complexity to sustain the constraints set by the environment and its growth parallel to the increase of the web of constraints ? Economic history has shown an everlasting trend toward a highly connected network of labyrinthine — connectionist — complexity. In this sense, complexity arose with the apparition of life and seems necessary to the pursuit of its evolution at all levels of organization.

We regard here *connectionism* — a less normative and more neutral term than *cooperation* whenever the system arises in economics or biology — as an answer to adapt to more and more viability constraints, which implies the emergence of links between the agents of a system and their evolution. A system is **disconnected (or autonomous, free, autarchic, decentralized, etc.)** if the connection matrix is the identity (or unit) matrix. In some loose words, the distance[7] between the **connection matrix** — which is the matrix linking each agent of the system to the others — and the identity matrix should capture and measure the concept of an **index of static connectionist complexity**. The larger such a connectionist complexity index, the more complex — or labyrinthine or intricate — the connectionist feature of the system.

[6]Physicists have attempted to measure "complexity" in various ways, through the concept of Clausius's entropy, Shannon's information, the degree of regularity instead of randomness, "hierarchical complexity" in the display of level of interactions, "grammatical complexity" measuring the language to describe it, temporal or spatial computational, measuring the computer time or the amount of computer memory needed to describe a system, etc.

The problem is that living systems being open, there is no way to describe them entirely and consequently, to demonstrate the truth of any proposition. We are left with metaphors, and among them, mathematical ones, to validate a consensual definition of one meaning — actually, one submeaning, so to speak — of complexity.

Therefore, scientific activity begins by dividing a system into two classes, the system under study and its environment. This division is always arbitrary, but often justified by the scientists in quest of explanation. Once conceptually isolated from its environment, a living system fuels itself in the last analysis on solar energy through the consumptions of wastes of the other components of the open system set apart in the description of the environment. Each component of the system which can evolve independently in the absence of constraints, must interact each other in order to maintain the viability of the system imposed by its environment.

[7]or the velocity $W'(t)$ of a time dependent connection matrix $W(t)$ starting from the identity matrix as we shall do in the case of heavy evolution. One can also measure other features of connectionnist complexity through the sparsity of the connection matrix, i.e., the number — or the position — of entries which are equal to zero or "small". The sparser such a connectionist matrix, the less complex the system.

1.2.3 Decentralization Mechanisms

Compounded with this issue of connectionist complexity — that is naturally involved in neural networks — is the question of whether these links relating one agent to another can be subsumed by regulation parameters — which we call in short regulees — and whether and when they provide the same evolutions of the state of the system. Such regulees are messages sent to the consumers, and the simplest ones are the prices.

In a **decentralized mechanism** — actually, we should say in an "a-centralized mechanism", since we do not assume the existence of a "center" — the information on the alloaction problem is split and mediated by, say, a "message" which summarizes part of the information. In our case, we use the "price" p as a main example of message (actually, the message with the smallest dimension). Knowing the price p, consumers are supposed to know how to choose their consumption bundle, **without**

- knowing the behavior of their fellow consumers
- knowing the set of scarce resources

Then the problem is to find what is the message which carries the relevant information.

Actually, we have to ask whether it is possible to find such a relevant message and then, **how to find it.**

If it is possible to answer the first type of question, it is much more difficult to investigate the second, leaving such problems to mythical players such as Adam Smith's "Invisible Hand"[8], the "Market" (with an upper case "M"), even the "Self-Regulation" mechanism, etc. We shall bethink that these players are not really operating on the price system, which we shall propose to regard as a **regulatory control** (a "regulee") to help the consumers to respect the scarcity constraints by delivering them proper informations on the behavior of all consumers and the set of available resources.

There are many other decentralized models, such as "rationing" mechanisms which involve shortages (and lines, queues, unemployement), or "frustration" of consumers, or "monetary" mechanisms, or others.

Naturally, there is no "pure" decentralization, since the choice of the decentralization message is in some sense centralized. The prices help consumers to make their choice in a decentralized way, but the difficulty is postponed to explain the evolution of prices.

[8] Adam Smith, together with Thomas Malthus, did influence Charles Darwin, where the biological reproductive success played the role of the economic profit, and where no outside mechanism is at work to regulate the system. One may go deeper in this analogy by saying that phenotypes play the role of commodities and genotypes the role of prices. This borrowing of economic metaphors by biologists is refunded those days by economists looking for metaphors in the theory of biological evolution.

Decentralization has a meaning in the context of a hierarchical organization in two levels (at least). We shall also briefly investigate the more complex situation where there are three (or more) levels of hierarchical organization. In this case, we shall distinguish among the messages the ones which provide informations to a given layer from the ones which provide message for all levels. In this way, one can conceive "cascades" of decentralization mechanisms using several kinds of messages (prices, quantities, lines or queues, advertisement, rumors, etc.) at each level of more complex hierarchical organizations in Russian dolls or Chinese boxes. At each level, an adequate class of messages allows us to decentralize the level below through a specific "institution".

In summary, since the nature of the constraints is collective and the dynamics are individual, so to speak, the "**viability property**" requiring that from any initial state satisfying the constraints starts at least one solution obeying them forever is not generally met. Hence, we have to couple or connect the dynamics of the system

1. either by connecting the agents of the original dynamics through connecting matrices $W := (w_i^j)_{i,j} \in \mathcal{L}(X,X)$:

$$\forall\, i = 1,\ldots,n, \;\; x_i'(t) \;=\; \sum_{j=1}^{n} w_i^j(t) g_j(x_j(t))$$

2. or by subtracting prices (or fluxes, controls, in short, regulees) to each agent of the original dynamical system:

$$\forall\, i = 1,\ldots,n, \;\; x_i'(t) \;=\; g_i(x_i(t)) - p(t)$$

We shall investigate these two ways and compare them: We shall see that **slow evolution evolution of prices and evolution through connecting matrices $W(t)$ as close as possible to the identity matrix $\mathbf{1}$ provide the same evolution of commodities, justifying mathematically the relevance of the price decentralization mechanism.** Indeed, the striking result we obtain in this framework is that these modes of regulation are equivalent by providing the same evolutions of commodities. In particular, this is the case of evolutions which minimize at each instant "complexity indices" of the form

$$\forall\, \alpha \in [0,1], \;\; \alpha\|\mathbf{1} - W(t)\|^2 + (1-\alpha)\|p(t)\|^2$$

which govern evolutions of commodities independent of the choice of the weight α.

We shall also investigate the behavior of such system when we bound both inflation and the "dynamical complexity index", which is the norm of the velocity of the connection matrices, by minimizing at each instant "dynamical complexity indices" of the form

$$\forall\, \alpha \in [0,1], \;\; \alpha\|W'(t)\|^2 + (1-\alpha)\|p'(t)\|^2$$

1.2.4 Adam Smith's Invisible Hand

From now on, we shall proceed by considering the mechanism of decentralization by prices only. This is the context in which Adam Smith, Léon Walras, Kenneth Arrow, Gérard Debreu and many others designed their contribution.

Indeed, there is no doubt that Adam Smith is at the origin of what we now call **decentralization**, i.e., the ability for a complex system moved by different actions in pursuit of different objectives to achieve an allocation of scarce resources[9]. The difficulty to grasp such a disordered way of regulation of economic processes, contrary to apparently more logical (or simple minded?) attractive organizational processes based on several varieties of planning procedures[10] led him to express it in a poetic manner. Let us quote the celebrated quotation of his book AN INQUIRY INTO THE NATURE AND THE CAUSES OF THE WEALTH OF NATIONS published in 1776, two centuries ago:

"Every individual endeavours to employ his capital so that its produce may be of greatest value. He generally neither intends to promote the public interest, nor knows how much he is promoting it. He intends only his own security, only his own gain. And he is in this led by an invisible hand *to promote an end which was no part of his intention. By pursuing his own interest, he frequently thus promotes that of society more effectually that when he really intends to promote it".*

We had to wait a century more for Léon Walras, a former engineer, to propose that this invisible hand "operates" on economic agents through prices, gaining enough information on the desires of the agents and the available commodities for guaranteeing their consistency, or the viability of the allocation system.

He presented in 1874 the general equilibrium concept in ÉLÉMENTS D'ÉCONOMIE POLITIQUE PURE as a solution to a system of nonlinear equations. At that time, when only linear systems were understood, the fact that the number of equations was equal to the number of unknowns led him and his immediate followers to make the optimistic assumption that a solution should necessary exist[11].

[9] With his compatriot David Hume, one can say that he is also at the origin of cybernetics and that he deeply influenced Charles Darwin. This is this cybernetical tradition that we chose to pursue here. In chapter 7 of his most famous book, he proposes an explicit dynamical process which is viable in spirit because transactions among agents are carried out at every step, contrary to the Walrasian tâtonnement.

[10] in favor among military organizational schemes.

[11] But it took another century, until 1954, for Kenneth Arrow and Gérard Debreu to find a mathematical solution to this problem. This solution, however, could not have been obtained without the fundamental Brouwer Fixed Point Theorem in 1910, which in turn required much modification to tailor it to this specific problem — by proving theorems whose assumptions could bear the same degree of economic interpretation as the conclusion. See an historical and epistemological account in [237, Ingrao & Israel].

1.2.5 Walras' Choice

In modern terms, the behavior of each consumer is described by a demand function $d_i(p,r)$ associating with a price p and an income r the ith consumption $x_i = d_i(p,r)$ of Mrs. i. A demand function subsumes a **passive way** to choose automatically a commodity knowing the price and the income. By contrast, a dynamical system describes rather an **active way**, in which the consumer **changes** her consumption and thus, acting on the system, becomes an economic **agent** in the real sense of the word.

The description of the consumer's behavior by demand functions furnishes a decentralized mechanism, since Mrs. i ignores the behavior of her fellow consumers and the size and the nature of the set M of scarce resources.

A **price** p associates linearly with any commodity y its value denoted by $\langle p, y\rangle$ (expressed in numbers of monetary units). It thus associates with the set M of available commodities its **total income** $\sigma_M(p) := \sup_{y\in M}\langle p, y\rangle$, which is the largest value of the available commodities y of M at price p. In Walrasian economies, it is assumed that this total income is allocated among consumers. Each consumer receives a share $r_i(p)$ of the total income $\sigma_M(p)$:

$$\forall\, p,\ \sum_{i=1}^{n} r_i(p) \ \le\ \sigma_M(p)$$

Once such an income allocation among consumers is made, Mrs. i chooses the commodity $x_i = d_i(p, r_i(p))$ knowing only the price p. The problem is then to find a price $\bar{p}$ (the Walrasian equilibrium price) such that

$$(d_1(\bar{p}, r_1(\bar{p})), \ldots, d_n(\bar{p}, r_n(\bar{p})))$$

forms an allocation.

This is a **decentralized model** because consumers do not need to know neither the choices of other consumers nor the set M of available commodities.

The basic Arrow-Debreu Theorem states in this case that such an equilibrium exists whenever a **budgetary rule** — *it is forbidden to spend more monetary units than earned* — is obeyed by consumer's demand functions:

$$\forall\, i = 1, \ldots n,\ \ \forall\, p,\ \ \langle p, d_i(p,r)\rangle \ \le\ r$$

Actually, if the demand functions obey the (collective) Walras law

$$\sum_{i=1}^{n}\langle p, d_i(p, r_i)\rangle \ \le\ \sum_{i=1}^{n} r_i$$

there exists at least a solution $\bar{p}$ to the allocation problem

$$\sum_{i=1}^{n} d_i(\bar{p}, r_i(\bar{p})) \ \in\ M$$

Such a solution $\bar{p}$ is called a **Walrasian equilibrium**.

Example The classical example of a Walrasian demand function is the one which is derived from the maximization of a consumer's utility function under the budgetary constraint $\langle p, x\rangle \leq r$, in such a way that the individual Walras law is automatically satisfied. When a unique solution to such an optimization problem exists, it provides an example of a demand function satisfying the individual Walras law. If several solutions do exist, one has to extend the problem to the case when demand functions are replaced by set-valued demand maps (see Chapters 12 and 13). □

But why did Walras call it an equilibrium ? He had in mind an underlying dynamical process regulating the evolution of prices for which the solution $\bar{p}$ of the above problem is an equilibrium in the mechanical sense, i.e., for which $\bar{p}$ is a constant evolution. Furthermore, if the solution $p(t)$ of this dynamical system converges to some value $\bar{p}$, then this asymptotic price is an equilibrium. Therefore, such dynamical process can be regarded as a (continuous) algorithm to obtain an equilibrium: *"Voyons à présent comment ce même problème de l'échange de plusieurs marchandises entre elles, dont nous venons de trouver la solution scientifique, est aussi celui qui se résout empiriquement sur le marché par le mécanisme de la concurrence... Cela se fait après réflexion, sans calcul, mais exactement comme cela se ferait par le calcul en vertu du système des équations d'équivalence des quantités demandées et offertes et de la satisfaction maxima complété par les restrictions convenues. ... Que faut-il donc prouver pour établir que la solution théorique et la solution du marché sont identiques ? Tout simplement que la hausse ou la baisse sont un mode de résolution par* tâtonnement *du système des équations d'égalité de l'offre et de la demande."*, he wrote in the twelfth lecture of his ELÉMENTS D'ÉCONOMIE THÉORIQUE.

In our framework, the *"système des équations d'égalité de l'offre et de la demande"* is described by the **excess demand map** E associating with any price p the difference

$$E(p) := \sum_{i=1}^{n} d_i(p, r_i(p)) - M$$

between the total demand $\sum_{i=1}^{n} d_i(p, r_i(p))$ and the total supply (the set M of scarce resources).

The **Walrasian tâtonnement**[12], in its continuous version, is defined by the

[12]Tâtonnement means "tentative process", "trial and error" — literally, cumbersomely walking in obscurity by touch (tâter). The translation of the verbal description by Walras into the framework of differential equations has been made by Kenneth Arrow and Léon Hurwicz.

differential inclusion

$$p'(t) \in E(p(t))$$

Hence, according to this law of supply and demand, the price increases whenever the excess demand is positive and decreases in the opposite case.

Furthermore, a Walrasian equilibrium price $\bar{p}$ is indeed an equilibrium of this underlying dynamical process because

$$0 \in E(\bar{p})$$

We observe that if $p(t)$ is a price supplied by the Walras tâtonnement process and if it is not an equilibrium, it cannot be implemented because the associated total demand $\sum_{i=1}^{n} d_i(p, r_i(p))$ is not necessarily available.

Hence, this model forbids consumers to transact as long as the prices are not equilibria. It is as if there was a super auctioneer calling prices and receiving offers from consumers. If the offers do not match, he calls another price according to the above dynamical process, but does not allow transactions to take place as long as the offers are not consistent, and this happens only at equilibrium!.

Tâtonnement is therefore not viable.

But the question which was asked despite this drawback is the convergence of solutions $p(t)$ of the above tâtonnement process to a limit, which then is an equilibrium. This issue, as well as either the Lyapunov stability[13] or the asymptotic stability[14] of a Walrasian equilibrium, require the use of adequate Lyapunov functions (introduced by Lyapunov in 1892) which decrease along the price paths $p(t)$ in such a way that one can derive the convergence of $p(t)$ when t goes to infinity. For example, an excess demand satisfying the property

$$\forall\, p \neq \bar{p}, \ \left\langle p - \bar{p}, \sum_{i=1}^{n} d_i(p, r_i(p)) \right\rangle \ < \ \sigma_M(p - \bar{p})$$

is shown to converge to an equilibrium. It had been difficult to provide a satisfying economic interpretation to such additional assumptions (to the Walras law). Furthermore, Scarf proposed in 1960 an example of excess demand function for which there exists a unique equilibrium which basin of attraction is empty. Hence, convergence of the Walras tâtonnement holds only for restrictive classes of tâtonnement processes.

On the other hand, in a series of papers by Debreu, Mantel, Sonnenschein, among many others, showed that whenever the number of agents is larger than or equal to the number of commodities, any continuous function satisfying the Walras law and homogeneous of degree zero can be regarded as

[13]This means that for any neighborhood of the equilibrium, there exists a smaller neighborhood from which the solutions remain in the given neighborhood.

[14]The means that there exists a subset, called the basin of attraction, from which the solutions converge to the equilibrium.

an excess demand function for some underlying Walrasian economic model associated with a continuous strictly convex function, destroying the hope that **every possible tâtonnement process** should converge to an equilibrium. These considerations have been studied further by Saari, Simon, Chiappori and Ekeland.

Finally, it may be too much to ask the entity which regulates the price (the Market, the Invisible Hand, the Gosplan, ...) to behave as a real decision-maker whereas consumers act passively according to their demand functions.

For these reasons, and also because these issues are treated in depth in many other books and papers, we shall let aside the Walrasian tâtonnement, the existence of a Walrasian equilibrium and the asymptotic stability theorems.

However, the legitimate admiration that Léon Walras deserves should not imply a dogmatic respect of his contribution by his followers: the equilibrium concept was a simplifying step in the attempt to grasp some essential economic feature in an otherwise complex maze of concepts. This concept had its use, as a first approximation, despite the fact that it rarely happened in economic history. The Walrasian tâtonnement still keeps its attractiveness among some economists despite its failure to explain the achievement of a (viable) allocation, let it be an equilibrium.

So, its **dépassement**, as well as the observation that the Walrasian tâtonnement is not viable and should be replaced by a viable dynamical system, should not be regarded by the faithfuls as a crime of **lèse majesté**. On the other hand, these shortcomings should not be used to claim that any decentralized mechanism using prices is merely a fantasy dreamed by theoreticians from their ivory towers — an "empty box", as it has been written — and even, to reject the relevance of mathematical metaphors in economics[15].

1.2.6 Nontâtonnement Models

Since the Walrasian **tâtonnement** is not viable, many authors proposed **nontâtonnement** dynamical processes (in particular, Arrow-Hahn, Negishi, Smale, Uzawa, among many others[16]).

Here, **both** the consumers act on their consumptions and Adam Smith's invisible hand act on the prices, according to a system of differential equations of the form

$$\begin{cases} \forall\, i = 1,\ldots,n, \;\; x_i'(t) \;=\; f_i(x_1(t),\ldots,x_n(t),p(t)) \\ \\ p'(t) \;=\; g(x_1(t),\ldots,x_n(t),p(t)) \end{cases}$$

[15]This is a typical instance of impatience and the totalitarian desire for monist explanations.

[16]We refer to [333, Smale], [305, Saari & Simon] and the books of [8, Arrow-Hahn] and [228, Hildenbrand & Kirman] for more detailed treatment of this issue.

Second, this dynamical system must yield viable solutions, in such a way that at each instant, consumers may transact and consume available resources.

In such models, the price dynamics described by g is regarded as a **supply and demand law**, assumed to describe how the invisible hand governs the evolution

But, in order to keep the consumptions viable, the Nagumo theorem requires to assume conditions on the rules f_i which:

1. **guarantee the viability for all prices**, so that such nontâtonnement models can no longer provide **explanations** of the role of prices in decentralization mechanisms,

2. and **are not decentralized** because the maps f_i have to depend upon the consumptions of the other consumers.

These requirements can be regarded as "rationing" conditions imposed on the consumers for respecting the scarcity conditions, providing "nonprice" decentralization mechanisms.

Actually, the issue studied by the proponents of nontâtonnement mechanisms is the **asymptotic convergence** of the solutions to Pareto optimal allocations (which are Walrasian equilibria for adequate allocations of the total income).

One naturally may add to the price mechanism any "rationing" mechanism, which is valid any time that decentralization by prices only is not possible. But before complying to that strategy, one may ask whether this is due to the fact that the supply and demand law described by the dynamics g is imposed *a priori* independently of the behavior of the consumers and the set M of scarce resources. We shall attempt to answer positively this question by characterizing the family of supply and demand laws which are consistent with the behavior of the consumers and the set of scarce resources. In other words, to answer the questions whether one can find price dynamics $g(x_1, \ldots, x_n, p)$ which can regulate the evolution of prices in such a way that **visible** consumers modify continuously their consumption **in a decentralized way**.

1.2.7 Visible Consumers and Viable Processes

We suggest to go a step further from the Walrasian tâtonnement by

1. conserving the dynamic (active) behavior of consumers as in nontâtonnement models, but in a decentralized way,

2. abandoning any *a priori* supply and demand rule regulating the evolution of prices.

Table 0.1: Comparison between Walrasian tâtonnement and viable process

Process:	Walrasian	Viable
Description of the behavior of consumers	demand functions $d_i(p,r)$ $x_i = d_i(p, r_i(p))$	change functions $c(x_i, p)$ $x_i'(t) = c_i(x_i(t), p(t))$
Demand map derived from change function		$x \in D_i(p)$ if and only if $c_i(x,p) = 0$
Derivation from utility function	$d_i(p,r)$ maximizes U_i under $\langle p, x\rangle \leq r$	$c_i(x,p)$ $= U_i'(x) - p$ $= \partial \left(U_i\vert_{\langle p,x\rangle \leq r}\right)(x)$
Equilibrium: stationarity and (static) viability	$\forall\, i,\ \bar{x}_i = d_i(\bar{p}, r_i(\bar{p}))$ such that $\sum_{i=1}^{n} \bar{x}_i \in M$	$\forall\, i,\ c_i(\bar{x}_i, \bar{p}) = 0$ such that $\sum_{i=1}^{n} \bar{x}_i \in M$
Budget rule	$\langle p, d_i(p,r)\rangle \leq r$	$\langle p, c_i(x,p)\rangle \leq 0$
(dynamic) Viability		$\exists\, p(t)$ such that $\forall\, t \geq 0,\ \sum_{i=1}^{n} x_i(t) \in M$
Characterization of the viability		$\forall\, (x_1, \ldots, x_n)$, $\Pi_M(x_1, \ldots, x_n) \neq \emptyset$
Regulation law		$p(t) \in \Pi_M(x_1(t), \ldots, x_n(t))$
Supply and Demand	$p'(t) \in \sum_{i=1}^{n} d_i(p(t), r_i(p(t))) - M$	$p'(t) \in G_M(x_1(t), \ldots, x_n(t), p(t))$
	a priori	*a posteriori*

In other words, instead of taking a supply and demand rule as a primitive of the model as it is done in both tâtonnement and nontâtonnement models, we shall derive them (and compute them) from the knowledge of the decentralized dynamics describing the behavior of the consumers and the knowledge of the set M of scarce resources.

It may be wise indeed to let the real decision-makers, the consumers in our case, to govern the evolution of their consumption through differential equations

$$x_i'(t) = c_i(x_i(t), p(t))$$

parametrized (or controlled) by the price $p(t)$, so that consumers change their consumptions knowing only the price $p(t)$ at each time t, without taking into account neither the behavior of the other consumers nor the knowledge of the set M of scarce resources.

Hence, a consumer is an economic agent, whose dynamical behavior is described by the function $(x,p) \mapsto c_i(x,p)$, called change function[17].

Therefore, it shares with the Walrasian static model its decentralization property.

The problem is then to find a time-dependent price $p(t)$ such that the associated solutions $x_i(t)$ of the above differential equations do form an allocation at each time t:

$$\forall\, t \geq 0, \ \sum_{i=1}^{n} x_i(t) \in M$$

We prove that this viability property holds true under a budget rule which is a dynamical version of the static Walras law. We shall even prove under the same assumption the existence of a viable equilibrium $(\bar{x}_1, \ldots, \bar{x}_n, \bar{p})$ of this dynamical system, which is a solution to

$$\begin{cases} \forall\, i = 1, \ldots, n, \ c_i(\bar{x}_i, \bar{p}) = 0 \\ \sum_{i=1}^{n} \bar{x}_i \in M \end{cases}$$

Example As in classical Walrasian economics, one can derive change functions from the consumer's wish to increase its utility under the budgetary constraint $\langle p, x\rangle \leq r$. The idea then is to use the steepest ascent method to the restriction of the utility function U to the budget set (which is no longer differentiable). Using nonsmooth analysis, i.e., replacing gradients by generalized gradients, we may take

$$c_i(x,p) := \partial\left(U_i|_{\langle p,x\rangle \leq r}\right)(x) = U_i'(x) - p$$

(when the utility function U_i is assumed to be differentiable. Otherwise, one can take the generalized gradient of U_i, and obtain a set-valued change map. We thus have to replace differential equations by differential inclusions to describe the behavior of consumers, and this is possible.)

---

[17]One can associate with any change function c_i a demand (set-valued) map D_i defined by $x \in D_i(p)$ if and only if $c_i(x,p) = 0$.

Actually, under adequate convexity assumptions, the static maximization mechanism: Find an allocation $\bar{x} = (\bar{x}_1, \ldots, \bar{x}_n)$ such that $\sum_{i=1}^{n} \bar{x}_i \in M$ and

$$\sum_{i=1}^{n} U_i(\bar{x}_i) = \max_{\sum_{i=1}^{n} x_i \in M} \sum_{i=1}^{n} U_i(x_i)$$

providing Pareto-optimal allocations conceals the three kind of algorithms: a Walrasian tâtonnement, a nontâtonnement process and a viable one. Indeed, convex analysis provides three characterizations of such an optimal solutions $(\bar{x}_1, \ldots, \bar{x}_n)$ and its Lagrange multiplier $\bar{p}$:

1. $(\bar{x}_1, \ldots, \bar{x}_n)$ maximizes the restriction of the function $\sum_{i=1}^{n} U_i$ to the set of allocations
2. $\bar{p}$ minimizes the dual problem
3. $(\bar{x}_1, \ldots, \bar{x}_n, \bar{p})$ is a saddle point of the associated Lagrangian

The (generalized) gradient method of each of the above problems are

1. a viable process

$$\begin{cases} x_i'(t) & = & U_i'(x(t)) - p(t) \\ \text{where} & & \\ p(t) \in N_M(x_1(t), \ldots, x_n(t)) & & \end{cases}$$

2. a nontâtonnement process of the form

$$\begin{cases} x_i'(t) & = & U_i'(x(t)) - p(t) \\ \text{where} & & \\ p'(t) & \in & \sum_{i=1}^{n} x_i(t) - S_M(p(t)) \end{cases}$$

(where the supply map S_M associates with any price p the set of resources yielding the maximal income)

3. a Walrasian tâtonnement process

$$p'(t) \in E(p(t))$$

where $E(p) = \sum_{i=1}^{n} D_i(p) - S_M(p)$ is the excess demand when $D_i(p)$ denotes the demand map derived from the utility function.

The Crandall-Pazy theorem implies the existence of solutions to each one of these three mechanisms (since those three dynamics are maximal monotone) and the convergence of both the viable process and the Walrasian tâtonnement, since the dynamics are (generalized) gradient methods.

This provides at least a class of instances where the Walras tâtonnement converges, offering positive examples countering Scarf's counter-example. □

Actually, we shall prove that the budgetary rule is just a sufficient condition, easy to implement for the viability of such a decentralized dynamical system and the existence of an equilibrium.

The viability theorem allows us to characterize the above viability property in terms of a pricing map Π_M associating with any allocation $(x_1, \ldots, x_n)$ a subset $\Pi_M(x_1, \ldots, x_n)$ of prices. This pricing map is built from the knowledge of M and the change functions c_i of the consumers.

The basic viability theorem implies that *the above dynamical system is viable if and only if for any allocation* $(x_1, \ldots, x_n)$*, the subset* $\Pi_M(x_1, \ldots, x_n)$ *is not empty. In this case, the existence of an equilibrium is also guaranteed.*

Furthermore, the prices which govern the evolution of allocations evolve according the (set-valued) "regulation law"

$$\forall\, t, \;\; p(t) \;\in\; \Pi_M(x_1(t), \ldots, x_n(t))$$

The above law provides at each instant the set of "viable" prices which govern the evolution of allocations of scarce resources, stating *a posteriori* how the invisible hand should choose the prices[18].

Once this has been established, one can raise other questions. Adapt, for instance, the above results when consumers would like their preference ordering increase along the evolution of their consumptions. This can be solved by bridling the initial regulation map Π_M, replacing it by a smaller one which associates the set of prices $p(t)$ providing allocations satisfying this supplementary requirement. We can take into account a nonstationary environment when both the change functions, the pricing map and the set of allocations depends either upon the time, or the history of the evolution of the allocations. We can take into account the case in which the sets of available commodities obey an evolution law (mutational equations, that are the analogues of differential equations governing the evolution of vectors).

Even in a stationary environment, we shall answer more traditional issues. For instance, whether the allocations $(x_1(t), \ldots, x_n(t))$ converges asymptotically to a limit $(\bar{x}_1, \ldots, \bar{x}_n)$, which is then an allocation, whether this convergence is at each instant Pareto-improving. Whether there is an attractor, whether there is a repeller around which the evolution fluctuates, whether the system is chaotic à la Saari. If a given loss function does not decrease along the evolution, can we reveal the concealed loss function which does have this property ? These are questions we shall attempt to answer affirmatively.

Prices offer a first source of variability — or flexibility — to allow the consumers to make their choice in a decentralized way by offering them as many velocities as potential prices. In some sense, this very variability offers a first source of contingent uncertainty, which has to be ultimately tamed by the scarcity of resources. However, a change function still describes a "price-dependent" deterministic behavior of a consumer because, when at

[18] such prices can be regarded as "open loop" controls in the terminology of systems theory.

time t, her consumption level and the price are known, the change function determines univoquely the velocity with which she modifies her consumption pattern. But she can make errors, after all, and not really wish or desire this prescribed velocity. We then can introduce a set of velocities[19] instead of one unique velocity associated with her consumption level and the price. In other words, we may describe the dynamical behavior of a consumer by a **differential inclusion** instead of a differential equation. This does not complicate that much the mathematical treatment, since differential inclusions offer the mathematical tools used in the first place to treat the initial viability problem. For instance, one source of uncertainty can be "frustrations" around a change map $c_i(x,p)$, which measures the difference between the desired (*a priori*) $c_i(x,p)$ and an actual (*a posteriori*) modification $c_i(x,p) - u_i$, where $\|u_i\|$ measures the frustration.

However, we shall answer before our basic question: Find the viable laws of supply and demand.

1.3 Deriving Decentralizing Supply and Demand Laws

The above regulation law is not yet a "supply and demand" law, because it does not tell **how the velocities of prices vary**. By supply and demand law, we mean here an adjustment law which is a dynamical process of the form

$$p'(t) \;=\; g(x_1(t), \ldots, x_n(t), p(t))$$

as in nontâtonnement models[20].

Naturally, if the pricing map $\Pi_M = \pi_M$ were actually single-valued and differentiable, the chain rule allows us to derive from the regulation law

$$\forall\, t,\;\; p(t) \;=\; \pi_M(x_1(t), \ldots, x_n(t))$$

[19]or even, a **fuzzy set of velocities**, since there are theorems on fuzzy viability for fuzzy differential inclusions (see [27,28, Aubin] for instance.). This allows us to put a kind of cost paid by the velocity to belongs to the fuzzy set of velocities. I uphold the view that differential inclusions or fuzzy differential inclusions provide more valid metaphors of the kind of uncertainty one find in economics than stochastic differential equations. However, there is a whole range of problems where some stochastic regularity makes sense, and which can be studied through stochastic differential equations. Since viability issues are imperative in economics and finance, G. da Prato and I have adapted the basic viability theorems for stochastic differential equations, which are presented in chapter 15 of the Tool-Box.

[20]for which we do not require *a priori* that price goes up when demand exceeds supply. One can not exclude that for certain economies prices of certain commodities interfere with the modification of prices of other commodities in order to maintain continuous transactions (the viability condition).

the law

$$\forall t,\ p'(t) = \pi'_M(x_1(t),\ldots,x_n(t))(c_1(x_1(t),p(t)),\ldots,c_n(x_n(t),p(t)))$$

which can be regarded as a supply and demand law. For such a law, the nontâtonnement system

$$\begin{cases} \forall i=1,\ldots n, & x_i'(t) = c_i(x_i(t),p(t)) \\ p'(t) & = \pi'_M(x_1(t),\ldots,x_n(t))(c_1(x_1(t),p(t)),\ldots,c_n(x_n(t),p(t))) \end{cases}$$

yields at each instant allocations of scarce resources (in a decentralized way).

When the pricing map is set-valued, the differential calculus of set-valued maps[21] allows also to differentiate the regulation law. It is possible to define a concept of derivative of set-valued maps. Denote by $\Pi'_M(x_1,\ldots,x_n,p)$ the derivative of the pricing map[22] Π_M at a point $(x_1,\ldots,x_n,p)$ of its graph, which is still a set-valued map associating with $(v_1,\ldots,v_n)$ a set

$$\Pi'_M(x_1,\ldots,x_n,p)(v_1,\ldots,v_n)$$

of marginal prices.

From now on, we shall set

$$G_M(x_1,\ldots,x_n,p) := \Pi'_M(x_1,\ldots,x_n,p)(c_1(x_1,p),\ldots,c_n(x_n,p))$$

One can prove a "set-valued chain rule" which allows to differentiate the regulation law and yields

$$\forall t,\ p'(t) \in G_M(x_1(t),\ldots,x_n(t),p(t))$$

We shall regard this differential inclusion as a "Supply and Demand Map" specifically designed to guarantee the viability of the decentralized nontâtonnement mechanism

$$\begin{cases} \forall i=1,\ldots n, & x_i'(t) = c_i(x_i(t),p(t)) \\ & \\ p'(t) \in G_M(x_1(t),\ldots,x_n(t),p(t)) & \end{cases}$$

The set-valued map $G_M(x_1,\ldots,x_n,p)$ is a kind of envelope of all single-valued supply and demand maps $g(x_1,\ldots,x_n,p)$ for which the **decentralized nontâtonnement process**

$$\begin{cases} \forall i=1,\ldots n, & x_i'(t) = c_i(x_i(t),p(t)) \\ & \\ p'(t) = g(x_1(t),\ldots,x_n(t),p(t)) & \end{cases}$$

[21] which was motivated by this very problem in the first place. For an exposition of set-valued analysis, we refer to the monograph SET-VALUED ANALYSIS by Hélène Frankowska and the author.

[22] Actually, we shall differentiate a "smaller" regulation map which governs the evolution of almost everywhere differentiable prices. This is a technical issue which is too long to explain here.

provides at each instant viable allocations of scarce resources.

In the language of viability theory, such selections g of G_M are called "adjustment laws.

The question arises how to find such selection procedures, and among them, to find the ones which retain some economic meaning.

We shall now advocate the ones which satisfy the **inertia principle**.

1.4 The Inertia Principle and Heavy Evolutions

Actually, if the behavior of the consumers is well defined, what about either the Market or the Planning Bureau, the task of which is now to find the prices $p(t)$ in $\Pi_M(x_1(t), \ldots, x_n(t))$? They do not behave as actual decision makers, knowing what is good or not (this is the case of even a planning bureau as soon as it involves more than three bureaucrats!). Hence, their role is only a **regulatory one**. If they are not able to optimize, we may assume that they only are able to correct the prices when the viability of the economic system is at stake, i.e., when the total consumption is no longer available.

Hence, we assume that Adam Smith's "Invisible Hand" or the Market or the Planning Bureau are able to "pilot" or "act" on the system by choosing such regulees according to the **inertia principle**:

> **Keep the price constant as long as the evolution provides allocations of available resources, and change them only when the viability is at stakes.**

Indeed, as long as the state of the system lies in the interior of the allocation set (the set of states satisfying scarcity constraints), any price will work. Therefore, the system can maintain the price inherited from the past. This happens if the system obeys the inertia principle (in its most extreme form, which can be ulteriorly be made more sophisticated).

Since the allocations may evolve while the price remains constant, the total consumption may reach the boundary of the set of scarce resources with an "outward" velocity. This event corresponds to a period of *crisis*: To survive, the system must find another price such that the new associated velocity forces the solution back inside the allocation set. Alternatively, if the scarcity constraints can evolve, another way to resolve the crisis is to relax the constraints (by technological progress, for instance) so that the state of the system lies in the interior of the new allocation set. When this is not possible, **strategies for structural change fail**: by design, this means that the solution leaves the allocation set and "dies".

This management by crisis or bankruptcy has been observed in economic history, so that we suggest to take these phenomena into account in the framework of this Inertia Principle[23]. Crisis could be defined here as either discontinuity of the price evolution or "fast" evolutions of prices.

The extreme form of the inertia principle leads to the question whether one can obtain the evolution of allocations of scarce resources **under a constant price**. We shall say that a price $\widehat{p}$ is a **punctuated equilibrium** if there exists a nonempty subset $N(\widehat{p})$ of allocations, called its **viability niche**, which is viable for a price:

$$\begin{cases} \forall\,(x_1^0,\ldots,x_n^0) \in N(\widehat{p}), \\ \text{the solutions to the system} \\ \forall\, i = 1,\ldots n, \;\; x_i'(t) \;=\; c_i(x_i(t),\widehat{p}) \\ \text{remain in the viability niche } N(\widehat{p}) \end{cases}$$

One can expect that the viability niches of most of the prices are empty. Naturally, **when the viability niche of a punctuated equilibrium is reduced to an unique allocation, this allocation is an equilibrium.**

Therefore, we have to select "supply and demand" functions (adjustment laws) $g(x_1,\ldots,x_n,p)$ which obey the inertia principle, thus providing rules for choosing prices when viability is at stakes in order to obtain allocations of scarce resources.

The simplest one (and most often, the most reasonable one) is to assume that at each instant, the prices are changed as slowly as possible. This is obtained by taking for map g the map ϖ_M defined by:

$$\begin{cases} \varpi_M(x_1,\ldots,x_n,p) \;\in G_M(x_1,\ldots,x_n,p) \\ \text{is the element of minimal norm in} \\ G_M(x_1,\ldots,x_n,p) \end{cases}$$

It obviously satisfies the inertia principle because if the velocity 0 of the price is available, it is picked by such a selection procedure, so that the price remains constant as long as 0 belongs to $G_M(x_1(t),\ldots,x_n(t),p(t))$.

Evolutions obeying this specific choice are called "heavy[24] evolutions", in the sense of heavy trends. Hence heavy evolutions are obtained by requiring

[23]This Inertia Principle provides an explanation of the concept **punctuated equilibrium** introduced in 1972 by Eldredge and Gould in paleontology. Excavations at Kenya's Lake Turkana have provided clear evidence of evolution from one species to another. The rock strata there contain a series of fossils that show every small step of an evolution journey that seems to have proceeded in fits and starts. Examination of more than 3,000 fossils by P. Williamson showed how 13 species evolved. The record indicated that the animals stayed much the same for immensely long stretches of time. But twice, about two million years ago and then, 700,000 years ago, the pool of life seemed to explode — set off, apparently, by a drop in the lake's water level. Intermediate forms appeared very quickly, new species evolving in 5,000 to 50,000 years, after millions of years of constancy, leading paleontologists to challenge the accepted idea of continuous evolution.

[24]This is justified by the fact that the velocity of the price is related to the acceleration of the consumptions, which, being minimal, has the maximal inertia.

at each instant the (norm of the) velocity of the price to be as small as possible.

Heavy solutions enjoy the property of "locking-in" punctuated equilibria.

Indeed, assume that for some instant $t_f > 0$, $p(t_f)$ is a punctuated equilibrium, then $p(t) = p_{t_f}$ for all $t \geq t_f$ and thus, $(x_1(t), \ldots, x_n(t))$ *remains forever in the viability niche of this punctuated equilibrium (and thus, an allocation of scarce resources).*

Therefore, for implementing this inertia principle, we have to provide conditions under which relevant prices $p(\cdot)$ are differentiable (almost everywhere), to build the differential inclusion which governs the evolution of differentiable relevant prices and then, select a differential equation in this differential inclusion (called an "adjustment law") which will obey the inertia principle.

Naturally, the division of the variables into two groups, commodities and prices, is too crude to stop here our investigation, even though it is better than no distinction at all between them.

We incidentally mentioned the introduction of frustrations, as a third group of variables to describe the individual uncertainty involved in the behavior of the consumers. It may be reasonable to assume that the evolution of frustrations depend upon the evolution of prices, whereas, in a non symmetric way, the evolution of prices is independent of the evolution of frustrations (while depending naturally upon the levels of consumptions, prices and frustrations). This leads to a **hierarchical game** where, in a sense made precise later, we attribute to the evolution of the price a heavier inertia than to the evolution of the frustrations, and to this one, a heavier inertia than the evolution of commodities. We can then define a "cascade" of niches of different orders, which lock-in successively the prices and the frustrations.

The problem of different time scales — a phenomenon of considerable importance in the articulation between economics and ecology — can be taken into account with the concepts of heavy solutions (and hierarchical heavy solutions). The traditional approach to these multiple time scales appeals to singular perturbation theory, where fast and slow evolutions are imposed *a priori* by imposing infinitely large or small parameters[25]. The inertia principle postulates an ordering of the variable of a system which provides the actual velocities of each variable instead of imposing these velocities a priori, and so, does not forbid slow solutions to evolve quickly when viability crisis happen.

[25] A new approach to the study of these slow and fast dynamics has been the use of "nonstandard analysis".

In summary, **given the decentralized behavior of the consumers described by the differential equations** $x_i' = c_i(x_i, p)$ **and the set of scarce resources, we can built the dynamics** ϖ_M **governing the behavior of the Market**, so that the evolution of the economic system is described by the system of differential equations

$$\begin{cases} i) & x_i'(t) = c_i(x_i(t), p(t)) \quad (i = 1, \ldots, n) \\ ii) & p'(t) = \varpi_M(x(t), p(t)) \end{cases}$$

Contrary to nontâtonnement models, this law governing the evolution of prices is not a modeling assumption, but a **consequence** of the modeling data of this elementary model satisfying the inertia principle.

In a nutshell, we assume implicitly that the "Invisible Hand" follows an "opportunistic" and "conservative" behavior of the system: A behavior which enables the system to allocate scarce resources among consumers as long as any price makes possible its regulation and to keep this price as long as it is possible. The "invisible hand" is an hand without qualities.

We thus attempt to explain the evolution of allocations and prices and to reveal the concealed feedbacks which allow the system to be regulated by prices.

1.5 Planning Procedures

Another way to find prices satisfying the regulation law is to obtain them by planning procedures.

This means that the planning bureau has to associate with any allocation $(x_1, \ldots, x_n)$ a price $\pi(x_1, \ldots, x_n)$ which it sends back to the consumer.

The viability theorem states that whenever the map

$$(x_1, \ldots, x_n) \mapsto \pi(x_1, \ldots, x_n)$$

is a **selection** of the pricing map Π_M in the sense that

$$\forall\, (x_1, \ldots, x_n), \quad \pi(x_1, \ldots, x_n) \in \Pi_M(x_1, \ldots, x_n)$$

then the evolution of consumption according

$$\forall\, i = 1, \ldots, n, \quad x_i'(t) = c_i(x_i(t), \pi(x_1(t), \ldots, x_n(t)))$$

yields allocations of scarce resources.

Planning procedures introduced by Drèze, de la Vallée-Poussin, Malinvaud, etc. fall in this category, and can be interpreted as closed loop controls.

One can obtain such selection by static optimization techniques (or, more generally, by game theoretical methods, or any other kind of technique). For instance, we can choose the element $\pi^0(x) \in \Pi(x)$ of minimal norm. Despite the lack of continuity of such a selection, we still can prove that the system of differential equations

$$x_i'(t) = c_i(x_i(t), \pi^0(x_1(t), \ldots, x_n(t)))$$

has viable solutions, which are called "slow allocations".

We shall prove that in the case when $c_i(x,p) := g_i(x) - p$, the slow evolution is the same than the evolution of the connection economy

$$\forall\, i = 1, \ldots, n, \;\; x_i'(t) = \sum_{j=1}^{n} w_j^i(t) g_j(x_j(t))$$

where the connecting matrices $w_i^j(t) \in \mathcal{L}(Y,Y)$ are as close as possible to the identity matrix $\mathbf{1}$.

This equivalence between slow regulation by prices and maximal decentralization by connecting matrices is quite striking.

However, this type of selection may not enjoy economic meaning, contrary to the heavy allocations we proposed earlier.

These selection procedures are **myopic** contrary to intertemporal optimization mechanisms.

Indeed, in the dynamical case, selection procedures split in two categories: we have to distinguish between "intertemporal optimization" problems and "myopic or instantaneous optimization" problems.

In intertemporal optimization (under infinite horizon), we maximize intertemporal discounted utility functions of the form

$$\int_0^\infty e^{at} W(t, x(t), p(t)) dt$$

under the constraint $(x(\cdot), p(\cdot)) \in \mathrm{Graph}(\Pi)$.

These are questions with which Calculus of Variations and Optimal Control Theory deal with. We shall briefly relate in Chapter 8 the problem of intertemporal optimization with a viability problem.

But **Optimal Control Theory** does require Adam Smith's invisible hand to "guide" the system by optimizing such an **intertemporal** optimality criterion, the choice of which is open to question even in static models, even when multicriteria or several decision makers are involved in the model.

Furthermore, the choice (even conditional) of the controls is made **once and for all** at some initial time, so that **they cannot be modified at each instant so as to take into account possible modifications** of the environment of the **system**, forbidding therefore **adaptation** to scarcity constraints.

Finally, *intertemporal optimization does require knowledge of the future* (even of a stochastic nature.) This postulates the possibility of experimentation or the belief that the phenomenon under study is periodic. Experimentation, by assuming that the evolution of the state of the system starting from a given initial state for a same period of time will be the same whatever the initial time, allows one to translate the time interval back and forth, and, thus, to "know" the future evolution of the system.

But in economics, as well as in biological evolution, experimentation is not possible[26]. Furthermore, **the dynamics of the system disappear and cannot be recreated.** Most economic systems do involve myopic behavior; while they cannot take into account the future, they are certainly constrained by the past.

We rather focus our attention to the problem of adaptation to the evolution of scarcity constraints and to the varying behavior of other economic agents on which no individual agent has a total control — and sometimes, no influence whatsoever.

Therefore, instead of using intertemporal optimization[27] that involves the future, we shall advocate here myopic selection procedures, either by adjustment laws (supply and demand maps) or by closed loop prices (planning procedures), for providing selection procedures of **viable evolutions** obeying, at each instant, scarcity or more generally, viability constraints which depend upon the **present or the past.** (This does not exclude **anticipations**, which are extrapolations of past evolutions, constraining in the last analysis the evolution of the system to be a function of its history.)

Viability theory deals with "dynamics under constraints", playing the role of "optimization under constraints" or "existence of equilibrium under constraints" in a dynamical manner. The viability theorem, which is an extension of Nagumo's theorem to differential inclusions, actually requires that the "regulees" involved in the nondeterministic dynamics play the role of Lagrange's or Kuhn-Tucker's multipliers, which is the reason why we call them "viability multipliers". In both cases, the concept of tangent cones to constained sets for "differentiating" the constraints, so to speak, and by duality, the concept of normal cones to the constraints are used, allowing the dual interpretation by prices and "budgetary" conditions to pop up to complement or replace conditions bearing on physical commodities. In the same way than optimization under constraints is much more difficult to handle than optimization without constraints, viability theory raises more obstacles to overcome than the study of ordinary systems of differential equations.

In some operational sense, the basic viability theorem can replace the Kakutani fixed point theorem each time it is used in a static model to make

[26]The twentieth century Soviet type (or military type) economic experimentation showed experimentally the limits of centralized operation of complex systems.

[27]which can be traced back to Sumerian mythology which is at the origin of Genesis: one Decision-Maker, deciding what is good and bad and choosing the best (fortunately, on an intertemporal basis, thus wisely postponing to eternity the verification of optimality), knowing the future, and having taken the optimal decisions, well, during one week...

it "dynamical". This is the strategy which we have tried to explain in the case of the problem of decentralized dynamical allocation of resources.

Chapter 2

Other Miscellaneous Illustrations

Introduction

We illustrate here the concept of viable solution, of viability kernels and niches, of the inertia principle and of heavy solutions in the case of simple explicit dynamical models for which one can exhibit the viability kernels.

The only advantage of these caricatural examples is to provide a graphical description of some results and to compare, *mutatis mutandis*, the evolutions of the state and of the regulee. These examples can tell many other stories, when one does not forget that the thousand remaining parameters which describe any reasonable model are frozen.

Several lessons can be inferred from these examples. For instance, a same recipe (keep the price constant) can have opposite consequences on the growth of consumption depending upon the initial state. A second one is that the intuition elaborated in a static framework may sometimes mislead us when evolution is present, by suggesting that such or such result, which makes sense in a static world, may have perverse effects in a dynamic one.

But it is dangerous to draw more conclusions than these modest models conceal. However, if a given reasoning is not validated in the case of such a simple model, it will not be valid in a more complex one.

1. We provide a presentation of the main concepts in the framework of the simplest model with one consumer, one sate — the consumption — and one regulee — the price. Naturally, replacing consumptions and prices by other items may provide other illustrations. We review the concepts of equilibrium, stable or not, of bifurcations and of viability niches of a regulee, which is the set of states regulated by a constant regulee. This leads to the inertia principle and to the problem of bounding inflation, which is handled by introducing "metasystems" from which we can obtain the heavy solutions.

2. We next provide more explicit analytical formula when the dynamics are linear or bilinear. Bilinear systems appear naturally in population dynamics, where the state is the population size and the regulee the Malthusian coefficient. The upper bound of the viability interval is called the **carrying capacity** of the population. We provide the explicit form of the regulation map under bounded inflation, which thus contains all the feedback maps expressing the evolution of viable Malthusian coefficients in terms of population size. The 1838 basic logistic Verhulst model naturally belongs to this class.

3. We then present a viability analysis of a simplified Keynesian model due to Jean Cartelier and Katharina Müllers, where they show that a low rate of interest is a most favorable condition not only because it is associated with a high equilibrium level of economic activity but also, and more significantly, because it increases the dynamical robustness of the economy.

4. The difficulties of collective decision-making models about climatic risks and technology motivated a viability analysis by Luc Doyen and Daniel Gabay taking into account the interactions between physical and economical requirements.

5. We also cite the same authors who used the viability approach for analyzing in the framework of the management of renewable resources the links between ecological models based on the logistic dynamics proposed by Verhulst and economic models by Graham and Schaeffer, which assume that the exploitation rate is proportional to the biomass and the economic activity.

6. These ideas have been further analyzed in the case of an ecological-economic model of the French Guyana shrimp fisheries by Christophe Béné and Luc Doyen. It gives an example of a time-dependent model, because both the evolution of the shrimps and the market are periodic.

7. We then expose two viability analysis of demographic models by Noël Bonneuil. The first one links the capital regarded as the state with regulee which are consumption and birth rates. The second one deals with the interactions between the Malthusian view, which propounds that the population stabilizes at a given income level (depending of the living standard and of technology) and the Boserupian one which suggests that technological change is triggered by demographic pressure.

8. The one dimensional model with which we began is now slightly modified for studying the case when the inflation bound is state dependent. This triggers the apparition of cycles, as Katharina Müllers showed.

9. We next present two illustrations of dynamical games. One is a viability study by Noël Bonneuil and Katharina Müllers on the predator-prey model with two states and two regulees, where the authors provide an analytical characterization of the viability kernel. It shows that the viability kernel is bounded, which implies that outside this viability kernel any bounded evolution violates the positivity constraints. Next, we report one of the viability analysis by Noël Bonneuil of the anthropological studies of Fredrik Barth dealing with the nomadization strategy of the Basseri tribes.

2.1 One Agent, One Commodity, One Price

Let $K := [a, b]$ the subset of a scarce commodity x, say hot chocolate (so that — being liquid — it is indivisible!). Assume that our consumer is greedy, her behavior amounts to **consuming more and more chocolate**. The velocity $g(x(t))$ with which she changes her consumption is then positive. In this case, her consumption increases according to the differential equation $x'(t) = g(x(t))$ and will eventually exceed the upper bound set by scarcity.

A mechanism is needed to slow down her covetousness. The one we retain here penalizes the consumption velocity by a nonnegative price: the consumption velocity is the difference between the above velocity and a function proportional to the price. In other words, the dynamical system corrected by prices is described by

$$x'(t) = g(x(t)) - h(x(t))p(t), \quad \text{where } p(t) \geq 0$$

and where $h(x) \geq 0$.

We shall present several examples[1] of such differential equations and use the viability kernel algorithms to present the actual representation of the concepts we present below as a hors d'œuvre.

2.1.1 Equilibria and Bifurcations

First, we can single out the prices associated with equilibria: they are the ones at which the velocity vanishes (since an equilibrium is a stationary state). If we suppose that $f(x) := \dfrac{g(x)}{h(x)}$ is defined for all $x \in K$, then $f : x \mapsto p = f(x)$ is the map associating with any consumption the price for which it is an equilibrium, the graph of which is called the **equilibrium curve**.

We are more interested by the inverse of this map, which associates with any price the (possibly empty) set of equilibria for this price.

[1] In some of the examples below, the signs of f and g will be changed to accommodate the mathematical metaphors to the story told.

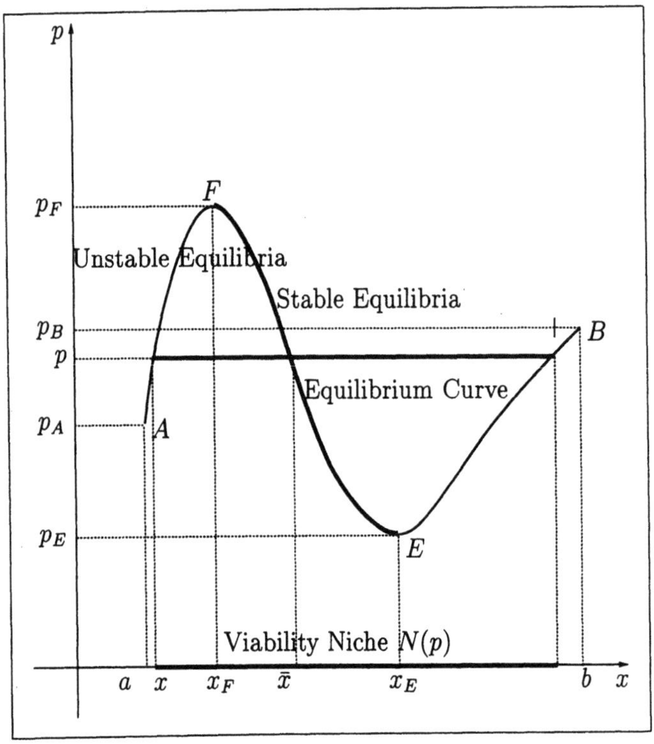

Figure 2.1: Stable and Unstable Equilibria, Bifurcations and Viability Niches

In the case of Figure 2.1, this set is empty if the price is either smaller than p_E or larger than p_F. At p_E appears a **bifurcation**: there exists an equilibrium x_E which gives birth to two equilibria when the price lies between p_E and p_A. At p_A, a third equilibrium appears, and there are three equilibria in this example when p ranges the interval $]p_A, p_B[$. We find again two equilibria when $p \in]p_B, p_F[$ and only one at p_F. In this example, there are four bifurcation prices, p_E, p_A, p_B and p_F.

Below the equilibrium curve, one notices that the consumption velocities are always positive, and negative above. Hence the two increasing branches AF and EB of the equilibrium curve are made of **unstable equilibria**. Indeed, for a given price, whether the initial consumption is slightly at the right or the left of the equilibrium, the consumption goes away from the equilibrium to the right and to the left respectively. Such equilibria are **repellers**.

Symmetrically, the decreasing branch FE of the equilibrium curve is made of **stable equilibria**. Indeed, for a given price, whether the initial consumption is slightly at the right or the left of the equilibrium, the consumption goes towards to the equilibrium from the left and from the right respectively. Such equilibria are actually **attractors**, since the consumptions starting nearby converge monotonically to such equilibria.

Equilibria E and F are called **saddle points**: attractors in one direction, repellers in the other.

2.1.2 The Regulation Map

Actually, in an evolutionary perspective, we are not interested by equilibria and their stability, but by the evolution of both prices and consumptions which comply the scarcity constraint.

The (set-valued) **pricing map** Π_K which associates with consumption x the set of viable prices: if the consumption is equal to a, a nonnegative velocity is required to renew with growth, and this is possible with prices ranging between 0 and p_A.

If the consumption belongs to the open interval $]a, b[$, then all prices are possible because, for short periods, the consumption can increase or decrease without leaving the scarcity interval.

Finally, when the consumption is equal to b, nonpositive velocities are required to slow it down, and this happens with prices larger than or equal to p_B.

In other words, the pricing map is defined by

$$\Pi_K(a) = [0, p_A],\ \Pi_K(x) = [0, \infty[\text{ when } a < x < b\ \&\ \Pi_K(b) = [p_B, +\infty[$$

(see Figure 2.2). This pricing map, being set-valued, is not deterministic: it leaves the possibility to errors. It imposes sufficiently low prices when the lower bound is reached and sufficiently high when the upper bound is achieved.

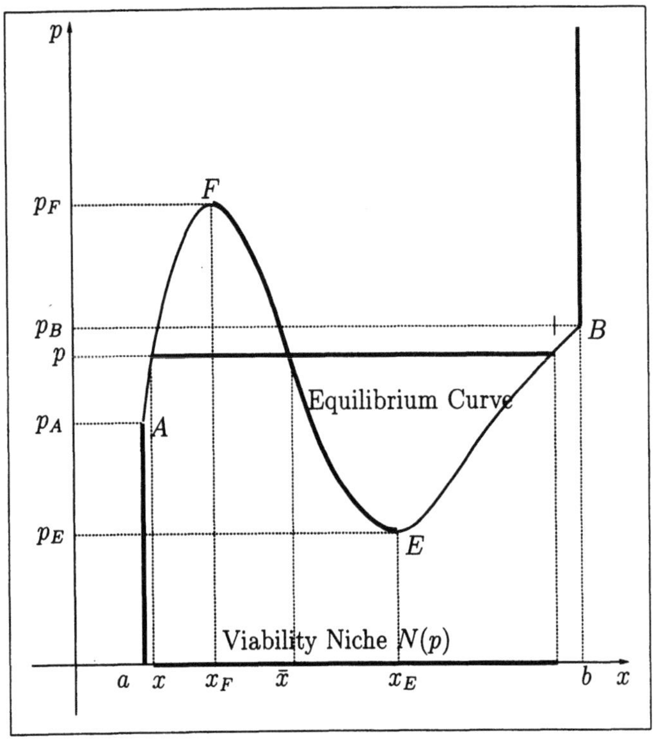

Figure 2.2: Regulation Map R_K

The pricing map guides the evolution of prices through the **regulation law**

$$\text{for all } t \geq 0, \quad p(t) \in \Pi_K(x(t))$$

It dictates what Adam Smith's invisible hand must at least do to choose the prices in order to comply the scarcity constraints.

On can devise a multitude of planning mechanisms which are consistent with this regulation map: **planning mechanisms** are selections $\pi(x) \in \Pi_K(x)$ of the pricing map used as **feedbacks** for which the differential equation

$$x'(t) = g(x(t)) - h(x(t))\pi(x(t))$$

has solutions (which are necessarily viable).

But using planning procedures is not the objective we pursue: we wish to show how the **inertia principle** works on this example.

2.1.3 Viability Niches

Consider an initial commodity-price pair below the equilibrium curve. Then, keeping the price constant, the velocity remains nonnegative and the consumption grows up.

At constant price (or at **zero inflation**), the scarcity interval $[a, b]$ is not necessarily viable. The **viability kernel** of our dynamical system regulated by prices is, by definition, the largest viability domain contained in the scarcity interval.

Take for instance a price smaller than p_E. The consumption grows to the upper bound b which it reaches with a positive velocity, so that it violates the scarcity constraint in **finite time**. The viability niche of such price is thus empty. When the price is p_E, the viability niche is reduced to the equilibrium x_E. When the price lies between p_E and p_A, its viability niche is the interval the lower bound of which is a and the upper bound is the unstable equilibrium associated with this price. The associated stable equilibrium is then an attractor of this system under constant price in the niche.

When the price ranges in $]p_A, p_B[$, its viability niche is the interval lying between the smallest unstable equilibrium and the largest unstable equilibrium, the stable equilibrium being an attractor of this system under zero inflation. Between p_B and p_F, the viability niche is the interval whose lower bound is the unstable interval and the upper bound is b. Finally, for p_F, the viability niche is made of the equilibrium x_F.

If the initial consumption is outside of the viability niche of a given price (which is always the case when its viability niche is empty), then the consumption evolving under constant price shall violate the scarcity constraint in finite time, either by traversing the upper bound if the price is below the equilibrium curve, or by going below the lower bound a if the price is above.

When the consumption reaches the upper bound b, Adam Smith's invisible hand must increase **at once** the price above the price p_B prescribed by the

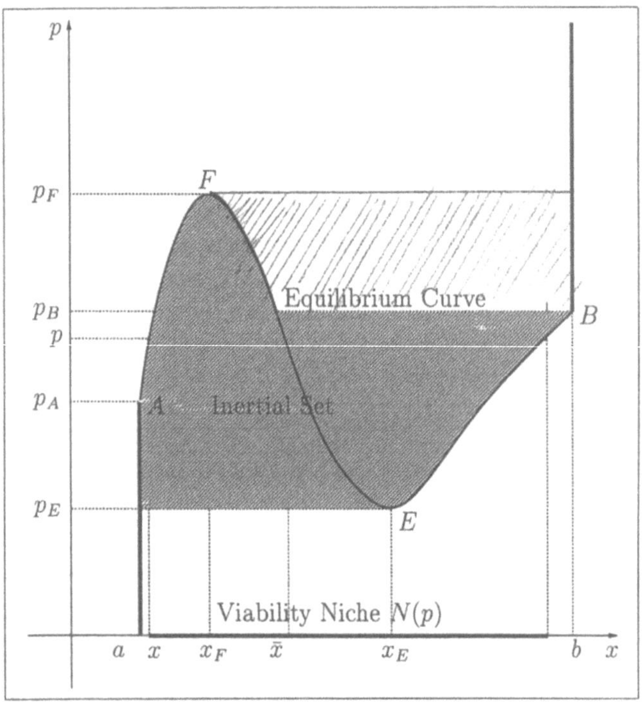

Figure 2.3: Inertial Set

pricing map. Or, if the price is above the equilibrium price, it must decrease ruggedly the price below p_A.

A first lesson can be drawn from this example: a same recipe — maintain the price constant — **can produce strictly contradictory consequences** — increase or decrease consumption.

A second lesson is the danger caused by the inertia principle: when the consumption reaches the boundary of the scarcity interval, only discontinuous changes (**impulses**) can maintain the viability of the system.

2.1.4 Bounding Inflation

Inflationist and Deflationist Solutions

When the initial consumption-price pair is below the equilibrium curve and the initial consumption outside the viability niche of the price, one can **slow down** consumption by increasing prices at a fixed velocity, say equal to some $c \geq 0$ (in other words, the price increases proportionally to time). Such solutions are called **inflationist**.

In this case, the consumption-price pair ranges over a trajectory such as the curve CD[2]. This curve is increasing below the equilibrium curve, cuts vertically this equilibrium curve and is decreasing above. Starting from the point C, both consumption and price increase until the consumption reaches its maximum and decreases, while the price continues to grow up.

All the trajectories of inflationist consumption-price pairs share the same behavior and do not cross.

We obtain the symmetric situation when the initial consumption-price pair is above the equilibrium curve and the initial consumption is outside its viability niche. The consumption is accelerated by diminishing prices at a

[2]These curves are the trajectories of the solutions $\rho_c^!$ to the differential equation

$$\frac{d}{dp}\rho_c^!(p) = \frac{g(\rho_c^!(p)) - h(\rho_c^!(p))p}{c} \tag{2.1.1}$$

As functions $p \mapsto x = \rho_c^!(p)$, they are increasing when the right-hand side is positive, i.e., below the equilibrium curve, vanish when they cross the equilibrium curve and decreasing above.

Indeed, when $p(t)$ is a solution to the differential equation $p'(t) = \psi(x(t), p(t))$, then the solution $x(t) := \rho(p(t))$ to the differential equation

$$x'(t) = g(x(t)) - h(x(t))p(t)$$

satisfies

$$x'(t) = \left(\frac{d\rho(p)}{dp}\right)p'(t) = \left(\frac{d\rho(p)}{dp}\right)\psi(x(t), p(t)) = g(\rho(p(t))) - h(\rho(p(t)))p(t)$$

In the case of inflationist solutions $\rho_c^!$, one takes $\psi(x, p) = c$.

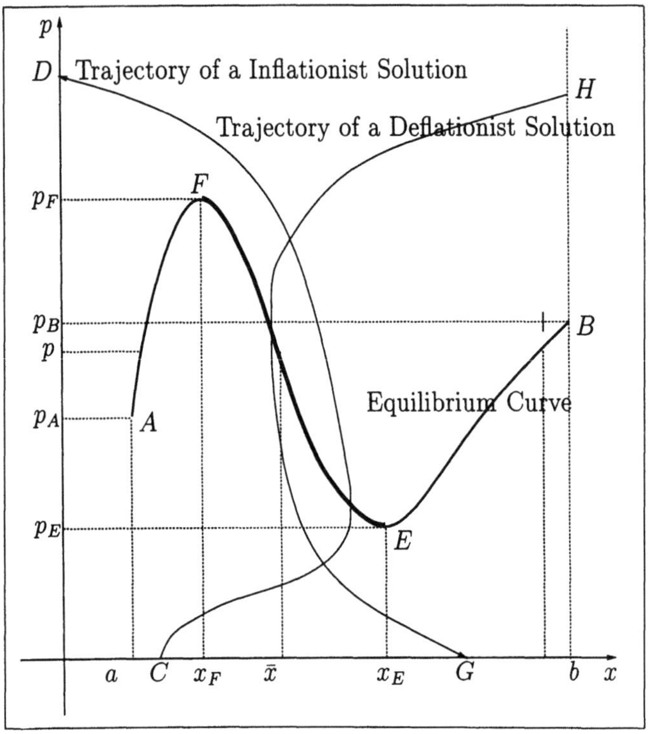

Figure 2.4: Trajectories of inflationist and deflationist solutions

constant velocity, say $-d$. Such solutions are called **deflationist**[3]. In this case, the consumption-price pair ranges a trajectory such as the curve GH. This curve is decreasing below the equilibrium curve, crosses it vertically and increasing above. Starting from H, both consumption and price decrease until the consumption achieves its minimum and increases again, while the price continues to go down.

2.1.5 The Metasystem

One may reject zero-inflation scenarios as exceptional and thus, the rigid form of the inertia principle which may lead to brutal discontinuities (impulses) of prices to maintain the viability of the system when the boundary of the scarcity interval is reached. But, when one think about it, it is not the inertia principle which has to be discarded, but the primitive aspect of it, which does not forbid impulses, i.e., discontinuous mutations of prices.

So, we add another assumption to this model: **prices must respect speed limits**. In other words, one imposes **bounds to inflation**: the price velocity must remain between bounds $-d$ and $+c$. Therefore, the evolution of consumption-price pairs is governed by the system of differential inclusions, called the **metasystem**

$$\begin{cases} i) \quad x'(t) = g(x(t)) - h(x(t))p(t) \ \text{ where } \ p(t) \geq 0 \\ ii) \quad \text{and} \quad -d \leq p'(t) \leq +c \end{cases} \tag{2.1.3}$$

Here, the former state and regulee become the two components (x, p) of the metasystem and the regulee of the metasystem is now the velocity of the regulee, which is left to evolve in a bounded interval.

The model becomes more complicated to solve, and one cannot guess what is the new pricing map. One can envision that if the consumption is near the boundary with a price such that the consumption-price is below the equilibrium curve, it would be impossible to keep the consumption inside the scarcity interval without prices violating the speed limit. This is because the old pricing map, valid for systems without speed limits, is no longer a pricing map for the metasystem (2.1.3) completed by speed limits on prices, i.e., under bounded inflation.

---

[3]In the case of deflationist solutions $\rho_d^\flat$, one takes $\psi(x,p) = -d$. These curves are then the trajectories of the solutions $\rho_d^\flat$ to the differential equation

$$\frac{d}{dp}\rho_d^\flat(p) = -\frac{g(\rho_d^\flat(p)) - h(\rho_d^\flat(p))p}{d} \tag{2.1.2}$$

As functions $p \mapsto x = \rho_d^\flat(p)$, they are decreasing above the equilibrium curve, vanish when they cross it and increasing below.

However, in this example, one can describe the new pricing map taking into account the bounded inflation[4].

For each consumption, the prices must range between the part of the inflationist trajectory below the equilibrium curve which goes through B and the part of the deflationist trajectory above the equilibrium curve which goes through A[5].

2.1.6 The Inertia Principle

Heavy Evolutions

A viable evolution is said to be **heavy** if the absolute value of the velocity of the viable price is the smallest one.

Let us begin by analyzing the case when the initial consumption-price pair $(x_0.p_0)$ is below the equilibrium curve, but above the inflationist curve going through B (which is the lower boundary of the graph of the new pricing rule).

If the initial consumption x_0 belongs to the viability niche of the initial price p_0, we already saw that it converges to the stable equilibrium associated with p_0, and thus, it is a heavy solution *par excellence*. If not, when the initial consumption is outside the viability niche of p_0, the consumption increases while the price remains constant, so that the trajectory of the consumption-price pair is horizontal. This works until it hits the inflationist curve going through B (for a consumption smaller than the upper bound b). Then, and only then, the price starts to evolve with the smallest velocity. A simple computation shows that there is only one viable price velocity, equal to $+c$. The consumption-price pair ranges now the inflationist curve until

[4]Its graph is the viability kernel of the rectangle $[a,b] \times [0,+\infty[$ for the system of differential inclusions (2.1.3). Therefore, the viability kernel algorithm and software allow us to compute them in the examples presented here.

[5]They are defined through the functions $\rho_c^{\sharp}$ and $\rho_d^{\flat}$ which are the solutions to differential equations (2.1.1) and (2.1.2) going through B and A respectively.

The equations of the upper curve is equal to

$$r_d^{\flat}(x) = p \text{ if and only if } x = \rho_d^{\flat}(p)$$

and the lower one is equal to

$$\begin{cases} r_c^{\sharp}(x) = 0 \text{ if } 0 \leq x \leq \rho_c^{\sharp}(0) \\ r_c^{\sharp}(x) = p \text{ if and only if } x = \rho_c^{\sharp}(p) \text{ when } \rho_c^{\sharp}(0) < x \leq b \end{cases}$$

The new pricing map associates with any consumption x the prices p such that

$$r_c^{\sharp}(x) \leq p \leq r_c^{\flat}(x)$$

The growth properties of the functions r and ρ are exchanged, since one is the inverse of the other.

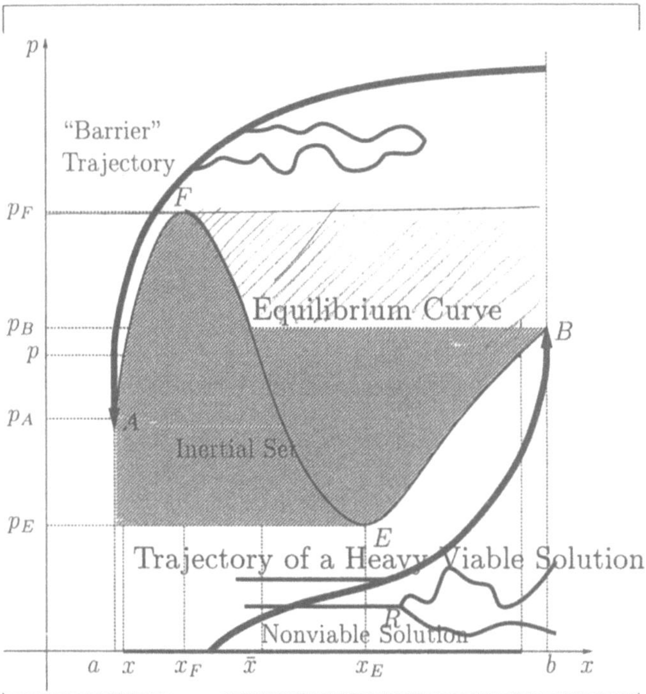

Figure 2.5: **Heavy Evolutions**

the moment when it achieves the equilibrium B. Being a heavy solution, it remains at this equilibrium.

We thus built the heavy solution starting at (x_0, p_0) to the **nondeterministic** system (2.1.3). Hence, the region comprised between the inflation curve going though B and the equilibrium curve is contained in the viability kernel of the system (2.1.3) under bounded inflation.

We also remark that the upper bound B is then **locked-in** by every heavy solution starting from a consumption-price pair below the equilibrium pair and above the inflationist curve going through B, when x_0 does not belong to the viability niche of p_0 (otherwise, the stable equilibrium associated with p_0 is locked-in).

If the price is maintained constant after crossing the inflationist curve going through B, it is impossible to maintain a viable consumption with an inflation below c. Indeed, let $R := (\bar{x}, p_0)$ be such a point. By braking the consumption taking the highest price velocity, the consumption-price pair ranges along an inflationist curve below the one which passes through B, and which does not cross it. Therefore, the consumption reaches b at finite time with a positive velocity, since the price is smaller than p_B. The solution

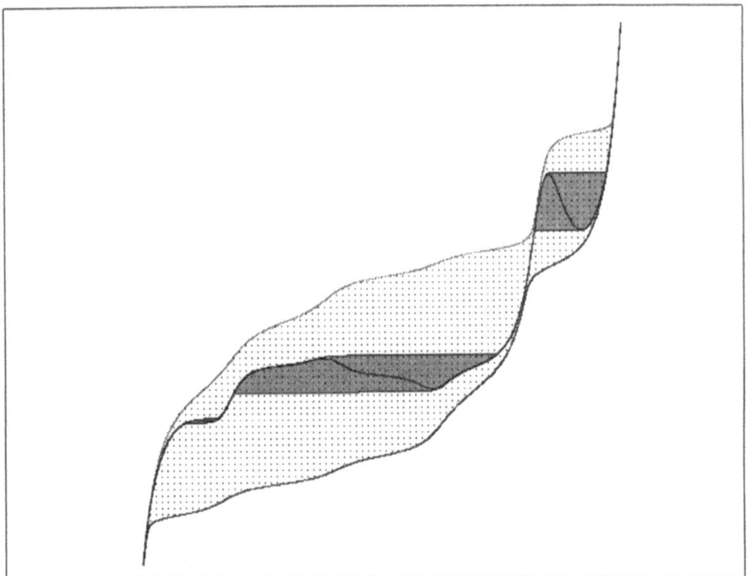

Viability kernel, inertial set and equilibrium curve for the system

$$\begin{cases} i. & x'(t) = x(t)(x(t)^2 - 1) - (\sin(2\pi x(t)) + 2)p(t) \\ ii. & p'(t) \in [-c, +c] \end{cases}$$

obtained by the viability kernel algorithm.

Figure 2.6: Viability Kernel of a Nonlinear System

is not viable: The situation is even worse for any other velocity $-d \leq p' \leq c$.

In other words, the speed limit on prices conceals a warning signal to prices, which must start to evolve with the highest velocity as soon as the inflationist curve going through B is reached. Otherwise, they need a higher inflation bound c_1 to remain viable (the inflation c_1 is the one such that the pair $R := (\bar{x}, p_0)$ lies in the trajectory associated with the inflationist curve associated with $c_1 \geq c$ and going through B).

The situation is completely symmetric if the initial consumption-price pair is above the equilibrium price and below the deflationist curve going through A. Consumption goes down under constant price until the consumption-price pair hits the deflationist curve going through A. Hence the price starts to decrease with velocity $-d$ until the consumption-price pair reaches the equilibrium A, at which it remains forever.

We observe a **hysteresis phenomenon**: if the price is too low, every heavy evolution starting outside its viability niche increases and reaches the equilibrium B, and if the price is too high, the heavy solution starting outside its viability niche reaches A.

We therefore checked that "viability" and "heaviness" suffice to determine a solution to the nondeterministic system (2.1.3).

Barrier Property

A general theorem due to Marc Quincampoix states that the inflationist curve going through B and the deflationist curve going through A enjoy the barrier property: once one of these curves is hit by a consumption-price pair, the evolution of the consumption-price pair cannot return into the interior of the viability kernel until it reaches one of the equilibria A or B. This is known under the name of the barrier property of the boundary of the viability kernel. It is also said that this boundary is semi-permeable. In other words, on these two curves, the price velocity is doomed to be equal either to c or $-d$. Otherwise, if the solution leaves the boundary of the viability kernel when it is contained in the interior of the constrained set, it also leaves it in finite time. Heavy solutions can travel only along these paths in order to maintain the viability of the system. This theorem shows how fragile and dangerous the viability of the system when the state approaches the boundary of the viability kernel.

If a Planner Would Exist

If a planner could command the evolution of prices, he could design other evolutions than the heavy ones which either converge to a stable equilibrium or locks-in one of the two equilibria A or B.

He could, for instance, want to reach the equilibrium B at any cost. Assume for instance that it starts from the consumption-price pair S (taken on the deflationist curve GH) above the equilibrium curve, where the consumption decreases. If the initial consumption does not belong to its viability niche, we saw that the heavy solution would arrive at the unwanted equilibrium A. In order to reach B, he can travel the deflationist curve GH with the price velocity $-d$ as long as the consumption-price pair does not reach the inflationist curve going through B. Despite price deflation, the consumption continues to decrease until it reaches its minimum when the deflationist curve GH crosses the equilibrium curve, contradicting the static intuition that a deflationist policy would increase consumption. The consumption starts to increase only when the equilibrium curve is traversed, and then our planner must maintain during some time his deflationist policy for increasing consumption, until the consumption-price pair reaches the inflationist curve going through B. Then, he must ruggedly change its policy by adopting an inflationist policy so that the consumption-price pair persists to grow while traveling this inflationist curve which leads the consumption-price pair to the desired equilibrium B.

2.2 Example: Affine and Bilinear Systems

2.2.1 Affine Systems

Let $K := [0,b]$ the subset of a scarce commodity x. Assume that the consumption rate of our greedy consumer is equal to $\alpha > 0$ and that her consumption is slowed down by a price: the evolution of its consumption is governed by the system

$$\text{for almost all } t \geq 0, \quad x'(t) = \alpha x(t) - p(t), \quad \text{where } p(t) \geq 0$$

subjected to the constraints

$$\forall\, t \geq 0, \quad x(t) \in [0,b]$$

We see at once that the viable equilibria of the system range over the equilibrium line $p = \alpha x$ and that they are unstable. The viability niche of the price $\bar{p}$ is reduced to its associated equilibria $\frac{\bar{p}}{\alpha}$.

The regulation map is given by the formula

$$\Pi_K(0) \;=\; \{0\}, \;\; \Pi_K(x) = \mathbf{R}_+ \;\text{ when }\; x \in]0,b[\;\;\&\;\; \Pi_K(b) = [\alpha b, +\infty[$$

When the inflation is bounded, the evolution of the consumption-price pair is governed by the system of differential inclusions

$$\left\{ \begin{array}{ll} i) & \text{for almost all } t \geq 0, \;\; x'(t) = \alpha x(t) - p(t) \\ ii) & \text{and } \;\; -c \leq p'(t) \leq c \end{array} \right. \tag{2.2.1}$$

which are viable in $[0,b] \times \mathbf{R}_+$.

We introduce the functions $\rho_c^\sharp$ and $\rho_c^\flat$ defined on $[0,\infty[$ by

$$\left\{ \begin{array}{lll} i) & \rho_c^\flat(p) & := \; \frac{c}{\alpha^2}(e^{-\alpha p/c} - 1 + \frac{\alpha}{c}p) \approx \frac{p^2}{2c} \\ ii) & \rho_c^\sharp(p) & := \; -ce^{\alpha(p-\alpha b)/c}/\alpha^2 + p/a + c/\alpha^2 \end{array} \right.$$

and the functions $r_c^\sharp$ and $r_c^\flat$ defined on $[0,b]^6$ by

$$\left\{ \begin{array}{ll} i) & r_c^\flat(x) = p \;\text{ if and only if }\; x = \rho_c^\flat(p) \\ ii) & r_c^\sharp(x) = 0 \;\text{ if }\; x \in [0, \rho_c^\sharp(0)] \;\; (\rho_c^\sharp(0) = \frac{c}{\alpha^2}(1 - e^{-\alpha^2 b/c})) \\ iii) & r_c^\sharp(x) = p \;\text{ if and only if }\; x = \rho_c^\sharp(p) \;\text{ when }\; x \in [\rho_c^\sharp(0), b] \end{array} \right.$$

[6]One can prove that the maps $r_c^\sharp$ and $r_c^\flat$ are solutions to the nonlinear "first-order partial differential inclusion"

$$0 \;\in\; r'(x)(\alpha x - r(x)) + [-c, +c] \tag{2.2.2}$$

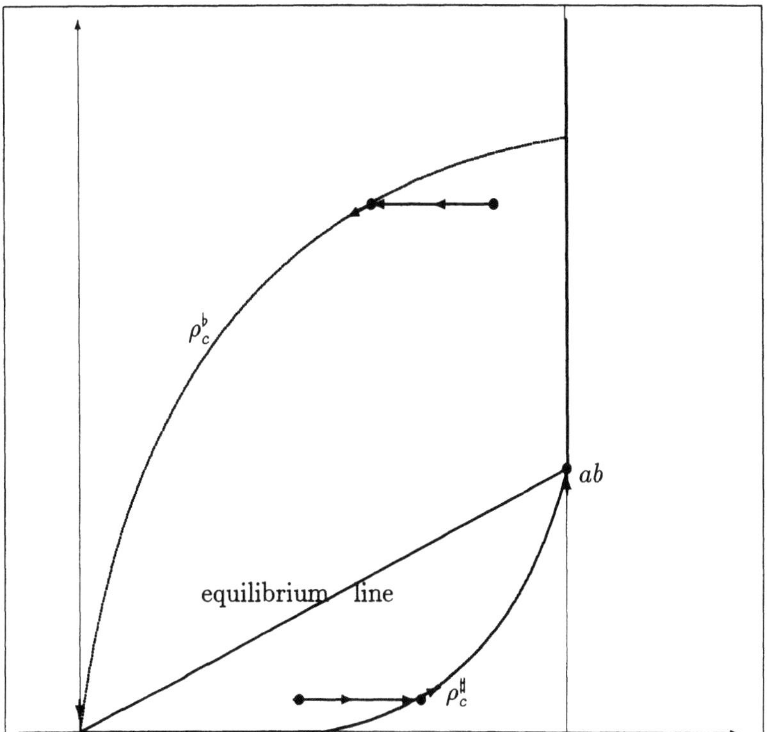

Viability kernel of a linear system. the equilibrium curve is a line made of unstable equilibria. other solutions where the velocities p' of the prices are randomly generated are shown. this computer simulation illustrates the barrier properties of the part of the boundary of the viability kernel contained in the interior of the square.

Figure 2.7: Regulation Map of an Affine System

We introduce now the set-valued map Π^c defined by[7]

$$\forall\, x \in [0,b],\;\; \Pi^c(x) = [r_c^\sharp(x), r_c^\flat(x)] \tag{2.2.4}$$

There exist solutions to (2.2.1) if and only if the initial state satisfies $p_0 \in \Pi^c(x_0)$. In this case, prices and commodities are related by the new pricing law:

$$\forall\, t \geq 0,\;\; p(t) \;\in\; \Pi^c(x(t))$$

Starting from a consumption-price pair within the viability kernel and below the equilibrium line, the heavy solution starts by evolving at zero inflation until the consumption-price pair hits the inflationist curve going through $B := (b, \alpha b)$. Then the price must increase with the highest velocity until the consumption-price pair reaches the equilibrium B.

2.2.2 Bilinear Systems

Bilinear systems are systems of the form

$$\begin{cases} i) & x'(t) \;=\; p(t)x(t) \\ ii) & p(t) \;\in\; [-\underline{p}, +\overline{p}] \end{cases}$$

where the regulee is simply chosen to be the growth rate of the system, only subjected to vary between lower and upper bounds.

The problem is to confine the state $x(t)$ to be viable in the interval $K := [a,b]$ where $0 \leq a < b$.

The equilibrium line is the horizontal line $p = 0$, and the viability niches of nonzero growth rates p are empty. For $p \in]0, +\overline{p}]$, the solutions are nondecreasing, and nonincreasing when $p \in [-\underline{p}, 0[$.

When $a > 0$, the regulation map is equal to

$$\Pi_K(x) \;:=\; \begin{cases} [0, +\overline{p}] & \text{if } \; x \;=\; a \\ \left[-\underline{p}, +\overline{p}\right] & \text{if } \; x \;\in]a,b[\\ \left[-\underline{p}, 0\right] & \text{if } \; x \;=\; b \end{cases}$$

and when $a = 0$, it is equal to

$$\Pi_K(x) \;:=\; \begin{cases} \left[-\underline{p}, +\overline{p}\right] & \text{if } \; x \;\in [a,b[\\ \left[-\underline{p}, 0\right] & \text{if } \; x \;=\; b \end{cases}$$

---

[7]By using tools of set-valued analysis, and in particular, the concept of contingent derivative $D\Pi(x,p)$ of a set-valued map Π, we shall see that Π^c is a set-valued solution to the first-order partial differential inclusion

$$\forall\,(x,p) \in \text{Graph}(\Pi),\;\; 0 \;\in\; D\Pi(x,p)(\alpha x - p) + [-c, +c] \tag{2.2.3}$$

and actually, the largest one with closed graph.

We require now that the price velocity must remain between bounds $-d$ and $+c$. Therefore, the evolution of consumption-growth rate pairs is governed by the system of differential inclusions, called the **metasystem**

$$\begin{cases} i) & x'(t) = p(t)x(t) \\ ii) & p'(t) \in [-d, +c] \end{cases}$$

subjected to metaviability constraints

$$\forall\, t \geq 0, \ \ (x(t), p(t)) \in [a, b] \times \left[-\underline{p}, +\overline{p}\right]$$

One can compute the new regulation map $\Pi_K^{(-d,c)}$ governing the evolution of viable solutions under bounded inflation. When $a > 0$, we find

$$\Pi_K^{(-d,c)}(x) = \left[-\min\left(\underline{p}\,,\sqrt{2c\log\left(\frac{x}{a}\right)}\right), +\min\left(\overline{p}\,,\sqrt{2d\log\left(\frac{b}{x}\right)}\right)\right]$$

and when $a = 0$,

$$\Pi_K^{(-d,c)}(x) = \left[-\underline{p}\,,\ +\min\left(\overline{p}\,,\sqrt{2d\log\left(\frac{b}{x}\right)}\right)\right]$$

The function

$$r^{\sharp}(x) := \sqrt{2d\log\left(\frac{b}{x}\right)}$$

limiting the upper part of the regulation map $\Pi_K^{(-d,c)}$ is the trajectory of the solutions with the lowest price velocity (equal to $-d$) passing through $(b, 0)$[8].

It is decreasing and has an inflection point $\left(\frac{b}{\sqrt{e}}, \sqrt{d}\right)$.

When $a > 0$, the function

$$r^{\flat}(x) := -\sqrt{2c\log\left(\frac{x}{a}\right)}$$

[8]Indeed, if $r(\cdot)$ is the trajectory of a solution with velocity equal to $-d$, we deduce by differentiating the relation $p(t) = r(x(t))$ that

$$-d = p'(t) = r'(x(t))x'(t) = r'(x(t))r(x(t))x(t)$$

Therefore the pair $(x(t), p(t))$ ranges over the graph of the solution $r(\cdot)$ to the differential equation

$$\frac{d}{dx}\left(\frac{r(x)^2}{2}\right) = r'(x)r(x) = \frac{-d}{x} = -d\frac{d\log(x)}{dx}$$

which is equal to

$$r(x)^2 + 2d\log(x) = \rho$$

where ρ is any constant. Requiring that $r^{\sharp}(b) = 0$ implies that $\rho = 2d\log(b)$. Replacing d by $-c$ and requiring that $r^{\flat}(a) = 0$ implies that $\rho = -2c\log(a)$.

limiting the lower part of the regulation map $\Pi_K^{(-d,c)}$ is the trajectory of the solutions with the upper price velocity (equal to $+c$) passing through $(a,0)$, which is also decreasing with no inflection point.

Starting from (x_0,p_0) where $p_0 \in \left[0\,,\sqrt{2d\log\left(\frac{b}{x_0}\right)}\right]$, the heavy solution $x(t) = x_0e^{p_0t}$ is associated with the constant growth rate $p(t) = p_0$ until the time $t_1 = \frac{\log\left(\frac{b}{x_0}\right)}{p_0} - \frac{p_0}{2d}$ when $x(t_1) = be^{\frac{-p_0^2}{2d}}$ because $r^\sharp(x(t_1)) = p_0$. Then $p(t) = p_0 - d(t-t_1)$ decreases with minimal velocity until the time $t_2 = t_1 + \frac{p_0}{d}$ when $x(t) = be^{\frac{-p(t)^2}{2d}}$ reaches the equilibrium $x(t_2) = b$.

When $a = 0$, the solution $x(t) = x_0e^{-p_0t}$ starting from $(x_0,-p_0)$ associated with the constant growth rate $-p_0$ where $p_0 \in [0,\underline{p}]$ is naturally viable in $[0,b]$, and thus, is the heavy solution associated with $-p_0$.

When $a > 0$ and when $p_0 \in \left[0\,,\sqrt{2c\log\left(\frac{x_0}{a}\right)}\right]$ the heavy solution $x^\flat(t) = x_0e^{-p_0t}$ starting at $(x_0,-p_0)$ is associated with the constant growth rate $p(t) = -p_0$ until the time $t_1^\flat = \frac{\log\left(\frac{x_0}{a}\right)}{p_0} - \frac{p_0}{2c}$ when $x^\flat(t_1^\flat) = ae^{\frac{p_0^2}{2c}}$ because $r^\flat(x^\flat(t_1^\flat)) = -p_0$. Then $p^\flat(t) = -p_0 + c(t-t_1)$ increases with maximal velocity until the time $t_2^\flat = t_1^\flat + \frac{p_0}{c}$ when $x^\flat(t) = ae^{\frac{p^\flat(t)^2}{2c}}$ reaches the equilibrium $x^\flat(t_2^\flat) = a$.

We deduce also that a necessary and sufficient condition for a differential equation

$$x'(t) \;=\; u(x(t))x(t)$$

to have a solution viable in the interval $[a,b]$ such that

$$-\underline{p} \;\leq\; u(x(t)) \;\leq\; +\overline{p} \;\;\&\;\; -d \;\leq\; \frac{d}{dt}u(x(t)) \;\leq\; +c$$

is that

$$\forall\, x \in [a,b], \;\; -\min\left(\underline{p}\,,\sqrt{2c\log\left(\frac{x}{a}\right)}\right) \;\leq\; u(x) \;\leq\; +\min\left(\overline{p}\,,\sqrt{2d\log\left(\frac{b}{x}\right)}\right)$$

2.2.3 The Verhulst Logistic Model

In population dynamics, where the state of the system represents the size of the population, the maximal population size b is called the carrying capacity. The growth rate p is said to be the Malthusian coefficient, which is the difference between fertility and mortality coefficients. Hence a feedback map $x \mapsto u(x)$ describes the effect of the population size x on the fertility and mortality. In those models, the Malthusian coefficients are assumed to satisfy the following phenomenological properties: for some $a_0 \in [a,b[$,

$$\begin{cases} i) & \forall\, x \in [a,a_0[, \;\; u'(x) \;>\; 0 \\ ii) & \forall\, x \in]a_0,b], \;\; u'(x) \;<\; 0 \\ iii) & u(b) \;=\; 0 \end{cases}$$

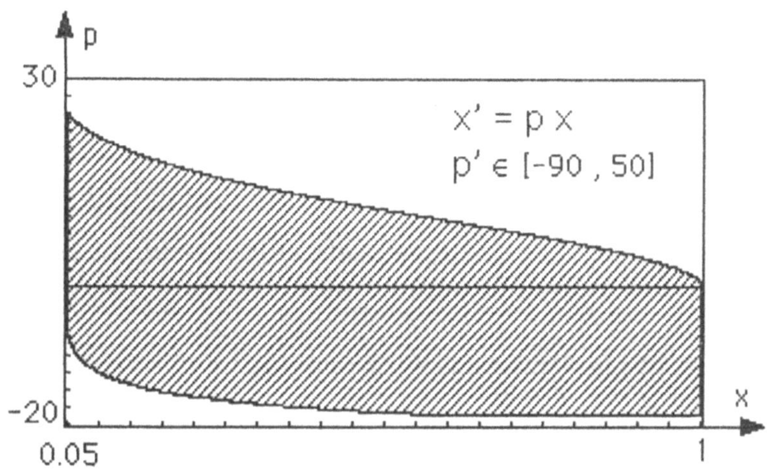

Regulation map of bilinear system $x'(t) = p(t)x(t)$ under bounded inflation $p'(t) \in [-d, +c]$. The constraints are $x(t) \in [0.05, 1]$ and $p(t) \in [-20, +30]$. We took $p'(t) \in [-90, 50]$.

Figure 2.8: Regulation Map of a Bilinear System

A constant positive growth rate $u(x) = p$, as in the original Malthusian model, implies exponential growth, and thus, a nonviable evolution. The increasing behavior of $u(x)$ on the interval $[a, a_0[$ is called the **Allee effect**, stating that at a low population size, an increase of the population size is desirable and has positive effects on population growth, whereas the decreasing behavior of $u(x)$ on the interval $]a_0, b]$ is called the **logistic effect**, stating that at high population, an increase of the size has a negative effect on the growth of the population.

If a bound on the evolution of the Malthusian coefficients is required, we have proved that actually $u(x)$ must be larger than or equal to $-\sqrt{2c\log\left(\frac{x}{a}\right)}$ to maintain the population above a size a and smaller than or equal to $+\sqrt{2d\log\left(\frac{b}{x}\right)}$ to maintain it below its carrying capacity b.

Writing the Malthusian coefficient as the difference $u(x) := \beta(x) - \mu(x)$ of fertility and mortality coefficients, we can rewrite the above condition in the form

$$\begin{cases} i) & \forall\, x \geq a,\ \beta(x) \leq \mu(x) - \min\left(\underline{p}\,, \sqrt{2c\log\left(\frac{x}{a}\right)}\right) \\ ii) & \forall\, x \leq b,\ \mu(x) \geq \beta(x) - \min\left(\overline{p}\,, \sqrt{2d\log\left(\frac{b}{x}\right)}\right) \end{cases}$$

We find as a special case the **purely logistic Verhulst model** with Malthusian coefficients of the form

$$u(x) := \beta(b - x)$$

proposed in 1838 by Verhulst in [343, Verhulst] and rediscovered in the thirties by Pearl in [286, Pearl]).

On can check that whenever $\beta \leq \frac{2\sqrt{d}}{b}$, we have

$$\forall\, x \leq b,\ u(x) := \beta(b - x) \leq \sqrt{2d\log\left(\frac{b}{x}\right)}$$

so that the viability of the Verhulst logisitc model under bounded inflation is guaranteed.

Remark — Let $f : [a, b] \mapsto [0, \infty[$ be a function such that the function $x \to \frac{1}{f(x)}$ has a primitive $x \mapsto g(x)$, which is then a nondecreasing function. Then we can prove in the same way that the set-valued map $\Pi_K^{(-d,c)}$

$$\Pi_K^{(-d,c)}(x) = \left[-\min\left(\underline{p}\,, \sqrt{2c\int_a^x \frac{dx}{f(x)}}\right)\,,\ \min\left(\overline{p}\,, \sqrt{2d\int_x^b \frac{dx}{f(x)}}\right)\right]$$

is the regulation map $\Pi_K^{(-d,c)}$ governing the evolution of viable solutions to

$$\begin{cases} i) & x'(t) = p(t)x(t) \\ ii) & p'(t) \in [-d, +c] \end{cases}$$

under bounded inflation subjected to metaviability constraints

$$\forall\, t \geq 0,\ \ (x(t), p(t)) \ \in \ [a, b] \times \left[-\underline{p}, +\overline{p}\right]$$

We observe that $r^\sharp(x) := \sqrt{2d \int_x^b \frac{dx}{f(x)}}$ satisfies

$$r^{\sharp'}(x) \ = \ -\frac{d}{f(x) r^\sharp(x)}$$

and

$$r^{\sharp''}(x) \ = \ d\frac{f'(x) r^{\sharp^2}(x) - d}{f(x) r^{\sharp^3}(x)}$$

In the same way, $r^\flat(x) := -\sqrt{2c \int_a^x \frac{dx}{f(x)}}$ satisfies

$$r^{\flat'}(x) \ = \ \frac{c}{f(x) r^\flat(x)}$$

and

$$r^{\flat''}(x) \ = \ -c\frac{f'(x) r^{\flat^2}(x) + c}{f(x) r^{\flat^3}(x)} \quad \square$$

2.3 An Elementary Keynesian Model

We present in this section a version of a Keynesian model proposed in [111, Cartelier & Müllers] **drastically** reduced in order to provide two-dimensional systems for illustrating some viability ideas at the price of economic relevance. Other choices of state-regulee pairs can be made. Here, the state is chosen to be the **stockpiling per unit of capital** and the regulee the **production per unit of capital** which depends on entrepreneurs within certain limits. It may be interpreted also as a rate of utilization of capital.

The interest rate is supposed to be fixed and the evolution of liquidities follows from the evolution of stocks and production.

Entrepreneurs own an amount of capital K, which is a quantity of the unique commodity held for the sake of production, and a proportion x of the same commodity which is the excess of production over sales, cumulative over time. It is equivalent to the gap between expected and realized profit. The proportion xK of the capital K differs from K in that xK reveals an involuntary stockpiling. A negative x describes a cumulate excess of orders over production.

Indeed a market economy keeps working only if entrepreneurs do not incur bankruptcy. This implies in turn that a state of confidence rules over the entire economy. Such a state of confidence may persist only if certain limits are not reached. For instance, the indebtedness of entrepreneurs must be moderate enough to guarantee debtors against insolvency. In the same way, the profitability of entrepreneurship must be high enough to convince people to commit themselves in such an activity. In our model, two thresholds are significant. The first one is $\overline{x}$, the maximum value for x, which is a rentability indicator. Beyond $\overline{x}$, it is no longer possible for entrepreneurs to remain active, because the difference between expected and realized profit is too large. The ratio x must not be too large if negative, because it would mean that entrepreneurs are not able to supply a sufficient fraction of the demand. We have hence the following constraint:

$$\forall\, t \geq 0,\ \ \underline{x} \ \leq\ x(t) \ \leq\ \overline{x} \tag{2.3.1}$$

It seems indeed sensible to admit that limited deviations have not dramatic consequences but also that some critic values must not be passed over.

Furthermore, the production p is also submitted to technological constraints summarized by

$$\forall\, t \geq 0,\ \ \underline{p} \ \leq\ p(t) \ \leq\ \overline{p}$$

Let us call *constrained set* the set

$$K \ =\ [\underline{x}\,, \overline{x}] \times [\underline{p}\,, \overline{p}]$$

defined by such limits on both the state x and the regulee p.

The constrained set is in general not a viability domain, but there exists a nonempty subset of the constrained set which satisfies the viability condition at every point, and we call the largest closed subset exhibiting this property the *viability kernel* of the constrained set.

This agrees with common sense, which suggests that even though we are not looking for equilibria (stationary states), the economy must "not be too far from equilibrium". This actually means that limits exist beyond which adaptive or innovative reactions cease to be possible. The crucial point is not whether economy is or is not in equilibrium but whether evolution "away of equilibrium" is or is not compatible with the fundamental data and ordinary rules of the economy. It becomes then natural to specify the minimal conditions under which an economy can work and to check if the dynamic evolution of the economy does not violate these bounds.

Jean Cartelier and Katharina Müllers conclude their study with those comments on such bounds: *"These bounds usually refer to ordinary and usual values of economic variables or economic ratios regarded as being acceptable whether close or not to equilibrium*[9]*. This is true for inflation rates, interest rates, etc. When an ordinary state of affairs, even if not very satisfactory, is*

[9]Being close or far to equilibrium does not really make sense since in actual economies equilibrium values are not known by agents. Market economies are characterized by uncertainty and private information. There is no clear reason why individuals should know the actions of others when they have to decide for themselves. They actually adopt *conventional* attitudes in order to make their life easier. Some of these *conventions* may be modeled by the constrained set.

dramatically changed, as would be the case if a period of low level of interest succeeds a durable period of high level, economic agents would have no reason to modify the rules by which they judge what is going on.

The constrained set may be viewed as more or less related to the conventions *by which people reduce the effects of uncertainty upon their behavior. As Keynes put it:* "Worldly wisdom teaches that it is better for reputation to fail conventionally than to succeed unconventionally" *([240, Keynes], p. 158). The strength of the relative stability of these conventions have been too often ignored by economic theory. Viability theory allows one to take a new view of this question and formally to deal with it."*

For that purpose, we turn now to the description of the dynamics governing the evolution of the ratio x. Entrepreneurs investment is ruled by two coefficients, $a \geq 0$, the "animal spirits" of the Keynesian tradition, i.e., the inducement to invest, and $b > 0$, the intensity of reaction to involuntary stockpiling, which measures the depletion of the investment rate by the amount of involuntary stockpiling: They enter the dynamics governing the evolution of the capital:

$$\frac{K'(t)}{K(t)} = a - bx(t)$$

The animal spirit depends upon the interest rate[10]. We refer to the original paper [111, Cartelier & Müllers] for the analysis of the dependence upon the interest rate.

The velocity of the stock $S = xK$ is assumed to be equal to the difference between the production and the velocity of the capital, so that the velocity of the stock $x(t)$ per unit of capital is equal to

$$x'(t) = p(t) - (1 + x(t))\frac{K'(t)}{K(t)}$$

Therefore, the dynamics of the stock $x(t)$ per unit of capital is governed by the differential equation

$$x'(t) = bx(t)^2 + (b - a)x(t) - a + p(t)$$

The equilibria are given by

$$\bar{x} = \left(-(b - a) \pm \sqrt{(b - a)^2 + 4b(a - p)}\right) /2b$$

Real roots exist if $((b-a)^2+4b(a-p) \geq 0$, which implies a maximum value for p equal to $p_{max} = (a+b)^2/4b$. For $p = p_{max}$, roots are equal, i. e., $\bar{x} = \frac{(a-b)}{2b}$.

[10] and is assumed to decrease when the interest rate increases. Usually, this dependence is affine: $a(i) = \alpha - \beta i$.

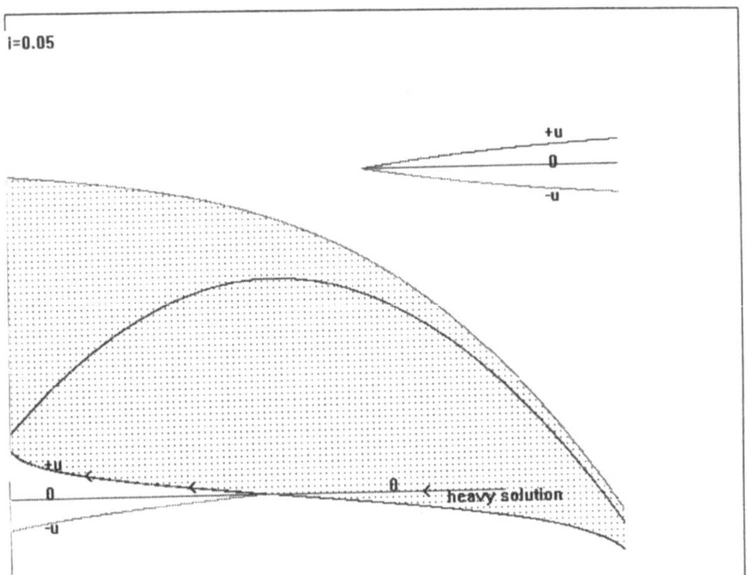

Viability kernel, the parabola curve and an example of heavy solution.

Figure 2.9: Viability Kernel of the Keynesian Model

In order to get $\bar{x} = 0$, i. e., a "full equilibrium" (in the sense that there is no stockpiling), we must have $a = b$. No demand pressure, coming either from investment or consumption exists when the economy evolves along the full equilibrium path. In this case we have $p_{max} = \bar{p} = a$, which is the usual **Keynesian multiplier formula**. Condition $a = b$ is never fulfilled but through the choice of an adequate rate by the banking system, and **nothing guarantees such a behavior of the banking system.**

The equilibrium curve is given by the formula

$$p = a - (b-a)x - bx^2 \tag{2.3.2}$$

This function defines a parabola dividing the x–p–plane in two monotonic cells: below the parabola, the velocity $x'(t)$ is negative and above, $x'(t)$ is positive. The equilibria $x < \frac{(a-b)}{2b}$ are stable and the equilibria $x > \frac{(a-b)}{2b}$ are unstable, while $\frac{(a-b)}{2b}$ is a saddle-point.

The relative position of the curve only depends on a and b. When animal spirits are strong enough, as compared with b, which is the intensity of reaction to involuntary stockpiling, the parabola's maximum is at the right of the vertical axis $x = 0$. It stands to the left in the opposite case. In both cases, full equilibrium does not correspond to p_{max}; in other words, full equilibrium and full employment do not coincide except for $a = b$.

This local stability analysis gives poor intuitions about the effective temporal path out of equilibrium. That an equilibrium is unstable or a saddle-point suggests only that the economy is pushed farther away from equilibrium, but nothing is known about the velocity of this process and the span of time during which economy is not too far from equilibrium.

We turn now our attention to the dynamics of the regulee p, the production per unit of capital.

Instead of reasoning with a law of adjustment *a priori* given, we fix limits to the speed of variation of the regulees but *the sign of variation is not preassigned.* These limits are more or less narrow, depending on the flexibility of the system. Obviously, to more flexible economies correspond wider viability domains or viability kernels. Thus, p is supposed to vary within certain limits but without any *a priori* law of variation[11].

We cite the authors of [111, Cartelier & Müllers]: *"This track unfortunately goes against the tradition of political economy, where the adjustment of some variables to disequilibria plays a prominent role. The "law of supply and demand" states that prices react in a determined direction in response to a difference between supply and demand in the market: the price of a*

[11] If the interest rate were allowed to evolve, they should be regarded as a second regulee, with a higher "inertia": when the constraints are about to be transgressed, the production per unit starts to evolve first while the interest rate remains constant. If this regulation process fails, then the interest rate evolves. This is the Müllers' cascades presented in Chapter 7.

particular commodity is assumed to vary according to the sign of the excess demand of this commodity. Another well-known law of adjustment concerns the variation of the quantity in reaction to the profitability of its production. For instance, in Walras' theory of production, the quantity varies according to the sign of the profit (sales minus costs). In our model, the production per unit of capital plays this role: It would seem natural to assume that p changes according to the sign of $x'(t)$ or $x(t)$. Hence, we only fix limits to the speed of variation of the regulees instead of assigning the sign of variation. Does it mean that we must renounce to the very notion of a law of adjustment? Absolutely not. On the contrary, we may hope to derive one from the study of the viability, which is far better than to postulate one beforehand".

We have to study whether the constrained set K is viable under the following "metasystem"

$$\begin{cases} i) & x'(t) = bx(t)^2 + (b-a)x(t) - a + p(t) \\ ii) & |p'(t)| \leq c \end{cases} \tag{2.3.3}$$

One can show that for this type of system, the viability kernel is limited by two trajectories starting on the respective intersection points of the equilibrium parabola (2.3.2) and the constrained set K such that p increases or decreases maximally, i. e.,

$$p'(t) = +c \text{ or } p'(t) = -c$$

The viability kernel $\text{Viab}(K)$ can be seen as the graph of a *regulation map* $R : [\underline{x}, \overline{x}] \longrightarrow [\underline{p}, \overline{p}]$,

$$p \in R(x) \text{ if and only if } (x,p) \in \text{Viab}(K)$$

This defines a (feedback) *regulation law* for the viable evolutions: a state–regulee solution $(x(\cdot), p(\cdot))$ of system (2.3.3) is viable if and only if it satisfies the regulation law

$$\forall t \geq 0, \ p(t) \in R(x(t)) \tag{2.3.4}$$

On the interior of the viability kernel, the velocity of the regulee can be arbitrarily chosen in the interval $[-c, +c]$. But on the boundaries, the velocity is univoquely determined. It is equal to $+c$ on the lower boundary and to $-c$ on the upper boundary of the viability kernel.

Instead of looking after the asymptotic consequences of *a priori* given "laws of adjustment" we are able to find what are the rules of adjustment which are compatible with viability. Moreover, even on the traditional question of the tendencies toward equilibrium, the regulation law (2.3.4) offers original answers. To mention only one example, it is clear that the *sign* of variation of p depends not only on the level of x but also on the level of the regulee p. For a given level of $x(t)$, the velocity $p'(t)$ must be positive (resp. negative) if the economy is on the lower (resp. upper) boundary of the viability kernel.

Of particular interest are heavy evolutions, where the regulee is constant as long as the viability of the solution is not at stake. It is only in highly unrealistic models that great speeds of adjustment are assumed. When adjustment costs and information problems are taken seriously, it appears that the economy is far less flexible than commonly admitted. New Keynesian economists (see [260, Mankiw & Romer]) have shown that prices, rate of interest or level of employment may be sticky as a result of agents's rational behavior because of menu costs or asymmetry of information. More generally, it seems realistic to assume that the psychological cost of changing rules and routines is sufficiently high to be incurred only in case of extreme need. Assuming heavy evolutions seems a convenient way of capturing inertia as an important feature of ecodynamics.

For example, if one starts in the interior of the viability kernel, an heavy solution is regulated by a constant regulee until reaching the boundary of the viability kernel. When the regulee p does not move, the solution leaves the viability kernel and hence finally the constrained set, because every solution starting from outside the viability kernel has to leave the constrained set in finite time. In this example, the regulee has to be increased with maximal velocity $+c$, so that the solution can remain on the boundary of the viability kernel and converge to a viable equilibrium.

Remark — One can show that **the smaller the interest rate, the larger the viability kernel** (when the animal spirits depends in an affine way of the interest rate). A permanently low rate of interest is then a most favorable condition. Not only because it is associated with a high equilibrium level of economic activity p but also, and more significantly, because it increases the dynamical robustness of the economy. A low rate of interest allows the economy to stay away from equilibrium without being condemned to collapse. *Ceteris paribus*, the larger the viability kernel the better the prospect of the economy. However this conclusion — not unexpected in a Keynesian model — ought to be somewhat qualified by more sophisticated models."

We quote the authors of [111, Cartelier & Müllers]: *The level of interest rate influences the ability of activity adjustment to maintain our economies in a viable state. Variations in the level of activity, whether decided "spontaneously" by the firms or "deliberately" by economic policy, are more or less successful in keeping the economy in a normal state of working. Their performance may crucially depend on the level of the rate of interest. Orthodox economists share the view that the level of the interest rate results from adjustment between saving and investment (or between supply and demand of funds on financial markets). A high level of interest reflects a relative saving scarcity. Nothing can be done for that except to promote incentives to save. Keynesians do not agree with this view: for them, "the rate of interest is a*

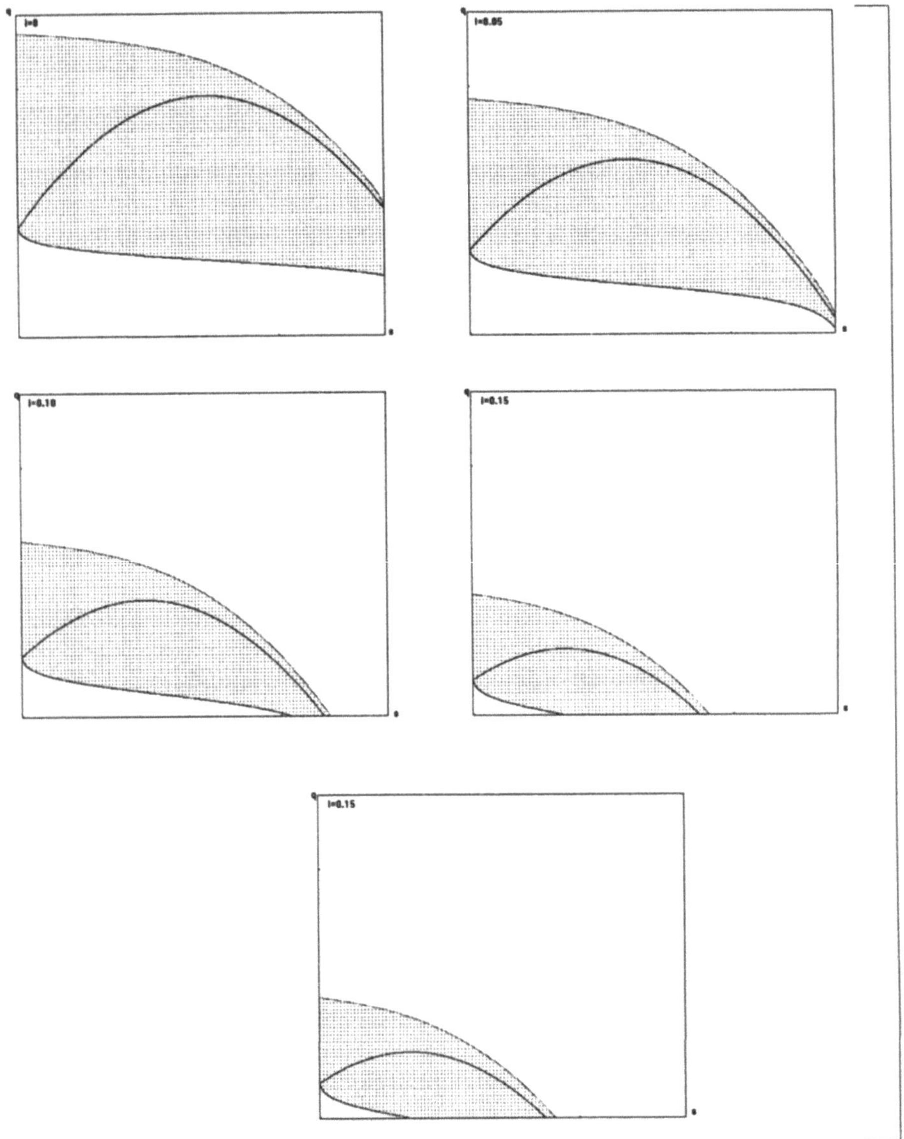

Figure 2.10: Viability Kernels for Increasing Interest Rates

highly conventional (...) phenomenon. For its actual value is largely governed by the prevailing view as to its value is expected to be" ([240, Keynes], p. 203). Policy can be effective in determining what the convention may be. About the interest rate, Keynesians maintain that "the lower the better". *There is no need to convince the reader of the importance of this issue. Day after day, financial markets and central banks decisions remind us of the strategic role played by the level of the rate of interest (and expectations about it)".* □

Remark — Households "own" their personal abilities, called labor, and they receive wages. They hold financial assets which are the counterparts of entrepreneurs indebtedness. These counterparts are either direct (bonds) or indirect, the banking system acting as an intermediary. In the balance–sheet of households, y means liquidity (demand – or term – deposits), expressed as a proportion of capital K (liquidity per unit of capital). Banking system's unique resources are households deposits yK. It lends all the deposits as short–term credits yK. Intermediation is supposed to be neutral.

In the model, the evolution of liquidity evolves according a differential equation of the form

$$y'(t) \ = \ bx(t)y(t) - ay(t) + cp(t)$$

where c depends upon the interest rate[12].

The indebtedness of entrepreneurs to banks cannot exceed the value $\overline{y}$. If it were the case, creditors would loose confidence and entrepreneurs would go into bankruptcy. For obvious reasons, y must be positive: otherwise, entrepreneurs would be creditors of banks and not the reverse! We have then a second constraint:

$$\forall t \geq 0, \ \ 0 \ \leq \ y(t) \ \leq \ \overline{l}$$

The interest rate i is another regulee of the system, which has been taken constant. The reason was that it has a higher inertia than the per-capita consumption for regulating the system. So, we have to study the viability of the system with two state variables x and y and either one regulee p when the interest rate is maintained constant or two regulees p and i. Using the two regulees p and b together, we have a hierarchical dynamical game which can be studied using the hierarchical cascade mechanism of Katharina Müllers presented in Chapter 7.

We refer the reader to the original paper [111, Cartelier & Müllers] for more details on this more complete model.

---

[12] in an affine way, usually: $c(i) = \gamma - \delta i$.

2.4 Technology and Climate Risk

Many difficulties of collective decision-making models about climatic risks are due to the interactions between physical and economical requirements (see [235,234,279,266,285] for a small sample of the literature). The precaution principle[13] and economic efficiency are often contradictory. In order to take into account the precaution principle[14], Luc Doyen, Daniel Gabay and Jean-Charles Hourcade have made a study of these questions in [174, Doyen, Gabay & Hourcade]) taking into account

1. the dynamic aspects of the processes involved[15],
2. exogenous uncertainties involved in decision and policy making,
3. endogenous uncertainties due to ignorance or the lack of control on Nature perturbations, which entices "worst case scenarios" instead of stochastic "fiability" reasoning,
4. the inertia due to a change adverse behavior of decision makers,
5. the viability of the system confronting physical and economical constraints.

In order to stylize the problems for providing a two-dimensional illustration, they isolated two variables:

1. the concentration of green-house gases — say, CO_2 — regarded as a state variable,
2. technological indexes of production processes polluting the atmosphere with the emission of these gases, regarded as regulees.

The constraints are both physical and economical, linking the cost of changing technology to the level of production (exogenous in this simple version).

In this model, the state $x \in [0,1]$ is the concentration of some green-house gas and the regulee $p \in [0,1]$ denotes a "technological index" of a polluting process, equal to 1 for the non polluting process and to 0 for the most pestilent one.

The viability constraints are split in

1. *Ecological constraints*

   $$\forall t \geq 0,\ 0 \leq x(t) \leq \overline{x}$$

   where $\overline{x}$ is a maximal concentration of green-house gases.

[13] stating that one should limit, bound or even forbid potential dangerous actions, without waiting for a scientific proof of their hazardous consequences, whatever the economic cost.

[14] see [34, Aubin] for certain comments on this topic.

[15] Demographic, cultural, economical and physical processes evolve and interact, and equilibrium or asymptotic properties are not sufficient, since transient evolutions do matter.

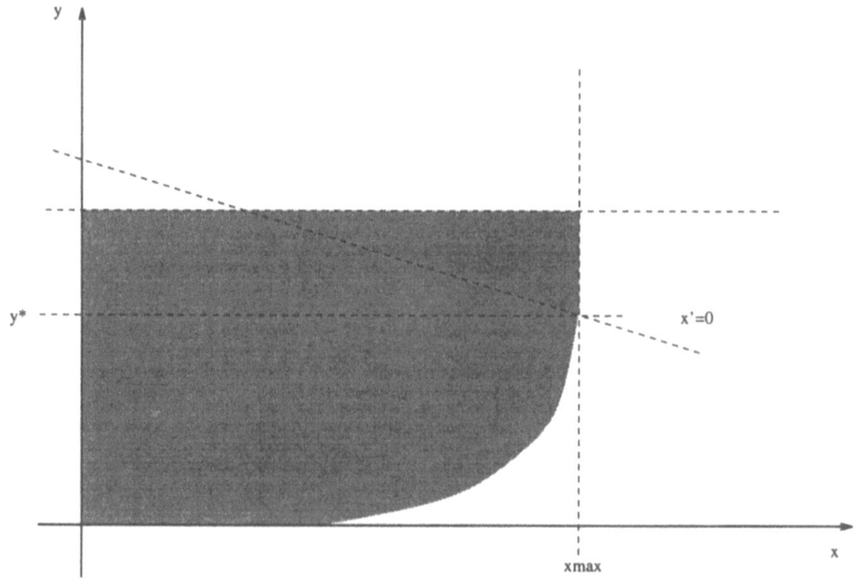

Equilibrium curve made of stable equilibria and viability kernel.

Figure 2.11: Viability Kernel

2. *Economic constraints*

$$\forall\, t \geq 0,\;\; 0 \;\leq\; p(t) \;\leq\; \overline{p}$$

expressing that the cost of operating a technology p should be less than the net profit (which depends upon the wealth derived from the production and from production costs, investments, etc., which are assumed to be constant for the sake of illustration.)

Hence the constrained set is defined by

$$K \;:=\; [0, \overline{x}] \times [0, \overline{p}]$$

We shall assume that $\overline{p} < 1$ in order to exclude the case when the least polluting technology $p = 1$ is viable.

The dynamics of the concentration of green-house gases is governed by

$$x'(t) \;=\; a(1 - p(t)) - bx(t)$$

That means that the variation of the concentration of the green-house gases depends upon a natural slow absorption phenomenon by the oceans ($-bx(t)$ with b "small") and is proportional to the short-term pollution generated when using the technology index $p(t)$ and the level of production (supposed to be constant in this illustrative example).

The rigidity and the inertia of technological changes is summarized by setting a bound to the velocity of the technology:

$$\forall\, t \geq 0, \;\; -d \;\leq\; p'(t) \;\leq\; +d$$

It is also assumed that beyond the standard bound on the velocity of the technological index obeys an economic constraint which reads

$$|p'(t)| \;\leq\; \delta(\overline{p} - p)$$

Therefore the metasystem governing the evolution of the state and the regulee are described by

$$\begin{cases} i) & x'(t) \;=\; a - bx(t) - ap(t) \\ ii) & |p'(t)| \;\leq\; \min(d, \delta(\overline{p} - p(t))) \end{cases} \tag{2.4.1}$$

The "equilibrium curve" is here the line $p = \dfrac{1 - bx}{a}$. Hence, all the equilibria are stable. The viability niches of the regulee p are empty if $p \in \left[0, \dfrac{1 - b\overline{x}}{a}\right[$ and equal to the interval $[0, \overline{x}]$ when $p \in \left[\dfrac{1 - b\overline{x}}{a}, \overline{p}\right]$, (this happens when $a\overline{p} + b\overline{x} \geq 1$).

It is easy to check that the constrained set K is viable under the metasystem whenever $\overline{x} \geq \dfrac{1}{b}$.

The viability kernel of K is empty when $a\overline{p} + b\overline{x} < 1$ because $\dfrac{1 - b\overline{x}}{a} > \overline{p}$. In this case, the constrained set is below the equilibrium line, so that all solutions $x(\cdot)$ decrease without being able to stop. If such is the situation, all solutions leave K in finite time (K is said to be a repeller).

Otherwise, the viability kernel is equal to

$$\mathrm{Viab}(K) \;=\; \{(x, p) \in K \mid x \leq \rho^{\flat}(p)\}$$

where $\rho^{\sharp}$ is the solution to the differential equation

$$\frac{d\rho^{\sharp}}{dp} \;=\; \frac{a - ap - b\rho^{\flat}(p)}{\min(d, \delta(\overline{p} - p))}$$

satisfying the initial condition $\rho^{\flat}\left(\dfrac{1 - b\overline{x}}{a}\right) = \overline{x}$.

In this case, there is danger when the technological index p is smaller than $\dfrac{1 - b\overline{x}}{a}$. One cannot use the same technology forever. The **heavy evolution** consists in keeping the same technology as long as the concentration of gas is smaller than $\rho^\flat(p)$. At this level, the technology has to be drastically changed with the velocity equal to $\min(d, \delta(\overline{p} - p(t)))$, while the concentration of gas increases until it reaches the level $\overline{x}$.

2.5 Management of Renewable Resources

Biologists and ecologists are rather interested to long horizons and survival — viability — problems whereas economists most of the time are preoccupied with short horizons, concentrating on efficiency and substitutability of commodities (see [215] for this basic point).

The simplest ecological models are based on the logistic dynamics proposed by Verhulst in [343, Verhulst] and Pearl in [286, Pearl] which govern the basic evolution of biological resources (see Section 2.2). This evolution is slowed down by economic activity which depletes it. The basic model was originated by Graham in 1935 (see [222, Graham]) and taken up by Schaeffer in 1954 (see [330, Schaeffer]), where it is assumed that the exploitation rate is proportional to the biomass and the economic activity. The equilibria of such dynamics are called in the literature the **sustainable yields**. The purpose of many investigations is to find an equilibrium yield maximizing some profit function. Beyond this static approach, optimal intertemporal optimization was used in economics for proposing "bang-bang" solutions, which may be not viable economically, and which bypass the inertia principle, translating many rigidities in the way of operating the capture of the renewable resource.

We present here a qualitative study by Luc Doyen and Daniel Gabay (see for more details) dealing with the viability of both economic and ecological constraints.

We denote by $x \in \mathbf{R}_+$ the biomass of the renewable resource and by $p \in \mathbf{R}_+$ the economic effort for exploiting it. The state and the regulee is subjected to viability constraints of the form

1. *Ecological constraints*

$$\forall\, t \geq 0,\;\; \underline{x} \;\leq\; 0 \;\leq\; \overline{x}$$

2. *Economic constraints*

$$\forall\, t \geq 0,\;\; cp(t) + C \;\leq\; \gamma p(t)x(t)$$

where $C \geq 0$ is a fixed cost, $c \geq 0$ the unit cost of economic activity and $\gamma \geq 0$ the price of the resource.

3. *Production constraints*

$$\forall\, t \geq 0,\;\; 0 \;\leq\; p(t) \;\leq\; \overline{p}$$

Actually, the economic constraint implies the positivity of $x(t)$. Hence the constrained set is defined by

$$K := \left\{(x,p) \mid \frac{cp+C}{\gamma p} \leq x \leq \overline{x} \ \& \ 0 \leq p \leq \overline{p}\right\}$$

It is not empty whenever

$$\gamma\overline{x} > c \ \& \ \frac{C}{\gamma\overline{x}-c} \leq \overline{p}$$

The dynamics involve the Verhulst logistic dynamics and the Schaeffer proposal:

$$x'(t) = rx(t)\left(1-\frac{x(t)}{\widehat{x}}\right) - p(t)x(t) \tag{2.5.1}$$

We set a bound to the velocity of the economic effort, which translates the rigidity of the economic behavior:

$$\forall\, t \geq 0, \ -d \leq p'(t) \leq +d$$

Therefore the metasystem governing the evolution of the state and the regulee are described by

$$\begin{cases} i) & x'(t) = rx(t)\left(1-\frac{x(t)}{\widehat{x}}\right) - p(t)x(t) \\ ii) & |p'(t)| \leq d \end{cases} \tag{2.5.2}$$

The "equilibrium curve" is here the union of the singleton $\{0,0\}$ and the line $p = r\left(1-\frac{x}{\widehat{x}}\right)$. They are called the "sustainable yields" in the literature. They are stable and viable if the lower boundary of K described by the graph of $p := \frac{C}{\gamma x - c}$ intersects the equilibrium line: viable equilibrium belong to the interval $[x_-, x_+]$ where $x_\pm$ are the real roots[16] of the equation

$$\gamma r x^2 - rx(\gamma\widehat{x}+c) + (C+rc)\widehat{x} = 0$$

Setting $p_\pm := r\left(1-\frac{x_\pm}{\widehat{x}}\right)$, we observe that the viability niches are not empty when $p \in [p_-, p_+]$ and that, in this case, they are equal to

$$N(p) = \left[\frac{C}{\gamma p - c}, \overline{x}\right]$$

[16] The discriminant is equal to $\Delta = \sqrt{r^2(\gamma\widehat{x}+c) - 4\gamma r\widehat{x}(C = rc)}$ so that the roots are equal to

$$x_\pm = \frac{r(c+\gamma\widehat{x}) \pm \Delta}{2\gamma r}$$

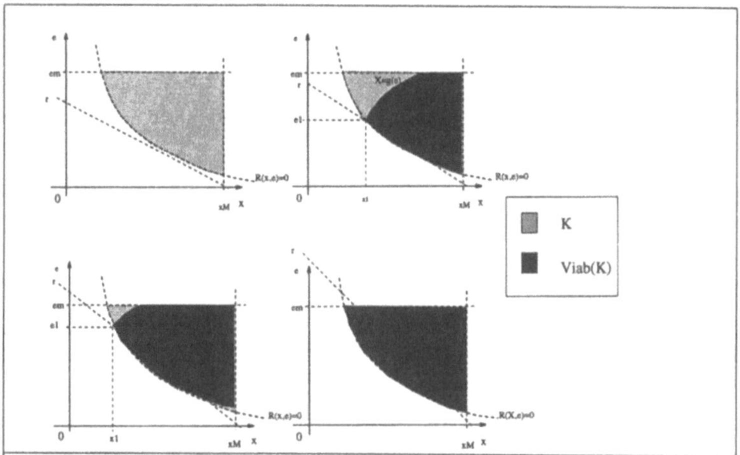

Viability kernels for different parameters.

Figure 2.12: Viability Kernel

One can see that K is viable whenever $p_- \geq \overline{p}$ because in this case, all the viability niches are nonempty. Otherwise, the viability kernel is equal to

$$\text{Viab}(K) = \{(x,p) \in K \mid x \geq \rho^\sharp(p)\}$$

where $\rho^\sharp$ is the solution to the differential equation

$$-d\frac{d\rho^\sharp}{dp} = r(1 - \frac{\rho^\sharp(p)}{\widehat{x}}) - p\rho^\sharp(p)$$

satisfying the initial condition $\rho^\sharp(\overline{p}) = x_-$.

In this case, danger happens when the economic effort p is larger than p_-. One cannot maintain the economic effort constant. The **heavy evolution** consists in keeping the economic effort constant as long as the biomass is larger than $\rho^\sharp(p)$. At this level, the economic effort has to be drastically reduced with the velocity d, i.e., using $p'(t) = -d$, while the biomass decreases until it reaches the level x_-.

2.6 Storage and Viability of a Fishery

The main examples of management of renewable resources is provided by fisheries, which goes back during the XIXth century to C.G. Petersen, a Dane, became a heated field when fifty years ago the world community started to regulate the exploitation of marine fauna. It started with the North-Sea herring (from 1.5 million tons in 1962 down to no exploitation in 1980), the

Peruvian anchovies, the sardine of South-Africa (from 1 million tons in 1962 down to no exploitation in 1980) to the control of seal and whale hunting. The control of these resources has triggered confrontations between fishers or even, their countries. Here, the viability analysis of a ecological-economic model of the French Guyana shrimp fisheries is taken from the paper [73, Béné & Doyen].

In classical models of firms, they adjust "endogenous" capital and labor input variable to an exogenous demand. In the case when resources are renewable — in fisheries, for instance, one of the inputs, the renewable resource, is also exogenous. This is an important difference, which leads to choose as a state variable of a production process using renewable resources the **storage**, which plays the role of a buffer between uncontrolled resources and controlled demand.

Also, contrary to purely biological predator-prey models, fisheries do have to adapt both to resource fluctuations and to cope with markets dynamics. This provides an example of time-dependent viability constraints.

For illustrating this point, we present here a simplified version of the study [73, Béné & Doyen] of the French Guyana shrimp fishery which offers the advantage of being a mono-species mono-market fishery, offering an empirical motivation to a theoretical study based on viability theory.

We isolate two variables, the state variable x which represents the storage, i.e., the difference between the catch quantity and the part of injected in the market. The regulee p is chosen to be the exportation flow.

An exogenous time dependent supply and demand analysis leads to viability constraints of the form

$$\forall\, t \geq 0, \;\; x(t) \;\in\; K(t) \;:=\; [0, s(t)]$$

where $s(t)$ is here a given time-dependent function derived from supply and demand through a detailed economic study exposed in the original [73, Béné & Doyen]. The set-valued map $t \rightsquigarrow K(t)$ is regarded as "tube".

The regulee is subjected to state-dependent time-dependent constraints, reflecting profit or budgetary constraints:

$$\forall\, t \geq 0, \;\; p(t) \;\in\; [c(x(t)), d(t)]$$

where $d(t)$ is the demand at time t and where the state-dependent $c(x)$ denotes the production cost (assumed to be affine in this example). The demand function is periodic (with a peak before Christmas and New-Year Eve periods, in this example). The evolution of the storage is governed by a simple time-dependent differential equation

$$x'(t) \;=\; g(t) - p(t)$$

where g is a periodic time-dependent function involving the evolution of the renewable resource (the shrimp), the fishing efforts and the catchability

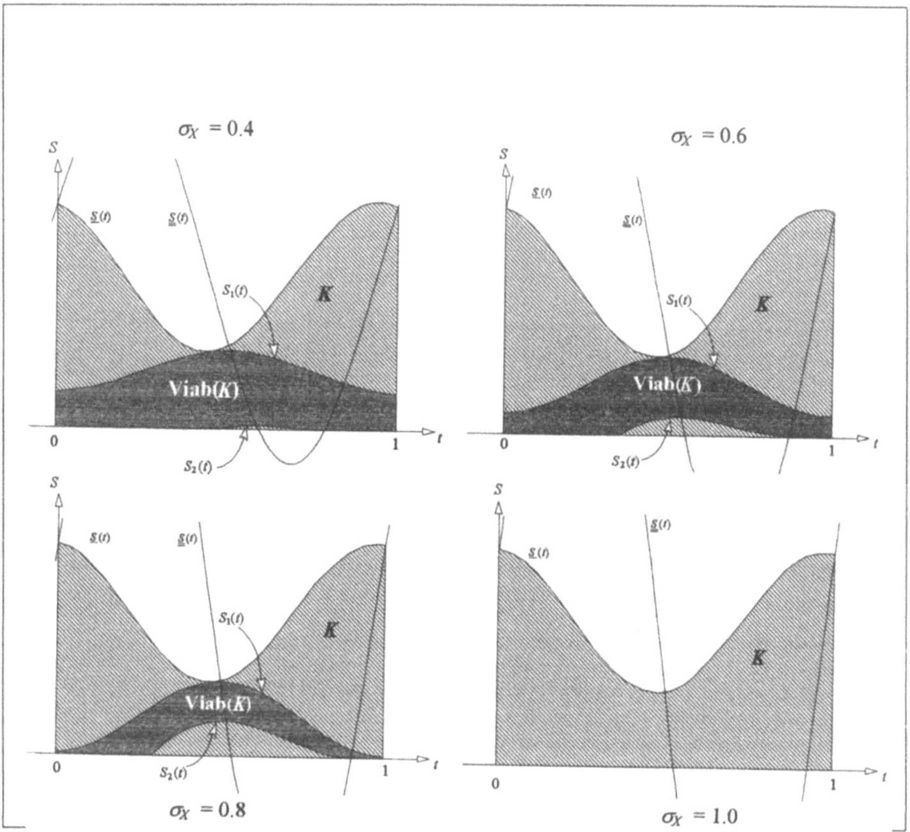

Figure 2.13: Viability Kernel for the Shrimp Management

coefficient, regarded as given in this study. The function g reflects dephazed seasonalities of both the resources — involving the life cycle of the shrimps — and the market — involving the cultural cycles of consumption pattern of shrimp eaters.

The usual addition of the auxiliary variable $w(t) := t$, solution to the differential equation $w'(t) = 1$, allows us to treat this time-dependent system as a time-independent system:

$$\begin{cases} x'(t) &= g(w(t)) - p(t) \\ w'(t) &= 1 \end{cases}$$

where

$$\forall\, t \geq 0,\;\; p(t) \;\in\; P(x(t), w(t)) \;:=\; [c(x(w(t))), d(w(t))]$$

subjected to the viability constraints

$$\forall\, t \geq 0,\;\; (x(t), w(t)) \;\in\; \mathcal{K} \;:=\; \{(x, w) \mid 0 \leq x \leq s(w)\}$$

The study [73, Béné & Doyen] provides the explicit analytical determination of the viability kernel of this problem as well as a qualitative analysis. They prove that

$$\text{Viab}(\mathcal{K}) \;:=\; \{(x, w) \mid \max(0, a_1(w) \;\leq\; a_2(w)\}$$

where $w \mapsto a_i(w)$ are solutions to differential equations which have been solved explicitly by the authors under conditions guaranteeing that the viability kernel is not empty. These conditions involve the amplitudes of both the resource and market oscillations. In this case, investments in storage capacities is worthless, since this strategy will not help to preserve the viability, which is maybe counterintuitive.

Being a subset of $\mathbf{R}^2$, this viability kernel can be regarded as the graph of a tube $t \rightsquigarrow K_\propto(t)$ defined by

$$K_\propto(t) \;:=\; [\max(0, a_1(t)), a_2(t)]$$

which provides the specification of the viable storage values: if at time t_0, the storage $x(t_0)$ belongs to $K(t_0)$, then it will be possible to find exportation flows which allows the storage to be confined between $c(t)$ and the demand $d(t)$ at each instant t, and actually, to be confined between the bounds $\max(0, a_1(t))$ and $a_2(t)$. Otherwise, no exportation strategy can maintain such constraint at each instant.

Further, having an explicit description of the viability kernel, one can compute the regulation law, providing the possible exportation flow p in terms of the time and the storage: Setting

$$R(t, x) \;:=\; \begin{cases} d(t) & \text{if } x(t) = a_2(t) \\ c(t) & \text{if } x(t) = a_1(t) > 0 \\ [b_1(t), b_2(t)] & \text{if } x(t) = 0 \\ [c(t), d(t)] & \text{if } x(t) \in\,]\max(0, a_1(t)), a_2(t)[\end{cases}$$

The regulation law states that

$$\forall t \geq 0, \; p(t) \in R(t, x(t))$$

Let us point-out that the regulation map is single-valued on the part of the boundary of the viability kernel contained in the interior of the constrained set, illustrating the semi-permeability property of this "barrier" provided by the Quincampoix Theorem.

This regulation map allows the fishery to adjust their storage volume in order to remain in the viability kernel, i.e., to remain viable. It contradicts the "intuitive" rule stating that "the higher storage volume the fishery enjoys, the larger the exportation flow". The above regulation map does not always impose a direct relationship between storage volumes and exportation flows, except when the pair storage-exportation lies on the part of the boundary of the viability kernel satisfying the barrier property.

We refer to [73, Béné & Doyen] for a deeper analysis showing that when the viability kernel is not empty, there are instants when the exploiting costs exceed the commercial value of the catch. This confirms the empirical observation that firms will not cease their activities as soon as their production costs are higher than their production's commercial value, a phenomenon termed sometimes the stickiness of the economic actor. This behavior was not taken into account in the description of the above system, but is derived from the analysis of the regulation map.

2.7 Demodynamics

Noël Bonneuil has used viability theory in demographics and other domains of social and biological sciences to take into account phenomena which were not adequately described by standard deterministic dynamical models. We begin here with the presentation of two models of demographic studies and with two other models of dynamical games in another section.

2.7.1 A Demo-Economic Model

In this demo-economic model suggested by Noël Bonneuil to explain the modern fluctuations of fertility in the Western world, the state variable x represents the per capita capital and the regulee p the per capita consumption. The evolution of capital is governed by the differential equation

$$x'(t) = \sqrt{x(t)} - p(t) + (d - b)x(t)$$

where p denotes the per capita consumption, b the birth rate and d the mortality rate. The constraints are $x \geq 0$, $p \geq 0$ and $b \geq 0$.

The equilibrium curve is given by the equation

$$p = \sqrt{x} + (d - b)x$$

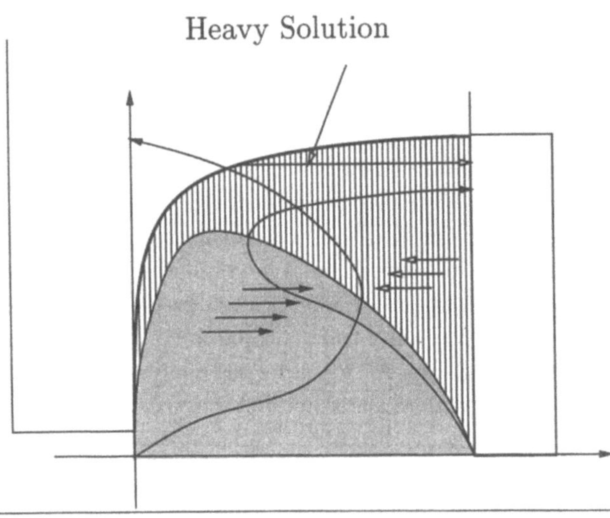

The axes represent the capital and the consumption. The equilibrium curve has two branches, the left one made of unstable equilibria and the right-one of stable equilibria.

Figure 2.14: Capital and Consumption

and shown in Figure 2.13. The left branch is made of unstable equilibria and the right branch of stable equilibria. The maximal consumption is denoted by $\bar{p}$.

At constant consumption larger than $\bar{p}$, the capital reaches 0 in finite time whatever the initial capital. If the initial capital-consumption pair is at the left of the equilibrium curve, the capital achieves 0 in finite time, but converges to an equilibrium if the initial capital-consumption pair is at the right of the left branch of the equilibrium curve.

One can consider the extreme cases when the consumption velocity is maximal (greedy solution) or minimal (frugal solution). The trajectory $p = \gamma(x)$ of the frugal solution going through $(0,0)$ determines the boundary of the regulation map of the system: it associates with any capital x the set of per-capita consumptions p below $\gamma(x)$.

The heavy solutions maintain the consumption constant as long as possible. If the solution does not converge to an equilibrium, the capital decreases and the capital-consumption pair hits the smallest viable frugal solution. In this case, the capital continues to decrease until it converges to 0.

The birth rate b is another regulee of the system, which has been taken constant. The reason was that it has a higher inertia than the per-capita consumption for regulating the system. Using the two regulees p and b together, we have a hierarchical dynamical game which can be studied using the hierarchical cascade mechanism of Katharina Müllers presented in Chapter 7.

2.7.2 Malthusian and Boserupian Scenarios

In demographics, two themes are debated for a consensus: the Malthusian one, which propounds that the population stabilizes at a given income level (depending of the living standard and of technology) whereas the Boserupian one suggests that technological change is triggered by demographic pressure.

In the Malthusian framework, the population grows naturally as long as the carrying capacity determined by resources and technology is not reached. According to Boserup, the pressure put by population on resources creates an incentive to use more productive technologies, so as to push farther away the equilibrium between population and resources.

Most of the mathematical models attempt to take into account these two phenomena in distinct deterministic models in which the links between demographic pressure and technological changes are univoquely fixed. Noël Bonneuil has succeeded in eluding the paradoxical conclusions of these points of view by considering technological progress as a regulee, allowing it to evolve in order to maintain the level of population positive, showing that due to this added flexibility, the two scenarios can be produced among many other possibilities. The pressure of the population on the use of new technologies is not necessarily deterministic whenever the population and the technology are in the viability kernel.

He proposed a simple model where the state is the logarithm of the population level x and the regulee is the logarithm of the technological progress denoted by p:

$$x'(t) \;=\; \alpha(w(t) - w^\star) - \theta p(t)$$

Noël Bonneuil inferred from simple and natural economic considerations that the *per capita* income w is equal to $w(t) = p(t) + \beta x(t) + \gamma x(t)^2$, so that the evolution of the population is governed by the differential equation

$$x'(t) \;=\; (\alpha - \theta)p(t) - \alpha w^\star + \alpha(\beta x(t) + \gamma x(t)^2)$$

The equilibrium curve is given by

$$p \;=\; \frac{\alpha}{\alpha - \theta}(w^\star - \beta x(t) - \gamma x(t)^2)$$

By putting a bound on the rate of technical level, Noël Bonneuil has computed the viability kernel and derived the map feeding back the population level to the technological level, which provides the regulation law prescribed by Boserup.

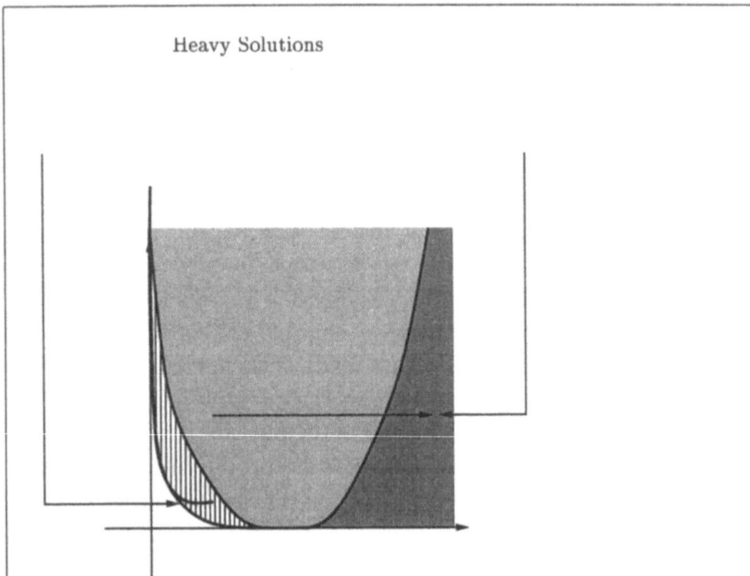

The axes represent the population and the technological level. The equilibrium curve. is a parabola. The left branch is made of unstable equilibria and the right branch of stable ones. Starting from the right of the right branch, the heavy solution decreases at constant production level and converges to the stable equilibrium associated with this technological level. Starting between the two branches, the heavy solution increases at constant technological level to a stable equilibrium. Starting from the left of the parabola, the population of the heavy solution decreases at constant technological level as long as it remains in the viability kernel, and thus, travels along the boundary of the viability kernel.

Figure 2.15: Malthus & Boserup

When the technological level is fixed, the population level converges either to a (stable) equilibrium or to extinction in finite time (when the technological level is too low).

But if we let the technological level free to evolve, other scenarios can emerge, whether the technological progress is maximal or minimal (see Figure 2.14).

Heavy solutions are obtained by keeping constant the technological level as long as possible. If the population does not converge to an equilibrium, it decreases and its trajectory hits the boundary of the viability kernel associated with the maximal technological progress. The population dies out in finite time if it does not make this revolution. Otherwise, the population still continues to decrease (and become extinct only at infinity) while technological level increases.

2.8 State-Dependent Bounds and Cycles

For simplicity, we started with constant bounds c and $-d$ on inflation. Naturally, these bounds may depend upon the state of the consumption.

$$\begin{cases} i) & x'(t) = g(x(t)) - h(x(t))p(t) \text{ where } p(t) \geq 0 \\ ii) & \text{and } -d(x) \leq p'(t) \leq +c(x) \end{cases} \tag{2.8.1}$$

The analysis becomes much more complicated, since we may observe oscillations which are killed by viability constraints, for instance. Still, the concepts of viability kernels and viability niches can be adapted.

We shall illustrate this equation with another story proposed by Philippe Lacoude for interpreting **business cycles**. Here, the "regulee" p does no longer represents the price, but the production of a producer. The differential equation (2.8.1)i) (with opposite signs) describes the behavior of the consumer, who modifies her consumption with a velocity proportional to p slowed down by an endogenous decay factor. The producer increases or decreases her production with respect to the demand she perceives with some uncertainty, measured by bound c:

$$\begin{cases} i) & x'(t) = p(t) - f(x(t)) \text{ where } p(t) \geq 0 \\ ii) & \text{and } -c \leq p'(t) - \varepsilon e(x(t)) \leq +c \end{cases}$$

We find the foregoing situation with $\varepsilon = 0$. In this example, the equation of the equilibrium curve is given by $p = f(x)$, the velocities are positive above[17] and negative below. Since the evolution of $p(\cdot)$ depends upon the consumption, the equilibrium "curve"[18] for the producer is made of the union

[17]We have changed the signs to accommodate the story...
[18]called the isocline

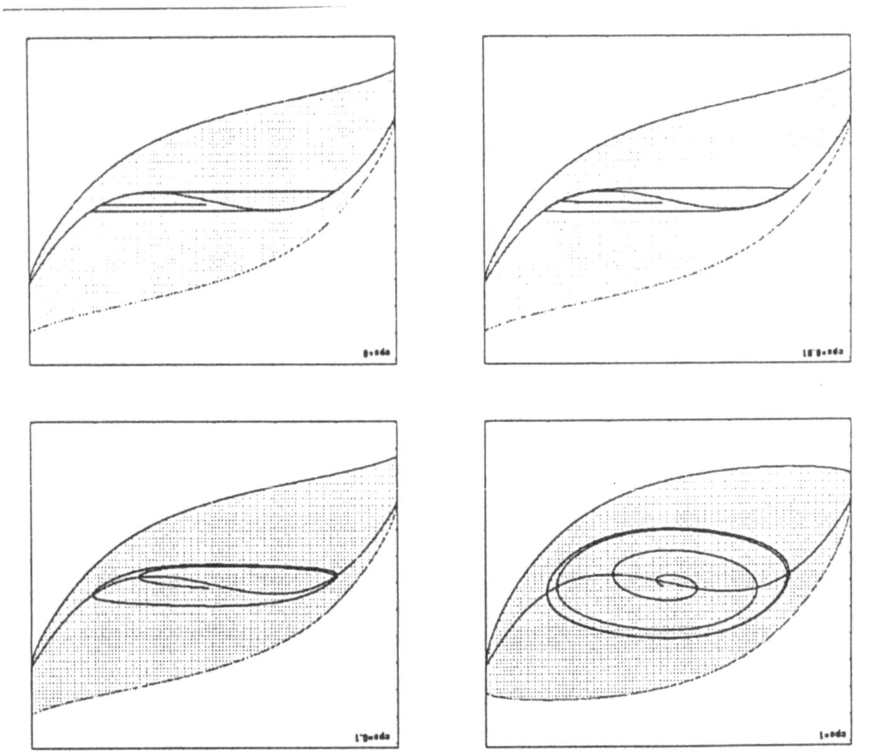

Equilibrium curve and viability kernel of the Van Der Pol oscillatory differential inclusion

$$\begin{cases} i) & x'(t) = a(\frac{x^3}{3} - x) - p \\ ii) & p'(t) - \varepsilon x(t) \in [-c, +c] \end{cases}$$

and viability kernel of

$$\begin{cases} i) & x'(t) = a(\frac{x^3}{3} - x) - p \\ ii) & p'(t) - \varepsilon x(t) = 0 \end{cases}$$

(oscillatory niche) for various values of ε.

Figure 2.16: Viability Kernel of an Oscillatory System

of the vertical lines $\{\alpha_i\} \times \mathbf{R}_+$ where the α_i denote the $n \geq 0$ distinct roots of the equation $e(x) = 0$ between the bounds a and b of the interval K. The equilibria of the foregoing system are the n points $(\alpha_i, f(\alpha_i))$.

We set $\alpha_0 = a$ and $\alpha_{n+1} = b$, so that the sign of the producer velocity is constant in each "cylinder" $M_i :=]\alpha_i, \alpha_{i+1}[\times \mathbf{R}_+$ $(i = 0, 1, \ldots, n)$. If we denote by L_- the area below the equilibrium curve and L_+ the region above, the $2n$ subsets $L_- \cap M_i$ and $L_+ \cap M_i$ are **monotonic cells**. In each of these cells, the monotonic behavior of the solutions is the same. One can prove easily that, starting from one monotonic cell, any solution of the foregoing system either leaves the monotonic cell in finite time or converges to an equilibrium.

Figure 2.15 shows the viability kernel of this "oscillatory" inclusion. The inertial set is the viability kernel of the system of differential equations

$$\begin{cases} i) & x'(t) = p(t) - f(x(t)) \text{ where } p(t) \geq 0 \\ ii) & p'(t) = \varepsilon e(x(t)) \end{cases}$$

It coincides with the region bounded by the limit cycle of this system. Starting outside of the inertial set, the heavy solution (with $c = 0$) starts to oscillate until it meets the boundary of the viability kernel. Then the evolution has to be slowed down by taking the evolution

$$p'(t) = \varepsilon e(x(t)) + c$$

which travels along the boundary of the viability kernel.

2.9 Examples of Dynamical Games

The metasystems of the one state-one regulee models presented earlier are dynamical systems with two "metastates" — the state and the regulee — and one "metaregulee" — the velocity of the regulee.

We turn now our attention to dynamical systems with two states and two regulee, which are still amenable to explicit analytical formulation, thanks to a study conducted by Noël Bonneuil and Katharina Müllers appeared in [86, Bonneuil & Müllers] in the ecological framework of the predator-prey model, which also enjoys several economic interpretations. Since predators and preys do exist in the economic jungle, we kept the original motivation of the authors.

We consider a system with two state variables denoted by x and y and two regulees denoted[19] by u and v. The evolution of the states in terms of the regulees is governed by the two-dimensional system of parametrized

[19] We do not use here the notations p or q for the regulees because in many models, velocities of regulees of former models are chosen as regulees of new systems, called metasystems.

differential equations:

$$\begin{cases} x'(t) &= f(x(t),y(t),u(t),v(t)), \quad u(t) \in [a_1,b_1] \\ y'(t) &= g(x(t),y(t),u(t),v(t)), \quad v(t) \in [a_2,b_2] \end{cases} \tag{2.9.1}$$

There are ambiguities in the vocabulary used. The vocable "differential games" coined by Rufus Isaacs in [239, Isaacs], the founder of the theory of differential games, refers both to the dynamics (2.9.1) alone and to the "rules of the game", so to speak. Such rule states that the first player maximizes a pay-off function and, in doing so, encounters the action of the second player, who himself seeks to maximize his pay-off function, which may be the opposite of the pay-off function of the first player. Such games are sometimes called "quantitative games".

Here, we are rather interested to what is called "qualitative games": A constrained set is given in which the state must remain, i.e., a viability constraint is given. The game consists in keeping at least one solution viable forever: the strategies chosen by each player are not directed towards any pay-off and winning is not an issue (the implied "aim" of each player is the ability to continue playing the game forever[20]). In mathematical terms, the game consists in safeguarding at least one solution of the system, starting from given initial conditions, which remains in a fixed closed constrained set forever. This is the reason why we chose to designate by "dynamical game" any dynamical system controlled by two or more players, without specifying the actual choice of the "game played", i.e., the scenario or the rules of the game to be played.

Playability is a cooperative concept, but there are several other "rules of a dynamical game" — other games to be played. In Chapter 7, we shall present Katharina Müllers' hierarchical games and Chapter 9 is devoted to noncooperative games played against an adverse "Nature".

2.9.1 The Prey-Predator Dynamical Game

We chose to illustrate such a dynamical game with the Lotka-Volterra prey-predator systems of the form

$$\begin{cases} x'(t) &= x(t)(a - u(t)x(t) - (v(t) - u(t))y(t)) \\ y'(t) &= y(t)(-b + v(t)x(t)), \end{cases} \tag{2.9.2}$$

where for all $t \geq 0$

$$u(t) \in [a_1, b_1], \quad v(t) \in [a_2, b_2] \tag{2.9.3}$$

Here, the states x and y are the average densities of the two species — or taxa — and the regulees are regarded as the "strategies" of the species.

[20]This is the "playability" concept introduced in Chapter 15 of VIABILITY THEORY, [28, Aubin].

The strategy u represents the "carrying capacity effect" and the strategy v summarizes the rate at which the predator captures the prey and the value of one unit of prey to the consumer.

The action of the predator on the prey depends on both strategies u and v. For the sake of simplicity, Noël Bonneuil and Katharina Müllers specify this action here as $v - u$, when $u < v$. When $u > v$, the predator cannot catch any prey, and dies. If strategies are associated to body sizes for instance, the higher u, the better the prey can escape the predator, but the lower its carrying capacity.

Here, they impose to the state to satisfy viability constraints of the form

$$\forall\, t \geq 0,\ \ (x(t), y(t)) \ \in\ K\ :=\ [0, \infty[\times[1, \infty[$$

where, after appropriate scaling, we consider a species to be extinct as soon as the density decreases below these bounds.

It was intuitively thought that the "nonlinearities" added to the model by subtracting to constant growth rates state dependent rates help in confining the nonnegative evolutions of the states in bounded subsets. We shall prove that this is not the case, at least when the strategies u and v are bounded. We shall show that the subset K is not viable, and that its viability kernel is bounded. Therefore, outside of it, and whatever the choice of the strategies u and v in a bounded set, the solutions leave K in finite time.

Several classical models specify a choice of strategy $u(x, y)$ and $v(x, y)$, of which equilibria and their stability properties are studied with respect to some relevant parameters (we refer to the wonderful book [275, Murray] for an exhaustive presentation of the topic.)

We adopt another point of view: We leave the choice of the strategies u and v open until the last moment, when we select the ones which maintain the viability of the system.

This point of view is shared for other reasons by several ecologists: For instance, in p. 199 of [287, Pianka], the author states that the theory of prey-predator interactions could be improved substantially simply by treating coefficients involved in Lotka–Volterra models "as variables". We quote Noël Bonneuil and Katharia Müllers: *"The assumption that strategies can vary, but within certain fixed bounds has a deep consequence on the prey-predator dynamics, onto which we would like here to shed another light: the set of possible co-evolutions is bounded, and its boundary can be determined. This boundary corresponds to a specific path of evolution.*

The problem is then to determine this set outside of which all trajectories are "deadly", and to determine the regulation of the strategies which allow the viability of the system. ... "The viability kernel plays a central role in understanding Darwinian selection, because the taxa who leave this viability kernel will surely be eradicated, whereas those who are still in it have at least one possibility to survive forever. To remain in the viability kernel, the species do not really "select" strategies which are viable. The viability kernel simply reveals that should the population densities be shifted outside of the viability kernel (by human intervention or some natural catastrophe), then there is no way the populations could avoid extinction, however adaptable they are, that is to say, however flexible they are at restructuring and hence redefining their taxa strategies within the constraints imposed. If it is impossible for infinitely flexible strategies (capable of varying instantaneously within given bounds) to rescue the situation, then adaptation under the constraints of natural selection can do no better.

Moreover, viability considerations deliberately center the debate on transient changes, regardless of what can happen asymptotically, and subsequently not necessarily in the neighborhood of equilibria or attractors. We have no worry of predicting the future trajectory the system will take; rather, we delineate which evolutions guaranteeing the existence of the taxa are possible, and if so, which strategies can regulate the system to make their existence possible".

For describing analytically the viability kernel of K, Noël Bonneuil and Katharina Müllers use the **monotonic cells** introduced by Olivier Dordan[21] in [165, Dordan] which help to clarify the qualitative behavior of dynamical game (2.9.1): The indexes δ_i, $i = 1, 2$ take their values in $\{+, -, 0, @\}$. The notation "@" represents indifferently either $+, -$ or 0.

Definition 2.9.1 *The set*

$$\begin{array}{rcl} K_{\delta_1\delta_2} & = & \{x \geq 0, y \geq 0 \mid \forall u \in [a_1, b_1],\ v \in [a_2, b_2]\} \\ & & \mathrm{sign}(f(x,y,u,v)) \ = \ \delta_1 \ \& \ \mathrm{sign}(g(x,y,u,v)) \ = \ \delta_2 \end{array} \tag{2.9.4}$$

is called monotonic cell *of the dynamical game (2.9.1).*

For example,

$$\begin{array}{rcl} K_{@@} & = & [0, \infty[^2, \\ K_{00} & = & \{(0,0)\}, \\ K_{-@} & = & \{(x,y) \in [0,\infty[^2; y > (a - b_1 x)/(a_2 - b_1) \text{ and } x \leq a/a_2\} \cup \\ & & \{(x,y) \in [0,\infty[^2; y > (a - a_1 x)/(a_2 - a_1) \text{ and } x > a/a_2\}, \end{array}$$

because $f(x,y,u,v) < 0$ for all $u \in [a_1, b_1], v \in [a_2, b_2]$ is equivalent to

$$y > \max_{u \in [a_1,b_1], v \in [a_2,b_2]} \frac{a - ux}{v - u} \tag{2.9.5}$$

where for all $(x,y) \in [0,\infty[^2$

$$\max_{u \in [a_1,b_1], v \in [a_2,b_2]} \left\{0, \frac{a - ux}{v - u}\right\} = \begin{cases} \dfrac{a - b_1 x}{a_2 - b_1} & \text{when } x \leq \dfrac{a}{a_2}, \\ \dfrac{a - a_1 x}{a_2 - a_1}, & \text{when } \dfrac{a}{a_2} < x \leq \dfrac{a}{a_1} \\ 0, & \text{when } x > \dfrac{a}{a_1} \end{cases} \tag{2.9.6}$$

[21] see Chapter 4 below.

Similarly,

$$\begin{array}{rcl} K_{@-} & = & [0, b/b_2[\times[0, \infty[\\ K_{--} & = & K_{-@} \cap K_{@-} \end{array}$$

Theorem 2.9.2 *Assume that for $x = b/b_2$*

$$1 \; < \; \max\left\{\frac{x(a-a_1x)}{b-a_1x}, \frac{x(a-b_1x)}{b-b_1x}\right\} \tag{2.9.7}$$

Then the viability kernel of K is not empty, and is delimited by the curves $\sigma^\sharp$ and $\rho^\sharp$:

$$\begin{array}{rcl} \mathrm{Viab}(K) & = & \{(x,y) \in K; \; y \leq \rho^\sharp(x) \text{ if } (x, \rho^\sharp(x)) \in K_{-@} \text{ and} \\ & & \qquad\qquad x \geq \sigma^\sharp(y) \text{ if } (\sigma^\sharp(y), y) \in K_{@-}\} \end{array} \tag{2.9.8}$$

where the functions $\sigma^\sharp$ and $\rho^\sharp$ are the solutions of the differential equations

$$\frac{d\sigma^\sharp}{dy}(y) \;=\; \min_{u\in[a_1,b_1], v\in[a_2,b_2]} \frac{f(\sigma^\sharp(y), y, u, v)}{g(\sigma^\sharp(y), y, u, v)} \qquad \sigma^\sharp(y_0) = x_0 \tag{2.9.9}$$

$$\frac{d\rho^\sharp}{dx}(x) \;=\; \max_{u\in[a_1,b_1], v\in[a_2,b_2]} \frac{g(x, \rho^\sharp(x), u, v)}{f(x, \rho^\sharp(x), u, v)} \qquad \rho^\sharp(x_1) = y_1, \tag{2.9.10}$$

with the initial conditions

$$(x_0, y_0) = (b/b_2, 1) \quad \text{and} \quad (x_1, y_1) \in K_{--} \cap Graph(\sigma^\sharp) \tag{2.9.11}$$

Noël Bonneuil and Katharina Müllers determine the viable feedback strategies associated to the viability kernel $\mathrm{Viab}(K)$, that is to say, for every state (x, y) the set $R_{\mathrm{Viab}(K)}(x, y)$ of strategies (u, v) which allow the system to remain viable.

Theorem 2.9.3 *For all $(x, y) \in K_{-@}$, let $\gamma(x, y)$ be the set of strategies (u, v) maximizing $\frac{g(x,y,u,v)}{f(x,y,u,v)}$, and for all $(x, y) \in K_{@-}$, let $\delta(x, y)$ be the set of strategies minimizing $\frac{f(x,y,u,v)}{g(x,y,u,v)}$. The regulation map for the predator–prey system associated to $\mathrm{Viab}(K)$ is given by*

$$R_{\mathrm{Viab}(K)}(x, y) \;=\; \begin{cases} \gamma(x, y), & \text{if } y = \rho^\sharp(x) \\ \delta(x, y), & \text{if } x = \sigma^\sharp(y) \text{ and } y > 1 \\ [a_1, b_1] \times [\max\{a_2, b/x\}, b_2], & \text{if } y = 1 \\ [a_1, b_1] \times [a_2, b_2], & \text{otherwise} \end{cases}$$

for all $(x, y) \in \mathrm{Viab}(K)$.

We refer to the original paper [86, Bonneuil & Müllers] for the proof of these theorems. Here are their conclusion: *"Interpreting coefficients in a dynamical game such as the Lotka-Volterra prey-predator game as varying*

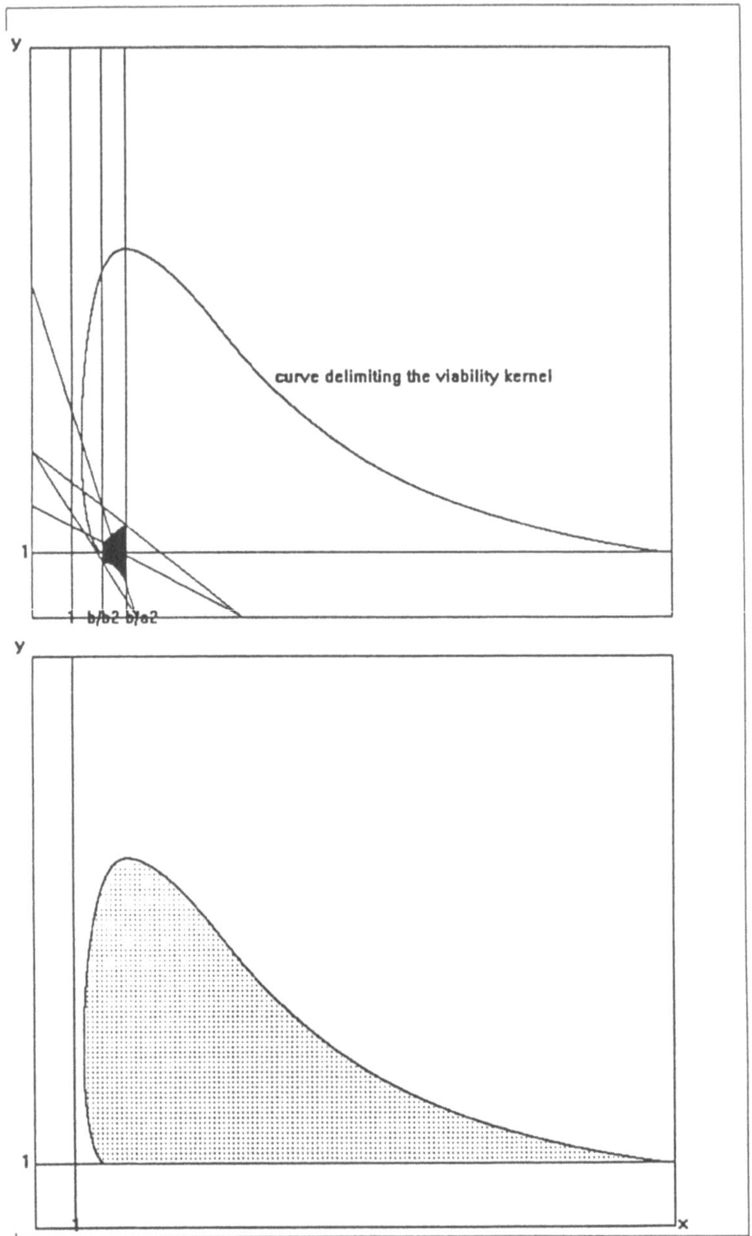

Equilibrium curve and viability kernel for the predator-prey model.

Figure 2.17: Viability Kernel for the Lotka-Volterra Model

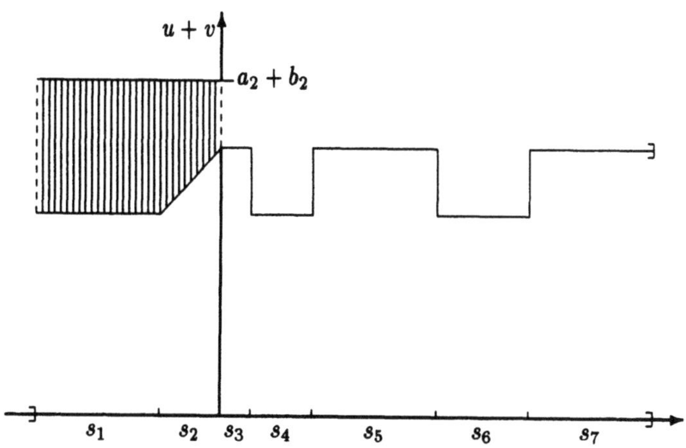

The figure represents symbolically the values of the regulation map on the boundary of the viability kernel. The horizontal axis represents the "flattened" boundary of the viability kernel, and the vertical axis the values $u + v$. The regulation map is deterministic on the segments in the interior of the constrained set, but set-valued on the boundary and on the part of the boundary of the viability kernel intersecting the boundary of the constrained set.

Figure 2.18: Regulation Map for the Lotka-Volterra Model

strategies is a different way of thinking ecological and economical interactions. Instead of guessing some law of change (frequency-dependent as presented usually, or some specified stochastic law), we have delimited which temporal paths are viable, that is to say those which can be sustained forever under given constraints.

Within these viability kernels, strategies can be picked up from a whole nonempty set, giving place to a great variety of possible predator-prey interactions. On the boundaries of these kernels, when the survival of species is at stake, only one path of interactions between prey and predator is observable, *because it is viable. All the other paths lead to extinction and disappear. The prey-predator interaction on the boundaries thus appears unequivocal between prey and predator, and strategies can be expressed as unequivocal (temporal) functions, possibly involving densities. An exogenous perturbation in this situation can pull the system back into the viability kernel, where strategies are supposed to be flexible enough to enable further survival, or, on the contrary, push it outside the viability kernel, where no strategy exists, however flexible, to rescue the situation and expect a long term existence."*

2.9.2 Barth's Nomads of South Persia

Noël Bonneuil has proposed in [85, Bonneuil] several mathematical models of Fredrik Barth's studies in social sciences and in particular, anthropology. He recognized in these studies modes of reasoning analogous to the ones used in viability theory. In Chapter 5 of Barth's [66, Barth] entitled "Models Reconsidered", Fredrik Barth insists on the evolutionary perspectives around the concept of social process, that he opposed to the then prevailing structuralist view, which is static by nature. He uses the terms "viability" and "controls" as well as "approximate equilibrium", not in their mathematical sense, naturally, but conveying the same meaning. His "approximate equilibria", for instance, correspond to our viability kernels. In his study of nomads of South Persia, Fredrik Barth specified the difficulty of forecasting the decision for a tribe to sedentarize and the failure of "mechanicist" models to give an account to the decisions taken by the nomads. We summarize here one of the Bonneuil's mathematical model of Barth's studies, choosing the one on nomads of South Persia taken from [65, Barth].

In this oversimplified version of Bonneuil's study, we isolate two variables, the state variable x which is the number of Basseri nomads of an independent economic unit, and the second state variable, which is the amount y of cattle of this economic unit. The regulation variables are the sedentarization rate u and the consumption (or predation) rate v. A positive sedentarization rate describes a sedentarization process while a negative one means a nomadization process.

The viability constraints state that an economic unit needs a minimum of cattle (about 60 animals) to survive. Hence the constrained set is defined by

$$K = \{(x,y) \in \mathbf{R}^2_+ \mid y \geq wx\}$$

where $w > 0$ is the minimal income, a threshold under which the survival of the tribe is at stakes.

The sedentarization and the predation rates are subjected to constraints of the form

$$(u,v) \in [-\underline{u}\,, +\overline{u}] \times [\underline{v}\,, +\overline{v}]$$

The lower limit $\underline{v}$ denotes the minimum predation rate necessary for survival. The evolution of the population is assumed to obey dynamics of the form

$$x'(t) = x(t)\left(\alpha \log\left(\frac{y(t)}{wx(t)}\right) - u(t)\right)$$

The cattle population obeys a Verhulst logistic dynamics corrected by a predation term analogous to (2.5.1):

$$y'(t) = ry(t)\left(1 - \frac{y(t)}{\tilde{y}}\right) - v(t)x(t)$$

The viability kernel of the constrained set K is computed by the viability kernel algorithm. From outside the viability kernel, all solutions leave the constrained set in finite time.

For small values of population and cattle, it coincides with the constrained set. When both the population and the cattle become large, then they must obey a new constraint, and to be above the round curve in the north-east part of the boundary. On the lower boundary of this viability kernel, the sedentarization rate is at its highest level. In this case, both the population and the cattle dwindle to reach the constraint $y = wx$. Quincampoix's Theorem also states that from this part of the boundary viability kernel above the line $y = wx$, no solution can enter the interior of the viability kernel. Therefore, if the population does not sedentarize fast enough, it is either doomed to vanish, or else, to sedentarize at an even higher sedentarization rate than $\overline{u}$. The later the decision to sedentarize, the highest the sedentarization rate needed to reinstate viability.

Starting with regulee u and v whose viability niches are not empty provide naturally heavy solutions par excellence: There is no need to change policy.

Otherwise, starting from the viability kernel, the solutions associated with constant regulee leave the constrained set in finite time. One can change them at the last moment, either with infinite velocity (impulses), shifting brutally the regulee in order to make the situation viable, or with bounded velocity for taking into account inertia constraints.

Actually, the two regulee do not enjoy the same inertia. As Noël Bonneuil shows, Barth's study suggests that when the viability of the tribe is at stakes, the nomads first modify their rate of predation. Only if this strategy is not

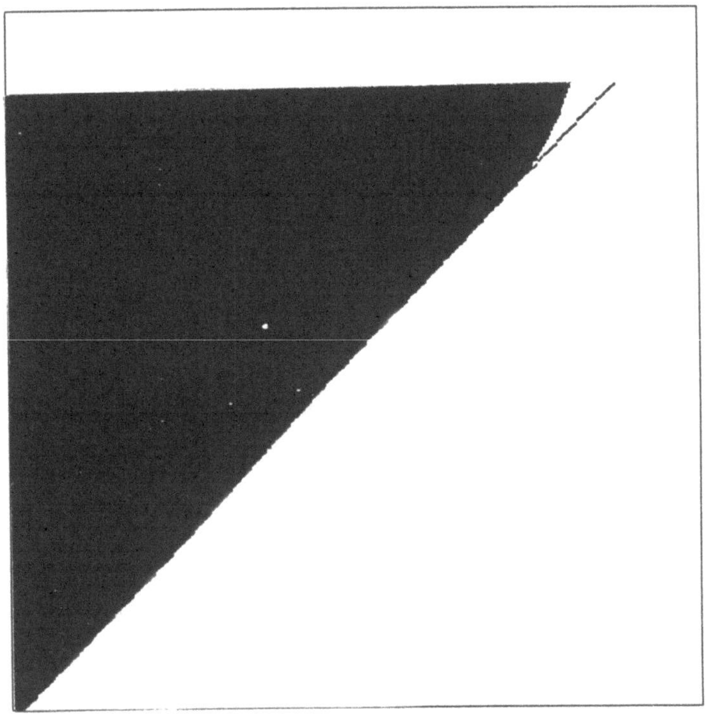

Figure 2.19: Viability Kernel for the Nomads of South Persia

sufficient to reestablish viability do the nomads start to sedentarize. The sedentarization regulee v has thus a higher inertia than the predation regulee u. This is the hierarchical cascade mechanism of Katharina Müllers presented in Chapter 7.

II

Ecodynamics

Chapter 3

Dynamical Resource Allocation

Introduction

We shall devote this chapter to the study of the basic problem of dynamical allocation of scarce resources ranging over a closed set $M \subset Y$ of the commodity space $Y := \mathbf{R}^l$. Its dual $Y^\star$, identified with $\mathbf{R}^l$, is interpreted as the price space.

1. Viability (or scarcity) constraints are described by the closed subset[1] K of allocations $x := (x_1, \ldots, x_n) \in X := Y^n$ of scarce resources:

$$K := \left\{ x \in Y^n \mid \sum_{i=1}^{n} x_i \in M \right\}$$

These are intended to describe the "viability" — or "sustainability", in a more ecological sense — of the dynamical economy.

The decentralized dynamical behavior of the consumers is described by n change functions $c_i : Y \times Y^\star \mapsto X$ and by the "*a priori* pricing map", which is a set-valued map $P : X \rightsquigarrow Y^\star$ associating with any allocation the set of possible prices $p(\cdot)$. Then the consumptions evolve in function of prices according to the dynamical laws

$$\begin{cases} i) \quad x_i'(t) = c_i(x_i(t), p(t)) \; (i = 1, \ldots, n) \\ ii) \quad p(t) \in P(x(t)) \end{cases}$$

A solution to this system is made of n functions $t \to x_i(t)$ satisfying this system for some time dependent price $t \to p(t)$.

[1]We shall investigate the cases when K depends upon the time or the history of the evolution of the commodities at the end of the chapter. We shall also cover the case of solutions which improve a reference preorder when time evolves.

The allocation set K is said to be **viable** if from every initial allocation $x_0 \in K$ starts at least one solution which is **viable** in the sense that

$$\forall\, t \geq 0,\ \sum_{i=1}^{n} x_i(t) \ \in\ M$$

The first task is to characterize the sets of scarce resources having this property. To be of value, this task must be done without solving the dynamical system and then, without checking the existence of viable solutions from each initial commodity.

An immediate intuitive idea jumps to the mind: at each point of the boundary of the set M, where the viability is at stake, there should exist velocities the sum of which is in some sense **tangent** to M and serves to allow the total consumption to bounce back and remain inside M. This is, in essence, what the Viability Theorem states in this basic problem of allocation of scarce resources. But, first, the mathematical implementation of the concept of tangency must be made.

We cannot be content with viability sets that are smooth manifolds, because most of the sets defined through inequality constraints which we need in economics would thereby be ruled out.

We shall show how to adapt the definition of tangent directions in the case of convex sets. But we can "implement" the concept of a direction v tangent to **any subset** K at $x \in K$, which should mean that starting from x in the direction v, we do not go too far from K.

To convert this intuition into a mathematical definition, we shall choose among the many ways that have been designed to translate what it means to be "not too far", the one suggested by Bouligand fifty years ago: a direction v is **contingent to K at $x \in K$** if it is a limit of a sequence of directions v_n such that $x + h_n v_n$ belongs to K for some sequence $h_n \to 0+$. The collection of such directions, which are in some sense "inward", constitutes a closed cone $T_K(x)$, called the **contingent cone**[2] to K at x. Naturally, except if K is a smooth manifold, we lose the fact that the set of contingent vectors is a vector-space. However, we can go forward despite this loss of amenity.

2. We then associate with the change functions c_i and the pricing map P confronted to the scarcity constraints (described by M) the **(set-valued) regulation map** Π_M. It maps any allocation x to the subset $\Pi_M(x)$ consisting of prices $p \in P(x)$ which are *viable* in the sense that

$$\sum_{i=1}^{n} c_i(x_i, p) \quad \text{is contingent to } M \text{ at } \sum_{i=1}^{n} x_i$$

[2]replacing the linear structure underlying the use of tangent spaces by the contingent cone is at the root of **Set-Valued Analysis**.

If, for every $x \in K$, there exists at least one viable price $p \in \Pi_M(x)$, we then say that the allocation set K is a **viability domain**.

The Viability Theorem we mentioned earlier holds true for a rather large class of systems, called **Marchaud systems**: Beyond imposing some weak technical conditions, the only severe restriction is that the change functions are affine with respect to the price and that the pricing map P has convex images $P(x)$.

The basic viability theorem states that **the allocation set K is viable if and only if K is a viability domain**. It also provides a **regulation law**:

$$\text{for almost all } t, \;\; p(t) \;\in\; \Pi_M(x(t))$$

in order to maintain the viability of a solution.

The multivaluedness of the regulation map — this means that several prices $p(t)$ may exist in $\Pi_M(x(t))$ — is an indicator of the "robustness" of the dynamical economy: **The larger the set $\Pi_M(x(t))$, the larger the set of disturbances which do not destroy the viability of the economy !**

3. We shall also characterize this property by **budgetary rules** and evidence that in certain cases, instantaneous Walras laws (individual and even collective ones) warrant that this condition is satisfied. They follow from a dual formulation of the tangential condition characterizing the viability the set of allocations. It allows us to find sufficient conditions for viability. Actually, we shall devote the whole Chapter 5 to the exploitation of this dual formulation for designing dynamical economies the regulees of which are "viability multipliers". In our example of allocating scarce resources, such a sufficient condition is a budgetary rule translating the fact that at each instant, it is forbidden to spend more monetary units than earned.

4. Once these basic results have been described, we embark upon adding new features to this very simple model. First, we allow consumers to be equipped with preference orders which must increase along at least one viable allocation originating from an initial allocation. This happens to be a very simple adaptation since the set of allocations preferred to a given one can be regarded as "state-dependent viability constraints".

   Simplicity was the main reason why we began to present the viability problem in a stationary environment. We shall then introduce time dependent dynamical economies for which the change functions $c_i(t, x, p)$, the pricing map $P(t, x)$ and the set $M(t)$ of scarce resources **depend upon the time**. At this point, we have to introduce **contingent derivatives** of the set-valued map $t \rightsquigarrow M(t)$, which is the **tube** describing the evolution of scarce resources, in order to describe the regulation law. This can be easily done, so that we can adapt the Viability Theorem for time-dependent sets of scarce resources and change functions.

5. However, the introduction of time is not sufficient: The environment of a dynamical economy not only depends upon the time and the allocation of the system at this instant, **but also upon the whole history of the evolution up to that time.** For instance, the change functions, the pricing map and the constrained set often depend upon the cumulated consequences of past evolutions.

   The viability problem still makes sense in this case: Find at each instant prices which provide consumers an information allowing them to evolve under history-dependent constraints. Therefore, we have to adapt the viability theorem to this case in order to provide the regulation map which governs the evolution of prices in terms of the past evolution of allocations.

6. We finally explore another evolutionary problem. We already mentioned the case of time-dependent dynamical economies confronted to a time-varying set of resources $M(t)$. However, this evolving set is given. The problem arises whether the evolution of the tube $M(t)$ can be governed by an evolutionary law in the same way as the evolution of a vector $x(t)$ can be governed by a differential equation. This makes sense — and even, is mandatory — when the set of resources is obtained from "technological sets" of producers. We thus may describe the behavior of producers who can govern the evolution of their technological sets.

   The answer is affirmative: It is possible to define what are called **mutational equations**

$$\overset{\circ}{M}(t) \ni \Phi(M(t))$$

   which, given an initial condition $M(0) = M_0$, governs the evolution of a unique tube $M(t)$ starting at M_0.

   In order to write down such a mutational equation, we have to define an adequate concept of **velocity of a tube.** For usual functions $t \mapsto x(t)$, the velocity is defined by the derivative $x'(t)$ of the function $x(\cdot)$. For tubes, the forward contingent derivative $DM(t,x)(1)$ we already introduced does not provide a concept of velocity, because it depends upon a state $x \in M(t)$. Therefore, we propose to use the concept of "mutation" $\Phi \in \overset{\circ}{M}(t)$, which may be for instance a convex compact subset satisfying the property

$$M(t) + h\Phi \subset B(M(t+h), \beta(h)h) \;\&\; M(t+h) \subset B(M(t) + h\Phi, \beta(h)h)$$

   where $\beta(h)$ converges to 0 with h. When $M(t) := \{x(t)\}$ is single-valued, we recognize the definition of the derivative.

   With this concept at hand, we can

(a) prove the existence and the uniqueness of a solution $M(\cdot)$ to the above mutational equation where Φ maps compact sets M to nonempty convex compact sets $\Phi(M)$,

(b) prove the formula

$$DM(t,x)(1) = \Phi + T_{M(t)}(x)$$

linking the forward contingent derivative and a mutation Φ belonging to $\overset{\circ}{M}(t)$.

Therefore, we can associate with a mutational equation $\overset{\circ}{M}(t) \ni \Phi(M(t))$ governing the evolution of the set of resources the regulation map Π_Φ defined by

$$\Pi_\Phi(x,M) := \left\{ p \in P(x) \mid \sum_{i=1}^{n} c_i(x_i,p) \in T_M(\sum_{i=1}^{n} x_i) + \Phi(M) \right\}$$

If the values $\Pi_\Phi(x,M)$ of the regulation map are not empty, then from any initial condition $x_0 \in M_0$, there exist a viable solution to the dynamical economy regulated by prices $p(t)$ obeying the regulation law

$$p(t) \in \Pi_\Phi(x(t),M(t))$$

3.1 Allocations of Scarce Resources

3.1.1 The Commodity Space

An **economic commodity** is by definition a good or a service supplied with a **measure unit**. Commodities can be dated, localized, contingent, etc. In this case, they are different. For example, two dated commodities with different dates and otherwise the same characteristics are different commodities.

In summary, we start with l commodities labeled $h = 1, \ldots, l$ and we denote by

$$e^h := (0,\ldots,1,\ldots,0) \in Y := \mathbf{R}^l$$

the **unit commodity** h (where 1 is at the h place).

We begin by assuming that **the commodities are indefinitely divisible**. This is a quite rough approximation of economic reality, but an imperative one which allows to describe mathematically the space of commodities as a finite dimensional vector-space. Actually, one can characterize commodities by the **services** which they produce, which are more "divisible" than the goods they yield.

A **commodity bundle**, or, in a more descriptive way, a **commodity basket**, is a basket made of x_1 units of commodity 1, x_2 units of commodity 2, ... ,

x_l units of commodity l. It is represented by the vector

$$x := \sum_{h=1}^{l} x_h e^h = (x_1, x_2, \ldots, x_l) \in Y$$

For the sake of simplicity, we shall speak from now on of commodities instead of commodity bundles or baskets.

So, the familiar finite dimensional vector-space $Y := \mathbf{R}^l$ is regarded as the **commodity space**, the canonical basis of which is made of the units of goods.

Naturally, some commodities will be eliminated, such as, for instance, commodities with negative units, which, at first glance, do not make sense.

Actually, it may be wise to accept negative goods if they are adequately interpreted. In exchange economics for instance, we distinguish among **produced** commodities and **consumed** commodities. One may represent for instance consumed commodities with a positive sign and produced commodities with a negative sign. Or, as another example, **labor** can be regarded as a negative **leisure**.

Why do we represent the commodity space by the finite dimensional vector-space $\mathbf{R}^l$? The reason is that we can indeed add commodities and multiply them by scalars, i.e., perform linear combinations of commodities. Therefore, we shall be able to exploit the rich structure of linear spaces.

3.1.2 The Set of Resources

We denote by $M \subset Y$ the set of physical **scarce resources** to be allocated among n consumers.

Scarcity is the key word, the basic requirement without which there would be no need of economics[3].

When producers are taken into account, the commodities of M to be allocated among consumers are **produced** by producers. The set M thus depends upon past and present decisions of producers. At least for the beginning, we shall assume in this chapter that there is a constant set M of scarce resources available at each time to consumers. Later on, when evolution will be taken into account, it will be possible to have M depend upon the time and cumulated consequences of past allocations, in order to take into account depletion, investments, pollution, etc.).

The first law to which economics must comply states that **it is impossible to consume more (physical) resources than available**[4]. We are mainly interested to **sustainable evolution**.

---

[3]Actually, ecology, since many economists restrict the scope of their investigations to exchange processes.

[4]by opposition to fiduciary goods, for which, despite the theoretical aim that their supply is limited, monetary creation allows the scarcity constraints to be relaxed.

Throughout this book, we assume mainly that M is **closed** and that it satisfies the **free disposal assumption**[5]:

$$M = M - \mathbf{R}^l_+ \tag{3.1.1}$$

This means that any commodity $y \leq x$ smaller than or equal to an available commodity x is still available.

Since M is a set of resources, it should be **bounded above** in the sense that

$$\exists\, \overline{y} \in Y \quad \text{such that } M \subset \overline{y} - \mathbf{R}^l_+ \tag{3.1.2}$$

(It cannot be bounded below because of the free disposal assumption).

We shall assume sometimes, for simplicity, that M is **convex**: convex combinations of scarce resources are still available. This is interpreted by economists by saying that **decreasing return to scale** prevails. If further more M is a cone, they say that **constant return to scale** prevails. (Actually, we shall be able to bypass this assumption in the dynamical case[6], but not in the static case, where convexity is a mathematical requirement for proving the existence of an equilibrium. In the dynamical case, it is just an amenity making life easier).

The particular case when $M := \{0\}$ or $M := -\mathbf{R}^l_+$ is called an **exchange economy**. In this case, there is no real consumption (since there are no resources), but a continuous exchange of commodities, with a possible destruction in the case when $M := -\mathbf{R}^l_+$.

3.1.3 Introducing the Consumers

We begin now the mathematical description of the n consumers $i = 1, \ldots, n$.

It starts by Mrs. i's **consumption set** $L_i \subset Y$, which represents the set of her potential consumptions. Actually, it is better to say that she will never accept a commodity outside her consumption set L_i. Most often, L_i is chosen to be the orthant $\mathbf{R}^l_+$.

Throughout this book, we assume mainly that the consumption sets L_i are **closed**.

Consumers are often assumed to have no **satiation**: this means that there is no limit to their desire to squander[7]. We describe it in mathematical terms by stating that $L_i = L_i + \mathbf{R}^l_+$.

---

[5]When K and L are subsets of a vector space X, we denote by $\alpha K + \beta L$ the subset

$$\alpha K + \beta L := \{\alpha x + \beta y\}_{x \in K,\, y \in L}$$

It is called the **Minkowski linear combination** of the subsets K and L. We always have $K \subset K + L$ whenever $0 \in L$.

[6]and replace it by a regularity assumption called **sleekness**. See Definition 11.2.6 in the Tool-Box.

[7]Despite its first glance appeal, this assumption is not always sensible.

We shall assume also that L_i is **bounded below**, i.e., that

$$\exists\, \overline{x}_i \in Y \quad L_i \subset \overline{x}_i + \mathbf{R}^l_+$$

Again, for simplicity, we shall sometimes assume that the consumptions sets are **convex**.

3.1.4 The Set of Allocations

We translate now the first economic law: **it is impossible to consume more (physical) commodities than available** by introducing the **set** K **of allocations of scarce resources among** n **consumers**.

We denote by $X := Y^n = \mathbf{R}^{ln}$ the Consumption Space of the n consumers. We set

$$x \;:=\; (x_1, \ldots, x_n) \;\in\; X$$

where x_i does no longer denote a component of a commodity bundle, but the commodity bundle of consumer i[8].

Therefore, the set of allocations is equal to

$$K \;:=\; \left\{ x \;:=\; (x_1, \ldots, x_n) \;\in \prod_{i=1}^{n} L_i \;\middle|\; \sum_{i=1}^{n} x_i \;\in\; M \right\} \tag{3.1.3}$$

Observe that it is not empty if and only if

$$0 \;\in\; \sum_{i=1}^{n} L_i - M \tag{3.1.4}$$

Conforming to what we called the first economic law amounts to evolving in the allocation set K (sustainable evolution) or to choosing elements (optimal ones or equilibria) in this allocation set.

Consequently, to proceed further, we need to make novel assumptions on the nature of the questions to answer and the behavior of consumers.

If everyone may easily agree on accepting the first economic law, the consensus about the behaviors of consumers and the way to describe them mathematically is far to be perfect and bound to evolve.

3.1.5 "Differentiating" Scarcity Constraints

The rule we imposed is that the total consumption $y(t) := \sum_{i=1}^{n} x_i(t)$ must remain in the set M of scarce resources. There is no problem as long as it remains in the **interior**[9] of M.

[8]It is hoped that this slight abuse of notation is forgiven by the reader. The context should efface any ambiguity.

[9]An element y belongs to the interior $\mathrm{Int}(M) := \overset{\circ}{M}$ of M if there exists a ball centered at y and contained in M (which then, "protects" this element from the complement of M).

Problems may arise when the total consumption reaches the boundary of M. Roughly speaking, one may feel that if the velocity $y'(t)$ "points outside" M, then the total consumption will leave the set M of scarce resources, destroying the viability (or sustainability) of the system. Otherwise, if the velocity $y'(t)$ is in some sense "tangent" to M, one may hope that the total consumption may bounce back inside the set M of scarce resources, keeping the system viable (or maintaining it sustainable).

Therefore, we need to implement the concept of tangency to any subset. The adequate choice to obtain viability theorems is to use the concept of **contingent cone**[10] introduced for the first time by Georges Bouligand in the thirties (and which happens to be the cornerstone to set-valued analysis, as we shall see soon).

Definition 3.1.1 *When K is a subset of X and x belongs to K, the* **contingent cone** $T_K(x)$ *to K at x is the closed cone of elements v*

$$\begin{cases} v \in T_K(x) \text{ if and only if } \exists\, h_n \to 0+ \text{ and } \exists\, v_n \to v \\ \text{such that } \forall\, n, \ \ x + h_n v_n \in K \end{cases}$$

In other words, starting from x in the direction v, one may leave K but not too much, since, for a countable number of "steps" h_n and approximate directions v_n, the point $x + h_n v_n$ lies in K.

It is very convenient to use the following characterization of this contingent cone in terms of distances: the **contingent cone** $T_K(x)$ to K at x is the closed cone of elements v such that

$$\liminf_{h \to 0+} \frac{d(x + hv, K)}{h} = 0$$

We also observe that

$$\text{if } \ x \in \operatorname{Int}(K), \ \text{ then } T_K(x) = X$$

and that

$$\text{if } \ K := \{\bar{x}\}, \ \ T_{\{\bar{x}\}}(\bar{x}) = \{0\}$$

It is convenient to posit the following

Definition 3.1.2 *Let $K \subset X$. A function $x(\cdot) : I \mapsto X$ is said to be* **viable in K on the interval I** *if and only if*

$$\forall\, t \in I, \ \ x(t) \in K$$

The **closure** $\overline{N}$ of a subset N is the set of limits of converging sequences of elements of N. The complement of the interior of M is the closure of the complement of M. The **boundary** ∂M is the intersection of the closure of M and of its complement: every ball centered at a point $y \in \partial M$ contains at least one element of M and at least one element of its complement.

[10] For a presentation of the ménagerie of tangent cones, we refer to chapter 4 of [54, Aubin & Frankowska].

The following proposition explains the importance of the role of contingent cones in "evolution under constraints", which is what viability theory is about:

Proposition 3.1.3 *Let $x(\cdot) : [0,T] \mapsto X$ be a differentiable viable function in K. Then*

$$\forall\, t \in]0,T[, \;\; x'(t) \in T_K(x(t))$$

There is a calculus of contingent cones which allows us to "compute them", and thus, to "apply" the forthcoming viability theorems[11]. For the time being, let us mention only that the contingent cone coincides with the tangent space $T_K(x)$ of differential geometry when K is a "smooth manifold". Also, when K is convex, one can prove that the contingent cone coincides with the **tangent cone** $T_K(x)$ to K at $x \in K$ of convex analysis, which is the closed cone spanned by $K - x$:

$$T_K(x) \;=\; \overline{\bigcup_{h>0} \frac{K-x}{h}}$$

In this case, the tangent cone is convex[12] for every $x \in K$.

The calculus of contingent cones is simplified when the subsets are closed and convex, since in this case, all the formulas we dream of are true (under adequate assumptions).

Since we aim at characterizing viable evolutions of allocations of scarce resources, we shall state right now the formula which expresses the contingent cone to K in terms of contingent cones to the consumption sets L_i and the set M of scarce resources:

Proposition 3.1.4 *Assume that the resource set M and the consumption sets L_i are closed and convex and satisfy the* constraint qualification assumption*:*

$$0 \;\in\; \text{Int}\left(\sum_{i=1}^{n} L_i - M\right) \tag{3.1.5}$$

Then

$$T_K(x) \;:=\; \left\{ v \;:=\; (v_1,\ldots,v_n) \;\in\; \prod_{i=1}^{n} T_{L_i}(x_i) \;\middle|\; \sum_{i=1}^{n} v_i \;\in\; T_M\left(\sum_{i=1}^{n} x_i\right)\right\} \tag{3.1.6}$$

[11] as well as the "equilibrium theorems under constraints" and "optimization theorems under constraints" (see Theorem 6.1.15 below, for instance).

[12] One can prove that the contingent cone $T_K(x)$ is convex whenever K "is **sleek** at x", which means that the set-valued map $T_K(\cdot)$ is lower semicontinuous at x. Convex subsets are sleek at every elements. Proposition 3.1.4 can be adapted to the case of sleek subsets when the constraint qualification assumption is replaced by a "transversality condition" (which, by the way, is shown to imply the constraint qualification property when the sets are convex.) See Theorem 11.2.8 and Table 11.1 below for more details.

The constraint qualification assumption (3.1.5) implies obviously the "transversality condition"[13]

$$\forall\, x \in K, \;\; \sum_{i=1}^{n} T_{L_i}(x_i) - T_M\left(\sum_{i=1}^{n} x_i\right) \;=\; Y$$

Observe that the constraint qualification assumption (3.1.5) is a slight strengthening of property (3.1.4) characterizing the nonemptiness of the set K of allocations. Without such an assumption, the above formula is no longer true, as the following counter-example shows:

Counter-Example — Property

$$T_{K_1 \cap K_2}(x) \;=\; T_{K_1}(x) \cap T_{K_2}(x)$$

is false when assumption $0 \in \text{Int}(K_1 - K_2)$ is not satisfied. Take for instance two balls K_1 and K_2 tangent at a point x. The tangent cone to the intersection $\{x\}$ is reduced to $\{0\}$, whereas the intersection of the tangent cones is a hyperplane. This shows that we cannot dispense with the **constraint qualification** assumptions in the calculus of tangent cones to inverse images and intersections. □

In order to check the constraint qualification assumption holds true, one has to solve the following equations with constraints: there exists $\eta > 0$ such that

$$\forall\, z \in \eta B, \;\; \exists\, x_i \in L_i, \;\; \exists\, y \in M \quad \text{such that } z = \sum_{i=1}^{n} x_i - y$$

This happens in particular under the (stronger) "Slater's condition"

$$\exists\, \tilde{x}_i \in L_i \quad \text{such that } \sum_{i=1}^{n} \tilde{x}_i \;\in\; \text{Int}(M)$$

Indeed, it is enough to take $\tilde{y} := \sum_{i=1}^{n} \tilde{x}_i$ and η equal to the radius of the ball centered at y and contained in M. This Slater condition thus requires that the interior of M is not empty, which excludes the case of equality constraints where $M := \{w\}$ is reduced to a singleton (as in exchange economies), and all the cases in which M actually lives in a subspace of smaller dimension than the one of Y.

When L_i and M are cones (centered at 0), then the constraint qualification assumption is equivalent to the "surjectivity condition"

$$\sum_{i=1}^{n} L_i - M \;=\; Y$$

since the only cone for which 0 is interior is the whole space.

We just recall here the following examples of tangent cones to standard sets:

[13] which is actually equivalent, thanks to the Constrained Inverse Function Theorem.

- When $M := \{w\}$ is reduced to a singleton, then $T_{\{w\}}(w) = 0$.
- When $M := w - \mathbf{R}^l_+$, then $v \in T_{w-\mathbf{R}^l_+}(y)$ if and only if $v_h \leq 0$ whenever $y_h = w_h$, $h = 1, \ldots, l$.
- When M is the probability simplex

$$S^l := \left\{ x \in \mathbf{R}^l_+ \mid \sum_{h=1}^{l} x_h = 1 \right\}$$

then the tangent cone $T_{S^l}(x)$ to S^l at $x \in S^l$ is the cone of elements $v \in \mathbf{R}^l$ satisfying

$$\sum_{h=1}^{l} v_h = 0 \quad \& \quad v_h \geq 0 \quad \text{whenever} \quad x_h = 0 \tag{3.1.7}$$

Indeed, let us take $v \in T_{S^l}(x)$. There exist sequences $h_p > 0$ converging to 0 and v_p converging to v such that $y_p := x + h_p v_p$ belongs to S^l for any $p \geq 0$. Then

$$\sum_{k=1}^{l} v_{p_k} = \frac{1}{h_p}\left(\sum_{k=1}^{l} y_{p_k} - \sum_{k=1}^{l} x_{p_k}\right) = 0$$

so that $\sum_{k=1}^{l} v_k = 0$. On the other hand, if $x_k = 0$, then $v_{p_k} = y_{p_k}/h_p \geq 0$, so that $v_k \geq 0$.

Conversely, let us take v satisfying (3.1.7) and deduce that $y := x + hv$ belongs to the simplex for h small enough. First, the sum of the y_k is obviously equal to 1. Second, $y_k \geq 0$, either when $x_k = 0$ because in this case v_k is nonnegative, or when $x_k > 0$, because it is sufficient to take $h < x_k/|v_k|$ for having $y_k \geq 0$. Hence y does belong to the simplex.

3.1.6 Preference Preorders

We may complete further the description of the behavior of the consumers by its **preference** preorder. It allows one consumer to compare two commodities in a **rational way**. This means here that every commodity is preferred to itself (this is called reflexivity) and that if one commodity is preferred to a second one, which is itself preferred to a third one, then the first one is preferred naturally to the third one (this is called transitivity).

Two commodities are **indifferent** if each is preferred to the other one. Then a commodity is **strictly preferred** to another one if it is preferred and not indifferent.

So, preference preorders are just particular cases of a **preorder** $\succeq$ relation. This is a binary relation which is both **reflexive** ($x \succeq x$ for every x) and **transitive** ($x \succeq y$ and $y \succeq z$ imply $x \succeq z$).

We can associate with a preorder relation the **equivalence relation**

$$x \sim y \text{ if and only if } x \succeq y \ \& \ y \succeq x$$

and the **order** relation

$$x \succ y \text{ if and only if } x \succeq y \text{ and not } y \succeq x$$

It is useful to characterize a preorder by the set-valued map L defined by

$$\forall\, x, \;\; L(x) := \{y \mid y \succeq x\}$$

the graph of which is the **graph of a preorder**. The subsets $L(x)$ are called the **sections** of the preorder.

Conversely, any set-valued map L **reflexive** (in the sense that $x \in L(x)$ for every x) and **transitive** (in the sense that $L(y) \subset L(x)$ for every $y \in L(x)$) defines the preorder $\succeq$ defined by $x \succeq y$ if and only if $x \in L(y)$.

Hence, from now on, we shall represent a preorder by a reflexive and transitive set-valued map. We shall say that a preorder is convex (resp. closed) if the subsets $L(x)$ are convex (resp. closed).

We say that a given preorder $\succeq$ is consistent with the order relation of the commodity space $\mathbf{R}^l$ if $y \succeq x$ whenever $y \geq z$ and $z \succeq x$. This can be translated in the form

$$\forall\, x, \;\; L(x) + \mathbf{R}^l_+ \subset L(x)$$

Assume now that each consumer not only wants to choose an allocation in her consumption set L_i, but that she wants to heighten her preference along a preorder $\succeq_i$ characterized by a set-valued map $L_i(\cdot)$ in the following sense:

Definition 3.1.5 *We shall say that a function $t \mapsto x(t)$ is* **increasing along a preference preordering** $\succeq_i$ *or* **monotone** *if*

$$\forall\, t \geq s \geq 0, \;\; x(t) \succeq_i x(s)$$

or, equivalently,

$$\forall\, t \geq s \geq 0, \;\; x(t) \in L_i(x(s))$$

So, monotone evolutions are "viable" in state dependent consumption sets $L_i(x)$, so to speak, since for each instant s, the consumptions must remain in the consumption set $L_i(x(s))$.

Therefore, it is easy to deduce from the definition of the contingent cones the following property of monotone functions.

Proposition 3.1.6 *Let $x_i(\cdot) : [0,T] \mapsto X$ be a differentiable function monotone along a preference preordering $\succeq_i$. Then*

$$\forall\, t \in]0,T[, \;\; x_i'(t) \in T_{L_i(x_i(t))}(x_i(t))$$

When consumptions sets of consumers are replaced by preference relations, the set $K(x)$ of preferred allocations is defined by

$$K(x) := \left\{ x := (x_1, \ldots, x_n) \in \prod_{i=1}^{n} L_i(x_i) \;\middle|\; \sum_{i=1}^{n} x_i \in M \right\} \quad (3.1.8)$$

Monotone allocations of resources are functions $t \mapsto x(t)$ such that

$$\forall\, t \geq s, \;\; x(t) \in K(x(s))$$

i.e.,

$$\begin{cases} \forall\, i = 1, \ldots, n, \;\; \forall\, t \geq s, \;\; x_i(t) \succeq_i x_i(s) \\ \forall\, t \geq 0, \;\; \sum_{i=1}^{n} x_i(t) \in M \end{cases}$$

Proposition 3.1.7 *Assume that the resource set M and the preference preorders $\succeq_i$ are closed and convex and satisfy the* constraint qualification assumption*:*

$$\forall\, x, \;\; 0 \in \operatorname{Int}\left(\sum_{i=1}^{n} L_i(x) - M \right) \quad (3.1.9)$$

Then

$$T_{K(x)}(x) := \left\{ v := (v_1, \ldots, v_n) \in \prod_{i=1}^{n} T_{L_i(x_i)}(x_i) \;\middle|\; \sum_{i=1}^{n} v_i \in T_M\left(\sum_{i=1}^{n} x_i \right) \right\} \quad (3.1.10)$$

3.2 Viable Evolutions of Allocations

By Proposition 3.1.3, we have an intuitive knowledge of what will be the basic viability theorem allowing consumers to allocate continuously scarce resources among themselves: their velocities $x'(t) := (x_i'(t))_{i=1,\ldots n}$ must remain tangent to K at $x(t) := (x_i(t))_{i=1,\ldots n}$. By Proposition 3.1.4 stated above, the dynamics of the consumers must obey the constraints

$$\begin{cases} \forall\, i = 1, \ldots n, \;\; x_i'(t) \in T_{L_i}(x_i(t)) \\ \sum_{i=1}^{n} x_i'(t) \in T_M\left(\sum_{i=1}^{n} x_i(t) \right) \end{cases}$$

Therefore, the dynamical behavior of consumers, describing the right hand sides of the differential equations governing the evolution of their consumptions, must obey the above conditions. They split in two: the first ones are decentralized to the level of each consumer. Each of them has to evolve according the rule

$$x_i'(t) \in T_{L_i}(x_i(t))$$

The second condition

$$\sum_{i=1}^{n} x_i'(t) \in T_M\left(\sum_{i=1}^{n} x_i(t)\right)$$

is "centralized", requiring the knowledge of the dynamical behavior of each consumer and the set M of scarce resources.

In the case of monotone allocations, the consumer has to evolve according to the rule

$$x_i'(t) \in T_{L_i(x_i(t))}(x_i(t))$$

instead.

We shall restrict ourselves to the design of dynamical behavior of consumers decentralized by messages sent to each consumers which summarize this knowledge of the dynamical behavior of each consumer and the set M of scarce resources. These messages can be chosen to be prices.

3.2.1 Prices as Messages for Decentralization

We assume now that the dynamic behavior of a consumer uses adequate informations — prices — to decide how to increase, decrease or keep constant her current consumption.

Very early economic activity, actually, trading activity, required the comparison of two commodities before an exchange, or a barter or swap. This is done by associating with each commodity its **value** expressed in **accounting or monetary units**, such as the EURO, the Franc, the Dollar, etc.

Even though barter is still used (in international trade), the idea to use a specific commodity, easy to handle and sufficiently divisible (liquid), as a unique mean of comparison appeared quite early in history. In order to compare two arbitrary commodities, the value of each commodity is expressed in terms of monetary units.

The mechanism which associates with a commodity this amount of monetary units is what is called the **price system** or simply, the **price**. A price p is then a map from the commodity space Y to the "value space" $\mathbf{R}$, associating to each commodity its value.

Since we have represented the commodity space by a vector space, in which one can perform linear combination of commodities, it is natural to continue to accept the relevance of the linear structure and to assume that the price is linear: the value of the sum of two commodities is the sum of the values and the value of λ times a commodity is λ times its value. In other words,

$$p(\lambda_1 x_1 + \lambda_2 x_2) = \lambda_1 p(x_1) + \lambda_2 p(x_2)$$

The Price Space is then the **dual**[14] $Y^\star := \mathcal{L}(Y, \mathbf{R})$ of the Commodity Space.

We shall supply the Price Space with the **dual basis**

$$(e_1^\star, e_2^\star, \ldots, e_l^\star)$$

where $e_h^\star$ associates with any commodity (basket) $x := (x_h)_{h=1,\ldots,l} \in Y$ the amount of units of the hth commodity: $e_h^\star(x) = x_h$.

We then deduce that

$$p(x) = \sum_{h=1}^{l} x_h p(e^h) = \sum_{h=1}^{l} e_h^\star(x) p(e^h) = \left(\sum_{h=1}^{l} e_h^\star p(e^h)\right)(x)$$

This shows that $p := \left(\sum_{h=1}^{l} e_h^\star p(e^h)\right)$ is a linear combination of the elements of the canonical basis. The component $p^h := p(e^h)$ of p in this dual basis is the **value of the unit of commodity** h, what is meant in the day to day language by the **price of** h.

We then write

$$p(x) = \sum_{h=1}^{l} p^h x_h =: \langle p, x\rangle$$

This nondegenerate bilinear form

$$(p, x) \in Y^\star \times Y \mapsto \langle p, x\rangle := p(x) \in \mathbf{R}$$

is called the **duality pairing**.

In general, we shall be led to choose **nonnegative prices**, i.e., prices in the positive cone

$$\mathbf{R}_+^{l\star} := \left\{p \in \mathbf{R}^{l\star} \mid p^h \geq 0\right\}$$

This is not always judicious. In instances when one consumer is forced to consume all available goods, it is sensible to accept negative prices. A glass of water in the desert may be attributed by someone a positive price, whereas in a dirty basement of a torture building, a victim is ready to attribute the last glass of water a negative price. Remember your childhood when a excessively caring mother forced you to finish your soup ...

So, to summarize, the first role played by a price is to compare two commodities x and y by comparing their values $\langle p, x\rangle$ and $\langle p, y\rangle$.

3.2.2 Change Functions of Visible Consumers

We shall capture the essence of the dynamical behavior of a consumer by describing her as a **change function**

$$(x, p) \mapsto c_i(x, p)$$

[14]The dual $Y^\star$ of a finite dimensional vector space Y is the space $\mathcal{L}(Y, \mathbf{R})$ of linear forms from Y to $\mathbf{R}$. It is isomorphic to the vector space Y, and in particular, has the same dimension.

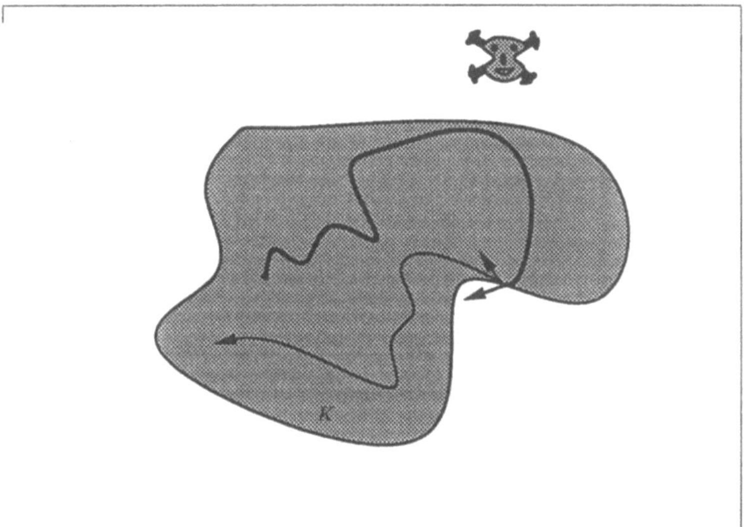

The subset K of the plane represents the constrained subset. At a point of the boundary contingent and not contingent velocities are indicated. Both the trajectories of a viable and a nonviable are shown.

Figure 3.1: Viable Solutions

associating with each commodity x owned by consumer i and the price p she observes on the market the velocity with which she will change her commodity.

Therefore, the dynamical behavior of consumer i is described by the differential equation

$$x_i'(t) = c_i(x_i(t), p(t))$$

It is decentralized in the sense that the decision of consumer i does neither involve the knowledge of the set M of available resources nor the behavior of her fellow consumers.

Her decision depends only upon her current consumption $x_i(t)$ and the "current price" (also called "spot price") $p(t)$ at time t: at each time t, she decides to increase, decrease or maintain constant her consumption according to her consumption and to the price she observes. Keeping constant her consumption of a commodity x_h means that she is satisfied.

Change functions should not be confounded with "demand functions" describing the behavior of a consumer in a static world. Actually, one can associate a demand map $D_i : \mathbf{R}^{l\star} \rightsquigarrow \mathbf{R}^l$ with a change function c_i by setting

$$x \in D_i(p) \text{ if and only if } c_i(x,p) = 0$$

In other words, the associated demand map of consumer i associates with any price p the set of equilibria x of her own dynamical system defined by her change function[15].

We shall provide later the examples of change maps derived from the problem of optimal allocations of scarce resources through utility functions of consumers: $c_i(x,p) = U_i'(x) - p$.

Another important class of change maps are the ones derived from the Hessian of a function U_i either by:

$$c_i(x,p) := U_i'(x) + \frac{1}{2}\frac{\partial^2 U_i}{\partial^2 x}(x)p$$

which satisfy

$$\langle p, c_i(x,p)\rangle = \langle U_i'(x), p\rangle + \frac{1}{2}\langle U_i''(x)p, p\rangle$$

We recall the following definition:

Definition 3.2.1 *We set*

$$\alpha(A) := \inf_{x \neq 0} \frac{\langle Ax, x\rangle}{\|x\|^2}$$

and we say that a linear operator $A \in \mathcal{L}(X,X)$ is semi positive definite if $\alpha(A) \geq 0$ It is said to be positive definite *if $\alpha(A) > 0$.*

[15] Conversely, one could associate with any demand map d_i the change function c_i defined by $c_i(x,p) := d_i(p) - x$. However, for several reasons which we shall elucidate later, we do not start with classical Walrasian demand functions to build our dynamical models, but the other way around, deriving demand maps from change maps taken as primitives of the behavior of consumers.

When the utility function U_i is concave, then $G(x) := -U_i''(x)$ is semi positive definite, and positive definite when U is strictly concave.

3.2.3 The Viability Problem

Now, we have to introduce an a priori law for price behavior. We can take into consideration external laws or regulations, and for that purpose, introduce a set-valued map $P : K \rightsquigarrow \mathbf{R}^l_+$, called a **a priori pricing map**, associating to each allocation $x \in K$ a subset $P(x) := P(x_1, \ldots, x_n) \subset \mathbf{R}^l_+$ of feasible prices (allowed by external regulations, for instance).

Hence, the prices are requested to obey the evolution law: $\forall\, t \geq 0,\;\; p(t) \in P(x(t))$.

In summary, the dynamics of the evolution of the consumption is described by

$$\forall\, t \geq 0, \quad \begin{cases} i) & x_i'(t) \;=\; c_i(x_i(t), p(t)) \;\; (i = 1, \ldots, n) \\ ii) & p(t) \;\in\; P(x(t)) \;:=\; P(x_1(t), \ldots, x_n(t)) \end{cases}$$

Equilibria of this dynamical system are solutions $(\bar{x}, \bar{p})$ to the system

$$\begin{cases} i) & c_i(\bar{x}_i, \bar{p}) \;=\; 0 \;\; (i = 1, \ldots, n) \\ ii) & \bar{p} \;\in\; P(\bar{x}) \;:=\; P(\bar{x}_1, \ldots, \bar{x}_n) \end{cases} \tag{3.2.1}$$

At equilibria, consumers are happy and keep the same allocation $\bar{x}_i \in D_i(\bar{p})$ of scarce resources forever.

We shall also address the problem of finding viable evolutions satisfying

$$\forall\, t \geq 0, \quad \begin{cases} i) & x_i(t) \;\in\; L_i \;\; (i = 1, \ldots, n) \\ ii) & \sum_{i=1}^n x_i(t) \;\in\; M \end{cases}$$

and viable equilibria, i.e., equilibria satisfying

$$\bar{x}_i \;\in\; L_i \;\; (i = 1, \ldots, n) \;\; \& \;\; \sum_{i-1}^{n} \bar{x}_i \;\in\; M$$

3.2.4 The Regulation Map

We use now the concept of tangent cone to associate with the behavior of the consumers described by change functions and the set M of scarce resources the **regulation map**:

Definition 3.2.2 *The* **regulation map** Π_M *is the set-valued map*

$$\forall\, x \in K,\; \Pi_M(x) \;:=\; \left\{ p \in P(x) \mid \sum_{i=1}^{n} c_i(x_i, p) \;\in\; T_M\left(\sum_{i=1}^{n} x_i\right) \right\}$$

built from the confrontation of the changes claimed by consumers and the set of scarce resources.

This regulation map relates each allocation $x \in K$ to prices such that the associated total change of consumptions is tangent to the set of scarce resources. The objective is to bring back the total consumption inside the set of scarce resources.

We posit now the assumptions we need to prove our viability theorem.

— ASSUMPTIONS ON THE CONSUMPTION AND RESOURCE SETS:

$$\left\{ \begin{array}{ll} i) & M = M - \mathbf{R}^l_+ \text{ is a closed convex subset} \\ ii) & \forall\, i = 1, \ldots, n, \;\; L_i = L_i + \mathbf{R}^l_+ \text{ is closed and convex} \\ iii) & 0 \in \operatorname{Int}\left(\sum_{i=1}^n L_i - M \right) \end{array} \right. \qquad (3.2.2)$$

These are the standard assumptions, the most severe one being the convexity of the consumption and resource sets, which is needed to guarantee the existence of an equilibrium (but not needed for the existence of viable evolution of allocations).

Assumptions (3.2.2)i) and ii) imply that

$$\forall\, x \in L_i, \;\; \mathbf{R}^l_+ \subset T_{L_i}(x) \;\; \& \;\; \forall\, y \in M, \;\; -\mathbf{R}^l_+ \subset T_M(y)$$

Assumption (3.2.2)iii) is the constraint qualification needed to compute the tangent cone to the set of allocations thanks to Proposition 3.1.4. It also implies that the set of allocation is not empty.

We shall not assume necessarily that the set K of allocations is compact, since this assumption is needed only for obtaining the existence of an equilibrium. However, we mention standard sufficient conditions guaranteeing the compactness of the set of allocations: There exist $\underline{y}$ and $\underline{x_i}$, $i = 1, \ldots, n$ such that

$$M \subset \underline{y} - \mathbf{R}^l_+ \;\; \& \;\; \forall\, i = 1, \ldots, n, \;\; L_i \subset \underline{x}_i + \mathbf{R}^l_+ \qquad (3.2.3)$$

or there exist $\underline{y}$ and $\underline{x_i}$, $i = 1, \ldots, n$ such that

$$\bar{y} \in M, \;\; M \cap (\bar{y} + \mathbf{R}^l_+) = \{\bar{y}\} \;\; \& \;\; \forall\, i = 1, \ldots, n, \;\; L_i \subset \underline{x}_i + \mathbf{R}^l_+$$

— ASSUMPTIONS ON THE CHANGE FUNCTIONS:

$$\left\{ \begin{array}{ll} i) & p \mapsto c_i(x, p) \; := \; g_i(x) - G_i(x)p \text{ is affine, where} \\ ii) & g_i : L_i \mapsto Y \text{ is continuous with linear growth} \\ & \text{(there exists } \gamma_i \text{ such that } \|g_i(x)\| \; \leq \; \gamma_i(\|x\| + 1)\text{)} \\ iii) & G_i : L_i \mapsto \mathcal{L}(Y^\star, Y) \text{ is continuous and bounded} \\ iv) & \forall\, x_i \in L_i, \; p \in \operatorname{Im}(P), \;\; c_i(x_i, p) \; \in \; T_{L_i}(x_i) \end{array} \right. \qquad (3.2.4)$$

Assumption (3.2.4)i) looks very restrictive[16]. However this assumption is satisfied for change functions derived from utility functions.

Assumption (3.2.4)iv) is made to simplify the presentation of the example. It is justified because it assumes that each consumer has a behavior consistent with her consumption set, in the sense that for any price, the velocity pushes the evolution in her consumption set.

We shall provide later on other viability theorem which do not require the assumption that the change functions are affine with respect to prices.

— ASSUMPTIONS ON THE PRICING MAP:

We set

$$\|P(x)\| := \sup_{y \in P(x)} \|y\|$$

and we say that F has **linear growth** if there exists a positive constant γ such that

$$\forall\, x \in \mathrm{Dom}(P),\ \ \|P(x)\| \ \le\ \gamma(\|x\| + 1)$$

Definition 3.2.3 (Marchaud Map) *We shall say that P is a* Marchaud map *if*

$$\begin{cases} i) & \textit{the graph and the domain of } P \textit{ are nonempty and closed} \\ ii) & \textit{the values } P(x) \textit{ of } P \textit{ are convex} \\ iii) & \textit{the growth of } P \textit{ is linear} \end{cases}$$

We shall posit the assumption:

$$P \text{ is a Marchaud map} \tag{3.2.5}$$

Constant pricing maps $P(x) \equiv \mathbf{R}^l_+ \cap cB$ and pricing maps of the form $P(x) \equiv \mathbf{R}^l_+ \cap \gamma(x)B$ where $\gamma(\cdot)$ is continuous (or even, just upper semicontinuous) with linear growth are obviously Marchaud maps.

3.2.5 Dynamical Allocation of Resources

We now address the problem of finding viable allocations, i.e., solutions to the controlled system

$$\begin{cases} i) \ \ x_i'(t) \ = \ c_i(x_i(t), p(t)) \ \ (i = 1, \ldots, n) \\ \\ ii) \ \ p(t) \ \in \ P(x(t)) \end{cases} \tag{3.2.6}$$

[16]We shall see that it is mandatory for proving the viability theorem, as the counter-example we shall present just before the proof of the Viability Theorem in the Tool-Box. Actually, what is needed is the convexity of the set of velocities. However, this assumption will be dropped when we shall set bounds to inflation for obtaining continuous prices instead of measurable prices.

satisfying

$$\forall\, t \geq 0, \quad \begin{cases} i) & x_i(t) \in L_i \quad (i = 1, \ldots, n) \\ ii) & \sum_{i=1}^{n} x_i(t) \in M \end{cases} \tag{3.2.7}$$

Recall that the **regulation map** Π_M is defined from the set of scarce resources M and the behavior of the consumers by

$$\forall\, x \in K,\; \Pi_M(x) := \left\{ p \in P(x) \mid \sum_{i=1}^{n} c_i(x_i, p) \in T_M\left(\sum_{i=1}^{n} x_i\right)\right\}$$

The Viability Theorem 4.1.5 that we shall state and prove later on implies that whenever this regulation map is strict[17], the **allocation set is viable under this system**. This means that from any initial allocation $x_0 = (x_{01}, \ldots, x_{0n})$ starts at least one viable allocation:

Theorem 3.2.4 *We posit assumptions (3.2.2), (3.2.4) and (3.2.5). Then the two following conditions are equivalent:*

$$\begin{cases} a) & \forall\, x \in K, \;\; \Pi_M(x) \neq \emptyset \\ b) & \forall\, x_0 \in K, \text{ there exists at least one allocation starting from } x_0 \\ & \text{evolving according to (3.2.6) and satisfying the constraints (3.2.7)} \end{cases}$$

In this case, the viable allocations are governed by the regulation law

$$\text{for almost all } t \geq 0, \quad p(t) \in \Pi_M(x(t)) \tag{3.2.8}$$

In other words, by replacing the *a priori* pricing map given independently of the set M of scarce resources by the regulation map Π_M, one could say that the set of allocations became **invariant** instead of being **viable**, in the sense that **all solutions governed by the regulation map are viable.** However, we have to point out that the regulation map Π_M is not Marchaud: Whenever inequality constraints are involved in the description of the set M, the graph of the tangent cone — and thus, the graph of the regulation map Π_M — is not closed ! This is one of the mathematical reasons why the viability theorem, even though it looks intuitive, is not trivial. In chapter 7, we shall study subregulation maps which have closed graphs. One way to find a closed subregulation map in the case of inequality constraints is to use the subset of the contingent cone introduced by Nina Maderner, as we shall show in one example below.

Remark — Naturally, we can extend this basic result in many directions and relax some of the assumptions.

---

[17]A set-valued map is said to be strict if its values are not empty.

For instance, if we are not interested in the existence of an equilibrium, but only in the evolutionary side of the allocation problem, we can dispense of the convexity assumptions and replace it by sleekness[18]. In this case, we replace the tangent cone to a convex subset by the contingent cone and assume instead that

$$\left\{\begin{array}{ll} i) & M = M - \mathbf{R}^l_+ \text{ is closed and sleek} \\ ii) & \forall\, i = 1,\ldots,n,\;\; L_i = L_i + \mathbf{R}^l_+ \text{ is closed and sleek} \\ iii) & \text{the transversality conditions:} \\ & \forall\, x \in K,\; \sum_{i=1}^n T_{L_i}(x_i) - T_M(\sum_{i=1}^n x_i) \;=\; Y \end{array}\right.$$

These assumptions imply that formula (3.1.5):

$$T_K(x) \;:=\; \left\{ v \;:=\; (v_1,\ldots,v_n) \;\in\; \prod_{i=1}^n T_{L_i}(x) \;\middle|\; \sum_{i=1}^n v_i \;\in\; T_M\left(\sum_{i=1}^n x_i\right)\right\}$$

still holds true in the nonconvex case. □

3.2.6 Examples

- When $M := \{w\}$ is reduced to a singleton, then

$$\Pi_M(x) \;=\; \left\{p \in P(x) \mid \sum_{i=1}^n c_i(x_i,p) \;=\; 0\right\}$$

- When $M := w - \mathbf{R}^l_+$, then $p \in P(x)$ belongs to $\Pi_M(x)$ if and only if

$$\sum_{i=1}^n c_i(x_i,p)_k \;\leq\; 0 \text{ whenever } \sum_{i=1}^n x_{i_k} \;=\; w_k$$

The drawback which appears as soon as inequality constraints do appear is that the graph of the above regulation map is not closed. This follows from the fact that the graph of the set-valued map $x \in \mathbf{R}_+ \rightsquigarrow T_{\mathbf{R}_+}(x)$ is not closed, not even locally compact (there is no compact neighborhood of 0 in $\text{Graph}(T_{\mathbf{R}_+}(\cdot))$.)

This the reason why Nina Maderner introduced a set-valued map $T_K^\diamond(\cdot)$ contained in $T_K(\cdot)$ with closed graph (see Theorem 11.2.15). In the simple case when $M := w - \mathbf{R}^l_+$, this subset of the tangent cone is defined by

$$T^\diamond_{w-\mathbf{R}^l_+}(y) \;:=\; \left\{v \in \mathbf{R}^l \mid \forall\, k = 1,\ldots,l,\;\; v_k + y_k \;\leq\; w_k\right\}$$

We associate with such a choice the regulation map $\Pi_M^\diamond(\cdot)$ defined by

$$\Pi^\diamond_M(x) \;=\; \left\{p \in P(x) \mid \sum_{i=1}^n c_i(x_i,p) + \sum_{i=1}^n x_i \;\leq\; w\right\}$$

The graph of such a "subregulation" map is then closed.

[18]because this property is sufficient for having the calculus of contingent cones we need.

3.2.7 Hierarchical Decentralization of a Dynamical Economy

The behavior of consumers described by the change functions

$$(x, p) \mapsto c_i(x, p)$$

uses the minimal information available to consumers in a decentralized mechanism: their owned commodity and the price p observed on the market. This is more an "ecological model" (allocation of scarce resources) than an "economic model" where transactions among consumers are taken into actions.

Indeed, the viability constraints

$$\forall\, i = 1, \ldots, n, \quad x_i(t) \in L_i \;\;\&\;\; \sum_{i=1}^{n} x_i(t) \;\in\; M$$

can be decomposed in the form

$$\sum_{i=1}^{n} x_i(t) \;=\; y(t) \;\;\&\;\; y(t) \;\in\; M$$

The first constraint is an "exchange constraint" of an economic nature whereas the second one is a "scarcity constraint" the nature of which is ecological.

One can "hierarchically decentralize" viable solutions of the dynamical economy

$$\forall\, i = 1, \ldots, n, \quad x_i'(t) = c_i(x_i(t), p(t)) \;\text{ where }\; p(t) \in P(x(t)) \tag{3.2.9}$$

by imposing the following "cascade" of constraints

$$\begin{cases} i) & \forall t \geq 0, \;\; x_i(t) \in L_i \\ ii) & \forall t \geq 0, \;\; \sum_{i=1}^{n} x_i(t) \;=\; y(t) \\ iii) & \forall t \geq 0, \;\; y(t) \;\in\; M \end{cases} \tag{3.2.10}$$

and by introducing a new agent governing the evolution of scarce resources according to an evolution law of the form

$$\begin{cases} i) & \text{for almost all } t \geq 0, \;\; y'(t) \;= d(y(t), q(t)) \\ ii) & \text{where } \; q(t) \;\in\; Q(y(t)) \end{cases} \tag{3.2.11}$$

It is obvious that the *first component* $x(\cdot)$ of any pair of solutions $(x(\cdot), y(\cdot))$ to the system ((3.2.9),(3.2.11)) satisfying viability constraints (3.2.10) is a solution to the initial dynamical economy (3.2.9) viable in the allocation set K.

On the other hand, solutions to the system (3.2.9) viable in K can be obtained in two steps:

— First, find a solution $y(\cdot)$ to the new dynamical economy on scarce resources (3.2.11) *viable in* M (ecological step) and then,

— second, find a solution $x(\cdot)$ to the initial dynamical economy (3.2.9) satisfying the viability constraints

$$\begin{cases} i) & \forall t \geq 0, \quad x_i(t) \in L_i \\ ii) & \forall t \geq 0, \quad \sum_{i=1}^n x_i(t) = y(t) \end{cases} \tag{3.2.12}$$

which does no longer involve the subset $M \subset Y$ *of constraints* (economical step).

In order to solve this new hierarchical allocation model, we need to introduce the "economic regulation map"

$$\begin{aligned} &\Pi_H(x_1, \ldots, x_n, y; q) \\ &:= \{p \in P(x) \mid \sum_{i=1}^n c_i(x_i, p) = d(y, q)\} \end{aligned}$$

The "ecological regulation map" of the dynamical economy on scarce resources (3.2.11) on the subset M is defined by

$$\Theta_M(y) \quad = \quad \{q \in Q(y) \mid d(y, q) \in T_M(y)\}$$

Definition 3.2.5 *We shall say the two dynamical economies (3.2.9) and (3.2.11) are* hierarchically decentralizable *if*

$$\forall\, y \in M,\ \forall\, q \in Q(y),\ \ \Pi_H(x_1, \ldots, x_n, y; q) \neq \emptyset$$

Theorem 3.2.6 *We posit assumptions (3.2.2), (3.2.4) and (3.2.5). If the two dynamical economies (3.2.9) and (3.2.11) are hierarchically decentralizable and if the ecological regulation map* Θ_M *is strict, then from any initial allocation* $y_0 := \sum_{i=1}^n x_{i0} \in M$ *starts at least one viable evolution of scarce resources* $y(t)$ *evolving according to (3.2.11) and governed by the* ecological regulation law

$$\text{for almost all } t \geq 0, \quad p(t) \in \Theta_M(y(t)) \tag{3.2.13}$$

and one allocation starting from $x_0 \in K$, *evolving according to (3.2.12) and governed by the* economical regulation law

$$p(t) \ \in\ \Pi_H(x_1(t), \ldots, x_n(t), y(t); q(t)) \tag{3.2.14}$$

This can illustrate *hierarchical organization* which is found in the evolution of so many economic systems. The decomposition of the viability constraints as a product of successive levels of constraints may induce an evolution at each level obeying the constraint binding its state to the state of the lower level. It is regulated by prices determined (in a set-valued way) by the evolution of the price and the state of the lower level.

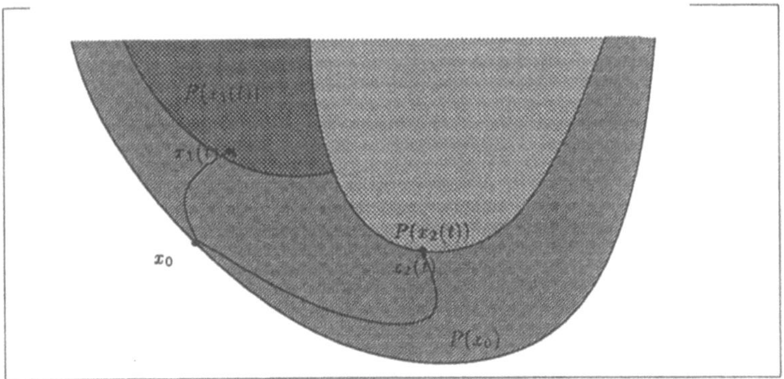

Each subset represents the subset of commodities preferred to a given commodity according to the preorder. Trajectories of some solutions increasing along which the preorder are indicated.

Figure 3.2: Monotone Solutions

3.2.8 Monotone Allocation of Resources

We now address the problem of finding monotone allocations along preference preorders of consumers, i.e., solutions to (3.2.6) satisfying

$$\begin{cases} i) & \forall\, s \geq 0,\ \ \forall\, t \geq s,\ \ x_i(t)\ \in\ L_i(x_i(s))\ \ (i = 1,\ldots,n) \\ & \\ ii) & \forall\, t \geq 0,\ \ \sum_{i=1}^{n} x_i(t)\ \in\ M \end{cases}$$

One then can prove:

Theorem 3.2.7 *Let us assume that M and the preference preorders L_i satisfy*

$$\begin{cases} i) & M = M - \mathbf{R}^l_+ \text{ is a closed convex subset} \\ ii) & \forall\, i = 1,\ldots,n,\ \ L_i(x) = L_i(x) + \mathbf{R}^l_+ \text{ is closed and convex} \\ iii) & 0\ \in\ \mathrm{Int}\left(\sum_{i=1}^{n} L_i(x_i) - M\right) \end{cases} \tag{3.2.15}$$

and

$$\begin{cases} i) & c_i(x,p)\ :=\ c_i(x) - G_i(x)p \text{ is affine, where} \\ ii) & c_i : L_i \mapsto Y \text{ is continuous with linear growth} \\ iii) & G_i : L_i \mapsto \mathcal{L}(Y^\star, Y) \text{ is continuous and bounded} \\ iv) & \forall\, x_i,\ p \in \mathrm{Im}(P),\ \ c_i(x_i,p)\ \in\ T_{L_i(x_i)}(x_i) \\ & \text{instead of assuming only that } c_i(x_i,p)\ \in\ T_{L_i}(x_i) \end{cases} \tag{3.2.16}$$

and that the pricing map P is Marchaud.

Then the two following conditions are equivalent:

$$\begin{cases} a) & \forall\, x \in K,\ \Pi_M(x) \neq \emptyset \\ b) & \forall\, x_0 \in K, \text{ there exists at least one } \textbf{monotone} \text{ allocation starting from} \\ & x_0 \text{ evolving according to (3.2.6) and satisfying the constraints (3.2.7)} \end{cases}$$

In this case, the monotone allocations are governed by the regulation law

$$\text{for almost all } t \geq 0, \quad p(t) \in \Pi_M(x(t))$$

3.2.9 Existence of an Equilibrium

We now state that under one of these equivalent conditions for viability, there exists at least one viable equilibrium $(\bar{x}_1, \ldots, \bar{x}_n, \bar{p})$: The following Theorem is an existence result of an equilibrium of the viable process which shares the same features of the Arrow-Debreu Theorem on the existence of a Walras equilibrium[19]:

Theorem 3.2.8 *We posit assumptions (3.2.2), (3.2.4) and (3.2.5). If the* regulation map $\Pi_M(\cdot)$ *is strict:*

$$\forall\, x \in K,\ \Pi_M(x) \neq \emptyset \tag{3.2.17}$$

and if the set K of allocations is compact, then there exists one viable equilibrium $(\bar{x}_1, \ldots, \bar{x}_n, \bar{p}) \in K \times \mathbf{R}^l_+$, i.e., a solution to the system of equations:

$$c_i(\bar{x}_i, \bar{p}) \;=\; 0 \quad (i = 1, \ldots, n)$$

satisfying the scarcity constraints

$$\forall\, i = 1, \ldots, n, \;\; \bar{x}_i \;\in\; L_i \;\; (i = 1, \ldots, n) \;\; \& \;\; \sum_{i-1}^{n} \bar{x}_i \;\in\; M$$

This Theorem follows from the Equilibrium Theorem 4.1.7 we shall prove later on. But before proving it, we emphasize the fact that the existence of an equilibrium is derived not only from the nonemptiness of the images of the regulation map Π_M, but also from the **convexity and the compactness of the allocation set** (assumptions which are **not needed** for the viability theorem).
□

[19] which is an equilibrium of the tâtonnement process.

3.3 Budgetary Constraints

As in optimization theory, duality theory plays an important role to justify the introduction of prices.

We shall begin by characterizing the necessary and sufficient condition for viability (which is a sufficient condition for the existence of an equilibrium) in terms of budgetary constraints. Next, we shall deduce that the dynamical equivalent of the Walras law, that may be called the **instantaneous Walras law**, guarantees that the regulation map is strict.

For that purpose, we need to introduce the concept of **normal cone** to a set K at x, which is the **polar cone** of the contingent cone, and to use the Separation Theorem, which will be couched in terms of **support functions**. We devote the next subsection to recall the definitions and to fix the notations.

3.3.1 Notes on Normal Cones

We begin by introducing support functions:

Definition 3.3.1 *Let K be a nonempty subset of a finite dimensional vector-space X. We associate with any continuous linear form $p \in X^\star$*

$$\sigma_K(p) \; := \; \sigma(K,p) \; := \; \sup_{x\in K}\langle p,x\rangle \; \in \; \mathbf{R}\cup\{+\infty\}$$

The function $\sigma_K : X^\star \mapsto \mathbf{R}\cup\{+\infty\}$ is called the **support function of** K[20]. *We also call*

$$\sigma_K^\flat(p) \; := \; \sigma^\flat(K,p) \; := \; \inf_{x\in K}\langle p,x\rangle \; \in \; \mathbf{R}\cup\{-\infty\}$$

the **lower support function**. *We say that the subsets of $X^\star$ defined by*

$$\begin{cases} i) & K^- \; := \; \{p \in X^\star \mid \sigma_K(p) \leq 0\} \\ ii) & K^\perp \; := \; \{p \in X^\star \mid \forall\, x \in K, \; \langle p,x\rangle = 0\} \end{cases}$$

are the **(negative) polar cone** *and* **orthogonal** *of K respectively.*

Examples

- When $K = \{x\}$, then $\sigma_K(p) \; = \; \langle p,x\rangle$
- When $K = B_X$, then $\sigma_{B_X}(p) \; = \; \|p\|_\star$
- If K is a cone, then

$$\sigma_K(p) \; = \; \begin{cases} 0 & \text{if} \;\; p \in K^- \\ +\infty & \text{if} \;\; p \notin K^- \end{cases} \quad \square$$

[20] It can be regarded as an income function associating with any price the maximal income produced by commodities of K.

When $K = \emptyset$, we set $\sigma_\emptyset(p) = -\infty$ for every $p \in X^\star$.

We observe that

$$\forall \lambda,\ \mu > 0,\ \sigma_{\lambda L+\mu M}(p) \ = \ \lambda\sigma_L(p) + \mu\sigma_M(p)$$

and in particular, that if P is a cone, then

$$\sigma_{M+P}(p) \ = \ \begin{cases} \sigma_M(p) & \text{if} \quad p \in P^- \\ +\infty & \text{if} \quad p \notin P^- \end{cases} \quad \square$$

The Separation Theorem[21] can be stated in the following way:

Theorem 3.3.2 (Separation theorem) *Let K be a nonempty subset of a Banach space X. Its* closed convex hull *is characterized by linear constraint inequalities in the following way:*

$$\overline{co}(K) \ = \ \{x \in X \mid \forall p \in X^\star,\ \langle p, x\rangle \ \le \ \sigma_K(p)\}$$

Furthermore, there is a bijective correspondence between nonempty closed convex subsets of X and nontrivial lower semicontinuous positively homogeneous convex functions on $X^\star$.

In particular,

1. *If K is a cone, then*

$$\overline{co}(K) \ = \ K^{--} \ \text{is its bipolar}$$

2. *If K is a vector space, then*

$$\overline{co}(K) \ = \ K^{\perp\perp} \ \text{is its biorthogonal}$$

Definition 3.3.3 *The* [22] *to K at x is defined by*

$$N_K(x) \ := \ (T_K(x))^-$$

When K is convex, we obtain the following characterization:

[21] This Separation Theorem is one corner stone of linear and convex functional analysis. It was discovered by the German mathematician Minkowski at the beginning of this century in finite dimensional spaces and extended independently in the 30's by Hans Hahn, an Austrian mathematician, the founder of the *Vienna Circle*, and Stefan Banach, the Polish founder of Linear Functional Analysis, to Banach spaces and to Hausdorff locally convex spaces, including weak topologies of Banach spaces. It is then known under the name of the Hahn-Banach Theorem.

[22] Since there are many normal cones for nonconvex subsets, this normal cone is known under several names in the literature: the Bouligand normal cone, or the subnormal cone, or the regular normal cone, as in [299, Rockafellar & Wets, Variational Analysis]. Since we shall use only this normal cone in this book, we shall call it simply the normal cone.

Proposition 3.3.4 *Assume that $K \subset X$ is a convex subset. Then*

$$N_K(x) \;=\; \{p \in X^\star \mid \langle p, x\rangle = \sigma_K(p)\}$$

or, equivalently,

$$N_K(x) \;=\; \{p \in X^\star \mid \forall\, y \in K, \langle p, y - x\rangle \;\leq\; 0\}$$

One can derive from Proposition 3.1.4

Proposition 3.3.5 *Consider the set of allocations*

$$K \;:=\; \left\{x \;:=\; (x_1, \ldots, x_n) \;\in \prod_{i=1}^{n} L_i \;\middle|\; \sum_{i=1}^{n} x_i \;\in\; M\right\} \tag{3.3.1}$$

Assume that the resource set M and the consumption sets L_i are closed and convex and satisfy the constraint qualification assumption *(3.1.5). Then*

$$N_K(x) \;:=\; \{(p + q_i)\}_{p \in N_M(\sum_{i=1}^{n} x_i),\, q_i \in N_{L_i}(x_i)} \tag{3.3.2}$$

When K is not convex, the normal cone can be characterized in the following way:

Proposition 3.3.6 *Let K be a subset of a finite dimensional vector-space X. Then $p \in N_K(x)$ if and only*

$$\begin{cases} \forall\, \varepsilon > 0,\ \exists\, \eta > 0 \ \text{ such that } \ \forall\, y \in K \cap B(x, \eta), \\ \\ \langle p, y - x\rangle \;\leq\; \varepsilon \|y - x\| \end{cases} \tag{3.3.3}$$

Proof — Let p satisfy the above property (3.3.3) and $v \in T_K(x)$. Then there exist h_n converging to 0 and v_n converging to v such that

$$y \;:=\; x + h_n v_n \;\in\; K \cap B(x, \eta)$$

for n large enough. Consequently, inequalities $\langle p, v_n\rangle \leq \varepsilon$ imply by taking the limit that $\langle p, v\rangle \leq \varepsilon$ for all $\varepsilon > 0$. Hence $\langle p, v\rangle \leq 0$, so that any element p satisfying the above property belongs to the polar cone of $T_K(x)$, even in the case of normed spaces.

Conversely, assume that p violates property (3.3.3): There exist $\varepsilon > 0$ and a sequence of elements $x_n \in K$ converging to x such that

$$\langle p, x_n - x\rangle \;>\; \varepsilon \|x_n - x\|$$

We set $h_n := \|x_n - x\|$, which converges to 0, and $v_n := (x_n - x)/h_n$. These elements belonging to the unit sphere, a subsequence (again denoted) v_n converges to some v. By definition, this limit belongs to $T_K(x)$, so that $\langle p, v\rangle \leq 0$. But our choice implies that $\langle p, v_n\rangle > \varepsilon$, so that $\langle p, v\rangle \geq \varepsilon$, a contradiction. □

3.3.2 Budget Characterization of the Regulation Map

Using the dual concept of normal cones, we deduce from the Separation Theorem 3.3.2 the following dual characterization of viability of the dynamical decentralized allocation mechanism.

Proposition 3.3.7 *We posit assumptions (3.2.2), (3.2.4) and (3.2.5). Then the regulation map* $\Pi_M(\cdot)$ *is strict if and only if the "budgetary condition"*

$$\sup_{q\in N_M(\sum_{i=1}^n x_i)} \inf_{p\in P(x)} \left\langle q, \sum_{i=1}^n c_i(x_i,p)\right\rangle \leq 0 \tag{3.3.4}$$

holds true.

Proof — The budgetary condition being obviously necessary, let us check that it is sufficient by using of the Separation Theorem 3.3.2. Indeed, assume that $\Pi_M(x) = \emptyset$ for some $x \in K$. This means that

$$0 \notin T_M\left(\sum_{i=1}^n x_i\right) - \left\{\sum_{i=1}^n c_i(x_i,p)\right\}_{p\in P(x)}$$

Since this subset is closed and convex, there exists $q \in Y^\star$ such that

$$\sup_{v\in T_M(\sum_{i=1}^n x_i),\, p\in P(x)} \left\langle q, v - \sum_{i=1}^n c_i(x_i,p)\right\rangle < 0$$

By fixing $p \in P(x)$, this implies first that

$$q \in N_M\left(\sum_{i=1}^n x_i\right) := \left(T_M\left(\sum_{i=1}^n x_i\right)\right)^-$$

By taking $v = 0 \in T_M\left(\sum_{i=1}^n x_i\right)$, we infer that

$$\sup_{p\in P(x)} \left\langle q, -\sum_{i=1}^n c_i(x_i,p)\right\rangle < 0$$

and therefore, that

$$\inf_{q\in N_M(\sum_{i=1}^n x_i)} \sup_{p\in P(x)} \left\langle q, -\sum_{i=1}^n c_i(x_i,p)\right\rangle < 0$$

a contradiction of the assumption. □

3.3.3 The Instantaneous Walras Law

The dual condition (3.3.4) involves "budget constraints" $\langle q, c_i(x,p)\rangle \leq 0$. Naturally, if we could associate with any $q \in N_M\left(\sum_{i=1}^n x_i\right)$ some $\mu > 0$ such that $p := \mu q \in P(x)$ and $\langle q, c_i(x, \mu q)\rangle \leq 0$, this dual condition would be satisfied.

We observe that the condition $M = M - \mathbf{R}^l_+$ implies that

$$N_M\left(\sum_{i=1}^n x_i\right) \subset \mathbf{R}^l_+$$

Definition 3.3.8 *We shall say that the change functions c_i satisfy the* (collective) instantaneous Walras law *if :*

$$\forall\, x \in K,\ \exists\, \lambda(x) > 0 \ \ \textit{such that}\ \forall\, p \in \mathbf{R}^l_+ \ \mid\ \|p\| \geq \lambda(x),\ \sum_{i=1}^n \langle p, c_i(x_i, p)\rangle \ \leq\ 0 \tag{3.3.5}$$

We then deduce from Theorem 3.2.4 and Proposition3.3.7 the following consequence:

Theorem 3.3.9 *We posit assumptions (3.2.2), (3.2.4) and (3.2.5). Assume that change functions c_i satisfy the instantaneous Walras law (3.3.5), that the pricing map satisfies*

$$\forall\, q \in \mathbf{R}^l_+,\ \exists\, \mu > 0 \ \ \textit{such that}\ \ \mu p \in P(x)$$

and that they are related by the condition

$$\forall\, x \in K,\ \ \beta_P(x) \ :=\ \inf_{p \in P(x)} \|p\| \ \geq\ \lambda(x)$$

Then, from every initial allocation $x_0 \in K$ starts at least one allocation evolving according (3.2.6), satisfycing

$$\text{for almost all } t \geq 0,\ \ \sum_{i=1}^n \langle p(t), x_i'(t)\rangle \ \leq\ 0 \tag{3.3.6}$$

which is regulated by the regulation law

$$p(t) \in \Pi_M(x(t))$$

Furthermore, when K is compact, there exists a viable equilibrium $(\bar{x}_1, \ldots, \bar{x}_n, \bar{p})$.

Proof — Indeed, if we fix $q \in \mathbf{R}^l_+$, there exists $\mu > 0$ such that $\mu q \in P(x)$. Therefore,

$$\mu\|q\| \geq \beta_P(x) \geq \lambda(x)$$

so that the instantaneous Walras law implies that

$$p := \mu q \in P(x) \ \& \ \langle q, c_i(x,p)\rangle = \frac{1}{\mu}\langle \mu q, c_i(x, \mu q)\rangle \leq 0$$

Hence Proposition 3.3.7 implies that K is a viability domain so that our theorem is a consequence of Theorem 3.2.4. Furthermore, since the instantaneous Walras law holds true, we deduce that property (3.3.6) is satisfied. □

The task we have now to fulfill is to provide some sufficient conditions for the collective instantaneous Walras law (3.3.5) to hold true.

Let S^l denote the unit simplex.

Proposition 3.3.10 *We write the change map in the more explicit form* $c_i(x,p) := g_i(x) - G_i(x)p$. *Assume that*

$$\text{the linear maps } G_i(x) \text{ are positive definite} \tag{3.3.7}$$

and set $\alpha(G_i(x)) := \inf_{p\in S^l}\langle G_i(x)p, p\rangle > 0$. *Then the change maps* c_i *satisfy the instantaneous Walras law with* $\lambda(x) = \max_{i=1,\ldots,n} \frac{\|g_i(x)\|}{\alpha(G_i(x))}$

Proof — Indeed, we observe that

$$\frac{\langle p, g_i(x)\rangle}{\langle G_i(x)p, p\rangle} \leq \frac{1}{\|p\|}\frac{\|g_i(x)\|}{\alpha(G_i(x))}$$

Therefore,

$$\forall\, q \in \mathbf{R}^l_+, \ \forall\, \|q\| \geq \frac{\|g_i(x)\|}{\alpha(G_i(x))}, \ \langle q, g_i(x) - G_i(x)q\rangle \leq 0$$

Hence the instantaneous Walras law is satisfied. □

Examples of positive definite operators are obtained by taking $G_i(x) := -U_i''(x)$ where $U_i''(x)$ is the Hessian of a strictly concave utility function U_i.

In this case, the collective Walras law has been decentralized further since the change functions c_i are to obey the individual instantaneous Walras law (or budgetary constraint)

$$\forall\, q \in \mathbf{R}^l_+ \ \text{ such that } \|q\| \geq \frac{\|g_i(x)\|}{\alpha(G_i(x))}, \ \langle q, g_i(x) - G_i(x)q\rangle \leq 0$$

Remark — We can portray property $\langle p, c_i(x,p)\rangle \leq 0$ by saying that it is forbidden to spend more monetary units than earned in continuous transactions. To be persuaded by this interpretation, it is enough to observe that the Walras law implies that along solutions to the system of differential equations (3.2.1), we have $\langle p(t), x_i'(t)\rangle \leq 0$, and thus, for all h small enough,

$$\langle p(t), x_i(t)\rangle \;\leq\; \langle p(t), x_i(t-h)\rangle + \varepsilon h$$

As we also can see, another advantage of the instantaneous Walras law is that it does not depend upon the set M of scarce resources.

The instantaneous Walras law guarantees that the images of the regulation map Π_M are not empty, without the knowledge of the set M of resources. The (collective) instantaneous Walras law allows the consumers to balance their "monetary transactions" at each instant.

3.4 Time-Dependent Resource Allocation

For simplicity, we began the mathematical description of a metaphoric economy in a stationary environment, in which the behavior of the consumers — described by the change functions and the consumption sets — and the set of scarce resources did not depend upon the time. It is only in this stationary context that the concepts of equilibria, absorbing states and attractors make sense. Unfortunately, stationary dynamical economies simply do not exist. Still, the concept of viability still makes sense in a dynamical environment, and the adaptation of the viability theorems can be made. *In a evolving environment, the concept of* adaptation *to viability constraints prevails over concepts of equilibrium and intertemporal optimality.*

In a nonstationary environment, we continue to begin the description of an economy by defining the set of allocations, keeping our basic economic law: it is impossible to consume more (physical) commodities than available.

But now, the set of scarce resources $M(t) \subset Y$ depends upon time: it is what we call a **tube**, i.e., a set-valued map $M : \mathbf{R}_+ \rightsquigarrow Y$ which associates with any time t the subset $M(t)$ of resources available at time t.

In a first step, we shall assume that this tube is given. But, in the extent where the resources are produced — and not only available —, we see that we have to answer the question of defining evolution laws governing the evolution of such tubes, as differential equations govern the evolution of trajectories. This is now possible thanks to the concept of mutational equations we shall present in the next section.

In the same way, the description of each consumer begins with the specification of her time-dependent consumption set $L_i(t)$.

Therefore, the sets $K(t)$ of allocations at each instant define an **allocation tube**:

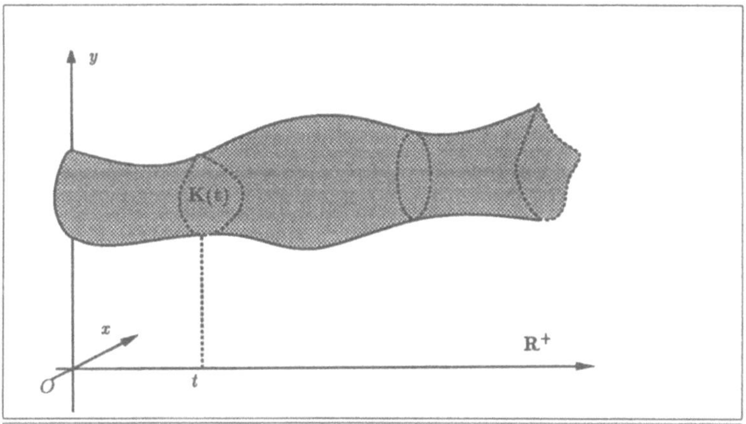

When the viability constraints evolve, the viability domain becomes a viability tube.

Figure 3.3: Viability Tubes

$$K(t) := \left\{ x := (x_1, \ldots, x_n) \in \prod_{i=1}^{n} L_i(t) \;\middle|\; \sum_{i=1}^{n} x_i \in M(t) \right\} \tag{3.4.1}$$

The first question which emerges is the following: *can we still differentiate these scarcity constraints ?*

We saw that we could do so by using the concept of contingent cone to the set K of allocations. When this set evolves with time, we need the concept of **contingent derivative** of the set-valued map $t \rightsquigarrow K(t)$ (see Definition 11.3.2 in the Tool-Box).

For this very purpose, a differential calculus of set-valued maps has been designed since the late seventies. By now, there is an explosion of competing concepts. In a proper Darwinian way, after this early explosion of opportunities, a selection process eliminated the non viable ones. Still, it survived two classes of "derivatives" of set-valued maps which by now are related by a fundamental formula: **contingent derivatives and mutations**. Contingent derivatives are needed to characterize the viability of an allocation tube under a dynamical economy and mutations are required to define **mutational equations** which govern the evolution of tubes, and thus, in our case, production processes.

Therefore, we can differentiate the viability constraint by using the concept of contingent derivative $DK(t,x)(1)$ of the tube at (t,x) in the forward direction 1: We recall that when $x \in K(t)$, $u \in DK(t,x)(1)$ if and only if there exist sequences $h_n > 0$ and u_n converging to 0 and u respectively such that

$$\forall\, n, \;\; x + h_n u_n \;\in\; K(t + h_n)$$

and that

$$\text{Graph}(DK(t,x)) \;=\; T_{\text{Graph}(K)}(t,x)$$

Proposition 3.4.1 *Let $x(\cdot) : [0,\infty[\mapsto X$ be a differentiable function viable in the tube in the sense that*

$$\forall\, t \geq 0, \;\; x(t) \;\in\; K(t)$$

Then the velocity belongs to the forward contingent derivative of the tube:

$$\forall\, t \geq 0, \;\; x'(t) \;\in\; DK(t, x(t))(1)$$

So, we need to compute the contingent derivative of the allocation tube. Since the graphs of the tubes are not convex in general, we have to replace the convexity assumptions of the earlier chapters which were needed both for having simpler computations and for guaranteeing the existence of equilibria by sleekness assumptions to which we already alluded and which are described in Definition 11.2.6 of the Tool-Box.

Proposition 3.4.2 *Assume that the resource tube $M(\cdot)$ and the consumption tubes $L_i(\cdot)$ are sleek and closed and that they satisfy the* transversality assumption*: For every $t \geq O$, $x \in K(t)$,*

$$Y \;=\; \sum_{i=1}^{n} T_{L_i(t)}(x_i) - T_{M(t)}\left(\sum_{i=1}^{n} x_i\right) \tag{3.4.2}$$

Then

$$\left\{ \begin{array}{l} DK(t,x)(1) \;:= \\ \{v \;:=\; (v_1, \ldots, v_n) \in \prod_{i=1}^{n} DL_i(t, x_i)(1) \mid \sum_{i=1}^{n} v_i \in DM\left(t, \sum_{i=1}^{n} x_i\right)(1)\} \end{array} \right.$$

Proof — The proof follows from Theorem 11.3.7 of the Tool-Box with $X := \mathbf{R}$, $Y := \mathbf{R}^{ln}$, $Z = \mathbf{R}^{ln} \times \mathbf{R}^{l}$, A defined by $A(x_1, \ldots, x_n) = (x_1, \ldots, x_n, \sum_{i=1}^{n} x_i)$ and F defined by

$$F(t) \;:=\; \prod_{i=1}^{n} L_i(t) \times M(t)$$

Since

$$T_{F(t)}(Ax) \;:=\; \prod_{i=1}^{n} T_{L_i(t)}(x_i) \times T_{M(t)}\left(\sum_{i=1}^{n} x_i\right)$$

we observe that assumption (3.4.2) implies the transversality condition (11.3.1) of Theorem 11.3.7. Since the tubes $L_i(\cdot)$ and $M(\cdot)$ are sleek, we also derive that

$$DF(t, x_1, \ldots, x_n, y)(1) \;=\; \prod_{i=1}^{n} DL_i(t, x_i)(1) \times DM(t, y)(1)$$

so that the conclusion of Proposition 3.4.2 follows from the conclusion of Theorem 11.3.7. □

With this formula at our disposal, we can apply the Viability Theorem 12.3.2 on tubes of the Tool-Box.

The dynamics of the evolution of the consumption is described by

$$\forall\, t \geq 0, \quad \begin{cases} i) & x_i'(t) \; = \; c_i(t, x_i(t), p(t)) \quad (i = 1, \dots, n) \\ ii) & p(t) \in P(t, x(t)) \; := \; P(t, x_1(t), \dots, x_n(t)) \end{cases} \tag{3.4.3}$$

We shall address the problem of finding viable evolutions satisfying

$$\forall\, t \geq 0, \quad \begin{cases} i) & x_i(t) \in L_i(t) \quad (i = 1, \dots, n) \\ ii) & \sum_{i=1}^n x_i(t) \; \in \; M(t) \end{cases} \tag{3.4.4}$$

Definition 3.4.3 *The* regulation map Π_M *is the set-valued map*

$$\begin{cases} \forall\, t \geq 0, \; x \in K(t), \; \Pi_M(t, x) \\ := \; \{p \in P(t, x) \mid \sum_{i=1}^n c_i(t, x_i, p) \in DM\left(t, \sum_{i=1}^n x_i\right)(1)\} \end{cases}$$

built from the confrontation of the changes claimed by consumers and the tube of scarce resources.

This regulation map associates with each instant $t \geq 0$ and with each allocation $x \in K(t)$ the prices such that the related total change of consumptions lies in the forward contingent derivative of the tube of scarce resources. The objective is to bring back the total consumption inside the tube of scarce resources.

We posit now the assumptions we need to prove the nonstationary version of the viability theorem.

— ASSUMPTIONS ON THE CONSUMPTION AND RESOURCE TUBES:
For every $t \geq 0$,

$$\begin{cases} i) & M(t) = M(t) - \mathbf{R}_+^l \text{ and the tube is closed and sleek} \\ ii) & \forall\, i = 1, \dots, n, \;\; L_i(t) = L_i(t) + \mathbf{R}_+^l \text{ and the tube is closed and sleek} \\ iii) & \forall\, x \in K(t), \;\; Y \; = \; \sum_{i=1}^n T_{L_i(t)}(x_i) - T_{M(t)}\left(\sum_{i=1}^n x_i\right) \end{cases} \tag{3.4.5}$$

— ASSUMPTIONS ON THE CHANGE FUNCTIONS:
For every $t \geq 0$,

$$\left\{\begin{array}{ll} i) & p \mapsto c_i(t,x,p) \;:=\; g_i(t,x) + G_{t,i}(x)p \text{ is affine, where} \\ ii) & g_i : \text{Graph}(L_i) \mapsto Y \text{ is continuous with linear growth uniform in } t \\ iii) & G_i : \text{Graph}(L_i) \mapsto \mathcal{L}(Y^\star, Y) \text{ is continuous and bounded} \\ iv) & \forall\, x_i \in L_i(t),\ p \in \text{Im}(P),\ \ c_i(t,x_i,p) \;\in\; DL_i(t,x_i)(1) \end{array}\right. \tag{3.4.6}$$

— ASSUMPTIONS ON THE PRICING MAP:

$$P : \mathbf{R}_+ \times Y^n \leadsto Y \text{ is a Marchaud map} \tag{3.4.7}$$

Theorem 3.4.4 *We posit assumptions (3.4.5), (3.4.6) and (3.4.7). Then the two following conditions are equivalent:*

1. *the nonstationary dynamical economy is viable in the sense that from every initial time t_0 and any initial state $x_0 \in K(t_0)$ starts at least one allocation evolving according (3.4.3) and satisfying the constraints (3.4.4)*

2. $$\forall\, t \geq 0,\ \forall\, x \in K(t),\ \ \Pi_M(t,x) \;\neq\; \emptyset$$

In this case, the viable allocations are governed by the regulation law

$$\text{for almost all } t \geq 0, \quad p(t) \in \Pi_M(t, x(t)) \tag{3.4.8}$$

In the next section, we shall consider the case when the tubes $M(\cdot)$ are obtained through kinds of differential equations called mutational equations.

3.5 Mutational Dynamical Economies

In the preceding section, we have let the resource set $M(t)$ evolve with time, but in a prescribed way, so to speak.

The question arises whether the evolution of the resource set $M(t)$ can be governed by a kind of differential equation, which, instead of ruling the evolution of vectors, do determine the evolution of sets. Such "differential equations" are called **mutational equations**. We devote Section 12.5? of the Tool-Box to a short presentation of mutational equations[23].

The way to adapt the concept of velocity of a time-dependent vector $x(t)$ to the case of a time -dependent subset $M(t)$ (which we have called a tube) is quite natural. The velocity $v := x'(t)$ of $x(t)$ can be defined in the following way:

$$x(t+h) \;\in\; B(x(t), \beta(h)h) \;\&\; x(t) \;\in\; B(x(t+h), \beta(h)h)$$

[23]which is the topic of the forthcoming monograph [35, Aubin].

where $\beta(h)$ converges to 0 with h and $B(x,\varepsilon)$ denotes the ball centered in x of radius ε. Written in this way, replacing the vector $x(t)$ by the subset $M(t)$ and the velocity $v \in Y$ by a convex compact subset $\Phi \subset Y$ — that we shall call a mutation —, we can propose a formal

Definition 3.5.1 *We shall say that a nonempty convex compact subset $\Phi \subset Y$ is a (forward)* mutation of a tube *$M(\cdot)$ at t if there exists a function $\beta(h)$ converging to 0 with h satisfying*

$$B(M(t)+h\varphi,\beta(h)h)\ \&\ M(t+h)\subset M(t)+h\Phi\subset B(M(t+h),\beta(h)h)$$

We denote by $\overset{\circ}{M}(t)$ the subset of mutations of the tube $M(\cdot)$ at t.

Remark — Actually, we shall enlarge the set of possible mutations by allowing Φ to depend upon x, i.e., to take for mutations set-valued maps $y \rightsquigarrow \Phi(y)$. This will be done in more details in the Tool-Box. □

We shall derive from Theorem 12.5.2 of the Tool-Box the following result linking mutations and contingent derivatives of tubes:

Theorem 3.5.2 *Assume that $M(\cdot) : \mathbf{R} \rightsquigarrow Y$ is a tube with closed values. Let $\Phi \in \overset{\circ}{M}(t)$ be a mutation of M at x. Then:*

$$\forall y \in M(t),\quad DM(t,y)(1) \;=\; \Phi + T_{M(t)}(y)$$

This theorem allows us to characterize a viability problem in tubes evolving according a mutational equation.

We denote by $\mathcal{K}(Y)$ the family of nonempty compact subsets (supplied with the Hausdorff topology). Let us consider a tube $M(\cdot) : \mathbf{R}_+ \rightsquigarrow Y$, solution to a mutational equation

$$\overset{\circ}{M}(t) \ni \Phi(M(t))$$

Theorem 3.5.3 *We posit assumptions (3.4.5), (3.4.6) and (3.4.7) of Theorem 3.4.4. Assume also that $\Phi : \mathcal{K}(Y) \mapsto \mathcal{K}(Y)$ is a continuous map satisfying*

$$\forall\, M,\ \ \|\Phi(M)\| := \sup_{v\in\Phi(M)} \|v\| \;<\; +\infty$$

The two conditions are equivalent:

1. *For every compact set M and every $x \in M$,*

$$\Pi_\Phi(x,M) \;:=\; \left\{p \in P(x) \mid \sum_{i=1}^n c_i(x_i,p) \in T_M\left(\sum_{i=1}^n x_i\right) + \Phi(M)\right\} \;\neq\; \emptyset$$

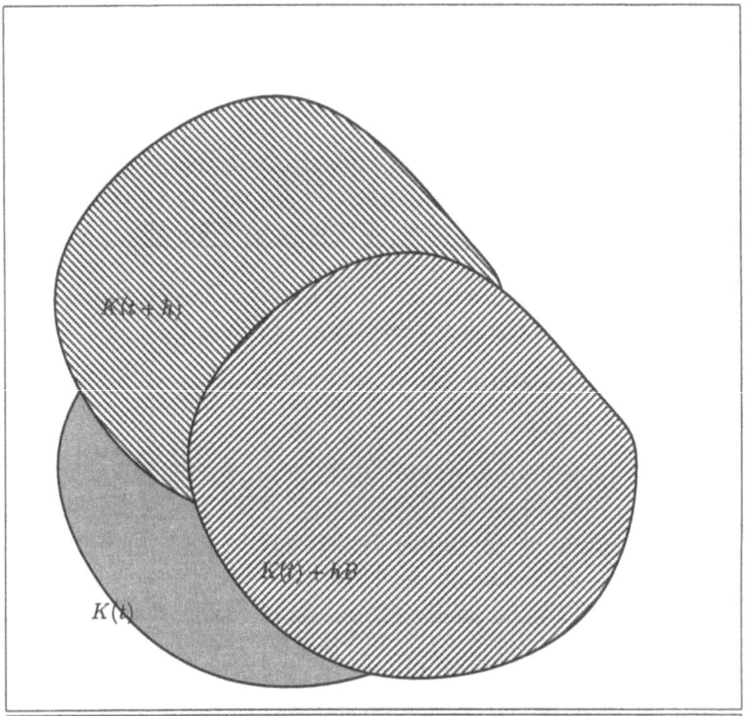

The mutation of a tube at time t is a compact convex subset B such that $K(t) + hB$ is close to $K(t + h)$.

Figure 3.4: Mutations of Tubes

2. *From every M_0 and from every $x_0 \in K_0$, there exists a solution to the dynamical economy*

$$\begin{cases} x_i'(t) = c_i(x_i(t), p(t)) \; (i = 1, \ldots, n) \\ p(t) \in P(x(t)) \end{cases}$$

which is viable in the sense that

$$\forall\, t \geq 0, \; \sum_{i=1}^{n} x_i(t) \in M(t)$$

where $M(t)$ is a solution to the mutational equation $\overset{\circ}{M}(t) \ni \Phi(M(t))$ starting at M_0.

In other words, whenever the set of resources is obtained through production processes, represented by technological subsets, as it is done in the Arrow-Debreu tradition, mutational equations can be used to provide mathematical metaphors for the dynamical behavior of producers.

3.6 History-Dependent Resource Allocation

Until now, we have assumed that the set of scarce resources was either constant or time dependent. But the most important economic phenomenon is that **the set of scarce resources depends upon delays, anticipations, and other cumulated consequences of the past, etc., i.e., on history.**

This is in particular the case when the scarcity constraints can be written in the form

$$\forall\, t \geq 0, \; \sum_{i=1}^{n} x_i(t) \in M\left(\sum_{i=1}^{n} \int_{-\infty}^{t} C_i(t-s)x_i(s)d\mu(s)\right) \tag{3.6.1}$$

where $M : Y \rightsquigarrow Y$ is a set-valued map, $C_i \in L^1(0, \infty; \mathcal{L}(Y, Y), d\mu)$ and $d\mu$ is either the usual Lebesgue measure dt for Volterra type of dependence on the history of the evolution up to time t or a discrete measure[24]. For the sake of simplicity, we do assume here that the consumption sets L_i are all equal to the commodity space $Y := \mathbf{R}^l$.

Hence, this type of scarcity constraints involves the history of the evolution in the form of cumulated damped consequences of past allocations.

It is then agreeable to translate the concept of history of the evolution of the state up to the instant $t > 0$ by using the operator $T(t)$ from the space of continuous functions $\mathcal{C}(-\infty, +\infty; X)$ to the **history space** $\mathcal{C} := \mathcal{C}(-\infty, 0; X)$

[24] If $x_i(\cdot)$ is continuous and bounded and if $C_i \in L^1(0, \infty; \mathcal{L}(Y, Y), dt)$ is (a class of) measurable integrable matrices for the measure $d\mu$, then $s \mapsto C_i(t-s)x_i(s)$ is (a class of) measurable integrable vectors of Y.

Figure 3.5: Translation $T(t)$

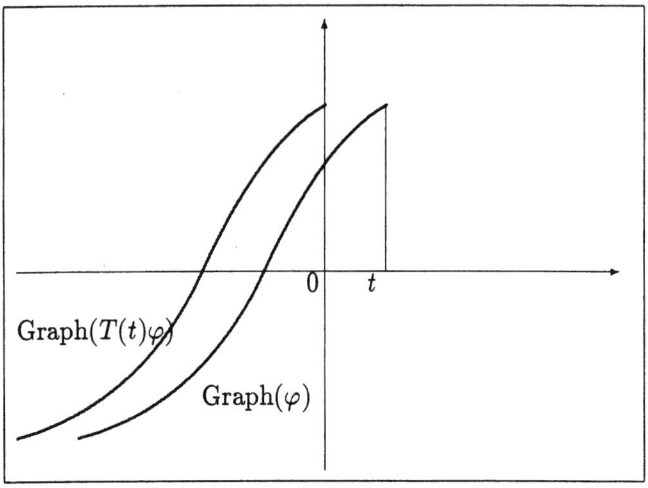

which associates with any continuous function $x(\cdot)$ its *history* $T(t)x$ *up to time* t defined by:

$$\forall\,\tau \in]-\infty, 0],\;\; T(t)x(\tau) \;:=\; x(t+\tau)$$

The history is also denoted by some authors by the notation $x_t := T(t)x$.

Hence, the set of allocations $K(T(t)x) \subset X := Y^n$ depends on the history $T(t)x$ of the allocation up to time t: It is given by:

$$K(T(t)x) \;:=\; \left\{ y \in \mathbf{R}^{l^n} \mid \sum_{i=1}^{n} y_i \;\in\; M\left(\sum_{i=1}^{n}\int_{-\infty}^{0} C_i(-s)x_i(t+s)d\mu(s)\right)\right\}$$

so that the history-dependent constraints can be written in the form:

$$\forall\, t \geq 0,\;\; x(t) \;\in\; K(T(t)x)$$

Observe that we can write the foregoing constraint (3.6.1) in the form

$$\forall\, t \geq 0,\;\; \sum_{i=1}^{n} x_i(t) \;\in\; M\left(\sum_{i=1}^{n}\int_{-\infty}^{0} C_i(-s)x_i(t+s)d\mu(s)\right)$$

When the measure is discrete, we obtain "delay scarcity constraints" of the type

$$\forall\, t \geq 0,\;\; x(t) \;\in\; M\left(\sum_{i=1}^{n}\sum_{j=-\infty}^{j=0} C_i(-s_j)x_i(t+s_j)\right)$$

where $s_j \leq 0$ constitute a decreasing sequence of delays and where the series of $\|C(s_j)\|$ is convergent.

Since we shall allow the set of available commodities to depend upon cumulated consequences of the past, there are no reason not to include this kind of memory in the dynamical economy. We thus consider other time dependent linear operators $A_i \in L^1(0,\infty;\mathcal{L}(Y,Y),d\mu)$ and $B \in L^1(0,\infty;\mathcal{L}(X,X),d\mu)$[25], change maps $c_i : Y \times Y^\star \mapsto Y$ and a pricing map $P : X \leadsto Y^\star$.

The evolution of commodities and prices is governed by the **Volterra dynamical economy**

$$\begin{cases} i) & x_i'(t) \;=\; c_i\left(\int_{-\infty}^t A_i(t-s)x_i(s)d\mu(s), p(t)\right) \\ \\ ii) & p(t) \;\in\; P\left(\int_{-\infty}^t B(t-s)x(s)d\mu(s)\right) \end{cases} \tag{3.6.2}$$

This makes sense only when at the initial time $t = 0$, the functions $x_i(\cdot)$ are already given bounded continuous functions φ_i defined for $s \leq 0$ (initial data are past allocations in this context):

$$\forall\, s \in]-\infty,0],\;\; x_i(s) \;=\; \varphi_i(s)$$

We would like to extend the characterization of the viability of a dynamical economy to the case of a Volterra dynamical economy (3.6.2) confronted to Volterra scarcity constraints (3.6.1).

For that purpose, we need to define a pricing map Π_M which is defined on $\mathcal{C}$.

We regard this set as the **history set** of "evolution histories" from $-\infty$ up to the initial time. We recall that a Lipschitz function is almost everywhere differentiable.

$$\begin{cases} \Pi_M(\varphi) \;:=\; \Big\{ p \in P\left(\int_{-\infty}^0 B(-s)\varphi(s)d\mu(s)\right) \;| \\ \\ \sum_{i=1}^n c_i\left(\int_{-\infty}^0 A_i(-s)\varphi_i(s)d\mu(s), p(t)\right) \\ \\ \in\; DM\left(\sum_{i=1}^n \int_{-\infty}^0 C_i(-s)\varphi_i(s)d\mu(s), \sum_{i=1}^n \varphi_i(0)\right)\left(\sum_{i=1}^n \int_{-\infty}^0 C_i(-s)\varphi_i'(s)d\mu(s)\right)\Big\} \end{cases} \tag{3.6.3}$$

We shall deduce from the Corollary 12.4.8 of Functional Viability Theorem 12.4.4 of the Tool-Box the following result.

Theorem 3.6.1 *We assume that the change functions c_i are continuous with linear growth and affine with respect to p, that P is Marchaud and that the*

[25]Recall that $X := Y^n$ is the allocation space.

time dependent linear operators A_i, B & C_i satisfy

$$\sup_{t\in[0,1]} \max(\|A_i(t)\|, \|B(t)\|, \|C_i(t)\|) \leq \gamma < +\infty$$

Assume that M is sleek, that for all $(y,z) \in \text{Graph}(M)$, $\text{Dom}(DM(y,z)) = Y$ and that

$$\sup_{(y,z)\in\text{Graph}(M)} \|DM(y,z)\| \leq \beta < +\infty$$

If the Volterra pricing map Π_M has nonempty values, then for any initial Lipschitz function satisfying

$$\sum_{i=1}^{n} \varphi_i(0) \in M\left(\sum_{i=1}^{n} \int_{-\infty}^{0} C_i(-s)\varphi_i(s)d\mu(s)\right)$$

there exists a solution $x(\cdot), p(\cdot)$ to the Volterra dynamical economy (3.6.2) obeying the initial condition

$$\forall\, s \in]-\infty, 0],\ \ x_i(t) = \varphi_i(s)$$

and the Volterra scarcity constraints (3.6.1).

The regulation law

$$\text{for almost all } t \geq 0,\ \ p(t) \in \Pi_M(x(t+\cdot))$$

governs the evolution of viable allocations.

This follows from a theorem where the dependence upon the history of the evolution is governed by differential inclusions with memory we shall present in the Tool-Box.

Chapter 4

Dynamical Economies

Introduction

1. The basic problem of dynamical allocation of scarce resources is a particular case of what we shall call a **dynamical economy** (P, c) governing the evolution of an abstract **commodity** and an abstract **price**. The commodities evolve according to the laws

$$\begin{cases} i) & x'(t) = c(x(t), p(t)) \\ ii) & p(t) \in P(x(t)) \end{cases}$$

where the commodity $x(\cdot)$ ranges over a finite dimensional vector-space X, the price $p(\cdot)$ ranges over another finite dimensional vector-space $\mathcal{M}$, $c : X \times \mathcal{M} \mapsto X$ describes the dynamics and where the set-valued map $P : X \rightsquigarrow \mathcal{M}$ may be regarded as an "*a priori* pricing map".

Here, the first equation describes how the price — regarded as a **message, or regulation control (in short, regulee)**, or again an **input** to the system — yields the commodity of the dynamical economy[1] — regarded as an **output** — whereas the second inclusion shows how some **commodity-dependent constraints act on the prices**. It depicts how the commodity-output "feeds back" to the price-input.

A solution to this system is a function $t \to x(t)$ satisfying this system for some time dependent price $t \to p(t)$.

Viability (or scarcity) constraints are described by a closed subset K of the commodity space: These are intended to describe the "viability" of the dynamical economy (in the basic model of resource allocation, K is the set of allocations of scarce resources to n consumers).

A subset K **is viable** under the dynamical economy described by c and P if from every initial commodity $x_0 \in K$ starts at least one solution

[1] once the initial commodity is fixed.

to the dynamical economy which is **viable** in the sense that

$$\forall\, t \geq 0, \quad x(t) \in K$$

We shall formulate in this more general framework the characterization of subsets viable under a dynamical economy. For that purpose, we associate with the dynamical economy (described by c and P) and with the viability constraints (described by K) the **(set-valued) regulation map** Π_K. It maps any commodity x to the subset $\Pi_K(x)$ consisting of prices $p \in P(x)$ which are *viable* in the sense that

$$c(x,p) \quad \text{is contingent to } K \text{ at } x$$

If, for every $x \in K$, there exists at least one viable price $p \in \Pi_K(x)$, we then say that K is a **viability domain** of the dynamical economy described by both c and P.

The Viability Theorem we mentioned earlier holds true for a rather large class of dynamical economies, called **Marchaud economies**[2].

The basic viability theorem states that for such systems K is viable under the dynamical economy (P, c) if and only if the regulation map Π_K has nonempty values.

The Viability Theorem also provides a **regulation law** for regulating the dynamical economy in order to maintain the viability of a solution: The viable solutions $x(t)$ are regulated by viable prices $p(t)$ through the **regulation law**:

$$\text{for almost all } t, \quad p(t) \;\in\; \Pi_K(x(t))$$

At this point arises the problem of finding sufficient conditions for guaranteeing that the pricing map has nonempty values. Since this is an existence problem — find prices in a given set — one can think of using existence theorems provided by nonlinear analysis, and we shall do that to some extent.

But it happens that we have at our disposal a dual formulation of the viability conditions involving **normal cones**, which are the polar (dual) cones to the tangent cones.

---

[2]It contains the class of dynamical economies of the form

$$x'(t) \;=\; g(x(t)) - G(x(t))p(t)$$

where $G(x)$ are linear operators from the price space to the commodity space and when the pricing map P has compact convex images $P(x)$.

Therefore, one can couch this necessary and sufficient condition for viability in terms of dual elements, usually accounted as prices. We face a situation analogous to the one in optimization under constraints, where the Lagrange multipliers, which belong to the normal cone to the set of allocations, are regarded as prices.

We shall then interpret this dual characterization as a **reservoir of budgetary rules** which we shall exploit in Chapter 5 through the use of "viability multipliers". Once written, the dual characterization provides sufficient conditions analogous to the Walras law in the classical Arrow-Debreu setting: As for "Walrasian demand functions", such budgetary law requires that one cannot spend more monetary units than earned during (instantaneous) transactions. We shall call these conditions "instantaneous Walras laws".

Within this framework of dynamical economies, we can go beyond the basic problem of allocation of scarce resources by involving other ingredients. We shall consider some of them once at a time, in this chapter as well as in Chapter 5, knowing that with some expertise, they all could be integrated in some kinds of "monster models".

In a nutshell, a dynamical economy is described by the change function and the pricing map. Given a constrained set, the Viability Theorem provides the necessary and sufficient conditions — couched in term of strictness of the regulation map — for the constrained set to be viable under the dynamical economy.

Many of the traditional interesting subsets such as **equilibrium points, trajectories of periodic solutions, limit sets of solutions, are examples of closed viability domains** (see Definition 6.2.13 and Proposition 6.2.14 below.) Actually, equilibrium points $\overline{x}$, which are solutions to

$$c(\overline{x},\overline{p}) = 0 \text{ for some } \overline{p} \in P(\overline{x})$$

are the smallest viability domains, the ones reduced to a single point. This is because being **stationary states**, the velocities $c(\overline{x},\overline{p})$ are equal to zero. Furthermore, we shall emphasize that there exists a basic and curious link between viability theory and general equilibrium theory: the General Equilibrium Theorem — an equivalent version of the 1910 **Brouwer Fixed Point Theorem**, the cornerstone of nonlinear analysis — states that **every compact convex viability domain contains an equilibrium point**. It finds here a particularly relevant formulation: **viability implies stationarity**.

We next provide the computation of the equilibrium map associating with each price the set of viable equilibria, which allows us to study the sensitivity of equilibra to perturbations of prices, known under the umbrella of "comparative statics". This result follows from the extension to set-valued maps of the Inverse Function Theorem stated in Chapter 14 of the Tool-Box.

2. After studying the sensitivity of equilibra to perturbations, we devote few pages to the problem of "dynamic qualitative economics".

   The **qualitative state** of a solution to the dynamical economy at a given time t is the knowledge of the monotonicity property of each component $x_i(t)$ of a solution $x(\cdot)$ to this dynamical economy, i.e., the knowledge of the sign of the derivatives $x_i'(t)$. Hence the **qualitative behavior** is the evolution of the qualitative states of the solution, i.e., the evolution of the vector of signs of the components of $x'(t) = c(x(t), p(t))$, which must be determined **without solving this dynamical economy.**

   Instead of studying the individual properties of the solutions to a dynamical property, we characterize the areas where each components either increases or decreases in a prescribed manner: they are the so-called **qualitative cells.** Each cell describes "qualitative properties" of the dynamical economy. In each of these cells, either all solutions leave the cell in finite time or at least one of them converges to an equilibrium. We provide sufficient conditions showing that from the boundary separating two qualitative cells, all solutions leave one cell to reach the other one, regarded as a "successor". In this case, we can derive a kind of qualitative discrete system on the qualitative cells indicating which are the successors if any of each qualitative cell.

   We shall study the qualitative behavior of the dynamical economy, i.e., the evolution of the functions $t \mapsto \text{sign}(x'(t))$ associated to solutions $x(\cdot)$ of the dynamical economy. Furthermore, we shall track down the "**landmarks**", i.e., the states at which the monotonic behavior of the solutions is modified. But, instead of finding them *a posteriori* by following the qualitative behavior of a given solution, we shall find them *a priori*, without solving the dynamical system, neither qualitatively nor analytically.

   To the extent that qualitative cells describe economic phenomena in the framework of a dynamical economy, this discrete dynamical system provides **causality** relations, by specifying what are the phenomena caused by a given one. In this sense, we are able to deduce from the model some "laws". This is one of the main motivations for borrowing the name of **qualitative economics** coined by Samuelson and extensively studied by Lancaster in the framework of "sign matrices".

   This is not always possible and we shall provide in Chapter 6 an extension of a result of Donald Saari on "chaos". Chaos here means the following property: Given any arbitrary infinite sequence of qualitative cells, there is always one solution which visits these cells in the prescribed order.

3. We then define **planning mechanisms** as single-valued maps $\varpi(\cdot)$ which are **selections** of the regulation map in the sense that $\varpi(x) \in \Pi_K(x)$ for all $x \in K$, since they associate with commodities prices "computed" by a planning bureau. Since the prices chosen this way depend only upon the commodities in a timeless manner, involving neither the past[3] nor the future (and this seems highly reasonable), one can say that such mechanisms describe a **myopic behavior** of the economic agents.

   Then the solutions to the differential equation

   $$x'(t) = c(x(t), \varpi(x(t)))$$

   (when they exist) are viable since the implemented prices $p(t) := \varpi(x(t))$ obey the regulation law by construction. Existence is guaranteed when $\varpi(\cdot)$ is continuous, but may still holds true for discontinuous, but explicit, feedbacks.

   This is case when we choose the price $\varpi^\circ(x) \in \Pi_K(x)$ with minimal norm. Viable solutions obtained with this planning mechanism are called **slow viable solutions**.

   We provide an explicit formula of the planning mechanisms driving slow solutions for dynamical economies of the form

   $$x'(t) = g(x(t)) - p(t)$$

   However, as it was explained in Chapter 1, one can regulate dynamical economies by connection matrices instead of prices:

   $$x'(t) = W(t)g(x(t))$$

   These connection matrices describe the dynamic relations among the economic agents needed to guarantee the viability of the system.

   We shall observe that one can associate with prices regulating such a dynamical economy a class of connection matrices providing the same evolutions, revealing the actual bilateral relations relating the economic agents when they use such a price as a decentralizing message. This provides a first explanation of the decentralizing role of prices.

   We next define the **index of (connectionist) complexity** of a connection matrix W — which is the matrix linking each agent of the system to the others — as the distance between this **connection matrix** W and the

[3]that is a flaw of this model which can be corrected by using history-dependent models later on.

identity matrix which describes the decentralized situation. We then prove that in this context, the connection matrix which minimizes this connectionist complexity index is associated with the planning mechanism driving slow viable evolutions, i.e., that the least complex connection matrix governs slow evolutions. This justifies further the decentralizing role of prices. On the mathematical side, the best framework to study these connection matrices is tensor algebra of matrices, which, despite its mysterious character, is very easy to use at the level required by our study. Chapter 10 of the Tool-Box provides what is needed to know for tackling these issues. The least complex connection matrix involves the tensor product of the minimal planning mechanism and the original dynamics. If one adopts this static measure of connectionist complexity and the relevance of such a mathematical metaphor, we derive that *the least complex connectionist matrix is not sparse* at all, since the entries of the tensor product of a vector $y \in \mathbf{R}^n$ and of a vector $p \in \mathbf{R}^m$ are $y_j p^i$ $(j = 1, \ldots n,\ i = 1, \ldots, m)$.

We do not address here the possibility of regulating dynamical economies with connection matrices obeying a given "sparsity structure" (for which some connection weights are required to be equal to 0).

4. Hence, we have to find **selection procedures** of the regulation map which provide either continuous selections or discontinuous selections for which the differential equation

$$x'(t) \;=\; c(x(t), \varpi(x(t)))$$

still has solutions. We saw that the lack of continuity of the regulation map forbids the viable price with minimal norms to be continuous, so that we cannot *a priori* use the existence theorems for differential equations which assume the continuity of the dynamics. This is also the case for other selection mechanisms of a selection in the regulation map $x \rightsquigarrow \Pi_K(x)$ using minimization or static game theoretical methods for choosing one price or a smaller subset of prices in $\Pi_K(x)$.

Hélène Frankowska found a way to overcome this obstacle by introducing a class of "selection procedures" which provides evolutions of feedback dynamics which are not necessarily continuous, provided that the regulation map is lower semicontinuous with closed convex images. We shall use this technique also at the end of this chapter to study the slow evolution of technological matrices, in Chapter 5 for solutions of dynamical economies with minimal frustration and in Chapter 7 for obtaining heavy evolutions and for studying hierarchical games.

We conclude this section by stating the celebrated Michael Theorem on continuous selections which states that under such an assumption, there exist continuous selections of the regulation map providing continuous feedback maps, as well as stating the less known Steiner Selection Theorem providing an explicit selection of a point of a compact convex subset. We shall need these two theorems in our short presentation of dynamical games against nature.

5. With these selection procedures in our hands, we investigate several planning mechanisms.

6. We finally take up again the complexity issue in another context. We introduced connection matrices in the dynamics and used them as regulees for sustaining the viability of the dynamical economy.

   But we could also introduce a connection matrix in the constraints

$$\forall t \geq 0, \ \ h(x(t)) \in M$$

   imposed upon the dynamical economy by relaxing them through the use of connection matrix W

$$\forall t \geq 0, \ \ h(W(t)x(t)) \in M$$

   which is nothing else that a technological matrix in a standard economic context.

   We shall derive the slowest evolution of technological matrices which guarantee the viability of the dynamical economy. We shall briefly return to the use of one-stage or multi-stage technological matrices in Chapter 5.

4.1 Toward Dynamical Economies

Instead of proving directly Theorems 3.2.4 and 3.2.8 (and thus, Theorem 3.3.9), we shall show that they are particular cases of viability theorems for dynamical economies.

First, we introduce the map $c : X \times \mathbf{R}^l_+ \mapsto X$ defined by

$$c(x,p) := (c_1(x_1,p), \ldots, c_n(x_n,p))$$

and we set

$$K := \left\{ x := (x_1, \ldots, x_n) \in \prod_{i=1}^n L_i \;\middle|\; \sum_{i=1}^n x_i \in M \right\}$$

Therefore, the dynamical laws

$$\begin{cases} i) & x_i'(t) = c_i(x_i(t), p(t)) \ \ (i = 1, \ldots, n) \\ ii) & p(t) \in P(x(t)) \end{cases}$$

governing the evolution of allocations are the laws governing a dynamical economy

$$\begin{cases} i) & x'(t) = c(x(t), p(t)) \\ ii) & p(t) \in P(x(t)) \end{cases}$$

Any viable equilibrium is a solution $(\overline{x}, \overline{p}) \in K \times \mathbf{R}^l_+$ to the system

$$c(\overline{x}, \overline{p}) = 0 \ \& \ \overline{p} \in P(\overline{x})$$

4.1.1 Dynamical Economy

Hence our dynamical model of allocation of scarce resources can be written in a semi-abstract form, between the explicit description and its translation as a differential inclusion that we will introduce later.

As it was advocated in the introduction of this chapter, it is convenient to choose a middle ground inspired from systems theory which allows us to incorporate more examples of dynamical economic models than the one of allocation of scarce resources.

We introduce two finite dimensional vector-spaces:

1. — the (abstract) commodity space X

2. — the (abstract) price space $\mathcal{M}$ and a *pricing set-valued map* $P : X \rightsquigarrow \mathcal{M}$ associating with any commodity x the (possibly empty) subset $P(x)$ of feasible prices associated with the commodity x. In other words, we assume that the available *prices of the system are required to obey constraints which may depend upon the commodity.* We shall investigate later the cases when the prices depend also upon the time and/or the history of the solution to the system.

The dynamics of the system are further described by a (single-valued) change map $c : \text{Graph}(P) \mapsto X$ which assigns to each commodity-price pair $(x, p) \in \text{Graph}(P)$ the *velocity* $c(x, p)$ with which the commodity evolves.

Hence the set

$$F(x) := \{c(x, p)\}_{p \in P(x)}$$

is the set of available velocities to the system when its commodity is x.

Definition 4.1.1 (Dynamical Economy) *A* dynamical economy (P, c) *is defined by*

— *a "pricing" set-valued map* $P : X \rightsquigarrow \mathcal{M}$

— *a map* $c : \text{Graph}(P) \mapsto X$ *describing the dynamics of the system.*

The evolution of the commodity and of the price is governed by

$$\begin{cases} i) & \text{for almost all } t, \ x'(t) = c(x(t), p(t)) \\ ii) & \text{where } p(t) \in P(x(t)) \end{cases} \quad (4.1.1)$$

We associate with any subset $K \subset \mathrm{Dom}(P)$ *the* regulation map $\Pi_K : K \leadsto \mathcal{M}$ *defined by*

$$\forall x \in K, \; \Pi_K(x) := \{p \in P(x) \mid c(x,p) \in T_K(x)\}$$

We shall say that the dynamical economy (P, c) *is a* **Marchaud dynamical economy** *if it satisfies the following conditions:*

$$\left\{\begin{array}{ll} i) & \mathrm{Graph}(P) \textit{ is closed} \\ ii) & c \textit{ is continuous} \\ iii) & c \textit{ is affine with respect to } p \textit{ and } P \textit{ has convex values} \\ iv) & c \textit{ and } P \textit{ have linear growth} \end{array}\right. \quad (4.1.2)$$

4.1.2 Viability Domains

We shall concentrate on the confrontation of the dynamical economy with constraints bearing on commodities. We denote by K the set of commodities satisfying these constraints.

Definition 4.1.2 *Let* K *be a subset of the domain of the pricing map* P. *We shall say that* K *is* **viable under the dynamical economy** (P, c) *if from any initial state* x_0 *in* K *starts* **at least** *one solution on* $[0, \infty[$ *to the dynamical economy (4.1.1) which is* **viable** *in* K.

We shall say that a subset $K \subset \mathrm{Dom}(P)$ *is a* **viability domain** *of the dynamical economy* (P, c) *if and only if*

$$\forall\, x \in K, \; \exists\, p \in P(x) \;\; \textit{such that} \;\; c(x,p) \in T_K(x)$$

i.e., if and only if the associated regulation map Π_K *is strict (has nonempty values).*

In particular, *viability domains reduced to an element* $\bar{x}$ *are equilibria.* Or we can say that closed viability domains are thick equilibria, so to speak. In particular, we shall mention later that "limit sets" of solutions are also closed viability domains. Since Adam Smith and above all, Léon Walras, the concept of equilibrium was closely related to the underlying dynamical process providing hopefully a way to achieve it, in finite or infinite time (see Chapter 6).

Therefore, the existence of an equilibrium is not the ultimate issue, the real one, when it can be solved, being the convergence of allocations when time goes to infinity, i.e., what is called **asymptotic convergence**.

Naturally, a given solution $x(\cdot)$ may not have a limit when $t \to +\infty$. However, one can always introduce the **limit set** of $x(\cdot)$, which is by definition the set of its **cluster points**, or of the limits of subsequences $x(t_n)$ when $t_n \to +\infty$. We shall prove that the limit sets of the solutions to a Marchaud dynamical economy are **closed viability domains** (see Proposition 6.2.14).

As we saw in the example we started with, duality conditions can be used to provide a further characterization of viability conditions[4].

Proposition 4.1.3 *Assume that the dynamical economy (P, c) is Marchaud. Then K is a viability domain of this dynamical economy if and only if*

$$\forall\, x \in K, \quad \sup_{p \in N_K(x)} \inf_{q \in P(x)} \langle p, c(x,q)\rangle \ \leq\ 0$$

Therefore, this dual characterization can lead to budgetary conditions (generalized Walras laws), since it is enough to design a map associating with any pair $(x,p) \in \mathrm{Graph}(N_K(\cdot))$ a price $q \in P(x)$ such that

$$\langle p, c(x,q)\rangle \ \leq\ 0$$

We exploit this situation and analogous ones to design viable economies in Chapter 5.

Applying the viability theorem to a dynamical economy for characterizing its viability or for guaranteeing the existence of an equilibrium amounts to checking that for any $x \in K$, there exists a solution $p \in P(x)$ to the inclusion $0 \in c(x,p) - T_K(x)$.

We can then appeal to the Equilibrium Theorem 14.4.1 below to provide an arsenal of sufficient conditions for the existence of such a viable price $p \in \Pi_K(x)$. We deduce for instance the following criterion:

Proposition 4.1.4 *Assume that K is sleek (see Definition 11.2.6), that the images $P(x)$ of the pricing map P are convex and compact and that there are linear operators $B(p) \in \mathcal{L}(\mathcal{M}, X)$ depending continuously upon p such that*

$$\forall\, (x,p) \in \mathrm{Graph}(P), \quad 0 \ \in\ c(x,p) - T_K(x) - B(p)T_{P(x)}(p) \qquad (4.1.3)$$

Then the regulation map $\Pi_K(\cdot)$ is strict and thus, K is a viability domain of the economy.

Observe that the above sufficient condition for the strictness of the regulation map Π_K has a dual version which can be written

$$\forall\, (x,p) \in \mathrm{Graph}(P), \quad \forall\, q \in N_K(x) \cap B(p)^{\star^{-1}} N_{P(x)}(p), \quad \langle q, c(x,p)\rangle \ \leq\ 0$$

which can itself be used to design other budgetary rules.

[4]That we shall deduce from Theorem 6.1.3 below.

4.1.3 The Viability Theorem for Dynamical Economies

We shall deduce from the general Viability Theorem 6.1.9 and Lemma 6.1.4 below the following (equivalent) characterization of the viability property:

Theorem 4.1.5 *Let us consider a Marchaud dynamical economy (P, c). Then a closed subset $K \subset \mathrm{Dom}(P)$ is viable[5] under the dynamical economy (P, c) if and only if the regulation map $\Pi_K(\cdot)$ is strict.*

Furthermore, any price $p(\cdot)$ regulating a viable solution $x(\cdot)$ in the sense that

$$\text{for almost all } t, \; x'(t) = c(x(t), p(t))$$

obeys the regulation law

$$\text{for almost all } t, \; p(t) \in \Pi_K(x(t)) \tag{4.1.4}$$

Remark — The regulation law (4.1.4) does not specify any regularity property of the function $t \mapsto p(t)$. One minimal regularity property that can be required is measurability with respect to time, the Filippov Measurable Selection Theorem[6] actually allows us to choose price functions obeying the regulation law (4.1.4) which are *measurable*.

We shall also provide in Chapter 7 weaker[7] conditions implying the existence of continuous prices, obtained however through a "smaller" regulation map. □

4.1.4 Using Viability Multipliers

We shall introduce a general method to design viable economies to which we shall devote Chapter 5. We start with a dynamical behavior described by a differential equation

$$x'(t) = g(x(t))$$

under which a sleek closed subset is not necessarily viable. The idea is to correct it by subtracting the normal cone to the constraints:

Proposition 4.1.6 *Let K be a sleek closed subset and $g : K \mapsto X$ be a continuous function. Then the set K is viable under the dynamical economy*

$$\begin{cases} i) & x'(t) = g(x(t)) - p(t) \\ ii) & p(t) \in N_K(x(t)) \cap \|g(x(t))\|B \end{cases} \tag{4.1.5}$$

[5]This means that for any initial commodity $x_0 \in K$, there exists a solution on $[0, \infty[$ to the dynamical economy (4.1.1) *viable in* K.

[6]See Theorem 8.2.10 of SET-VALUED ANALYSIS, [54, Aubin & Frankowska] for instance.

[7]which do not require that $c(x, p)$ is affine with respect to p nor that the pricing map P is Marchaud.

where the prices range over the dual $X^\star := X$ of the state space X (identified with its dual). This change function satisfies the instantaneous Walras law with $\lambda(x) = \|g(x)\|$:

$$\forall \|p\| \geq \|g(x)\|, \ \langle p, g(x) - p \rangle \ \leq \ 0$$

Proof — Since K is sleek, the set-valued map $x \rightsquigarrow N_K(x)$ has a closed graph. Hence the pricing map P defined by $P(x) := N_K(x) \cap \|g(x)\|B$ is Marchaud. We recall that projection onto the closed convex cone $T_K(x)$ satisfies

$$\mathrm{Proj}_{T_K(x)}(g(x)) \ = \ g(x) - \mathrm{Proj}_{N_K(x)}(g(x)) \ \in \ P(x)$$

since the normal cone is by definition the polar cone of the contingent cone. Therefore it belongs to the regulation map:

$$\forall\, x \in K, \ \mathrm{Proj}_{T_K(x)}(g(x)) \ \in \ \Pi_K(x)$$

This implies that K is a viability domain of the dynamical economy (4.1.5), so that from any initial state starts a solution to (4.1.5) which remains in K. □

Remark — When K is closed and convex, we deduce that the dynamical economy (4.1.5) can be written

$$\left\{ \begin{array}{ll} i) & \forall\, y \in K, \ \langle x'(t) - g(x(t)), x(t) - y \rangle \ \leq \ 0 \\ ii) & \|p(t)\| \ \leq \ \|g(x(t))\| \end{array} \right. \tag{4.1.6}$$

because to say that $p(t) := g(x(t)) - x'(t)$ belongs to $N_K(x(t))$ amounts to saying that

$$\forall\, y \in K, \ \langle p(t), y - x(t) \rangle \ \leq \ 0$$

Problems (4.1.6) are called **differential variational inequalities.**

By analogy with optimization under constraints $x \in K$ in optimization theory, we can interpret the elements of the dual space $X^\star$ as **viability multipliers** (instead of Lagrange or Kuhn-Tucker multipliers). We shall come back to this problem in Chapter 5. □

4.1.5 Equilibria of a Dynamical Economy

The general Equilibrium Theorem 6.1.15 and Lemma 6.1.4 below provide the following sufficient condition for the existence of an equilibrium:

Theorem 4.1.7 *Let us consider a Marchaud dynamical economy (P, c). If K is a convex and compact viability domain, then there exists a viable equilibrium $\bar{x} \in K$ of the dynamical economy.*

At this point, let us also derive from Ekeland's Theorem another sufficient conditions for a dynamical economy to possess an equilibrium without convexity assumptions. We mention in particular this other criterion:

Theorem 4.1.8 *Let (P,c) be a Marchaud dynamical economy and $V : X \mapsto \mathbf{R}_+ \cup \{+\infty\}$ a nonnegative nontrivial extended lower semicontinuous function. We denote by*

$$D_\uparrow V(x)(u) := \liminf_{h\to 0+,\, v\to u} \frac{V(x+hv)-V(x)}{h}$$

its contingent derivative (see section 9.3.2 of the Tool-Box). Assume that for $\rho > 0$,

$$\forall\, x \in \mathrm{Dom}(V),\ \inf_{q\in P(x)} (D_\uparrow V(x)(c(x,q)) + \rho\|c(x,q)\| \leq 0 \qquad (4.1.7)$$

Then there exists an equilibrium of the dynamical economy.

Proof — By Ekeland's Theorem 14.6.4 below, there exists $x_\varepsilon \in K$ such that

$$\forall\, v \in X,\ 0 \leq D_\uparrow V(x_\varepsilon)(v) + \varepsilon\|v\|$$

Since V satisfies (4.1.7), we deduce that there exists $p_\varepsilon \in P(x_\varepsilon)$ such that

$$D_\uparrow V(x_\varepsilon)(c(x_\varepsilon,p_\varepsilon)) + \rho\|c(x_\varepsilon,p_\varepsilon)\| \leq 0$$

We thus infer that for all $\varepsilon > 0$, there exist $p_\varepsilon \in P(x_\varepsilon)$ such that

$$\rho\|c(x_\varepsilon,p_\varepsilon)\| \leq \varepsilon\|c(x_\varepsilon,p_\varepsilon)\|$$

By taking $\varepsilon < \rho$, we deduce that $c(x_\varepsilon,p_\varepsilon) = 0$, so that the pair $(x_\varepsilon,p_\varepsilon)$ is an equilibrium of the dynamical economy. □

4.1.6 Sensitivity and Local Uniqueness of the Equilibrium

We associate with each parameter p the set

$$E(p) := \{x \mid p \in P(x) \ \&\ c(x,p) = 0\}$$

of the equilibria of $c(\cdot,p)$ viable in $P^{-1}(p)$. We say that $E : \mathcal{M} \rightsquigarrow X$ is the *equilibrium map* of the dynamical economy.

We measure the **sensitivity of equilibria** to price variations through the contingent derivative of the equilibrium map, which associates with any marginal price modification the set of marginal variations of the equilibrium. Furthermore, we provide conditions under which it is locally unique (in the sense that there exists a neighborhood of the equilibrium which does not contain any other equilibrium).

We begin by stating this result when the pricing map $P(\cdot)$ is constant:

Theorem 4.1.9 *Let us assume that the change function c of a dynamical economy is continuously differentiable and that the pricing map is constant on the set* K*:* $\forall\, x \in K,\ P(x) := P$*. Assume that both* K *and* P *are convex*[8].

1. *The contingent derivative* $DE(x,p)$ *of the equilibrium map is defined by:*

$$\begin{cases} u \in DE(p,x)(\pi) \Longleftrightarrow \\ u \in T_K(x), \quad \pi \in T_P(p) \ \& \ \dfrac{\partial c(x,p)}{\partial x} u + \dfrac{\partial c(x,p)}{\partial p} \pi \ = \ 0 \end{cases}$$

2. *Assume that*

$$T_K(x) \cap \ker\left(\frac{\partial c(x,p)}{\partial x}\right) \ = \ \{0\}$$

i.e., that $u = 0$ *whenever* $u \in T_K(x)$ *satisfies* $\dfrac{\partial c(x,p)}{\partial x} u \ = \ 0$*. Then there exists a neighborhood* $B(x,\eta)$ *in which* x *is the only equilibrium of* $E(p)$*.*

Then the foregoing statement is a particular case of

Theorem 4.1.10 *Let us assume that the change function c of a dynamical economy* (P,c) *is continuously differentiable and that* P *is closed and sleek.*

1. *The contingent derivative* $DE(x,p)$ *of the equilibrium map is characterized by:*

$$u \in DE(p,x)(\pi) \Longleftrightarrow \pi \in DP(x,p)(u) \ \& \ \frac{\partial c(x,p)}{\partial x} u + \frac{\partial c(x,p)}{\partial p} \pi \ = \ 0$$

2. *Assume that*

$$\ker(DP(x,p)) \cap \ker\left(\frac{\partial c(x,p)}{\partial x}\right) \ = \ \{0\}$$

i.e., that $u = 0$ *whenever* $u \in X$ *satisfies*

$$0 \ \in \ DP(x,p)(u) \ \& \ 0 \ = \ \frac{\partial c(x,p)}{\partial x} u$$

then $u = 0$*. Then there exists a neighborhood* $B(x,\eta)$ *in which* x *is the only equilibrium of* $E(p)$*.*

This result is a particular case of Theorem 8.1.7 of [28, Aubin].

---

[8] or sleek.

4.2 Qualitative Economics

Let us consider a Marchaud *dynamical economy* (P, c) defined by

— a "pricing" set-valued map $P : X \leadsto \mathcal{M}$

— a map $c : \mathrm{Graph}(P) \mapsto X$ describing the dynamics of the system.

The purpose of this paper is to study the "qualitative" evolution of the commodities and of prices governed by

$$\begin{cases} i) & \text{for almost all } t, \ x'(t) = c(x(t), p(t)) \\ ii) & p(t) \in P(x(t)) \end{cases} \tag{4.2.1}$$

subjected to the constraint

$$\forall t \geq 0, \ x(t) \in K := \mathrm{Dom}(P)$$

We shall assume from now on that the graph and the domain of the pricing map P are closed.

The **qualitative state** of a solution to the dynamical economy at a given time t is given by the knowledge of the monotonicity property of each component $x_i(t)$ of the commodity $x(\cdot)$, i.e., the knowledge of the sign of the derivatives $x_i'(t) = c_i(x(t), p(t))$. Hence the **qualitative behavior** is the evolution of the qualitative states of the solution, i.e., the evolution of the vector of signs of the components of $x'(t) = c(x(t), p(t))$. The problem is to determine theses signs **without solving the differential inclusion** underlying the dynamical economy.

In order to denote the qualitative states and track down their evolution, we introduce the set $\mathcal{R} := \{-, 0, +, \}$ made of the three signs $+$ and $-$ and 0 with which we associate the three cones

$$\mathbf{R}_- :=]-\infty, 0[, \ \mathbf{R}_0 := \{0\}, \ \& \ \mathbf{R}_+ :=]0, +\infty[$$

and the three closed cones

$$-\mathbf{R}_+ :=]-\infty, 0], \ 0\mathbf{R}_+ := \{0\}, \ \& \ +\mathbf{R}_+ := [0, +\infty[$$

We next introduce the n-dimensional **confluence space** $\mathcal{R}^n$ defined by

$$\mathcal{R}^n := \{ -, 0, + \}^n$$

the convex cones

$$\mathbf{R}^n_a := \{ v \in \mathbf{R}^n \mid \text{sign of } (v_i) = a_i \}$$

and their closures

$$a\mathbf{R}^n_+ := \{ v \in \mathbf{R}^n \mid \text{sign of } (v_i) = a_i \text{ or } 0 \}$$

We set $X := \mathbf{R}^n$ and we denote by $a \in \mathcal{R}^n$ a sign vector. We then associate with such a sign vector the **monotonic cell** $\mathcal{P}_a \subset \mathcal{P} := \text{Graph}(P)$ defined in the following way: $\mathcal{P}_a$ is the subset of commodity-price pairs $(x,p) \in \mathcal{P}$ such that

$$c(x,p) \in \mathbf{R}^n_a \tag{4.2.2}$$

and the **large monotonic cell** $\overline{\mathcal{P}}_a \subset \mathcal{P} := \text{Graph}(P)$ defined in the following way: $\overline{\mathcal{P}}_a$ is the subset of commodity-price pairs $(x,p) \in \mathcal{P}$ such that

$$c(x,p) \in a\mathbf{R}^n$$

We see for instance that $\mathcal{P}_0$ *is the graph of the equilibrium map.*

We shall denote by $P_a : X \rightsquigarrow \mathcal{M}$ and $\overline{P_a}$ the set-valued maps the graphs of which are equal to $\mathcal{P}_a$ and $\overline{\mathcal{P}_a}$ respectively and by K_a and $\overline{K}_a$ their domains (i.e., the projections onto X of $\mathcal{P}_a$ and $\overline{\mathcal{P}}_a$).

A first property of these monotonic cells is stated in the following theorem:

Theorem 4.2.1 *Assume that $a \in \{-,0,+\}^n$ is a vector sign. If the monotonic cell $\mathcal{P}_a$ is not empty and bounded, then, from any initial state $(x_0,p_0) \in \mathcal{P}_a$, either all solutions leave $\mathcal{P}_a$ in finite time or one of them is such that $x(t)$ converges to an equilibrium $\bar{x}$.*

Proof — Assume that a solution $x(\cdot)$ is such that $(x(t),p(t))$ remains in $\mathcal{P}_a$ for all nonnegative $t \geq 0$. Let any i such that $a_i \neq 0$. Since

$$x_i(t) - x_i(0) = \int_0^t x_i'(\tau)d\tau$$

we deduce that $x_i(t)$ is monotone and bounded because the sign of $x_i'(t) = c_i(x(t),p(t))$ is constant in the monotonic cell $\mathcal{P}_a$. Therefore, it converges to some number $\bar{x}_i$ when $t \to +\infty$.

Consequently, each component of the solution $x(\cdot)$ either is constant or converges, so that $x(t)$ converges to a limit, which is then an equilibrium of the dynamical economy. □

We shall provide sufficient conditions for such monotonic cells $\mathcal{P}_a$ to be nonempty. It may happen that for some $T \geq 0$,

$$(x(T),p(T)) \notin \mathcal{P}_a \;\&\; (x(T),p(T)) \in \mathcal{P}_b$$

i.e., that the evolution enters a new monotonic cell $\mathcal{P}_b$.

We shall study whether such situations happen at every commodity-price pair in the "intersection" of two monotonic cells: In this case, one can say that the sign vector b is a monotonic successor of the sign vector a.

These laws thus **reveal causality relations** between qualitative phenomena concealed in the dynamical system, by specifying the successors of each monotonic cells (see (6.4.1) and the links with chaos à la Saari). They present a major interest in economics and game theory for making some sense out of the maze of quantitative properties.

The question we answer now is whether these qualitative cells are non empty and/or singular and to compute their contingent cones.

Theorem 4.2.2 *We assume that P is closed and sleek and that c is continuously differentiable.*

1. *The contingent cone to the monotonic cell $\overline{\mathcal{P}}_a$ at (x,p) is contained in the set of elements (v,π) such that $\pi \in DP(x,p)(v)$ and such that*

$$\text{sign of } \left(\frac{\partial c(x,p)}{\partial x}v + \frac{\partial c(x,p)}{\partial p}\pi\right)_i = a_i \text{ or } 0$$

whenever $c_i(x,p) = 0$. The converse is true whenever the transversality condition:

$$\left\{\frac{\partial c(x,p)}{\partial x}v + \frac{\partial c(x,p)}{\partial p}\pi\right\}_{\pi \in DP(x,p)(v)} - a\mathbf{R}^n_+ = \mathbf{R}^n$$

holds true.

2. *Let $(\bar{x},\bar{p})$ be an equilibrium. We posit the transversality condition:*

$$\left\{\frac{\partial c(\bar{x},\bar{p})}{\partial x}v + \frac{\partial c(\bar{x},\bar{p})}{\partial p}\pi\right\}_{\pi \in DP(\bar{x},\bar{p})(v)} - a\mathbf{R}^n_+ = \mathbf{R}^n$$

Then the monotonic cell $\mathcal{P}_a$ is nonempty and $(\bar{x},\bar{p})$ belongs to its closure. In particular, if

$$\left\{\frac{\partial c(\bar{x},\bar{p})}{\partial x}v + \frac{\partial c(\bar{x},\bar{p})}{\partial p}\pi\right\}_{\pi \in DP(\bar{x},\bar{p})(v)} = \mathbf{R}^n$$

then the 3^n monotonic cells $\mathcal{P}_a$ are nonempty. (We have a qualitatively unstable situation *since every qualitative behavior can be implemented from an initial qualitative state in the neighborhood of this equilibrium.)*

Proof — The first part of the theorem follows directly from the formula of the contingent cone to the inverse image of a set because

$$\overline{\mathcal{P}}_a := \mathcal{P} \cap c^{-1}(a\mathbf{R}^n_+)$$

and from the characterization to the contingent cone to the cone $a\mathbf{R}^n_+$:

$$T_{\overline{\mathcal{P}}_a} = \left\{(v,p) \in T_{\mathcal{P}}(x,p) \mid \frac{\partial c(x,p)}{\partial x}v + \frac{\partial c(x,p)}{\partial p}\pi \in T_{a\mathbf{R}^n_+}(c(x,p))\right\}$$

For proving the second part, we apply the Constrained Inverse Function Theorem 14.6.1 of the Tool-Box to the map $(x,p,z) \mapsto c(x,p) - z$ from $X \times \mathcal{M} \times X$ to X restricted to the closed subset $\text{Graph}(P) \times a\mathbf{R}^n_+$ at the point $(\bar{x}, \bar{p}, 0)$. Its tangent cone is equal to the product $T_{\text{Graph}(P)}(\bar{x},\bar{p}) \times a\mathbf{R}^n_+$ since

$$T_{a\mathbf{R}^n_+}(0) \;=\; a\mathbf{R}^n_+$$

Therefore, we know that there exists $\varepsilon > 0$ such that, for all $w \in \varepsilon[-1,+1]^n$, there exist an element $(x,p) \in \text{Graph}(P)$ and an element $z \in a\mathbf{R}^n_+$ satisfying $c(x,p) - z = w$ and $\|x - \bar{x}\| + \|p - \bar{p}\| + \|z\| \leq l\|w\|$. Taking in particular $w_i = a_i\varepsilon$, we see that $c_i(x,p) = a_i\varepsilon + z_i$ and thus, that the sign of $c_i(x,p)$ is equal to a_i for all $i = 1, \ldots, n$. Hence (x,p) belongs to $\mathcal{P}_a$ and $\|x - \bar{x}\| + \|p - \bar{p}\| \leq l\varepsilon$. □

We first distinguish among the monotonic cells the one which are **monotonic equilibria**:

Definition 4.2.3 *A qualitative cell $a \in \mathcal{R}^n$ is said to be a monotonic equilibrium (or a qualitative equilibrium) if from any $(x,p) \in \overline{\mathcal{P}}_a$ starts at least one solution to the dynamical economy which remains in the monotonic cell $\overline{\mathcal{P}}_a$.*

The first statement of Theorem 4.2.2 provides a characterization of monotonic equilibria of a dynamical economy:

Corollary 4.2.4 *Assume that for any $(x,p) \in \overline{\mathcal{P}_a}$, the transversality condition:*

$$\left\{ \frac{\partial c(x,p)}{\partial x} v + \frac{\partial c(x,p)}{\partial p} \pi \right\}_{\pi \in DP(x,p)(v)} - a\mathbf{R}^n_+ \;=\; \mathbf{R}^n$$

holds true. If there exists a constant $\gamma > 0$ such that for every $(x,p) \in \overline{\mathcal{P}_a}$ satisfying $c_i(x,p) = 0$, there exists $\pi \in DP(x,p)(c(x,p)) \cap \gamma B$ such that

$$\text{sign of } \left(\frac{\partial c(x,p)}{\partial x} c(x,p) + \frac{\partial c(x,p)}{\partial p} \pi \right)_i \;=\; a_i \text{ or } 0$$

then the sign vector a is a monotonic equilibrium.

Proof — We consider the system of differential inclusions

$$\left\{ \begin{array}{lrcl} i) & x'(t) & = & c(x(t),p(t)) \\ ii) & p'(t) & \in & \gamma B \end{array} \right. \tag{4.2.3}$$

The first statement of Theorem 4.2.2 and the assumptions of this corollary imply that $\overline{\mathcal{P}}_a$ is a viability domain of this system of differential inclusions. Hence, from any $(x,p) \in \overline{\mathcal{P}}_a$ starts at least one solution to the dynamical economy which remains in the monotonic cell $\overline{\mathcal{P}}_a$. □

When $a \in \mathcal{R}^n$ is no longer a monotonic equilibrium, what are the successors, if any, of a given transition state a?

This question does not always receive an answer, because, for a solution $x(\cdot)$ starting from some initial state $x \in K_a$ may oscillate around the boundary of K_a in the sense that there exist two sequences $t_n > 0$ and $s_n > 0$ converging to $0+$ such that $x(t_n) \in \overline{K}_a$ and $x(s_n) \notin \overline{K}_a$ (see Proposition 6.3.7).

We can exclude this pathological phenomenon in the following instance:

Lemma 4.2.5 *Assume that the change function c is continuously differentiable and that P is sleek, lower semicontinuously differentiable and satisfies*

$$\exists\, \delta > 0 \quad \textit{such that } \forall\, (x,p) \in \mathrm{Graph}(P),$$

$$\frac{\partial c(x,p)}{\partial x} c(x,p) \in \delta B - \frac{\partial c(x,p)}{\partial p} DP(x,p)(c(x,p))$$

Assume also that the constant $\gamma := \sup_{(x,p)\in\mathcal{P}} \|DP(x,p)(c(x,p))\|$ *is finite.*

Let $(x,p) \in \overline{\mathcal{P}}_a$. *We assume that*

- *either* $\mathrm{sign}\,(c_i(x,p)) \;=\; a_i \neq 0$
- *or else,*

$$\mathrm{sign}\,\left(\frac{\partial c(x,p)}{\partial x} c(x,p) + \frac{\partial c(x,p)}{\partial p}\pi\right)_i \;=\; b_i$$

if $\mathrm{sign}\,(c_i(x,p)) \;=\; 0.$

Then there exists $T > 0$ *such that, for all* $t \in]0,T[$, *the solution* $(x(t),p(t))$ *remains in the cell* $\mathcal{P}_{\Phi(a)}$ *where*

$$\Phi(a)_i \;:=\; \begin{cases} a_i & \text{when } \mathrm{sign}\,(c_i(x,p)) \;=\; a_i \neq 0 \\ b_i & \text{when } \mathrm{sign}\,(c_i(x,p)) \;=\; 0 \end{cases}$$

Proof — Let us consider the constants $\delta > 0$ and

$$\gamma := \sup_{(x,p)\in\mathcal{P}} \|DP(x,p)(c(x,p))\|$$

which is finite by assumption. We then introduce the Marchaud system of three differential inclusions

$$x'(t) \;=\; v(t),\;\; p'(t) \;\in\; \gamma B \;\;\&\;\; v'(t) \;\in\; \delta B$$

subject to the viability constraints

$$\forall\, t \geq 0,\;\; (x(t),p(t)) \;\in\; \mathcal{P} \;\;\&\;\; v(t) \;=\; c(x(t),p(t))$$

We deduce from the Viability Theorem 6.1.9 that for every initial condition $p \in P(x)$ such that

$$\frac{\partial c(x,p)}{\partial x}c(x,p) \in \delta B - \frac{\partial c(x,p)}{\partial p}DP(x,p)(c(x,p))$$

there exists a solution $(x(\cdot)), p(\cdot)$ to the dynamical economy $x(0) = x$, $p(0) = p$, $v(0) = c(x,p)$ and for almost all $t \geq 0$, $x''(t)$ belongs to the set

$$\frac{\partial c(x(t),p(t))}{\partial x}c(x(t),p(t)) + \frac{\partial c(x(t),p(t))}{\partial p}DP(x(t),p(t))(c(x(t),p(t)))$$

Consider now $(x,p) \in \mathcal{P}_a$. If the sign of $c_i(x,p)$ is equal to $a_i \neq 0$, there exists $\eta_i > 0$ such that the sign of $x_i'(t)$ remains equal to a_i when $t \in [0, \eta_i[$.

If the sign of $c_i(x,p)$ is equal to 0, we know that the sign of the second derivative $x_i''(\tau)$ is equal to the sign of

$$\left(\frac{\partial c(x,p)}{\partial x}c(x,p) + \frac{\partial c(x,p)}{\partial p}\pi\right)_i$$

for some $\pi \in DP(x,p)(c(x,p))$ thanks to the continuity requirements. By assumption, this sign is equal to b_i which is assumed to be different from 0. Hence there exists $\eta_i > 0$ such that the sign of $x_i''(t)$ remains equal to b_i when $t \in]0, \eta_i[$, so that the sign of

$$x_i'(t) \;=\; \int_0^t x_i''(\tau)d\tau$$

remains equal to b_i on the interval $]0, \eta_i[$.

Hence we have proved that there exists some $\eta > 0$ such that $(x(t), p(t)) \in \mathcal{P}_b$ for $t \in]0, T[$ where $T := \min_i \eta_i > 0$. □

Let $(\bar{x}, \bar{p})$ be an equilibrium. We shall say that the qualitative cell $\overline{\mathcal{P}}_a$ is "singular" at $(\bar{x}, \bar{p})$ if $(\bar{x}, \bar{p})$ is locally the only point of the monotonic cell $\overline{\mathcal{P}}_a$, i.e., if there exists a neighborhood $N(\bar{x})$ of $\bar{x}$ such that:

$$\forall\, (x,p) \in N(\bar{x},\bar{p}),\;\; (x,p) \neq (\bar{x},\bar{p}), p \in P(x) \;\text{ implies that }\; p \notin \overline{\mathcal{P}}_a(x)$$

Theorem 4.2.6 *Let $(\bar{x}, \bar{p})$ be an equilibrium. We posit the following assumption:*

$$\begin{cases} \text{whenever } \pi \in DP(\bar{x},\bar{p})(u) \text{ satisfies} \\ \dfrac{\partial c(\bar{x},\bar{p})}{\partial x}u + \dfrac{\partial c(\bar{x},\bar{p})}{\partial p}\pi \;\in\; a\mathbf{R}^n_+ \\ \text{then } \; u = 0 \;\&\; \pi = 0 \end{cases} \tag{4.2.4}$$

Then the qualitative cell $\overline{\mathcal{P}}_a$ is singular at $(\bar{x}, \bar{p})$.

Proof — Assume the contrary: for all $n > 0$, there exists $(x_n, p_n) \in B((\bar{x}, \bar{p}), 1/n)$, $(x_n, p_n) \neq (\bar{x}, \bar{p})$ such that $c(x_n, p_n)$ does belong to $a\mathbf{R}^n_+$. Let us set $h_n := \|x_n - \bar{x}\| + \|p_n - \bar{p}\| > 0$, which converges to 0, $u_n := \frac{x_n - \bar{x}}{h_n}$ and $\pi_n := \frac{p_n - \bar{p}}{h_n}$. Since (u_n, π_n) belongs to the unit ball, which is compact, a subsequence (again denoted) (u_n, π_n) converges to some element (u, π) of the unit ball. Furthermore, π belongs to $DP(\bar{x}, \bar{p})(u)$.

Finally, since

$$c_n(x_n, p_n) = h_n \left(\frac{\partial c(\bar{x}, \bar{p})}{\partial x} u + \frac{\partial c(\bar{x}, \bar{p})}{\partial p} \pi + \varepsilon_n \right) \in a\mathbf{R}^n_+$$

for all $n > 0$, we infer that the limit $\left(\frac{\partial c(x, p)}{\partial x} u + \frac{\partial c(x, p)}{\partial p} \pi \right)$ belongs to $a\mathbf{R}^n_+$. Since (u, π) is different from 0 and since $\pi \in DP(\bar{x}, \bar{p})(u)$, we obtained a contradiction of the assumption. □

4.3 Slow Viable Solutions

Consider a dynamical economy (P, c) described by

$$\begin{cases} i) \quad x'(t) = c(x(t), p(t)) \\ ii) \quad p(t) \in P(x(t)) \end{cases} \tag{4.3.1}$$

We have associated with each viability domain K the **regulation map** $\Pi_K \subset P$ associating with every state $x \in K$ the set

$$\Pi_K(x) := \{p \in P(x) \mid c(x, p) \in T_K(x)\}$$

of **viable prices**. We did prove that under adequate assumptions on P and c that K is a viability domain if and only if the images $\Pi_K(x)$ of the regulation map are not empty for all $x \in K$. In this case, the price functions $p(\cdot)$ which regulate viable solutions obey the **regulation law**

$$\text{for almost all } t \geq 0, \;\; p(t) \in \Pi_K(x(t))$$

But how can one find prices $p(t) \in \Pi_K(x(t))$?

We then intoduce **planning mechanisms** which are single-valued **selections** $\varpi(\cdot)$ of the regulation map: $\varpi(x) \in \Pi_K(x)$ for all $x \in K$.

Then the solutions to the differential equation

$$x'(t) = c(x(t), \varpi(x(t)))$$

(when they exist) are viable since the implemented prices $p(t) := \varpi(x(t))$ obey the regulation law by construction. Existence is guaranteed when $\varpi(\cdot)$ is *continuous*, but may still holds true for discontinuous, but explicit, feedbacks.

Hence, we have to find **selection procedures** of the regulation map which provide either continuous selections or discontinuous selections for which the above differential equation still has solutions.

By assuming a myopic behavior of price-takers in some models, we can assume for instance that prices are chosen in the regulation map with minimal norms. By doing that, we obtain discontinuous planning mechanisms, because, unfortunately, the lower semicontinuity of the regulation map is not sufficient for implying the continuity of this minimal norm. But we shall overcome this difficulty by observing that such selections can be obtained from the subset $\Pi_K(x)$ by "slicing out" the piece we desire with a **selection procedure** $S_{\Pi_K}(x)$, where the map $S_{\Pi_K}(\cdot)$ has "good" properties that the selection does not have[9]. For building planning mechanisms, we can for instance think of explicitly selecting some prices of the regulation map, for instance, the price $\varpi^\circ(x) \in \Pi_K(x)$ with minimal norm. Viable solutions obtained with this planning mechanism are called **slow viable solutions**. But since the minimal selection can be obtained by selection procedures, we shall be able to derive the existence of such slow solutions. We may use as well other selection procedures involving optimization or game theoretical mechanisms, and thus provide many instances of evolution obtained under a myopic behavior of the price-takers, as we shall see later on.

4.3.1 Slow Viable Solutions of Dynamical Economies

Definition 4.3.1 *We define the* minimal selection $\varpi_K^\circ(\cdot)$ *of the regulation map* $\Pi_K(\cdot)$ *by*

$$\varpi_K^\circ(x) \in \Pi_K(x) \ \& \ \|\varpi_K^\circ(x)\| = \min_{y \in \Pi_K(x)} \|y\|$$

whenever the values of $\Pi_K(\cdot)$ *are closed and convex.*

The solutions to differential equation

$$x'(t) \;=\; c(x(t), \varpi_K^\circ(x(t))) \qquad (4.3.2)$$

are called slow *viable solutions to dynamical economy (4.3.1).*

Unfortunately, since there is no hope of having continuous regulation maps Π_K in general (as soon as we have inequalities constraints), this minimal selection is not continuous. However, the differential equation has viable solutions (4.3.2) under adequate assumptions.

Indeed we derive from Proposition 4.4.2 and Theorem 4.5.2 below the existence of slow viable solutions:

---

[9]The idea of introducing and using these selection procedures is due to Hélène Frankowska.

Theorem 4.3.2 *Consider a Marchaud dynamical economy (P, c). If the regulation map is lower semicontinuous with nonempty convex values, then the dynamical economy (4.3.1) has slow viable solutions.*

Propositions 11.2.10 and 11.2.11 of the Tool-Box provide criteria for guaranteeing the lower semicontinuity of the regulation map.

4.3.2 Regulation by Slow Prices

As the simplest example, we look for ways of coordinating the decisions of the consumers by subtracting prices to their dynamical priceless behavior

$$x'(t) = g(x(t)) - p(t)$$

in order that viability constraints of the form

$$\forall t \geq 0, \ \ h(x(t)) \in M$$

are always satisfied.

We define the **regulation map** Π_M by

$$\Pi_M(x) := \{p \mid h'(x)(g(x) - p) \in T_M(h(x))\}$$

The prices $p(t)$ regulating viable solutions are given by the **regulation law**[10]

$$p(t) \in \Pi_M(x(t))$$

We recall in the Tool-Box that when $B \in \mathcal{L}(X, Y)$ is surjective, its orthogonal right inverse is equal to

$$B^+ = B^\star(BB^\star)^{-1}$$

that we can supply Y with the final norm μ^B defined by $\mu^B(z) := \|B^+ z\|$ and that we denote by π_K^B the projector of best approximation onto the closed subset K for this final norm. In this case, Proposition 10.1.4 below implies that the unique solution $\bar{x}$ to the minimization problem

$$\inf_{Bx \in K+v} \|x - u\|$$

[10]In the case of the problem of allocations of scarce resources

$$\forall t \geq 0, \ \sum_{i=1}^{n} x_i(t) \in M$$

for instance, we saw that

$$\Pi_M(x) := T_M\left(\sum_{i=1}^{n} x_i\right) - \frac{1}{n}\sum_{i=1}^{n} g_i(x_i)$$

is equal to

$$\bar{x} = u - B^+(\mathbf{1} - \pi_K^B)(Bu - v)$$

We finally recall that when M is sleek (and in particular, convex), then its tangent cones $T_M(y)$ are convex (see Theorem 11.2.8 of the Tool-Box). In this case, since the polar cone to the contingent cone $T_M(y)$ is the normal cone $N_M(y)$, we also know that the solution to the minimization problem

$$\inf_{Bx \in T_M(y)+v} \|x - u\|$$

is equal to

$$\bar{x} = u - B^+(\mathbf{1} - \pi_{T_M(y)}^B)(Bu - v) = u - B^\star \pi_{N_M(y)}^{B^\star}(BB^\star)^{-1}(Bu - v)$$

where $\pi_{N_M(y)}^{B^\star}$ denotes the projector onto the normal cone $N_M(y)$ when the dual $Y^\star$ is supplied with the "dual final norm" $\|B^\star q\|$.

Slow solutions to this dynamical economy are given by

Proposition 4.3.3 *Let us assume that the map g is continuous and bounded and that the resource map h is continuously differentiable and satisfies the uniform surjectivity condition:*

$$\forall\, x \in K, \ \ h'(x) \text{ is surjective \& } \sup_{x \in K} \|h'(x)^+\| < +\infty \tag{4.3.3}$$

and that M is closed convex (or more generally, sleek). Then the slow solution of the dynamical economy

$$x'(t) = g(x(t)) - p(t) \tag{4.3.4}$$

subjected to the viability constraints

$$\forall\, t \geq 0, \ \ h(x(t)) \in M$$

is the solution to the differential equation

$$x'(t) = g(x(t)) - \varpi^\circ(x(t))$$

where

$$\left\{ \begin{array}{lll} \varpi^\circ(x) & = & h'(x)^+ \left(\mathbf{1} - \pi_{T_M(h(x))}^{h'(x)}\right) h'(x)g(x) \\ & & \\ & = & h'(x)^\star \pi_{N_M(h(x))}^{h'(x)^\star}(h'(x)h'(x)^\star)^{-1}h'(x)g(x) \end{array} \right. \tag{4.3.5}$$

Proof — Proposition 11.2.10 implies that the regulation map Π_M is lower semicontinuous because the set-valued map $T_M(\cdot)$ is lower semicontinuous and because the surjectivity assumption (4.3.3) implies assumption (11.2.1) of Proposition 11.2.10 below. Slow viable solutions do exist and are obtained by the price $\varpi^\circ(x)$ of minimal norm among the prices satisfying $h'(x)p \in T_M(h(x)) - h'(x)g(x)$. Proposition 10.1.4 implies formula (4.3.5). □

4.3.3 Regulation by Connection Matrices

Instead of decentralizing the dynamical economy by prices according to the dynamical economy

$$x'(t) = g(x(t)) - p(t)$$

in order to comply to viability constraints of the form

$$\forall t \geq 0, \; h(x(t)) \in M$$

we look for the problem of sustaining viability of the evolution by connecting each consumer with the other ones through a connection matrix $W := (w_i^j) \in \mathcal{L}(X,X)$:

$$x'(t) = W(t)g(x(t)) \tag{4.3.6}$$

Naturally, we obtain a fully decentralized economy when the connection matrix is the identity matrix. But this is not always the case, and consumers may take into account the behavior of the other consumers in their dynamics. Then the new regulation parameter is no longer a price $p \in X^\star$, but a connection matrix $W \in \mathcal{L}(X,X)$.

We shall then prove a striking equivalence result: the regulation by slow solutions is the same than the regulation by connecting matrices with maximal decentralization, obtained with connection matrices as close as possible to the identity matrix. In other words, the regulation price is an efficient decentralization mechanism.

We introduce the new regulation map

$$R_M(x) := \{W \in \mathcal{L}(X,X) \mid h'(x)Wg(x) \in T_M(h(x))\}$$

Recall that whenever $p \in X^\star$ and $y \in Y$, we denote by $p \otimes y \in \mathcal{L}(X,Y)$ the rank one linear operator defined by $x \mapsto (p \otimes y)(x) := \langle p, x\rangle y$, the matrix of which is

$$\left(p^i y_j\right)_{\substack{i=1,\ldots,m \\ j=1,\ldots,n}}$$

Here, both finite dimensional vector spaces X and Y are identified with their duals.

We then observe the following

Proposition 4.3.4 *Let $\Gamma : X \mapsto X$ be any continuous map such that*

$$\inf_{x \in K} \langle \Gamma(x), g(x)\rangle > 0$$

Then a connection matrix of the form

$$W(x) := \mathbf{1} - \frac{\Gamma(x)}{\langle \Gamma(x), g(x)\rangle} \otimes \varpi(x)$$

the entries of which are equal to

$$w_{ij}(x) = \delta_{i,j} - \frac{\Gamma(x)_i \varpi(x)_j}{\langle \Gamma(x), g(x)\rangle}$$

belongs to $R_M(x)$ if and only if $\varpi(x)$ belongs to $\Pi_M(x)$ and the viable solutions to the differential equations $x' = W(x)g(x)$ and $x' = g(x) - \varpi(x)$ do coincide.

Proof — Indeed, we observe that

$$W(x)g(x) = g(x) - \frac{\langle \Gamma(x), g(x)\rangle}{\langle \Gamma(x), g(x)\rangle}\varpi(x) = g(x) - \varpi(x)$$

so that the two differential equations $x' = W(x)g(x)$ and $x' = g(x) - \varpi(x)$ are the same. Furthermore, to say that $W(x) \in R_M(x)$ amounts to saying that

$$h'(x)W(x)g(x) = h'(x)g(x) - h'(x)\varpi(x) \in T_M(h(x))$$

so that the viability conditions are the same. □

In particular, the regulation by connection matrices of the form

$$W(x) := \mathbf{1} - \frac{g(x)}{\|g(x)\|^2} \otimes \varpi(x)$$

is equivalent to a price decentralization mechanism whenever $g(x) \neq 0$ for every $x \in K$.

The problem of maximal decentralization is then to find connection matrices as close as possible to the identity matrix $\mathbf{1}$. It happens that such connection matrices maximizing decentralization are of the above form.

Proposition 4.3.5 *Let us assume that the map g is continuous and bounded and that the resource map h is continuously differentiable and satisfies the uniform surjectivity condition:*

$$\forall x \in K, \ h'(x) \text{ is surjective } \& \ \sup_{x \in K} \frac{\|h'(x)^+\|}{\|g(x)\|} < +\infty \tag{4.3.7}$$

Let $M \subset Y$ be a closed convex (or more generally, sleek) subset. Then the scarcity set $K := h^{-1}(M)$ is viable under the connection economy (4.3.6) if and only if for every $x \in K$, the image $R_M(x)$ of the regulation map is not empty.

The solution with maximal decentralization subjected to the viability constraints

$$\forall t \geq 0, \ h(x(t)) \in M$$

is governed by the differential equation

$$x'(t) = W^\circ(x(t))g(x(t))$$

where

$$\left\{\begin{array}{lcl} W^\circ(x) & = & \mathbf{1} - \frac{g(x)}{\|g(x)\|^2} \otimes h'(x)^+ \left(\mathbf{1} - \Pi^{h'(x)}_{T_M(h(x))}\right) h'(x)g(x) \\ & = & \mathbf{1} - \frac{g(x)}{\|g(x)\|^2} \otimes h'(x)^\star \pi^{h'(x)^\star}_{N_M(h(x))}(h'(x)h'(x)^\star)^{-1}h'(x)g(x) \end{array}\right. \tag{4.3.8}$$

Furthermore, the slow viable solutions to dynamical economy (4.3.4) regulated by prices and the viable solutions to the connection economy (4.3.6) under maximal decentralization coincide.

Proof — We can apply the same proof as the one of Proposition 4.3.3 where the connection economy is a dynamical economy for the dynamical system $c(x, W) := Wg(x)$ and where the role of the price p is played by the connection matrix W. Then, using the definition 10.2.1 of tensor products of linear operators of the Tool-Box, the regulation map can be written in the form:

$$R_M(x) := \{W \in \mathcal{L}(X, X) \mid (g(x) \otimes h'(x))W \in T_M(h(x))\}$$

so that viable solutions are regulated by the regulation law

$$(g(x(t)) \otimes h'(x(t)))W(t) \in T_M(h(x))$$

Since the map $g(x) \otimes h'(Wx)$ is surjective from the space $\mathcal{L}(X, X)$ of connection matrices to the resource space Y, we observe that

$$\|(g(x) \otimes h'(x))^+\| = \frac{\|h'(x)^+\|}{\|g(x)\|}$$

so that the assumptions of Proposition 11.2.10 are satisfied thanks to (4.3.7). Hence the regulation map R_M is lower semicontinuous.

Therefore, the viable connection matrices closest to the identity matrix $\mathbf{1}$ are the solutions W° minimizing the distance $\|W - \mathbf{1}\|$ among the connection matrices satisfying the above constraints.

We infer from Theorem 10.3.6 that since the map $g(x) \otimes h'(Wx)$ is surjective from the space $\mathcal{L}(X, X)$ of connection matrices to the resource space Y, the solution is given by formula (4.3.8). □

Remark: Mixed Regulation by Prices and Connection Matrices Let us consider the mixed regulation mode using both connection matrices W and prices p:

$$x'(t) = Wg(x(t)) - p(t)$$

subjected to the same viability constraints

$$\forall t \geq 0, \ h(x(t)) \in M$$

We still supply X with the Euclidian norm, identify X with its dual. Then the regulation map takes the form

$$R_M(x) := \{(W,p) \in \mathcal{L}(X,X) \times X \mid h'(x)Wg(x) - h'(x)p \in T_M(x)\}$$

We supply the space $\mathcal{L}(X,X) \times X$ with the norm $\nu_{\alpha,\beta}$ defined by

$$\nu_{\alpha,\beta}(W,p)^2 := \alpha\|W\|^2 + \beta\|p\|^2$$

Then the slow solutions are regulated by the pair $(W_{\alpha,\beta}(x), \varpi_{\alpha,\beta}(x))$ in the regulation map which minimizes the distance $\nu_{\alpha,\beta}(W - \mathbf{1}, p)$ in $R_M(x)$. By Theorem 10.3.7 below, the solution is given by

$$\begin{cases} W_{\alpha,\beta}(x) = \mathbf{1} - \dfrac{\beta g(x)}{\alpha + \beta\|g(x)\|^2} \otimes h'(x)^+ \left(\mathbf{1} - \pi^{h'(x)}_{T_M(h(x))}\right) h'(x)g(x) \\ \varpi_{\alpha,\beta}(x) = \dfrac{\alpha}{\alpha + \beta\|g(x)\|^2} h'(x)^+ \left(\mathbf{1} - \pi^{h'(x)}_{T_M(h(x))}\right) h'(x)g(x) \end{cases}$$

We then observe that, setting $\varpi^\circ(x) := h'(x)^+ \left(\mathbf{1} - \Pi^{h'(x)}_{T_M(h(x))}\right) h'(x)g(x)$,

$$\begin{cases} W_{\alpha,\beta}(x)g(x) - \varpi_{\alpha,\beta}(x) \\ = g(x) - \dfrac{\beta\|g(x)\|^2}{\alpha + \beta\|g(x)\|^2}\varpi^\circ(x) - \dfrac{\alpha}{\alpha + \beta\|g(x)\|^2}\varpi^\circ(x) \\ = g(x) - \varpi^\circ(x) \end{cases}$$

Therefore, for every pair (α, β), the slow viable solutions, which are solutions to the differential equations

$$x'(t) = W_{\alpha,\beta}(x(t))g(x(t)) - \varpi_{\alpha,\beta}(x(t))$$

are the slow solutions to the initial dynamical economy, i.e., solutions to

$$x'(t) = g(x(t)) - \varpi^\circ(x(t)) \quad \square$$

4.4 Selection Theorems

Slow viable evolutions are not the only choice offered, since we can choose other selections of the regulation map. We can for instance minimize a given function on the subset $\Pi_K(x)$ or solve a game on such a subset. Actually, these selection procedures are instances of a general concept of selection procedures introduced by Hélène Frankowska.

4.4.1 Selection Procedures

A selection $S(F(\cdot))$ of a set-valued map F is a set-valued map contained in F. We introduce here the class of selections obtained through a selection procedure $S_F(\cdot)$, which allows us to obtain the desired selection of F by "cutting pieces" out of its images $F(x)$.

Definition 4.4.1 (Selection Procedure) *A* selection procedure *of a set-valued map $F : X \rightsquigarrow Y$ is a set-valued map $S_F : X \rightsquigarrow Y$ satisfying*

$$\begin{cases} i) & \forall x \in \mathrm{Dom}(F), \quad S(F(x)) := S_F(x) \cap F(x) \neq \emptyset \\ ii) & \textit{the graph of } S_F \textit{ is closed} \end{cases}$$

The set-valued map $S(F) : x \rightsquigarrow S(F(x))$ is called the selection *of F.*

Then the selection is a single-valued map denoted $s(F(\cdot))$ whenever

$$\forall x \in \mathrm{Dom}(F), \quad s(F(x)) \ := \ S_F(x) \cap F(x) \ \text{ is a singleton}$$

4.4.2 Minimal Selections

When the values of a set-valued map F are closed and convex, we shall take minimal selection defined by

$$F^\circ(x) \ := \{u \in F(x) \mid \|u\| = \min_{y \in F(x)} \|y\|\} \tag{4.4.1}$$

The upper semicontinuity of F, even when it is closed convex valued, is not strong enough to imply the continuity of the minimal selection[11].

However, the selection procedure S_F° defined by

$$S_F^\circ(x) \ := \ \{y \in Y \mid \|y\| \ \leq \ d(0, F(x))\} \tag{4.4.2}$$

is the minimal selection procedure because the graph of the minimal selection is equal to:

$$\mathrm{Graph}(F^\circ) = \mathrm{Graph}(F) \cap \mathrm{Graph}(S_F)$$

Proposition 4.4.2 *Let us assume that $F : X \rightsquigarrow Y$ is closed and lower semicontinuous with convex values. Then the graph of the minimal selection is closed*[12].

Proof — Since F is lower semicontinuous, the function $x \mapsto d(0, F(x))$ is upper semicontinuous, so that the graph of $S_F^\circ(\cdot)$ is closed. □

---

[11]Consider the set-valued map $F : \mathbf{R} \rightsquigarrow \mathbf{R}$ defined by

$$F(x) := \begin{cases} \{2\} & \text{if} \quad x \neq 0 \\ [1,2] & \text{if} \quad x = 0 \end{cases}$$

It is upper semicontinuous with compact convex values and its minimal selection is obviously not continuous.

[12]If moreover F is upper hemicontinuous, then the minimal selection is continuous. See Theorem 9.3.4 of SET-VALUED ANALYSIS, [54, Aubin & Frankowska].

4.4.3 Optimal Selections

We can easily provide other examples of selection procedures through optimization thanks to the Maximum Theorem.

Proposition 4.4.3 *Let us assume that a set-valued map $F : X \leadsto Y$ is lower semicontinuous with compact values. Let $V : \mathrm{Graph}(F) \mapsto \mathbf{R}$ be continuous. Then the set-valued map S_F defined by:*

$$S_F(x) := \left\{ y \in Y \mid V(x,y) \leq \inf_{z \in F(x)} V(x,z) \right\}$$

is a selection procedure *of F which yields the selection $S(F)$ equal to:*

$$S(F(x)) = \left\{ y \in F(x) \mid V(x,y) \leq \inf_{z \in F(x)} V(x,z)) \right\}$$

Proof — Since F is lower semicontinuous, the function

$$(x,y) \mapsto V(x,y) + \sup_{z \in F(x)} (-V(x,z))$$

is lower semicontinuous thanks to the Maximum Theorem. Our proposition follows from :

$$\mathrm{Graph}(S_F) = \\ \left\{ (x,y) \mid V(x,y) + \sup_{z \in F(x)} (-V(x,z)) \leq 0 \right\} \quad \square$$

Most selection procedures through game theoretical models or equilibria are instances of this general selection procedure based on Ky Fan's Inequality (Theorem 14.1.2).

Proposition 4.4.4 *Let us assume that a set-valued map $F : X \leadsto Y$ is lower semicontinuous with convex compact values. Let $\varphi : X \times Y \times Y \mapsto \mathbf{R}$ satisfy*

$$\begin{cases} i) & \varphi(x,y,z) \text{ is lower semicontinuous} \\ ii) & \forall (x,y) \in X \times Y, \;\; z \mapsto \varphi(x,y,z) \text{ is concave} \\ iii) & \forall (x,y) \in X \times Y, \;\; \varphi(x,y,y) \leq 0 \end{cases}$$

Then the map S_F associated with φ by the relation

$$S_F(x) := \left\{ y \in Y \mid \sup_{z \in F(x)} \varphi(x,y,z) \leq 0 \right\}$$

is a selection procedure of F yielding the selection map $x \mapsto S(F(x))$ defined by

$$S_F(x) := \left\{ y \in F(x) \mid \sup_{z \in F(x)} \varphi(x,y,z) \leq 0 \right\}$$

Proof — Ky Fan's inequality (see Theorem 14.1.2) states that the subsets $S_F(x)$ are not empty since the subsets $F(x)$ are convex and compact. The graph of S_F is closed thanks to the assumptions and the Maximum Theorem because it is equal to the lower section of a lower semicontinuous function:

$$\text{Graph}(S_F) = \left\{(x,y) \mid \sup_{z\in F(x)} \varphi(x,y,z) \leq 0\right\} \quad \square$$

Proposition 4.4.5 *Assume that $Y = Y_1 \times Y_2$, that a set-valued map $F : X \rightsquigarrow Y$ is lower semicontinuous with convex compact values and that $a : X \times Y_1 \times Y_2 \to \mathbf{R}$ satisfies*

$$\begin{cases} i) & a \text{ is continuous} \\ ii) & \forall\,(x,y_2) \in X \times Y_2,\ \ y_1 \mapsto a(x,y_1,y_2) \text{ is convex} \\ iii) & \forall\,(x,y_1) \in X \times Y_1,\ \ y_2 \mapsto a(x,y_1,y_2) \text{ is concave} \end{cases}$$

Then the set-valued map S_F associating to any $x \in X$ the subset

$$\begin{array}{l} S_F(x) \ := \ \{(y_1,y_2) \in Y_1 \times Y_2 \text{ such that} \\ \forall (z_1,z_2) \in F(x),\ \ a(x,y_1,z_2) \leq a(x,z_1,y_2)\} \end{array}$$

is a selection procedure of F (with convex values). The selection map $S(F(\cdot))$ associates with any $x \in X$ the subset

$$\begin{array}{l} S(F)(x) \ := \ \{(y_1,y_2) \in F(x) \text{ such that} \\ \forall (z_1,z_2) \in F(x),\ \ a(x,y_1,z_2) \leq a(x,y_1,y_2) \leq a(x,z_1,y_2)\} \end{array}$$

of saddle-points *of $a(x,\cdot,\cdot)$ in $F(x)$.*

Proof — We take

$$\varphi(x,(y_1,y_2),(y_1',y_2')) \ := a(x,y_1,y_2') - a(x,y_1',y_2)$$

and we apply the above theorem. □

We derive from the Equilibrium Theorem 6.1.15 selection procedures yielding equilibria in $F(x)$.

Proposition 4.4.6 *Let us assume that a set-valued map $F : X \rightsquigarrow Y$ has nonempty compact convex values. Let us consider an upper semicontinuous set-valued map E with nonempty compact convex values from* Graph(F) *to Y satisfying:*

$$\forall (x,y) \in \text{Graph}(F),\ \ E(x,y) \cap T_{F(x)}(y) \neq \emptyset$$

Then the set-valued map S_F defined by

$$S_F(x) \ := \ \{y \in Y \mid 0 \in E(x,y)\}$$

is selection procedure of F. The selection map $S(F)$ associates with any $x \in$ Dom(F) *the set*

$$S(F)(x) \ := \ \{y \in F(x) \mid 0 \in E(x,y)\}$$

of equilibria of $E(x,\cdot)$ in $F(x)$.

4.4.4 Michael's Selection Theorem

We shall now state the celebrated Michael's theorem stating that lower semicontinuous convex-valued maps do have **continuous selections**.

Theorem 4.4.7 (Michael's Theorem) *Let F be a lower semicontinuous set-valued map with closed convex values from a compact metric space X to a Banach space Y. It does have a continuous selection.*

In particular, for every $\overline{y} \in F(\overline{x})$ there exists a continuous selection f of F such that $f(\overline{x}) = \overline{y}$.

We refer for instance to section 9.1 of SET-VALUED ANALYSIS for the proof of this Theorem. □

4.4.5 Steiner's Selection Theorem

Let us consider a smooth manifold Ω of boundary $\Gamma = \partial\Omega$ for which Stokes' formula is true.

For a nonempty convex compact subset K of $\mathbf{R}^n$, we define its *Steiner point* (or *Krümmungsschwerpunkte*, also called sometimes the *curvature centroid*) $s(K)$ by:

$$s(K) \;=\; \frac{\text{meas}(\Gamma)}{\text{meas}(\Omega)} \int_\Gamma p\sigma(K,p)\omega(dp)$$

where $\sigma(K,p) := \sup_{v\in K}\langle p, v\rangle$ is the support function of K, ω is the measure on Γ proportional to the Lebesgue measure normalized by $\int_\Gamma \omega(dp) = 1$.

The main example is naturally the unit ball $\Omega = B^n$ of $\mathbf{R}^n$ whose boundary is the unit sphere $\Gamma = \Sigma^{n-1}$ in $\mathbf{R}^n$. Then the original Steiner point is given by the formula

$$s(K) \;=\; n\int_{\Sigma^{n-1}} p\sigma(K,p)\omega(dp)$$

When the dimension $n = 1$ and $\Omega := [a, b]$, we define the Steiner point of K by

$$s(K) \;:=\; \frac{1}{b-a}(a\sigma(K,+1) + b\sigma(K,-1))$$

If Ω is symmetric, i.e., $\Omega = -\Omega$, , it follows that $s(K) = -s(-K)$, since $\sigma(K,p) = \sigma(-K,-p)$. Therefore if K is also symmetric, then $s(K) = 0$.

Theorem 4.4.8 *For any bounded subset K of $\mathbf{R}^n$,*

$$s(K) \;\in\; \overline{\text{co}}(K)$$

Furthermore $s(\cdot)$ is Lipschitz:

$$\forall\, K,\; L \text{ compact and convex,}\;\; \|s(K) \;-\; s(L)\| \;\le\; \frac{\text{meas}(\Gamma)}{\text{meas}(\Omega)}\|\Gamma\|\boldsymbol{d}(K,L)$$

where

$$\begin{cases} i) & h^\flat(K,L) = \sup_{x\in K} d(x,L) = \sup_{x\in K}\inf_{y\in L} d(x,y) \\ ii) & d(K,L) := \max(h^\flat(K,L), h^\flat(L,K)) \end{cases} \qquad (4.4.3)$$

denote the Hausdorff semi-distance and distance between the compact subsets K and L.

An immediate application of the above construction is the existence of a continuous selection for a continuous set-valued map:

Theorem 4.4.9 *Consider a set-valued map F from a metric space to nonempty compact convex subsets of $\mathbf{R}^n$. Then $f(x) := s(F(x))$ is a selection of the set-valued map F, continuous whenever F is continuous and Lipschitz whenever F is Lipschitz.*

See the proof of Theorem 9.4.3 of [54, Aubin & Frankowska].

Proof — Indeed, the set-valued map F with closed compact values is continuous (respectively Lipschitz) if and only if for every p, its support function $x \mapsto \sigma(F(x),p)$ is continuous (respectively Lipschitz) because one can check that

$$\sup_{p\in B}(\sigma_K(p) - \sigma_L(p)) = h^\flat(K,L)$$

where B is the unit ball of $X^\star$ (see for instance [12, Aubin]). □

4.5 Planning in Dynamical Economies

The lower semicontinuity of the regulation map Π_K of a dynamical economy provides not only slow solutions, but other planning mechanisms thanks to the Michael Selection Theorem and the other selection procedures of the regulation map introduced in the preceding section.

4.5.1 Continuous Planning Mechanisms

Continuous single-valued selections ϖ_K of the regulation map Π_K are **viable planning mechanisms**, since the Viability Theorem states that the differential equation

$$x'(t) = c(x(t), \varpi_K(x(t)))$$

enjoys the viability property:

Proposition 4.5.1 *Consider a Marchaud dynamical economy (P,c). If its regulation map is lower semicontinuous with nonempty convex values, then the dynamical economy can regulate viable solutions in K by* **continuous planning mechanisms.**

This result is not useful in practice, since Michael's Selection Theorem does not provide constructive ways to find those continuous planning mechanisms. We could instead use the Steiner selection of the regulation map, but it does not enjoy convincing economic interpretation.

However, other selection procedures of the regulation map introduced in the preceding section provide a whole family planning mechanisms using optimization or static game theoretical scenarios for choosing prices in the regulation map.

4.5.2 Planning Mechanisms

Theorem 4.5.2 *Consider a Marchaud dynamical economy (P, c) and suppose that K is a viability domain. Let S_{Π_K} be a selection of the regulation map Π_K. Suppose that the values of S_{Π_K} are convex. Then, for any initial commodity $x_0 \in K$, there exist a viable solution starting at x_0 and a viable price to dynamical economy (4.3.1) which are regulated by the selection $S(\Pi_K)$ of the regulation map Π_K, in the sense that* for almost all $t \geq 0$,

$$p(t) \in S(\Pi_K)(x(t)) := \Pi_K(x(t)) \cap S_{\Pi_K}(x(t))$$

If the selection $s(\Pi_K)(\cdot)$ is single-valued, it provides a planning mechanism *regulating viable solutions of the dynamical economy (4.3.1).*

Proof — Since the convex selection procedure S_{Π_K} has a closed graph and convex values, we can replace the dynamical economy (4.3.1) by the dynamical economy

$$\begin{cases} i) & x'(t) = c(x(t), p(t)) \\ ii) & \text{for almost all } t, \ p(t) \in P(x(t)) \cap S_{\Pi_K}(x(t)) \end{cases} \tag{4.5.1}$$

which satisfies the assumptions of the Viability Theorem 4.1.5. It remains to check that K is still a viability domain for this "smaller" system. But by construction, we know that for all $x \in K$, there exists $p \in S(\Pi_K)(x)$, which belongs to the intersection $P(x) \cap S_{\Pi_K}(x)$ and which is such that $c(x, p)$ belongs to $T_K(x)$.

Hence the new dynamical economy (4.5.1) enjoys the viability property, so that, from all initial states $x_0 \in K$ starts one viable solution and a viable price to the dynamical economy (4.5.1) which, for almost all $t \geq 0$, are related by

$$\begin{cases} i) & p(t) \in P(x(t)) \cap S_{\Pi_K}(x(t)) \\ ii) & c(x(t), p(t)) \in T_K(x(t)) \end{cases}$$

Therefore, for almost all $t \geq 0$, $p(t)$ belongs to the intersection of $\Pi_K(x(t))$ and $S_{\Pi_K}(x(t))$, i.e., to the selection $S(\Pi_K)(x(t))$ of the regulation map Π_K. □

In order to provide other examples of selection procedures, and in particular, continuous selections or minimal selections, we have to assume (and thus, to check) that the regulation map Π_K is **lower semicontinuous.**

We can now multiply the possible corollaries, since we have given several instances of selection procedures of set-valued maps. We shall just mention some of the examples. We begin by selecting viable solutions through minimization procedures:

Proposition 4.5.3 *Consider a Marchaud dynamical economy* (P, c) *and suppose that the regulation map is lower semicontinuous with nonempty convex images. Let us consider a loss function*

$$V : (x,p) \in \text{Graph}(P) \mapsto V(x,p) \in \mathbf{R}$$

be continuous and convex with respect to p. *Then, for any initial commodity* $x_0 \in K$, *there exists a viable solution* $x(\cdot)$ *to the dynamical economy (4.3.1) regulated by an open loop price* $p(\cdot)$ *satisfying for almost all* $t \geq 0$,

$$p(t) \in \Pi_K(x(t)) \;\; \& \;\; V(x(t),p(t)) \;=\; \inf_{q \in \Pi_K(t)} V(x(t),q)$$

Proof — This is a consequence of Theorem 4.5.2 and Proposition 4.4.3. □

When the price space $\mathcal{M} := \mathcal{M}_1 \times \mathcal{M}_2$ is the product of two price spaces, viable prices can be required to be **saddle-points** of two-person games:

Proposition 4.5.4 *Consider a Marchaud dynamical economy* (P, c), *where* $P(x) = P_1(x) \times P_2(x)$ *is the product of two price sets and* $c(x,p) := g(x) + G_1(x)p_1 + G_2(x)p_2$. *Assume that the regulation map*

$$\Pi_K(x) \;:=\; \{(p_1,p_2) \in P_1(x) \times P_2(x) \mid G_1(x)p_1 + G_2(x)p_2 \in T_K(x) - g(x)\}$$

is lower semicontinuous with nonempty convex values. Let $a : X \times \mathcal{M}_1 \times \mathcal{M}_2 \to \mathbf{R}$ *satisfy*

$$\begin{cases} i) & a \text{ is continuous} \\ ii) & \forall (x,p_2) \in X \times Y_2, \;\; p_1 \mapsto a(x,p_1,p_2) \text{ is convex} \\ iii) & \forall (x,p_1) \in X \times Y_1, \;\; p_2 \mapsto a(x,p_1,p_2) \text{ is concave} \end{cases}$$

Then, for any initial commodity $x_0 \in K$, *there exist a viable solution* $x(\cdot)$ *and open loop prices* $p_1(\cdot)$ & $p_2(\cdot)$ *satisfying for almost all* $t \geq 0$,

$$\begin{cases} i) & x'(t) \;=\; c(x(t)) + G_1(x(t))p_1(t) + G_2(x(t))p_2(t) \\ ii) & p_1(t) \in P_1(x(t)) \;\;\&\;\; p_2(t) \in P_2(x(t)) \\ iii) & \forall (q_1,q_2) \in \Pi_K(x), \\ & a(x(t),p_1(t),q_2) \leq a(x(t),p_1(t),p_2(t)) \leq a(x(t),q_1,p_2(t))\} \end{cases}$$

Proof — The proof follows from Theorem 4.5.2 and Proposition 4.4.7. □

As a last example, we state:

Proposition 4.5.5 *Consider a Marchaud dynamical economy (P, c) and suppose that the images of the regulation map are nonempty, convex and compact. Let us consider an upper hemicontinuous set-valued map E with nonempty closed convex values from* Graph(P) *to $\mathcal{M}$ satisfying:*

$$\forall (x,p) \in \text{Graph}(\Pi_K), \quad E(x,p) \cap T_{\Pi_K(x)}(p) \neq \emptyset$$

Then, for any initial commodity $x_0 \in K$, there exists a viable solution $x(\cdot)$ to the dynamical economy (4.3.1) regulated by an open loop price $p(\cdot)$ satisfying for almost all $t \geq 0$,

$$p(t) \in \Pi_K(x(t)) \ \& \ 0 \in E(x(t), p(t))$$

4.6 Slow Evolution of Technological Matrices

Let (P, c) describe a dynamical economy

$$\begin{cases} i) \quad x'(t) = c(x(t), p(t)) \\ ii) \quad p(t) \in P(x(t)) \end{cases} \tag{4.6.1}$$

subjected to the scarcity constraints

$$\forall\, t \geq 0, \quad h(x(t)) \in M$$

where $h : X \mapsto Y$ represents the resource map and $M \subset Y$ the set of available resources.

When the necessary and sufficient condition $h'(x)c(x,p) \in T_M(h(x))$ is not met, and when one cannot regulate the dynamics by using connection matrices as we did in Proposition 4.3.4, we propose another way of regulating the economy by to "correcting" the original constraints by utilizing technological matrices for obtaining new constraints of the form

$$x \in X \ \& \ W \in \mathcal{L}(X,X) \text{ must satisfy } h(Wx) \in M$$

We shall return in Chapter 5 to the problem of using technological matrices in the viability constraints.

The original dynamical economy is obtained when the technological matrix is equal to the identity matrix $\mathbf{1}$.

We now set a bound on the velocity of the technological matrix, since technological advance is bounded and suffers from a certain inertia:

$$\|W'(t)\| \leq \varphi(x(t), W(t))$$

where B denotes the unit ball of the space $\mathcal{L}(X,X)$.

Therefore, one can regulate viable solutions using both prices and connection matrices through the system of differential inclusions which governs the initial dynamical economy together with the technological matrix which guides it:

$$\begin{cases} i) & x'(t) = c(x(t),p(t)) \\ ii) & p(t) \in P(x(t)) \\ iii) & \|W'(t)\| \leq \varphi(x(t),W(t)) \end{cases} \tag{4.6.2}$$

subject to the viability constraints

$$\forall\, t \geq 0,\ \ h(W(t)x(t)) \in M$$

We set

$$\mathcal{K} := \{(x,W) \in X \times \mathcal{L}(X,X) \ \text{ satisfying } \ h(Wx) \in M\}$$

and

$$\begin{cases} \Pi_M(x,W) := \\ \{(V,p) \in \varphi(x,W)B \times P(x) \mid h'(Wx)(Vx + Wc(x,p)) \in T_M(h(Wx))\} \end{cases}$$

Theorem 4.6.1 *Let us assume that the dynamical economy (P,c) is Marchaud, that φ is continuous with linear growth and that the resource map h is continuously differentiable and satisfies the transversality condition:*

$$\forall\, x \in K,\ \ h'(Wx) \text{ is surjective} \tag{4.6.3}$$

and that M is convex (or more generally, sleek).

If for any $(x,W) \in \mathcal{K}$, $\Pi_M(x,W) \neq \emptyset$, then from any initial $(x_0,W_0) \in \mathcal{K}$ starts one viable solution $(x(\cdot),W(\cdot),p(\cdot))$ to the system (4.6.2) satisfying the regulation law

$$\text{for almost all } t \geq 0,\ \ (W'(t),p(t)) \in \Pi_M(x(t),W(t)) \tag{4.6.4}$$

In particular, for any selection $r(x,W)$ of $h'(x)^{-1}(T_M(h(Wx))$ and any positive-definite map $\Gamma \in \mathcal{L}(X,X^\star)$, the solutions to the system of differential equations

$$\begin{cases} i) & x' = c(x,p) \\ ii) & W' = -\dfrac{\Gamma x}{\langle \Gamma x, x\rangle} \otimes (r(x,W) - Wc(x,p)) \end{cases} \tag{4.6.5}$$

are viable in $\mathcal{K}$.

Proof — The system (4.6.2) is actually a differential inclusion

$$(x'(t), W'(t)) \in \{c(x(t), p)\}_{p \in P(x)} \times \varphi(x(t), W(t))B$$

where B denotes the unit ball of the space $\mathcal{L}(X, X)$. We are looking for solutions to this differential inclusion viable in the set $\mathcal{K}$. We set $H(x, W) := h(Wx)$. Since it is the inverse image of a closed subset, we know that the transversality condition

$$\mathrm{Im}(H'(x, W)) - T_M(H(x, W)) = X$$

(which is satisfied thanks to assumption (4.6.3)) implies that

$$\{(v, V) \in X \times \mathcal{L}(X, X) \mid h'(Wx)(Vx + Wv) \in T_M(h(Wx))\}$$

is contained in $T_K(x)$. Therefore, assumption (7.4.2) implies that the tangential condition of the Viability Theorem is satisfied, and thus, that from any initial condition $(x_0, W_0) \in \mathcal{K}$, starts one viable solution $(x(\cdot), W(\cdot))$ to the above differential inclusion.

It satisfies: for almost all $t \geq 0$,

$$h'(W(t)x(t))(W'(t)x(t) + W(t)c(x(t), p(t))) \in T_M(h(W(t)x(t)))$$

which is a regulation law specifying how prices and matrices evolve in terms of commodities. In particular, the dynamics of (4.6.5) satisfy

$$h'(Wx)(W'x + Wc(x, p)) := h'(Wx)r(x, W) \in T_M(h(Wx))$$

so that from any initial state $(x_0, W_0) \in \mathcal{K}$ starts a viable solution of (4.6.5).

Theorem 4.6.2 *Let us assume that the map g is continuous and bounded is linear and that the resource map h is continuously differentiable and satisfies the transversality condition:*

$$\forall (x, W) \in \mathcal{K}, \ \ h'(Wx) \text{ is surjective \& } \sup_{x \in K} \frac{\|h'(Wx)^+\|}{\|x\|} < +\infty \qquad (4.6.6)$$

and that M is convex (or more generally, sleek).

From any initial $(x_0, W_0) \in \mathcal{K}$ starts one viable solution $(x(\cdot), p(\cdot), W(\cdot))$ to the system (4.6.2) minimizing at each instant the velocity of the technological matrix, which is governed by the system of differential equations

$$\begin{cases} i) \ \ x' = c(x, p) \\ ii) \ \ W' = -\dfrac{x}{\|x\|^2} \otimes h'(Wx)^+ \left(1 - \pi^{h'(Wx)}_{T_M(h(Wx))}\right) h'(Wx)Wc(x, p) \\ \qquad\ \ = -\dfrac{x}{\|x\|^2} \otimes h'(x)^\star \pi^{h'(x)^\star}_{N_M(h(Wx))}(h'(Wx)h'(Wx)^\star)^{-1}Wc(x, p) \end{cases}$$

In particular, when $M := \{y_0\}$, this system boils down to

$$\begin{cases} i) & x' = c(x,p) \\ ii) & W' = -\frac{x}{\|x\|^2} \otimes h'(Wx)^+ h'(Wx)W(c(x,p)) \end{cases}$$

where $h'(Wx)^+h'(Wx)W(c(x,p))$ is the projection of $h'(Wx)Wc(x,p)$ onto the kernel of $h'(Wx)$.

Proof — We introduce the map R_M defined on $\mathrm{Graph}(P) \times \mathcal{L}(X,X)$ by

$$\begin{cases} R_M(x,p,W) \\ := \{V \in \varphi(x,W)B \mid h'(Wx)Vx \in T_M(h(Wx)) - h'(Wx)Wc(x,p)\} \end{cases}$$

Proposition 11.2.10 implies that the regulation map R_M is lower semicontinuous because the set-valued map $T_M(\cdot)$ is lower semicontinuous and the uniform surjectivity assumption (4.6.6) implies assumption (11.2.1).

Since the regulation law can be written in the form of the differential inclusion

$$W'(t) \in R_M(x(t),p(t),W(t))$$

we infer that the evolution of viable solutions are solutions to the system of differential inclusions

$$\begin{cases} i) & x'(t) = c(x(t),p(t)) \\ ii) & W'(t) \in R_M(x(t),p(t),W(t)) \end{cases} \tag{4.6.7}$$

Let

$$S^\circ_{R_M}(x,p,W) := \{V \in \mathcal{L}(X,X) \mid \|V\| \leq d(0,R_M(x,p,W))\}$$

the minimal selection procedure of this lower semicontinuous set-valued map, which is Marchaud. Then the solutions system

$$\begin{cases} i) & x'(t) = c(x(t),p(t)) \\ ii) & W'(t) \in S^\circ_{R_M}(x(t),p(t),W(t)) \end{cases}$$

viable in $\mathcal{K}$ are the solutions to with minimal norm because, for almost all $t \geq 0$,

$$W'(t) \in R_M(x(t),p(t),W(t)) \cap S^\circ_{R_M}(x(t),p(t),W(t))$$

Hence they are minimizing at each instant the dynamical complexity index because $W'(t)$ is the minimal velocity in this set. Setting $x := x(t)$ and $W := W(t)$, then $V := W'(t)$ is the matrix V solution to the quadratic minimization problem:

$$\inf_{h'(Wx)Vx \in T_M(h(Wx)) - h'(Wx)W(c(x,p))} \|V\|^2$$

We infer from Theorem 10.3.6 that since the map $x \otimes h'(Wx)$ is surjective from the space $\mathcal{L}(X, X)$ of technological matrices to the resource space Y, the solution is given by

$$V = -\frac{x}{\|x\|^2} \otimes h'(Wx)^+ \left(\mathbf{1} - \pi^{h'(Wx)}_{T_M(h(Wx))}\right) h'(Wx)W(c(x,p))$$

Then the solutions to the system

$$\begin{cases} i) \quad x'(t) = c(x(t)), p(t)) \\ ii) \quad W'(t) \in \dfrac{x(t)}{\|x(t)\|^2} \otimes \\ \quad h'(W(t)x(t))^+ \left(T_M(h(W(t)x(t))) - h'(W(t)x(t))W(t)(c(x(t), p(t)))\right) \end{cases}$$

minimize at each instant the dynamical complexity index.

When M is a singleton, the evolution of the technological matrix is governed by the differential equation

$$W'(t) = -\frac{x(t)}{\|x(t)\|^2} \otimes h'(W(t)x(t))^+ \left(h'(W(t)x(t))W(t)(c(x(t), p(t)))\right) \quad \square$$

Chapter 5

Designing Dynamical Economies

Introduction

The question arises how to design dynamical economies under which a given constrained set is viable. We shall motivate this method by introducing frustrations, then describe the general method and proceed with three classes of more and more complicated examples which can be omitted in a first reading.

1. Let us begin with the problem of dynamic allocation of scarce resources. In Chapter 3, we presented dynamical economies which are nondeterministic in the extent where their behavior depends only upon prices. In other words, their wishes are fulfilled whenever the evolution of the price is governed by the regulation law.

   But we could add a possibility for making mistakes, either in the behavior of the price or in the computations that a consumer makes when she determines a change function $c(x,p)$ when the state is x and when the price is p. Therefore, we introduce a parameter u, which we shall interpret as the **frustration**, which is allowed to evolve in a set F of frustration parameters. Instead of wanting to meet the wish $x'(t) = c(x(t),p(t))$ of the consumer, we shall enlarge the range of possible evolutions of the consumers by allowing evolutions of the form

   $$x'(t) \in c(x(t),p(t)) + u(t) \quad \text{where} \quad u(t) \in F$$

   In other words, the frustration parameter $u(t)$ represents the difference between the actual (algebraic) increase of consumption and her desired one. We thus add a new "dimension" to the nondeterministic nature of the dynamics of the economy, which now depends both upon the price and the frustration parameters (which play the role of regulees). In the framework of discrete time, the desired consumption is often (and soundly) identified with an *ex ante* wish and the actual consumption as an *ex post* state. In the continuous time version, the *ex ante* / *ex post* dichotomy can be replaced by the *a priori* / *a posteriori* dichotomy, so that the concept of frustration which encapsulated the two ideas remains valid.

But it happens that the introduction of prices and frustrations could have also been obtained through a general method that we shall exploit in this chapter.

2. In my opinion, when one has to design a mathematical metaphor for an evolutionary model of economic behavior, one should start by gathering the constraints of the problem which cannot — or should not — be violated.

   This requires first to delineate the endogenous states of the system under study and to discriminate them from the rest of the variables, regarded as exogenous, constituting in some sense the "environment" of the system under investigation. This partition between variables, which dictates the level of abstraction of a particular investigation, is the first source of constraints that the endogenous variable must obey.

   Usually, there are few disputes among "modelers" when they are listing these constraints. Serious disagreements may begin when behavioral assumptions have to be made.

   In order to weaken such controversies, or to prorogue the ultimate choice of a behavioral description of the economic agents, the strategy I suggest is to begin by characterizing the tangent cones to the sets of constraints, as we did with the simple example of allocation of scarce resources. Indeed, we know that any dynamic behavior of the economic agents has to prescribe velocities which are tangent to the set of constraints.

   This is a never-ending task, which becomes very early entangled. We shall perform this task here for a class of constraints. We abide with the convex case only for the sake of simplicity, the nonconvex case being reviewed in the Tool-Box (one needs however to replace convexity assumptions by sleekness assumptions in order to perform the computation of contingent cones).

   Once the computation of the tangent cone is done — as well as their dual version, the normal cones —, we can devise general strategies for designing dynamical behaviors of the economic agents under which the constraints are viable.

   Indeed, the tangent cone being computed, we obtain a kind of "envelope" in which one can choose the change functions consistent with the constraints. Now, the problem of choice of a behavior of the consumers is well circumscribed: one can propose, describe or suggest such and such shape of a change function and check whether or not a representative of this class belongs to this "envelope". Or one can propose a choice criterion and choose among viable dynamical economies of this "envelope" the ones which satisfy optimally this criterion (this question has been investigated in the framework of planning mechanisms). Or we can use the dual formulation involving the normal cones to the constrained set. As in optimization theory, where the normal cones to the set constrained furnish the spaces to which the Lagrange multipliers belong, the normal cone provides here the set in which to find prices as well as other regulatory controls, called regulees. This provides one way (but not the unique one) allowing **to design change functions of dynamical economies for which the constrained set is viable.** This strategy is familiar to economists, since it amounts to use for prices the very same Lagrange multipliers that are used in optimization under constraints for relaxing the constraints.

Therefore, the difficulty is confined to the elaboration of the list of constraints that in a given model the state of the system must obey and to the computation of the tangent and normal cones of the related constrained set, so that we can use the strategy outlined above. It is not foreign to the scheme followed in the static framework when optimization under constraints is exploited.

For that purpose, we shall derive from the characterization of tangent cones to a class of constraints the form of the regulation law. We shall in particular single out the case in which a given allocation set K described as an intersection of constrained sets K_j can be regulated by price systems for which each component p_j regulates the viability of the constrained set K_j. In this case, which is exceptional, one can say that the regulation map is "decoupled".

3. We then illustrate briefly the implementation of this strategy by integrating sociological effects through (fuzzy) coalitions and by adding producers to the consumers who had to share resources.

   Competitiveness means in our context that consumers can trade at any instant and with anyone. We shall attempt to translate noncompetitiveness by introducing coalitions $A \subset N := \{1, \ldots, n\}$ of the n consumers and by relaxing the allocation requirement by asking only that any instant t, there exist a coalition $A(t) \in \mathcal{P}(N)$ and an allocation such that

$$\sum_{i \in A(t)} x_i(t) \in M(A(t))$$

where $M(A) \subset Y$ denotes the set of commodities available to the members of the coalition A. We shall also derive in this case a regulation law ruling the evolution of coalitions in terms of the evolution of commodities and prices, and thus, "activity intervals" of each consumer i, which is made of the instants t for which $i \in A(t)$.

Actually, coalitions range over a discrete set, where it is impossible to define velocities, and thus, to envision a dynamical model governing the evolution of coalitions, adding a sociological touch to an otherwise purely economic behavior. One way to overcome this difficulty is to allow **fuzzy coalitions** of consumers to form. Fuzzy coalitions are described by "rates of participations" of consumers between 0 and 1, and thus, by vectors of $\mathbf{R}^n$. Therefore, we are formally able to define velocities of fuzzy coalitions and thus, to study their evolution through differential equations or inclusions.

In this context, the viability multiplier involves both prices and "coordination regulees", which allow the dynamical economy to remain viable. Whenever one can define and localize an institution allocating these "coordination regulees", one can suggest an account of the evolution of social consensus, of "herd behavior" of "mimetic" consumers, or of Panurgean consumer.

4. We continue this chapter with a program for future investigations of dynamical economies involving consumers and producers which can be omitted in a first reading. We face challenges analogous to the static case when too many constraints reveal Lagrange multipliers the structure of which becomes too intricate to be tracked down.

5. We finally briefly investigate the intricate case when the regulees are technological matrices involved in a one-stage or a multi-stage production process, when the constraints are of the form

$$W_{m-1} \cdots W_j \ldots W_0 x \ \in \ M$$

This describes a production process associating with the commodity x the intermediate input $x_1 := W_0 x$, which itself requires an input $x_2 := W_1 x_1$, and so on.

Instead of imposing a given sequence of technological matrices and studying the evolution of the commodities under dynamical systems of the form

$$x_j'(t) \ = \ g_j(x_j(t)) - p_j(t)$$

we want to derive also the evolution of technological matrices $W_j(t)$ in such a way that the viability constraint

$$\forall\, t \geq 0, \ \ W_{m-1}(t) \cdots W_j(t) \ldots W_0(t) x(t) \ \in \ M \ \subset \ Y$$

is satisfied at each instant. Using viability multipliers, one can prove that dynamical economies of the form

$$\begin{cases} x_0'(t) &= g_0(x_0(t)) - W_0^\star(t)p_1(t) \\ x_j'(t) &= g_j(x_j(t)) + p_j(t) - W_j^\star(t)p_{j+1}(t) \ \ (j = 1, \ldots, m-1) \\ x_m'(t) &= g_m(x_m(t)) + p_m(t) \end{cases}$$

govern viable solutions provided that the evolution of technological matrices $W_j(t)$ obeys dynamics involving the tensor product of $x_j(t)$ and the "intermediate" price $p^{j+1}(t)$.

The purpose of this chapter is then to illustrate that, knowing the constraints, one can design classes of dynamical economies susceptible to make these constraints viable.

5.1 Frustrations

5.1.1 Frustrated Consumers

We go back to the problem of dynamical allocation of scarce resources investigated in the preceding chapter. We shall now relax the assumption (3.2.4)iv) of Theorem 3.2.4 which requires that for any $x \in L_i$ and for every price p, the velocity of the consumption is tangent to the consumption set L_i (or $L_i(x_i)$ when the consumption sets L_i are replaced by preference preorders described by reflexive and transitive set-valued maps). But it may happen that a given consumer may not be happy with some prices in the sense that the velocity which is associated with it does not bring back the consumption in her consumption set. We therefore regard the distance between the velocity $c_i(x,p)$ and the tangent cone $T_{L_i}(x)$ as a kind of "minimal frustration".

Therefore, in order to maintain the viability of the economy without assuming property (3.2.4)iv), we have to add around the initial velocity $c_i(x,p)$ other velocities $u_i \in F_i$, which we shall regard as **frustrations**.

Hence the dynamic of the economy is governed by

$$x_i'(t) = c_i(x_i(t), p(t)) - u_i(t) \ \text{ where } \ u_i(t) \in F_i \ \ \& \ \ p(t) \in P(x(t)) \quad (5.1.1)$$

The former economy is the particular case when $F_i = \{0\}$.

We introduce now the regulation maps

$$\begin{cases} \Phi_i(x,p) & := F_i \cap (c_i(x,p) - T_{L_i}(x)) \\ \Pi_M(x,u) & := \{p \in P(x) \mid \sum_{i=1}^n c_i(x,p) \in T_M\left(\sum_{i=1}^n x_i\right) + \sum_{i=1}^n u_i\} \end{cases}$$

We observe that whenever the sum $\sum_{i=1}^n u_i$ of the frustrations is equal to 0, then

$$\Pi_M(x,u) \;=\; \Pi_M(x)$$

Theorem 5.1.1 *We posit assumptions (3.2.2), (3.2.5) and*

$$\left\{\begin{array}{ll} i) & c_i(x,p) \;:=\; g_i(x) - G_i(x)p \;\text{ where} \\ ii) & g_i : L_i \mapsto Y \;\text{ is continuous with linear growth} \\ iii) & G_i : L_i \mapsto \mathcal{L}(Y^\star, Y) \;\text{ is continuous and bounded} \\ iv) & F_i \;\text{ is convex, compact and contains } 0 \end{array}\right. \tag{5.1.2}$$

Then the two following conditions are equivalent:

$$\left\{\begin{array}{l} a) \;\; \forall\, x \in K, \;\; \exists\,(u,p) \;\; \textit{such that} \\ \\ \qquad p \in \Pi_M(x,u) \;\;\&\;\; \forall\, i = 1,\ldots n, \;\; u_i \in \Phi_i(x_i,p) \\ \\ b) \;\; \forall\, x_0 \in K, \;\text{starts one allocation evolving according (5.1.1)} \end{array}\right.$$

In this case, the viable allocations are governed by the regulation law*: for almost all* $t \geq 0$,

$$p(t) \in \;\Pi_M(x(t),u(t))\;\&\; u_i(t) \;\in\; \Phi_i(x_i(t),p(t)) \tag{5.1.3}$$

If the change functions satisfy the instantaneous Walras law, then the budget constraints become

$$\forall\, i = 1,\ldots n, \;\;\text{for almost all } t \geq 0, \;\; \langle p(t), x_i'(t)\rangle \;\leq\; \langle p(t), u_i(t)\rangle \tag{5.1.4}$$

Under these conditions, there exists at least a viable equilibrium, a solution $(\bar{x}_1,\ldots,\bar{x}_n,\bar{u}_1,\ldots,\bar{u}_n,\bar{p})$ *to*

$$\left\{\begin{array}{ll} i) & \forall\, i = 1,\ldots,n, \;\; 0 \;=\; c_i(\bar{x}_i,\bar{p}) + \bar{u}_i \\ ii) & \forall\, i = 1,\ldots,n, \;\; \bar{x}_i \;\in\; L_i, \;\; \bar{u}_i \;\in\; F_i \\ iii) & \bar{p} \in P(\bar{x}) \;\;\&\; \sum_{i=1}^n \bar{x}_i \;\in\; M \end{array}\right.$$

This theorem states that the viability of an economy allowing frustrated consumers amounts to solving at each state a $(n+1)$ person game whose players are the n consumers and the "Market", where the strategies are the frustrations u_i and the prices p and where the decision rules are the set-valued maps $p \rightsquigarrow \Phi_i(x_i,p)$ and $u \rightsquigarrow \Pi_M(x,u)$.

Indeed, Theorem 5.1.1 states that the set of allocations is viable under the frustrated dynamical economy if and only if for every allocation $x \in K$, there exists a fixed point $(u_1,\ldots,u_n,p)$ to the set-valued map

$$(u_1,\ldots,u_n,p) \;\rightsquigarrow\; \prod_{i=1}^n \Phi_i(x_i,p) \times \Pi_M(x,u)$$

Consequently, every adequate fixed point theorem provides not only a sufficient condition for the existence of such a fixed point, but also a feedback mechanism for associating with every allocation $x \in K$ a solution of the above $n+1$-person game through the regulation law (5.1.3). This is the case of the Kakutani Fixed Point Theorem 14.1.2 stated in the Tool-Box:

Corollary 5.1.2 *We posit assumptions (3.2.2), (3.2.5) and (5.1.2) of Theorem 5.1.1. If for every $x \in K$,*

$$\forall\, u \in F,\ \Pi_M(x,u) \neq \emptyset\ \ \&\ \forall\, i = 1,\ldots,n,\ \forall\, p \in P(x),\ \ \Phi_i(x_i,p) \neq \emptyset$$

then

$$\forall\, x_0 \in K, \text{ starts one allocation evolving according (5.1.1)}$$

and there exists an equilibrium.

Inequality (5.1.4) provides a budgetary interpretation whenever we assume that the change function obeys the instantaneous individual Walras law

$$\forall\, x \in L_i,\ \ \forall\, p,\ \ \langle p, c_i(x,p)\rangle\ \leq\ 0$$

In this case, we obtain

$$\langle p(t), x_i'(t)\rangle\ \leq\ \langle p(t), u_i(t)\rangle$$

One can regard the value of the frustration as a quantity of monetary units which compensates the consumer for modifying her current allocation according to a market price $p(t)$ which does not bring her consumption in her consumption set L_i.

Naturally, we can adapt to this context of frustrated economies Theorem 3.2.7 when the consumption sets L_i are replaced by preference preorders described by reflexive and transitive set-valued maps $x_i \rightsquigarrow L_i(x_i)$.

Actually, one can allow the frustration sets F_i to depend upon x_i. More generally, frustrated economies are particular cases of economies where change functions are replaced by set-valued change maps, which provide examples of **dynamical games** (see Chapter 9).

5.1.2 Minimal Frustration

The relaxation of assumption (3.2.4)iv) of Theorem 3.2.4 brings us to the question of finding evolutions under minimal frustration. In mathematical terms, we wish to provide assumptions under which the frustrations are minimal in the sense that they realize the minimum distance between the velocity $c_i(x_i,p)$ and the tangent cone $T_{L_i}(x_i)$.

Since $\Phi_i(x,p) := F_i \cap (c_i(x,p) - T_{L_i}(x))$ is closed and convex, there exists a unique frustration of minimal norm denoted by $u_i^0(x,p) \in \Phi_i(x_i,p)$.

Theorem 5.1.3 *We posit assumptions (3.2.2), (3.2.5) and (5.1.2) of Theorem 5.1.2.*

Assume furthermore that there exists a constant $\rho > 0$ *such that*

$$\forall x \in L_i, \ \forall p \in P(x), \ c_i(x,p) + \rho B \subset F_i + T_{L_i}(x) \tag{5.1.5}$$

We also assume that for every $p \in P(x)$, *the subsets* $\Phi_i(x_i, p)$ *are not empty. Let us we denote by* $u_i^0(x,p) \in \Phi_i(x_i,p)$ *the frustration of minimal norm and set*

$$\Pi_M^0(x) := \left\{ p \in P(x) \mid \sum_{i=1}^n c_i(x,p) \in T_M\left(\sum_{i=1}^n x_i\right) + \sum_{i=1}^n u_i^0(x,p) \right\}$$

If for any $x \in K$, *there exists a price* $p \in \Pi_M^0(x)$, *then from every initial allocation* $x_0 \in K$ *starts one viable allocation with minimal frustration evolving according to the system*

$$\begin{cases} \forall i = 1, \ldots n, \ \ x_i'(t) \ = \ c_i(x_i(t), p(t)) - u_i^0(x(t), p(t)) \\ \text{for almost all } t \geq 0, \ \ p(t) \ \in \ \Pi_M^0(x(t)) \end{cases}$$

Proof — Assumption (5.1.5) and Proposition 11.2.10 of the Tool-Box imply that the set-valued maps Φ_i are lower semicontinuous because the consumption sets L_i being assumed convex, and thus sleek, the set-valued map $T_{L_i}(\cdot)$ is lower semicontinuous.

The Maximum Theorem 11.2.5 implies that the function

$$x \mapsto d(c_i(x,p), T_{L_i}(x))$$

is upper semicontinuous, so that the set-valued map $x \rightsquigarrow S_{\Phi_i}(x)$ defined by

$$S_{\Phi_i}(x) \ := \ \{u \in Y \mid \|u\| \leq d(c_i(x,p), T_{L_i}(x))\}$$

has a closed graph and convex values.

We now set

$$C_i(x_i, p) \ := \ c_i(x_i, p) - F_i \cap S_{\Phi_i}(x_i)$$

These set-valued change maps are Marchaud maps, and by construction, the projection $u_i^0(x_i, p)$ of 0 onto $\Phi_i(x_i, p)$ is such that the velocity $c_i(x_i, p) - u_i^0(x_i, p)$ belongs to the subset $C_i(x_i, p)$ and is tangent to $T_{L_i}(x_i)$.

Therefore, we can apply Theorem 5.1.3 to deduce the existence of a viable allocation satisfying

$$\forall i = 1, \ldots n, \begin{cases} x_i'(t) \ = \ c_i(x_i(t), p(t)) - u_i(t) \\ u_i(t) \ \in \ F_i \ \& \ \|u_i(t)\| \ \leq \ \|u_i^0(x_i(t), p(t))\| \end{cases}$$

But since the solution is viable, we also know that for almost all $t \geq 0$,

$$c_i(x_i(t), p(t)) - u_i(t) \in T_{L_i}(x_i(t))$$

In other words, $u_i(t)$ belongs to $\Phi_i(x_i(t), p(t))$ and its norm is minimal. Therefore, $u_i(t) = u_i^0(x_i(t), p(t))$. □

Remark — We could have introduced the frustrations for mathematical reasons only, as "viability multipliers", as we shall do systematically in this chapter. Indeed, the set of allocations can be regarded as the inverse image of $\prod_{i=1}^{n} L_i \times M$ under the map

$$x \mapsto (x_1, \ldots, x_n, \sum_{i=1}^{n} x_i)$$

from Y^n to Y^{n+1}. Its transpose is the map from $Y^{\star^{(n+1)}}$ to $Y^{\star^n}$ defined by

$$(u_1, \ldots, u_n, p) \mapsto (p + u_1, \ldots, p + u_n)$$

In other words, prices and frustrations can be regarded as viability multipliers, the prices being associated with the collective scarcity constraints, the frustrations with the individual consumption constraints. □

5.2 Designing Economies through Viability Multipliers

5.2.1 Constraints Defined as Inverse Images

We can construct dynamical economies viable on constrained sets $K := h^{-1}(M)$ which are inverse images by continuously differentiable maps $h : X \mapsto Y$ by regulating them with prices belonging to the dual of the space Y. This is a situation analogous to the case of optimization under constrained sets $K := h^{-1}(M)$, where the Lagrange multipliers, regarded as prices, also belong to the dual $Y^\star$.

We posit the transversality condition

$$\forall\, x \in K, \quad \mathrm{Im}(h'(x)) - T_M(h(x)) := Y \tag{5.2.1}$$

It is satisfied whenever $h'(x)$ is surjective.

By transposition, this implies that

$$\ker(h'(x)^\star) \cap N_M(h(x)) = \{0\}$$

Therefore, we can associate with $h : X \mapsto Y$ and the closed convex set (or, more generally, sleek set) $M \subset Y$ satisfying the transversality condition (5.2.1) the positive constant

$$\gamma_M^h(x) := \inf_{q \in N_M(h(x)),\, q \neq 0} \frac{\|h'(x)^\star q\|}{\|q\|} > 0$$

since the unit sphere of a finite dimensional vector space is compact.

We recall that we associated with the pricing map the constant

$$\beta_P(x) := \inf_{p \in P(x)} \|p\|$$

Proposition 5.2.1 *Let us assume that $K := h^{-1}(M)$ is the inverse image by a continuously differentiable map $h : X \mapsto Y$ of a closed convex (or sleek) subset $M \subset Y$ satisfying the transversality condition (5.2.1). Assume that the pricing map $P : X \rightsquigarrow Y^\star$ is a Marchaud map satisfying*

$$\forall\, x \in K,\ \ \forall\, p \in N_M(h(x)),\ p \neq 0,\ \ \exists\, \mu > 0 \ \ \text{such that } \mu p \in P(x) \quad (5.2.2)$$

Consider a change map defined by $c(x,q) := g(x) - G(x)q \in X$ where $g : X \mapsto X$ is continuous with linear growth and where $G : X \mapsto \mathcal{L}(X^\star, X)$ is bounded and continuous and where the matrices $G(x)$ are positive-definite:

$$\alpha(G(x)) := \inf_{v \neq 0} \frac{\langle G(x)v, v\rangle}{\|v\|^2} > 0$$

If the maps g and G, the pricing map P and the constraints are linked by the consistency relation

$$\forall\, x \in K,\ \ \gamma_M^h(x)\beta_P(x) \geq \frac{\|g(x)\|}{\alpha(G(x))} \quad (5.2.3)$$

then the dynamical economy

$$\begin{cases} x'(t) = g(x(t)) - G(x(t))h'(x(t))^\star p(t) \\ p(t) \in P(x(t)) \end{cases}$$

is viable on K and the regulation map is defined by

$$\Pi_K(x) = (h'(x)G(x)h'(x)^\star)^{-1}\,(h'(x)g(x) - T_M(h(x)))$$

Proof — Since M is sleek, h is continuously differentiable and since the transversality assumption (5.2.1) holds true, then $N_K(x) = h'(x)^\star N_M(h(x))$ thanks to the calculus of normal cones to convex (and more generally, sleek) subsets. We have to prove that for every $x \in K$

$$\sup_{r \in N_K(x)} \inf_{p \in P(x)} \langle r, c(x, h'(x)^\star p)\rangle = \sup_{q \in N_M(h(x))} \inf_{p \in P(x)} \langle h'(x)^\star q, c(x, h'(x)^\star p)\rangle \leq 0$$

because, by Proposition 4.1.3, this implies that $\Pi_K(x)$ is not empty.

We know by assumption that for every $q \in N_M(h(x))$, there exists $\mu > 0$ such that $p := \mu q \in P(x)$. Therefore,

$$\|h'(x)^\star \mu q\| \;=\; \frac{\|h'(x)^\star q\|}{\|q\|}\|\mu q\| \;\geq\; \gamma_M^h(x)\,\beta_P(x)$$

and thus

$$\begin{cases} \langle h'(x)^\star q, g(x) - G(x)h'(x)^\star p\rangle \;=\; \langle h'(x)^\star q, g(x)\rangle - \mu\langle h'(x)^\star q, G(x)h'(x)^\star q\rangle \\ \leq\; \|h'(x)^\star q\|(\|g(x)\| - \gamma_M^h(x)\,\beta_P(x)\alpha(G(x))) \;\leq\; 0 \end{cases}$$

Hence K is a viability domain of the dynamical economy which is therefore viable. Furthermore, $p \in P(x)$ belongs to $\Pi_K(x)$ if and only if

$$h'(x)G(x)h'(x)^\star p \;\in\; h'(x)g(x) - T_M(h(x)) \;\square$$

Example Condition (5.2.2) means that the normal cone $N_M(h(x))$ to M at $h(x)$ is contained in the bipolar cone $P(x)^{--}$, which is the closed cone spanned by the compact convex subset $P(x)$.

When $0 \notin P(x)$, then $P(x)$ is called the "sole" of $P(x)^{--}$. One can prove that a cone has a compact sole if and only the interior of its polar cone is not empty.

Therefore, when the interior of the contingent cone $T_M(h(x))$ of a closed sleek subset M at $h(x)$ has a non empty interior and if

$$x \mapsto v_0(x) \;\in\; -\mathrm{Int}(T_M)(h(x))$$

the sole

$$P(x) \;:=\; \{p \in N_M(h(x)) \mid \langle p, v_0(x)\rangle \;=\; \|v_0(x)\|\beta_P(x)\}$$

is a compact convex subset spanning the normal cone $N_M(h(x))$ and satisfying

$$\forall\, p \in P(x),\;\; \beta_P(x) \;\leq\; \|p\| \;\leq\; \frac{\beta_P(x)\|v_0(x)\|}{\rho(x)}$$

where $\rho(x) \;=\; \inf_{y \notin T_M(h(x))} \|v_0(x) + y\|$ is the radius of the largest ball centered at $v_0(x)$ contained in $T_M(h(x))$. When both $x \mapsto v_0(x)$ and $x \mapsto \beta_P(x)$ are continuous, the set-valued map $x \rightsquigarrow P(x)$ is upper semicontinuous

Let us also mention that the Slater condition

$$\forall\, x \in K,\;\; \exists\, u_0(x) \mid h'(x)u_0(x) \;=\; v_0(x) \;\in\; -\mathrm{Int}(T_M(h(x)))$$

implies the transversality condition (5.2.1).

Finally, let us observe that if there exists a cone $S \subset Y$ such that

$$M \;=\; M + S$$

then $T_M(y) \supset S$ and $N_M(y) \subset S^-$. □

Remark — By the Lop-Sided Minimax Theorem 14.1.4 below, since $P(x)$ is compact and convex, we know that there exist $\bar{p} \in P(x)$ such that

$$\begin{cases} \sup_{q\in N_M(h(x))} \langle h'(x)^\star q, g(x) - G(x)h'(x)^\star \bar{p}\rangle \\ = \sup_{q\in N_M(h(x))} \inf_{p\in P(x)} \langle h'(x)^\star q, g(x) - G(x)h'(x)^\star p\rangle \\ = \inf_{p\in P(x)} \sup_{q\in N_M(h(x))} \langle h'(x)^\star q, g(x) - G(x)h'(x)^\star p\rangle \ \leq \ 0 \end{cases}$$

Therefore, any element achieving the supremum over the normals $q \in N_M(h(x))$ is achieved at $\bar{q} \in N_M(h(x))$ such that

$$\langle h'(x)^\star q, g(x) - G(x)h'(x)^\star \bar{p}\rangle \ = \ 0 \ \square$$

5.2.2 Anatomizing Constraints

Consider now the more explicit case when the dynamics and the viability constraints of a dynamical economy are anatomized in the following way:

$$\begin{cases} i) & X := \prod_{i=1}^n X_i \\ ii) & Y := \prod_{j=1}^m Y_j, \quad M := \prod_{j=1}^m M_j \\ iii) & \mathcal{M} := \prod_{j=1}^m Y_j^\star, \quad P(x) := \prod_{j=1}^m P_j(x) \end{cases} \tag{5.2.4}$$

and when

$$\forall x \in X, \ \ h(x) \ := \ (h_1(x), \ldots, h_m(x))$$

so that K is the intersection of the subsets K_j defined by:

$$K_j \ := \ \{x \in X \ | \ h_j(x) \in M_j\} \tag{5.2.5}$$

When the resource maps h_j are continuously differentiable, we set

$$h_j^{i'}(x) \ := \ \frac{\partial}{\partial x_i} h_j(x)$$

We assume that the pricing map $P(x) := \prod_{j=1}^m P_j(x)$ and that the change function can be decomposed in the form

$$c_i(x,p) \ := \ g_i(x) - G_i(x)p$$

where $G_i(x) \in \mathcal{L}(X_i^\star, X_i)$ are positive definite.

The constants $\gamma_M^h(x)$ and $\beta_P(x)$ are equal to

$$\gamma_M^{h^2}(x) \ := \inf_{q_j \in N_{M_j}(h_j(x))} \frac{\sum_{i=1}^n \left\|\sum_{j=1}^m h_j^{i'}(x) q_j\right\|^2}{\sum_{j=1}^m \|q_j\|^2}$$

and to

$$\beta_P^2(x) \ := \inf_{p_k \in P_k(x)} \sum_{k=1}^n \|p_k\|^2$$

respectively. We set

$$D_j^k(x) \;=\; \sum_{i=1}^{n} h_j^{i'}(x)G_i(x)h_k^{i'}(x)^\star \;\in\; \mathcal{L}(X_k^\star, Y_j)$$

and $b(x) := h'(x)g(x)$ the vector of components

$$b_j(x) \;=\; \sum_{i=1}^{n} h_j^{i'}(x)g_i(x)$$

In a game-theoretical framework, where the indexes j are interpreted as labels of players, the regulees $p := (p_j)_{j=1,...m}$ are split in the form $p := (p_j, p_{\hat{j}})$ where $p_{\hat{j}} := (p_k)_{k\neq j}$ denotes the components of the regulee p "controlled" by the other players. Hence, from the view point of a given player j, she has to choose her own regulee p_j in response to the state x of the system and the choice of the regulees $p_{\hat{j}}$ of the other players. We thus introduce the "discriminating regulation maps" Δ_j defined by

$$\left\{\begin{array}{lll} \Delta_j(x, p_{\hat{j}}) & := & \{p_j \in P_j(x) \text{ such that} \\ D_j^j p_j & \in & b_j(x) - \sum_{k\neq j} D_j^k(x)p_k - T_{M_j}(h_j(x))\} \end{array}\right.$$

Proposition 5.2.2 *We assume that*

$$\left\{\begin{array}{ll} i) & \forall k, \; P_k \text{ is Marchaud} \\ ii) & \forall i, \; g_i : \mathrm{Dom}(P) \mapsto X_i \text{ is continuous and bounded} \\ iii) & \forall i, \; G_i : \mathrm{Dom}(P) \mapsto \mathcal{L}(X_i^\star, X_i) \text{ is continuous, bounded and} \\ & \text{positive definite} \end{array}\right.$$

and that

$$\left\{\begin{array}{ll} i) & \text{the subsets } M_j \text{ are closed and sleek} \\ ii) & \text{the maps } h_j \text{ are continuously differentiable} \\ iii) & \forall\, x \in K, \; \forall q_j \in Y_j \; (j = 1, \ldots, n), \; \exists p_i \in X_i \text{ such that} \\ & q_j \in \sum_{i=1}^{n} h_j^{i'}(x)p_i + T_{M_j}(h_j(x)) \;\; (j = 1, \ldots m) \end{array}\right. \quad (5.2.6)$$

Then under consistency assumption

$$\forall\, x \in K, \;\; \gamma_M^{h^2}(x)\beta_P^2(x) \;\geq\; \max_{i=1,\ldots,n} \frac{\|g_i(x)\|^2}{\alpha(G_i(x))^2} \quad (5.2.7)$$

from any initial state $x_0 \in K$ starts one solution $(x_i(\cdot))_{i=1,...n}$ on $[0, \infty[$ to the dynamical economy

$$\left\{\begin{array}{lll} \forall i = 1, \ldots, m, \;\; x_i'(t) & = & g_i(x(t)) - \sum_{k=1}^{m} G_i(x(t))h_k^{i'}(x(t))^\star p_k(t) \\ & & \\ \forall\, j = 1, \ldots, m, \;\; p_j(t) & \in & P_j(x_1(t), \ldots, x_n(t)) \end{array}\right.$$

viable *in K and regulated by prices obeying the* discriminating regulation laws

$$\text{for almost all } t, \;\; \forall j = 1, \ldots, m, \;\; p_j(t) \in \Delta_j(x(t), p_{\hat{j}}(t)) \quad (5.2.8)$$

Proof — Transversality assumptions (5.2.6)iii) imply that

$$\begin{cases} i) & T_K(x) = \bigcap_{j=1}^m T_{K_j}(x), \\ ii) & T_{K_j}(x) = \{v \in X \mid \sum_{i=1}^n h_j^{i'}(x_i)v_i \in T_{M_j}(h_j(x))\} \end{cases}$$

We thus infer formulas (5.2.8) because $p \in \Pi_K(x)$ if and only if

$$\forall j = 1, \dots m, \ \ p_j \in \Delta_j(x, p_{\hat{j}})$$

In other words, the choice of viable regulees is obtained through decision rules $\Delta_j(x, \cdot)$ in a noncooperative way. □

Selection mechanisms using Nash equilibria can then be used for selecting such prices (see Proposition 4.4.4 and Theorem 4.5.2).

We shall say that the regulation map is *decoupled* if

$$\forall j \neq k, \ \ D_j^k(x) = 0$$

In this case, the discriminating maps Δ_j are independent of the choice of the other regulees:

$$\Delta_j(x, p_{\hat{j}}) = \Pi_{K_j}(x) := \{p_j \in P_j(x) \mid D_j^j(x)p_j \in b_j(x) - T_{M_j}(h_j(x))\}$$

so that each "partial" viability domain K_j is regulated by the ith component of the regulee in the sense that

$$\text{for almost all } t, \ \ \forall j = 1, \dots, m, \ \ p_j(t) \in \Pi_{K_j}(x(t))$$

We note that

$$\Delta_j(x, p_{\hat{j}}) = \Pi_{K_j}(x) - \sum_{k \neq j} D_j^j(x)^{-1} D_j^k(x) p_k$$

5.2.3 Case of Linear Constraints

This is in particular the case when we suppose that $K := H^{-1}(M)$ is the inverse image by linear operators $H \in \mathcal{L}(X, Y)$ where $H = (h_j^i)$ is written as a "block matrix":

$$\forall x \in X, \ \ Hx := (h_1 x, \dots, h_m x) \text{ where } h_j x := \sum_{i=1}^n h_j^i x_i$$

so that K is the intersection of the subsets K_j defined by

$$K_j := \left\{ x \in X \mid \sum_{i=1}^n h_j^i x_i \in M_j \right\}$$

We posit the **constraint qualification assumption**

$$0 \in \operatorname{Int}(\operatorname{Im}(H) - M) \tag{5.2.9}$$

which can be proved to be equivalent[1] to the **transversality condition**

$$\forall\, x \in K, \quad \mathrm{Im}(H) - T_M(Hx) \ := \ Y \tag{5.2.10}$$

We set

$$D_j^k(x) \ = \ \sum_{i=1}^n h_j^i G_i(x) h_k^{i\star} \ \in \ \mathcal{L}(X_k^\star, Y_j)$$

and the vector $b(x) :=:= hg(x)$ of components

$$b_j(x) \ = \ \sum_{i=1}^n h_j^i g_i(x)$$

and we introduce the discriminating regulation maps defined by

$$\begin{cases} \Delta_j(x, p_{\hat{j}}) \ := \ \{p_j \in P_j(x) \ \text{such that} \\ D_j^j p_j \ \in \ b_j(x) - \sum_{k \neq j} D_j^k(x) p_k - T_{M_j}(\sum_{i=1}^n h_j^i x_i)\} \end{cases}$$

Then viable evolutions are governed by the regulation laws

$$\text{for almost all } t, \ \forall j = 1, \ldots, m, \ \ p_j(t) \in \Delta_j(x(t), p_{\hat{j}}(t)) \ \ \square$$

5.2.4 Case of Multilinear Constraints

Assume now that the resource maps $h_j : \prod_{i=1}^n X_i \mapsto Y_j$ are multilinear. We set

$$\forall\, i = 1, \ldots n, \ \ \hat{i} \ := \ \{j = 1, \ldots, n, \mid j \neq i\}, \ \ X_{\hat{i}} \ := \ \prod_{j \neq i} X_j$$

so that any $x := (x_1, \ldots, x_n) \in \prod_{i=1}^n X_i$ can be written

$$x := (x_{\hat{i}}, x_i) \in X_{\hat{i}} \times X_i$$

We denote by $h_j^i(x_{\hat{i}}) \in \mathcal{L}(X_i, Y_j)$ the linear operator defined by

$$\forall\, u_i \in X_i, \ \ h_j^i(x_{\hat{i}})(u_i) \ := \ h_j(x_{\hat{i}}, u_i)$$

which naturally coincides with $h_j^{i'}(x)$. We then deduce the following

Corollary 5.2.3 *We assume that*

$$\begin{cases} i) & \forall k, \ \ P_k \ \textit{is Marchaud} \\ ii) & \forall i, \ \ g_i : \mathrm{Dom}(P) \mapsto X_i \ \textit{is continuous and bounded} \\ iii) & \forall i, \ \ G_i : \mathrm{Dom}(P) \mapsto \mathcal{L}(X_i^\star, X_i) \ \textit{is continuous and bounded} \end{cases}$$

[1]It is obvious that (5.2.9) implies (5.2.10), whereas the converse follows from the Constrained Inverse Function Theorem 14.6.1 of the Tool-Box.

and that

$$\left\{\begin{array}{ll} i) & \text{the subsets } M_j \text{ are closed and sleek} \\ ii) & \text{the maps } h_j \text{ are multilinear} \\ iii) & \forall\, x \in K,\ \forall q_j \in Y_j\ (j = 1, \ldots, n),\ \ \exists p_i \in X_i \ \text{ such that} \\ & q_j \in \sum_{i=1}^{n} h_j^i(x_{\hat{i}}) p_j + T_{M_j}(h_j(x))\ \ (j = 1, \ldots m) \end{array}\right. \tag{5.2.11}$$

Then under consistency assumption

$$\forall\, x \in K,\ \ \gamma_M^{h^2}(x)\beta_P^2(x) \ \geq \max_{i=1,\ldots,n} \frac{\|g_i(x)\|^2}{\alpha(G_i(x))^2} \tag{5.2.12}$$

from any initial state $x_0 \in K$ *starts one solution* $(x_i(\cdot))_{i=1,\ldots n}$ *on* $[0, \infty[$ *to the dynamical economy*

$$\left\{\begin{array}{lll} \forall i = 1, \ldots, m, \ \ x_i'(t) & = & g_i(x(t)) - \sum_{k=1}^{m} G_i(x(t)) h_k^i(x_{\hat{i}}(t))^\star p_k(t) \\ \\ \forall\, j = 1, \ldots, m, \ \ p_j(t) & \in & P_j(x_1(t), \ldots, x_n(t)) \end{array}\right.$$

viable *in* K *and regulated by prices obeying the* regulation laws

$$\text{for almost all } t,\ \ \forall j = 1, \ldots, m,\ \ p_j(t) \in \Delta_j(x(t), p_{\hat{j}}(t))$$

where the discriminating regulation maps Δ_j *are defined by*

$$\left\{\begin{array}{lll} \Delta_j(x, p_{\hat{j}}) & := & \{p_j \in P_j(x) \ \text{ such that} \\ D_j^j p_j & \in & b_j(x) - \sum_{k \neq j} D_j^k(x) p_k - T_{M_j}(h_j(x))\} \end{array}\right.$$

where

$$D_j^k(x) \ := \ \sum_{i=1}^{n} h_j^i(x_{\hat{i}}) G_i(x) h_k^i(x_{\hat{i}})^\star$$

Example 1: Value Economies Let us consider a continuously differentiable function $h : (x,p) \in X \times X^\star \mapsto h(x,p) \in \mathbf{R}$. Hamiltonian systems

$$\left\{\begin{array}{lll} x'(t) & = & h_p'(x(t), p(t)) \\ p'(t) & = & -h_x'(x(t), p(t)) \end{array}\right.$$

are designed to satisfy the constraint $h(x(t), p(t)) = 0$.

Actually, we introduce the constraint

$$\forall\, t \geq 0,\ \ (x(t), p(t)) \ \in \ \text{Graph}(P)$$

to which we add a value constraint of the form

$$\forall\, t \geq 0,\ \ h(x(t), p(t)) \ \in \ I \ \subset \ \mathbf{R}$$

Let us set

$$K \ := \ \{(x,p) \in \text{Graph}(P) \mid h(x,p) \in I\}$$

The subset K is the inverse image of Graph$(P) \times I$ under the map $(x,p) \mapsto (x,p,h(x,p))$. The transpose of its derivative is the map

$$(u,q,\lambda) \in X \times X^\star \times \mathbf{R} \;\mapsto\; (q + \lambda h'_x(x,p), v + \lambda h'_p(x,p)) \in X^\star \times X$$

If we posit the transversality condition

$$\forall\,(x,p) \in \text{Graph}(P),\;\; h'_x(x,p) + h'_p(x,p)DP(x,p) \text{ is surjective}$$

and if we assume that the graph of P is closed and sleek, then we deduce from the calculus of contingent cones that the contingent cone to the set of constraints is defined by $T_K(x,p) :=$

$$\{(u,\pi) \in \text{Graph}(DP(x,p)) \mid \langle h'_x(x,p),u\rangle + \langle h'_p(x,p),\pi\rangle \in T_I(h(x,p))\}$$

We can use viability multipliers in order to design viable **value economies**, which are dynamical economies the state of which is made of both commodities $x \in X$ and prices $p \in X^\star$, and governed by

$$\begin{cases} i) & x'(t) \;=\; g(x(t),p(t)) - G(x(t),p(t))(q(t) + \lambda(t)h'_x(x(t),p(t)) \\ ii) & p'(t) \;=\; e(x(t),p(t)) - E(x(t),p(t))(v(t) + \lambda(t)h'_p(x(t),p(t))) \\ iii) & p(t) \in P(x(t)) \end{cases}$$

where

$$\begin{cases} g : X \times X^\star \mapsto X \;\&\; G : X \times X^\star \mapsto \mathcal{L}(X^\star,X) \\ e : X \times X^\star \mapsto X^\star \;\&\; E : X \times X^\star \mapsto \mathcal{L}(X,X^\star) \end{cases} \tag{5.2.13}$$

and where the regulee are made of triples $(v,q,\lambda) \in X \times X^\star \times \mathbf{R}$.

When the linear operators $G(x,p)$ and $E(x,p)$ are positive definite, we know that the value constraints are viable under such value economies.

The regulation map associates with any $(x,p) \in \text{Graph}(P)$ the set

$$\begin{array}{l} R_P(x,p) \;:=\; \{(v,q,\lambda) \mid q \in DP(x,p)(v) \;\&\; \lambda \;\in\; T_I(h(x,p)) \\ -\dfrac{\langle h'_x(x,p), g(x,p) - G(x,p)q\rangle + \langle e(x,p) - E(x,p)v, h'_p(x,p)\rangle}{\langle h'_x(x,p), G(x,p)h'_x(x,p)\rangle + \langle E(x,p)h'_p(x,p), h'_p(x,p)\rangle} \left.\vphantom{\dfrac{a}{b}}\right\} \end{array}$$

If we assume that for every $(x,p) \in \text{Graph}(P)$, we have

$$e(x,p) - E(x,p)x \;\in\; DP(x,p)(g(x,p) - G(x,p)p)$$

then we can take $v = 0$ and $q = 0$ in the above characterization of the regulation map.

In the case when $h(x,p) := \langle p,x\rangle$, when we want to keep the value $\langle p,x\rangle$ in a given viability interval I, then such constraints are viable under value economies of the form

$$\begin{cases} i) & x'(t) = g(x(t),p(t)) - G(x(t),p(t))(q(t) + \lambda(t)p(t)) \\ ii) & p'(t) = e(x(t),p(t)) - E(x(t),p(t))(v(t) + \lambda(t)x(t)) \\ iii) & p(t) \in P(x(t)) \end{cases}$$

Viable solutions are govern according the regulation map associating with any $(x,p) \in \text{Graph}(P)$ the set

$$\begin{array}{l} R_P(x,p) := \{(v,q,\lambda) \mid q \in DP(x,p)(v) \ \& \ \lambda \in T_I(\langle p,x\rangle) \\ \quad - \dfrac{\langle p, g(x,p) - G(x,p)q\rangle + \langle e(x,p) - E(x,p)v, x\rangle}{\langle p, G(x,p)p\rangle + \langle E(x,p)x, x\rangle}\Big\} \end{array}$$

Example 2: Increasing Production Capacities Let us consider the case when the set of allocations

$$K := \left\{ x \in \prod_{i=1}^n L_i \mid \sum_{i=1}^n x_i \in M \right\}$$

of resources is empty. We regard here $M \subset Y := \mathbf{R}^n$ as the production set of one producer, which cannot meet the demand using only one unit of labor, say. The question arises whether one can revive the viability of the allocation mechanism by increasing the number of labor units : We introduce for that purpose a scalar $w \in [1,+\infty[$ measuring the number of labor units and consider the new set of allocations defined by

$$K := \left\{ (x,y,w) \in \prod_{i=1}^n L_i \times M \times [1,+\infty[\mid \sum_{i=1}^n x_i \leq wy \right\} \tag{5.2.14}$$

This is an example of constraints defined as inverse images by the bilinear map

$$(x,y,w) \in \prod_{i=1}^n L_i \times M \times [1,\infty[\mapsto \sum_{i=1}^n x_i - wy \in Y$$

the derivative of which is equal to

$$(u,v,\lambda) \in Y^n \times Y \times \mathbf{R} \mapsto \sum_{i=1}^n u_i - wv - \lambda y \in Y$$

Its transpose is equal to

$$p \in Y^\star \mapsto (p,\ldots,p,-pw,-\langle p,y\rangle)$$

We assume that the dynamical behavior of the consumers and the producer is described by the following simple change functions of the form

$$\begin{cases} x_i'(t) = g_i(x_i(t)) - p_i(t) \ (i = 1,\ldots,n) \\ y'(t) = g_0(y(t)) - q(t) \\ w'(t) := e(w(t)) - \lambda(t) \end{cases}$$

We posit the transversality condition

$$\sum_{i=1}^{n} T_{L_i}(x_i) - T_M(y) + y\mathbf{R}_+ \; = \; Y$$

We then deduce from Corollary 5.2.3 that *the constraints (5.2.14) involving labor units are viable under dynamical economies of the form*

$$\begin{cases} x_i'(t) & = & g_i(x_i(t)) - p(t) \;\; (i = 1, \ldots, n) \\ y'(t) & = & g_0(y(t)) + w(t)p(t) \\ w'(t) & := & e(w(t)) + \langle p(t), y(t) \rangle \end{cases}$$

One can compute the regulation map governing the evolution of the prices:

$$p \in R_K(x, y, w)$$

if and only if

$$\sum_{i=1}^{n} (g_i(x_i) - p) - w(h(y) + wp) - y(e(w) + \langle q, y, \rangle) \; = \; r$$

where

$$r_h \; \leq \; 0 \text{ whenever } \left(\sum_{i=1}^{n} x_i - wy \right)_h \; = \; 0$$

Therefore, setting $z := \sum_{i=1}^{n} g_i(x_i) - wh(y) - ye(w) - r$, if and only if

$$(n + w^2)p + \langle p, y \rangle y \; = \; z$$

The solution to this equation is given by

$$p \; = \; \frac{1}{n + w^2} \left(z - \frac{y \langle z, y \rangle}{n + w^2 + \|y\|^2} \right)$$

5.3 Dynamical Socio-Economies

In the basic dynamical problem of resource allocation, we assumed that all consumers can trade at any instant and with anyone. This is too strong an assumption, since we know that shortages imply queues in the consumption side and unemployment in the labor side.

One first attempt to take into account this phenomenon is to introduce coalitions $A \subset N := \{1, \ldots, n\}$ of the n consumers and to relax the allocation requirement by asking only that any instant t, there exist a coalition $A(t) \in \mathcal{P}(N)$ and an allocation such that

$$\sum_{i \in A(t)} x_i(t) \; \in \; M(A(t))$$

where $M(A) \subset Y$ denotes the set of commodities available to the members of the coalition A.

We shall begin by investigating this case and derive a regulation law linking the evolution of the prices to the evolution of allocations and coalitions. In particular, if at time t, $A(t) = N$ is the "grand coalition", then the viable prices are the ones obtained when all consumers were allowed to participate to the market. If the price is "wrong" in the sense that it is not viable in the nonshortage case, then the regulation law provides the coalitions that can form for which an allocation can be found in the subeconomy restricted to this coalition. This regulation law provides a first explanation of the evolution of coalitions for such a crude model, and also, activity intervals of each consumer i, which is made of the instants t for which $i \in A(t)$.

The first definition of a coalition which comes to mind is that of a subset of consumers $A \subset N$. Thus there are 2^n coalitions. However, although the number of coalitions very soon becomes important, it remains finite, which prevents us from using analytical techniques, and in particular, from tackling dynamical models of evolution of coalitions. One way to overcome this difficulty is to embed the family of coalitions, subsets of the set of consumers, in the subset of **fuzzy coalitions** which range over a finite dimensional vector space. This allows us to define **social mobility** by using velocities of fuzzy coalitions and thus, to couple "sociodynamics" with "ecodynamics".

5.3.1 Regulation by Shortages

We introduce a family $\mathcal{A} \subset \mathcal{P}(N)$ of coalitions $A \subset N$ of consumers containing the "grand coalition" N. We assume that each coalition can build a "subeconomy" by allocating among their participants a subset $M(A) \subset Y$ of allocations. We thus associate with any coalition $A \in \mathcal{A}$ the allocation set

$$K(A) := \left\{ x \in \prod_{i=1}^{n} L_i \mid \sum_{i \in A} x_i \in M(A) \right\}$$

of coalition A. Then, the allocation problem is relaxed. It requires that for every

$$\forall\, t \geq 0,\ \ \exists\, A(t) \in \mathcal{A} \ \text{ such that } \sum_{i \in A(t)} x_i(t) \ \in\ M(A(t)) \qquad (5.3.1)$$

This amounts to replace the former set K of allocation by the larger set

$$\mathcal{K} := \bigcup_{A \in \mathcal{A}} K(A)$$

We shall be able to proceed because we know how to compute the contingent cone to a finite union of closed subsets:

Lemma 5.3.1 *Denote by $\mathcal{A}(x)$ the set of coalitions $A \in \mathcal{A}$ such that $x \in K(A)$. Then*

$$T_{\mathcal{K}}(x) \;=\; \bigcup_{\mathcal{A}(x)} T_{K(A)}(x)$$

Then the velocity $x'(t)$ of an allocation viable in $\mathcal{K}$ instead of K satisfies

$$\forall\, t \geq 0,\; \sum_{i \in A(t)} x_i'(t) \;\in\; T_{M(A(t))}\left(\sum_{i \in A(t)} x_i(t)\right)$$

We adapt the dynamics of the evolution in the following way:

$$\forall\, t \geq 0, \quad \begin{cases} i) & x_i'(t) \;=\; g_i(x_i(t)) - G_i(x_i(t))p(t) \\ ii) & p(t) \;\in\; P(x(t)) \;:=\; P(x_1(t), \ldots, x_n(t)) \end{cases} \tag{5.3.2}$$

Definition 5.3.2 *The* regulation map $\Pi_{\mathcal{M}}$ *is the set-valued map associating with any $A \in \mathcal{A}$ and any $x \in K(A)$ the subset*

$$\Pi_{\mathcal{M}}(x, A) \;:=\; \left\{ p \in P(x) \mid \sum_{i \in A} c_i(x_i, p) \;\in\; T_{M(A)}\left(\sum_{i \in A} x_i\right)\right\}$$

built from the confrontation of the changes claimed by consumers of a coalition and the set of scarce resources allocated to this coalition.

This regulation map relates each coalition $A \in \mathcal{A}$ and allocation $x \in K(A)$ to prices such that the associated total change of consumptions is tangent to the set of scarce resources of the coalition. We shall adapt the assumptions to this case in the following way:

$$\begin{cases} i) & \forall\, A \in \mathcal{A},\; M(A) = M(A) - \mathbf{R}^l_+ \text{ is a closed convex subset} \\ ii) & \forall\, i = 1, \ldots, n,\;\; L_i = L_i + \mathbf{R}^l_+ \text{ is closed and convex} \\ iii) & \forall\, A \in \mathcal{A},\; 0 \;\in\; \text{Int}\left(\sum_{i \in A} L_i - M(A)\right) \end{cases} \tag{5.3.3}$$

Theorem 5.3.3 *We posit assumptions (5.3.3), (3.2.4) and (3.2.5). Then the two following conditions are equivalent:*

$$\begin{cases} a) & \forall\, x \in \mathcal{K},\; \exists\, A \in \mathcal{A}(x) \mid \Pi_{\mathcal{M}}(x, A) \;\neq\; \emptyset \\ b) & \forall\, x_0 \in K, \text{ starts at least one allocation evolving according (5.3.2)} \end{cases}$$

In this case, the viable allocations are governed by the regulation law

$$\text{for almost all } t \geq 0,\;\; p(t) \in \Pi_{\mathcal{M}}(x(t), A(t)) \tag{5.3.4}$$

Actually, this regulation law becomes the ternary relation:

$$\text{for almost all } t \geq 0, \quad (x(t), A(t), p(t)) \in \text{Graph}(\Pi_{\mathcal{M}})$$

which can be red in several ways: if two of its three items are chosen, we obtain the evolution of the third one in terms of the evolution of the two first ones. In particular, it explains the evolution of the coalitions of consumers in terms of the evolution of both consumptions and prices.

5.3.2 Social Mobility

In order to obtain a less passive evolution law of coalitions, we would like to design differential equations governing the evolution of coalitions for providing a mathematical metaphor of social mobility. For that purpose, one can embed the family of subsets of a (discrete) set N of n consumers (economic agents, players, decision makers, etc.) to the space $\mathbf{R}^n$ through the map χ associating with any coalition $A \in \mathcal{P}(N)$ its characteristic function[2] $\chi_A \in \{0,1\}^n \subset \mathbf{R}^n$.

By definition, the family of fuzzy sets[3] is the convex hull $[0,1]^n$ of the power set $\{0,1\}^n$ in $\mathbf{R}^n$. Therefore, we can write any fuzzy set in the form

$$\alpha = \sum_{A \in \mathcal{P}(N)} m_A \chi_A \text{ where } m_A \geq 0 \ \& \sum_{A \in \mathcal{P}(N)} m_A = 1$$

The memberships are then equal to

$$\forall i \in N, \ \alpha_i = \sum_{A \ni i} m_A$$

Consequently, if m_A is regarded as the probability for the set A to be formed, the membership of the player i to the fuzzy set α is the sum of the probabilities of the coalitions to which player i belongs. Player i participates fully in α if $\alpha_i = 1$, does not participate at all if $\alpha_i = 0$ and participates in a fuzzy way if $\alpha_i \in]0,1[$.

---

[2]This canonical embedding is more adapted to the nature of the power set $\mathcal{P}(N)$ than the universal embedding of a discrete set M to $\mathbf{R}^m$ by the Dirac measure associating with any $j \in M$ the jth element of the canonical basis of $\mathbf{R}^m$. The convex hull of the image of M by this embedding is the probability simplex of $\mathbf{R}^m$. Hence fuzzy sets offer a "dedicated convexification" procedure of the discrete power set $M := \mathcal{P}(N)$ instead of the universal convexification procedure of frequencies, probabilities, mixed strategies derived from its embedding in $\mathbf{R}^m = \mathbf{R}^{2^n}$.

[3]This concept of the fuzzy set was introduced in 1965 by L. A. Zadeh. Since then, it has been wildly successful, above all in many areas outside mathematics. Lately, we found the following quotation by A. Bercoff: *"Aujourd'hui, nous nageons dans la poésie pure des sous ensembles flous"* ...

François Mitterand

in "LA LUTTE FINALE", Michel Lafon (1994), p.69.

In a competitive dynamical economy, at each instant, the "grand coalition" N (or $\mathbf{1} := (1, \ldots, 1)$) is formed and all consumers are allowed to allocate the scarce commodities of the set $M \subset \mathbf{R}^l$.

The purpose of this section is to introduce imperfect competition by assuming that only some fuzzy coalition[4] α can form and allocate among their participants a specific set $M(\alpha) \subset \mathbf{R}^l$ of available commodities. Instead of requiring that at each instant,

$$\sum_{i=1}^{n} x_i(t) \in M(\mathbf{1}) = M(N)$$

as we did until now, we are content with the formation of a coalition $\alpha(t) \in [0,1]^n$ which "plays" a subeconomy described by

$$\sum_{i=1}^{n} \alpha_i(t) x_i(t) \in M(\alpha(t))$$

When $\alpha := \chi_A$ is a usual coalition, we find again the constraint (5.3.1).

Now, we introduce the socio-economic state $(x, \alpha) \in Y^n \times \mathbf{R}^n$ satisfying the constraints

$$\sum_{i=1}^{n} \alpha_i x_i \in M(\alpha)$$

where M denotes now a set-valued map from the set $[0,1]^n$ of fuzzy coalitions[5] to the commodity space $Y := \mathbf{R}^l$, called the **resource map**.

---

[4]Actually, this idea of using fuzzy coalitions has already been used in the framework of cooperative games with and without side-payments. For instance, it has been shown that in the framework of cooperative games with side payments involving fuzzy coalitions, the concepts of Shapley value and core did coincide with the (generalized) gradient of the "characteristic function" at the "grand coalition". The differences between these concepts for usual games is explained by the different ways one "fuzzyfy" a characteristic function defined on the set of usual coalitions. See [16,17, Aubin], [12, Aubin, Chapter 12] and [19, Aubin, Chapter 13].

[5]The choice of "cooperative" fuzzy coalitions $\alpha \in [0,1]^n$ is arbitrary.

We could, for instance, introduce negative memberships when players enter a coalition with aggressive intents. This is mandatory if one wants to be realistic ! A positive membership is interpreted as a cooperative participation of the player i in the coalition, while a negative membership is interpreted as a non-cooperative participation of the ith player in the generalized coalition. In what follows, one can replace the cube $[0,1]^n$ by any product $\prod_{i=1}^{n}[\lambda_i, \mu_i]$ for describing the cooperative or noncooperative behavior of the consumers.

Technically, the choice of the scaling $[0,1]$ inherited from the tradition built on integration and measure theory is not adequate for describing convex sets. When dealing with convex sets, we have to replace the characteristic functions by indicators taking their values in $[0,+\infty]$ and take their convex combinations to provide an alternative allowing to speak of "fuzzy" convex sets. Therefore, "toll-sets" are nonnegative cost functions assigning to each element its cost of belonging, $+\infty$ if it does not belong to the toll set (See Chapter 8). The set of elements with finite positive cost do form the "fuzzy boundary" of the toll set, the set of elements with zero cost its "core". This has been done to adapt viability theory to "fuzzy

So, the set of coalition-dependent allocations is described by a set-valued map $K : [0,1]^n \rightsquigarrow Y$ defined by

$$K(\alpha) := \left\{ x \in Y^n \mid \sum_{i=1}^n \alpha_i x_i \in M(\alpha) \right\}$$

In order to proceed, we need to compute the contingent derivative of the resource map:

Proposition 5.3.4 *Assume that the resource map $M : [0,1]^n \rightsquigarrow Y$ is sleek and closed and that they satisfy the* transversality assumption*: For every $x \in K(\alpha)$,*

$$Y = \sum_{i=1}^n \alpha_i T_{L_i}(x_i) + \sum_{i=1}^n x_i T_{[0,1]}(\alpha_i) - T_{M(\alpha)}\left(\sum_{i=1}^n \alpha_i x_i\right) \tag{5.3.5}$$

Then

$$\begin{cases} DK(\alpha, x)(\beta) \\ := \left\{ v \in Y^n \;\middle|\; \sum_{i=1}^n (\alpha_i v_i + \beta_i x_i) \in DM\left(\alpha, \sum_{i=1}^n \alpha_i x_i\right)(\beta) \right\} \\ \text{whenever } \beta_i \geq 0 \text{ if } \alpha_i = 0,\ \beta_i \leq 0 \text{ if } \alpha_i = 1 \text{ and } \emptyset \text{ if not} \end{cases} \tag{5.3.6}$$

Proof — Indeed, the graph $\mathcal{K}$ of the set-valued map K is the set of pairs $(\alpha, x) \in [0,1]^n \times Y^n$ such that

$$A(\alpha, x) := \sum_{i=1}^n \alpha_i x_i \in M(\alpha)$$

The proof follows from an adaptation of Theorem 11.3.7 of the Tool-Box. □

The socio-economic state must remain in the inverse image of the graph of M under the map

$$(x, \alpha) \mapsto (\alpha, \sum_{i=1}^n \alpha_i x_i)$$

---

viability theory" in which both the sets $P(x)$ of price and the set M of available commodities are toll sets. See Chapter 10 of VIABILITY THEORY, [28, Aubin] for more details.

Actually, the Cramer transform

$$C_\mu(p) := \sup_{x \in \mathbf{R}^n} \left(\langle p, x \rangle - \log \left(\int_{\mathbf{R}^n} e^{\langle x, y \rangle} d\mu(y) \right) \right)$$

maps probability measures into the family of toll sets. In particular, it transforms convolution products of density functions to inf-convolutions of extended functions, Gaussians to squares of norms, etc. See [47, Aubin & Dordan] for more informations on this topic.

the derivative of which is the map

$$(v,\gamma) \mapsto \left(\gamma, \sum_{i=1}^{n} \gamma_i x_i + \sum_{i=1}^{n} \alpha_i v_i\right)$$

It transpose is equal to

$$(p,\beta) \mapsto ((\alpha_i p)_{i=1,\ldots,n}, (\beta_i + \langle p, x_i\rangle)_{i=1,\ldots,n})$$

We recall that the normal cone $N_{\mathrm{Graph}(M)}(\alpha, y)$ defines the graph of the codifferential[6] $DM(\alpha,y)^\star : Y^\star \rightsquigarrow \mathbf{R}^n$ through the formula

$$\beta \in DM(\alpha,y)^\star(p) \text{ if and only if } (-\beta, p) \in N_{\mathrm{Graph}(M)}(\alpha, y)$$

Hence, using p and $\beta \in DM(\alpha,y)^\star(p)$ as viability multipliers, we design the dynamics of the evolution of the consumption and of the coalition in the following way:

$$\forall\, t \geq 0, \quad \begin{cases} i) & x_i'(t) = g_i(x_i(t)) - \alpha_i(t) G_i(x_i(t)) p(t) \quad (i = 1,\ldots,n) \\ ii) & \alpha_i'(t) = \gamma_i(\alpha_i(t)) + \beta_i(t) - \langle p(t), x_i(t)\rangle \quad (i = 1,\ldots,n) \\ iii) & (p(t), \beta(t)) \in P(x(t), \alpha(t)) \end{cases} \tag{5.3.7}$$

where $p(t) \in Y^\star$ is a price and $\beta(t) \in \mathbf{R}^{n^\star}$ is a coordination regulee. where $P(x,\alpha)$ is a compact sole of the normal cone

$$N_{\mathrm{Graph}(M)}\left(\alpha, \sum_{i=1}^{n} \alpha_i x_i\right)$$

We shall address the problem of finding viable evolutions satisfying

$$\forall\, t \geq 0, \; \sum_{i=1}^{n} \alpha_i(t) x_i(t) \in M(\alpha(t))$$

Definition 5.3.5 *The regulation map Π_M is the set-valued map*

$$\begin{cases} \forall\, \alpha \in [0,1]^n, \; x \in K(\alpha), \; \Pi_M(x,\alpha) := \\ \left\{(p,\beta) \in P(x,\alpha) \mid \sum_{i=1}^{n} \alpha_i^2 G_i(x_i) p - \sum_{i=1}^{n} \beta i x_i + \sum_{i=1}^{n} \langle p, x_i\rangle x_i \right. \\ \in \sum_{i=1}^{n} (\alpha_i g_i(x_i) + \gamma_i(\alpha_i) x_i) \\ \left. -DM\left(\alpha, \sum_{i=1}^{n} \alpha_i x_i\right) \left((\gamma_i(\alpha_i) + \beta_i - \langle p, x_i\rangle)_{i=1,\ldots,n}\right)\right\} \end{cases}$$

built from the confrontation of the changes claimed by consumers and "social mobility" described by the evolution law of the coalitions.

[6] See Definition 11.3.5 of the Tool-Box.

This regulation map associates with each fuzzy coalition $\alpha \in [0,1]^n$ and with each allocation $x \in K(\alpha)$ the prices such that the related total change of consumptions lies in the contingent derivative of the resource map in the direction of "social mobility". The objective is to bring back the total consumption of the members of the fuzzy coalition inside the set of scarce resources available to the fuzzy coalition.

We posit now the assumptions we need to prove the noncompetitive version of the viability theorem.

— ASSUMPTIONS ON THE CONSUMPTION AND RESOURCE MAPS:

$$\mathrm{Int}(T_{\mathrm{Graph}(M)}(\alpha, y)) \neq \emptyset \quad \text{and } M \text{ is closed and sleek} \tag{5.3.8}$$

— ASSUMPTIONS ON THE CHANGE FUNCTIONS:

$$g_i(x,p) \text{ and } \gamma(\alpha) \text{ are continuous and bounded} \tag{5.3.9}$$

— ASSUMPTIONS ON THE PRICING MAP:

$$P : Y^n \times \mathbf{R}^n \rightsquigarrow Y \times \mathbf{R}^n \text{ is a Marchaud map} \tag{5.3.10}$$

Theorem 5.3.6 *We posit assumptions (5.3.8), (5.3.9) and (5.3.10). Then the two following conditions are equivalent:*

1. *the dynamical socio-economy is viable in the sense that from every initial coalition α_0 and any initial state $x_0 \in K(\alpha_0)$ starts at least one fuzzy coalition and one allocation of $M(\alpha(t))$ evolving according (5.3.7)*

2. *the regulation map is strict:*

$$\forall\, \alpha \in [0,1]^n,\ \forall\, x \in K(\alpha),\ \Pi_M(x,\alpha) \neq \emptyset$$

In this case, the viable allocations are governed by the regulation law

$$\text{for almost all } t \geq 0, \quad (p(t), \beta(t) \in \Pi_M(x(t), \alpha(t)) \tag{5.3.11}$$

Example: Herd Behavior We associate with any $(p,\beta) \in \Pi_M(x,\alpha)$ the subset

$$A_+(x,\alpha;p,\beta) \ := \ \{i \in N \mid \gamma_i(\alpha_i) + \langle p, x_i\rangle - \beta_i \ > \ 0\}$$

of consumers whose rates of participations are increasing, and the subset

$$A_-(x,\alpha;p,\beta) \ := \ \{i \in N \mid \gamma_i(\alpha_i) + \langle p, x_i\rangle - \beta_i \ < \ 0\}$$

of consumers who are being excluded.

Therefore, one can deduce the evolution of the upward class of consumers by studying the evolution of $A_+(x(t),\alpha(t);p(t),\beta(t))$ as time evolve.

Whenever one can define and localize an institution allocating the coordination regulees β, one can solve the problem of increasing social consensus in another way, by associating with prices p coordination regulees β such that $A_+(x,\alpha;p,\beta)=N$, i.e., the following subset of the regulation map

$$\begin{cases} (p,\beta)\in \Pi_M(x,\alpha) \ \text{ satisfying} \\ \forall\, i=1,\ldots,n, \ \ \gamma_i(\alpha_i)+\langle p,x_i\rangle-\beta_i \ \geq \ 0 \end{cases}$$

Coalitions $A_+(x,\alpha;p,\beta)$ and $A_-(x,\alpha;p,\beta)$ describe the "herd behavior" of the consumers. They are formed of "mimetic" consumers.

A consumer i_0 is Panurgean for a coalition $A\subset N$ if for every consumer $j\in A$,

$$\gamma_{i_0}(\alpha_{i_0},\beta_{i_0}-\langle p,x_{i_0}\rangle)\gamma_i(\alpha_i)+\langle p,x_i\rangle-\beta_i \ \geq \ 0$$

Remark: Psycho-Socio-Economics — We can also enrich the description of the players by representing each player i by what psychologists call her "behavior profile".

We consider q "behavioral qualities" $k=1,\ldots,q$. For example, $k=1$: intelligence, $k=2$: patience, $k=3$: creativity, etc. We also suppose that a behavior can be measured (evaluated) in vector spaces B_k ($k=1,\ldots,q$). A behavior profile of a player is then defined as a linear operator $A:=(A_k)_{k=1,\ldots,q}\in\mathcal{L}(Y,\prod_{k=1}^q B_k)$ which specify the quantities $A_k(x)\in B_k$ of the q qualities k that the player implements when he receives a commodity $x\in Y$. Thus, instead of representing each player by a letter of the alphabet, she is described as a linear operator of $\mathcal{L}(Y,\prod_{k=1}^q B_k)$.

We then suppose that each player may implement all, none, or only some of her behavioral profile when she participates in a social coalition.

For example, suppose that we have retained the 2 qualities $k=1$: intelligence and $k=2$: patience. A player may implement these two qualities in different ways, according as to whether she is participating in the Fraternal Society of Social Psychology, the Cooking Club or the Association of Belote Players. In the first case, we assume that she implements the two qualities in full, while in the second case she does not use her intelligence at all but proves her patience to the full and in third case he uses half of the potential of each of these two qualities.

If the consumers are described by their behavior profiles

$$A^i=(A_1^i,\ldots,A_k^i,\ldots A_q^i)\in\mathcal{L}(Y,\prod_{k=1}^q B_k)$$

the qualities brought into play by a social coalition $\alpha\in[-1,+1]^{qn}$ are equal to

$$\forall i=1,\ldots,n,\ \ \{(\alpha_i^k A_k^i)(x)\}_{k=1,\ldots,q}\in\prod_{k=1}^q B_k$$

Consider n players represented by their behavior profiles in $\mathcal{L}(Y, \prod_{k=1}^{q} B_k)$. Any matrix $\alpha = (\alpha_i^k)$ describing the levels of participation $\alpha_i^k \in [-1, +1]$ of the behavioral qualities k for the n players i is called a **social coalition.** The set of all social coalitions which it is possible to construct is the qn-dimensional hypercube $[-1, +1]^{qn}$.

qualities / players	quality 1	...	quality k	...	quality q
player 1	α_1^1		α_1^k		α_1^q
⋮					
player i	α_i^1		α_i^k		α_i^q
⋮					
player n	α_n^1		α_n^k		α_n^q

The q rows represent the levels of behavioral participation of the q players.

The evolution of the behavior of the q qualities of each consumers i will be described by differential equations governing consumptions and the evolution of social coalitions :

$$\begin{cases} x_i'(t) = c_i(x_i(t), \sum_{k=1}^{q} \alpha_i^k(t) A^{k\star} p^k(t)) \\ (i = 1, \ldots, n) \\ \alpha_i^{k\prime}(t) = \Gamma_i^k(\alpha_i^k(t), \beta_i^k(t) - \langle p^k(t), A^k x_i \rangle) \\ (i = 1, \ldots n, \; k = 1, \ldots q) \end{cases}$$

It is a matter of routine to extend the results concerning fuzzy coalitions to the ones concerning "social groups".

5.4 Introducing the Producers

For providing an example of the above strategy, we subjoin m producers denoted by $j = 1, \ldots, m$ to our former n consumers $i = 1, \ldots, n$. We shall begin by itemizing the constraints and then, explaining the notations. The producer j provides a commodity $y^j \in Y := \mathbf{R}^l$ and the consumer i acquires a commodity $x_i^j \in Y$ at firm j.

We denote by $y := \{y^1, \ldots, y^m\} \in Y^m$ the sequence of commodities produced by the set of producers $k = 1, \ldots m$. The state of the consumer $i = 1, \ldots, n$ is described by $x_i := \{x_i^j\}_{j=1,\ldots,m} \in Y^m$. We shall denote by $x := \{x_i\}_{i=1,\ldots,n} \in Y^{mn}$ the "vector-matrix" of consumption bundles x_i^j secured by consumer i at producer j.

We chose a middle ground between too simple a model such as the allocation of scarce resources and a very complete model with too many constraints to handle. The choice of the level of abstraction is naturally a matter of convenience and taste. This one is presented as an illustration of the strategy we advocated more than a definitive framework for describing an economy.

Here is one conceivable list of constraints:

$$\left\{\begin{array}{l} x_i^j \in L_i^j \\ \\ \sum_{j=1}^m B_i^j x_i^j \in L_i \\ \\ y^j \in U^j(x_1^j, \ldots, x_n^j) \\ \\ \sum_{i=1}^n A^j y^j \in M\left(\left(x_i^j\right)_{i=1,\ldots,n,\, j=1,\ldots,m}\right) \\ \\ \sum_{i=1}^n x_i^j \in Q^j(y^j) \end{array}\right. \tag{5.4.1}$$

which can be divided in three groups, the first two constraints dealing with consumers, the second two ones with the producers and the last one with the exchange constraints.

Consumption The description of consumer i is made of the following items:

1. Consumption sets $L_i^j \subset Y$ of commodities x_i^j produced by j which can be consumed by consumer i. For instance, if there are no possible trade relations between consumer i and producer j, we set $L_i^j = \{0\}$
2. Service sets $L_i \subset Z$ where Z denotes a vector-space of services. For instance, one of the components of the service may represent labor, the associated service set L_i of consumer i describing the upper limit of labor that consumer i can bear with. Other components of the service space Z could describe savings, investments, etc.
3. operators $B_i^{j,k} \in \mathcal{L}(Y, Z)$ associating with any commodity x_i^k the service $B_i^{j,k} x_i^k$ delivered to firm j by consumer i. By setting $B_i^k := \sum_{j=1}^m B_i^{j,k}$, we deduce that B_i^k associates with the commodity x_i^k of consumer i the services he can provide when he acquires commodity x_i^k.

Therefore, the constraints on the consumer are represented by the consumption set $\vec{L}_i$ of consumer i defined by

$$\vec{L}_i := \left\{(x_i^j)_{j=1,\ldots,m} \in \prod_{j=1}^m L_i^j \mid \sum_{k=1}^m B_i^k x_i^k \in L_i\right\} \tag{5.4.2}$$

Production We denote by $R := \mathbf{R}^k$ the vector space describing the resource space available to producers. Each producer j is described by

1. a technological matrix $A^j \in \mathcal{L}(Y, R)$ associating with a commodity $y \in Y := \mathbf{R}^l$ the resource $A^j y \in R$ needed to produce it

2. a set-valued map $U^j : Y^n \rightsquigarrow Y$ associating with any commodity x_i^j of consumer i acquired at firm j some resources (for instance, labor, savings) used has an input by producer j

   In many examples, the map U_j is defined through

$$U^j(x_i) := V^j\left(\sum_{i=1}^{n} C_i^j x_i^j\right)$$

   where operators $C_i^j \in \mathcal{L}(Y^m, R^j)$ associate with any commodity bundle x_i of consumer i the resources $\sum_{i=1}^{n} C_i^j x_i^j \in R^j$ and a set-valued map $V^j : R^j \rightsquigarrow Y$ prescribing the potential production of firm j associated with resources attributed directly to him by the consumers

3. and a set-valued map $M : X^{mn} \rightsquigarrow R$ associating with any consumption bundle $x := (x_i^j)$ the set $M(x)$ of resources available to the producers.

   In numerous examples, the production map is defined by

$$M\left(\left(x_i^j\right)_{i=1,\ldots,n,\; j=1,\ldots,m}\right) := M_0\left(\sum_{i=1}^{n} D_i x_i\right)$$

   where the operators $D^i \in \mathcal{L}(Y^{mn}, Z)$ associate with any commodity bundle x_i of consumer i the services (for instance, labor, savings) $D_i x_i = \sum_{k=1}^{m} D_i^k x_i^k$ he can procure to the whole economy when it acquires the commodity bundle x_i and $M_0 : Z \rightsquigarrow R$ prescribes the set of available resources in resources to be allocated among the producers.

Then the production map of the productive system can be written in the form

$$\vec{M}(x) := \left\{ y \mid y^j \in U^j(x_1^j, \ldots, x_n^j) \mid \sum_{i=1}^{n} A^j y^j \in M\left(\left(x_i^j\right)_{i=1,\ldots,n,\; j=1,\ldots,m}\right)\right\} \tag{5.4.3}$$

Exchange We introduce a trade map $Q^j : Y \rightsquigarrow Y$ associating with any produced commodity y^j the set $Q^j(y^j)$ of available commodities to be consumed or saved or stocked. For instance, if $z^j \in \mathbf{R}_+^l$ denotes the maximal amount of storage of producer j, the vector interval $Q^j(y^j) := [y^j - z^j, y^j]$ provides an example of such a map Q^j.

Then the exchange constraint reads

$$\forall\, j = 1, \ldots, m, \quad \sum_{i=1}^{n} x_i^j \in Q^j(y^j)$$

For simplicity, we shall use only here prices as control parameters to design dynamical economies under which the set of allocations

$$\mathcal{K} := \left\{(x,y) \in Y^{nm} \times Y^m \mid x \in \prod_{i=1}^{n} \vec{L}_i,\ y \in \vec{M}(x)\ \&\ \sum_{i=1}^{n} x_i^j \in Q^j(y^j)\right\} \tag{5.4.4}$$

is viable.

Such results imply the viability of the evolution of an economy with only a minimal message (or semiotic) mechanism. But, in the examples we delineated above, we could add among the regulatory control (that we should call regulees) the storage z^j regulating the trade map Q^j, the linear operators C_i^j and D_i regulating the production maps U^j and M for providing redundant controls to make the economy more robust.

Once the list (5.4.1) of constraints written, we have to compute the tangent cone to this allocation set for obtaining the whole "envelope" of dynamical economies under which this allocation set is viable.

For that purpose, we have to posit the constraint qualification assumption which states that there exists $\eta > 0$ such that, for every perturbation

$$\left((a_i^j)_{i,j}, (b_i)_i, ((q_i^j)_i, -r^j)_j, ((s_i^j)_{i,j}, -p), (\rho^j, -\sigma^j)_j\right)$$

with norm smaller than or equal to η, there exist a solution (x,y) satisfying $\forall\, i = 1,\dots,n,\ j = 1,\dots,m$,

$$\begin{cases} x_i^j \in a_i^j + L_i^j \\ \sum_{j=1}^m B_i^j x_i^j \in b_i + L_i \\ y^j \in r^j + U^j(x_1^j - q_1^j, \dots, x_n^j - q_n^j) \\ \sum_{i=1}^n A^j y^j \in p + M\left(\left(x_i^j - s_i^j\right)_{i=1,\dots,n,\ j=1,\dots,m}\right) \\ \sum_{i=1}^n x_i^j = \rho^j + Q^j(y^j - \sigma^j) \end{cases}$$

We recall that the **codifferential of** F **at** (x,y) has been defined in the Tool-Box by

$$p \in DF(x,y)^\star(q) \text{ if and only if } (p,-q) \in N_{\text{Graph}(F)}(x,y)$$

Theorem 5.4.1 *Assume that the sets L_i^j, L_i and the graphs of the maps U^j and M are closed and convex and that the foregoing constraint qualification assumption holds true. Let us consider continuous maps g_i and h^j describing the autonomous dynamical behavior of consumers and producers when* *they are unconstrained. Then the set of allocation (5.4.1) is viable under dynamical economies of the form*

$$\begin{cases} \forall\, i = 1,\dots,n,\ j = 1,\dots,m, \\ x_i^{j'}(t) \ = \ g_i^j(x_i^j(t)) - \rho^j(t) - B_i^{j\star} b_i(t) - a_i^j(t) - q_i^j(t) - s_i^j(t) \\ \forall\, j = 1,\dots,m, \\ y^{j'}(t) \ = \ h^j(y^j(t)) + r^j(t) + \sigma^j(t) + A^{j\star} p(t)) \end{cases}$$

where the pricing map is given by

$$\begin{cases} a_i^j(t) \in N_{L_i^j}(x_i^j)(t) \ \&\ b_i(t) \in N_{L_i}\left(\sum^j B_i^j x_i^j(t)\right) \\ q_i^j(t) \ \in \ DU^{j\star}((x_i^j(t))_i, y^j(t))(r^j(t)) \\ s_i^j(t) \ \in \ DM^\star\left(\left(x_i^j(t)\right)_{i,j}, \sum_{j=1}^m A^j y^j(t)\right)(p(t)) \\ \rho^j(t) \ \in \ DQ^{j\star}\left(y^j(t), \sum_{i=1}^n x_i^j(t)\right)(\sigma^j(t)) \end{cases}$$

Proof — The tangent cone to the set of allocations is equal to

$$\begin{cases} T_{\mathcal{K}}(x,y) \ := \ \Big\{(u_i^j \in T_{L_i^j}(x_i^j), v^j \in DU^j((x_i^j)_i, y^j)((u_i^j)_i) \ | \\ \sum_{j=1}^m B_i^j v_i^j \in T_{L_i}\left(\sum_{j=1}^m B_i^j x_i^j\right), \\ \sum_{j=1}^m A^j v^j \in DM\left(\left(x_i^j\right)_{i,j}, \sum_{j=1}^m A^j y^j\right)\left(\left(u_i^j\right)_{i,j}\right), \\ \sum_{i=1}^n u_i^j \ \in \ DQ^j\left(y^j, \sum_{i=1}^n x_i^j\right)(v^j)\Big\} \end{cases} \tag{5.4.5}$$

Indeed, one can write the allocation set $\mathcal{K}$ in the form $H^{-1}(\mathcal{M})$ where

$$H(x,y) \ := \ \left(x, \left(\sum_{j=1}^m B_i^j x_i^j\right)_i, \left((x_i^j)_i, y^j\right)_j, \left(\left(x_i^j\right)_{i,j}, \sum_{j=1}^m A^j y^j\right), \left(\sum_{i=1}^n x_i^j, y^j\right)_j\right)$$

maps $Y^{mn} \times S^m$ to $Y^{mn} \times Y^n \times (Y^n \times Y)^m \times (Z^{mn} \times R) \times (Y^m \times Y^m)$ and where

$$\mathcal{M} \ := \ \prod_{i,j} L_i^j \times \prod_i L_i \times \prod_j \operatorname{Graph}(U^j) \times \operatorname{Graph}(M) \times \prod_{j=1}^m \operatorname{Graph}(Q^j)$$

Its transpose is defined on the dual of the space

$$Y^{mn} \times Y^n \times (Y^n \times Y)^m \times (Z^{mn} \times R) \times (Y^m \times Y^m)$$

of multipliers (prices) of the form

$$\left(a_i^j)_{i,j}, (b_i)_i, ((q_i^j)_i, -r^j)_j, ((s_i^j)_{i,j}, -p), (\rho^j, -\sigma^j)_j\right)$$

Each constraint defining the allocation set requires in some sense this sequence of prices to regulate viable evolutions.

The transpose is given by the formula:

$$\begin{cases} H^\star\left((a_i^j)_{i,j},(b_i)_i,((q_i^j)_i,-r^j)_j,((s_i^j)_{i,j},-p),(\rho^j,-\sigma^j)_j\right) \\ = \left(\left(\rho^j + a_i^j + B_i^{j\star}b_i + s_i^j + q_i^j\right)_{i=1,\ldots,n,j=1,\ldots m}, \left(-r^j - A^{j\star}p - \sigma^j\right)_{j=1,\ldots,m}\right) \end{cases}$$

We deduce that the normal cone $N_{\mathcal{K}}(x,y)$ to $\mathcal{K}$ is the set of

$$\left(\left(\rho^j + a_i^j + B_i^{j\star}b_i + s_i^j + q_i^j\right)_{i=1,\ldots,n,j=1,\ldots m}, \left(-r^j - A^{j\star}p - \sigma_i^j\right)_{j=1,\ldots,m}\right)$$

where

$$\begin{cases} a_i^j \in N_{L_i^j}(x_i^j) \;\&\; b_i \in N_{L_i}\left(\sum^j B_i^j x_i^j\right) \\ q_i^j \;\in\; DU^{j\star}((x_i^j)_i, y^j)(r^j) \\ s_i^j \;\in\; DM^\star\left(\left(x_i^j\right)_{i,j}, \sum_{j=1}^m A^j y^j\right)(p) \\ \rho^j \;\in\; DQ^{j\star}\left(y^j, \sum_{i=1}^n x_i^j\right)(\sigma^j) \end{cases}$$

When the pricing map is given *a priori*, we can design viable dynamical economies in the following way:

Theorem 5.4.2 *Let us consider n change functions c_i of the consumers and m change functions d^j describing the behavior of the consumers satisfying the instantaneous Walras condition. Then, under adequate consistency conditions which we do not write explicitly, the set of allocation (5.4.1) is viable under dynamical economies of the form*

$$\begin{cases} \forall\, i = 1,\ldots,n, \\ x_i^{j'}(t) \;=\; c_i^j(x_i^j(t),(\rho^j(t) + B_i^{j\star}b_i(t) + a_i^j(t) + q_i^j(t) + s_i^j(t)) \\ \forall\, j = 1,\ldots,m, \\ y^{j'}(t) \;=\; d^j(y^j(t), -r^j(t) - \sigma^j(t) - A^{j\star}p(t)) \end{cases}$$

For any initial allocation $(x_0, y_0) \in \mathcal{K}$, there exist prices

$$\left(a_i^j(t))_{i,j},(b_i(t))_i,((q_i^j(t))_i,-r^j(t))_j,((s_i^j(t))_{i,j},-p(t)),(\rho^j(t),-\sigma^j(t))_j\right)$$

such that the solutions of the foregoing differential equations have solutions which form at each instant an allocation.

5.5 Splitting Production Processes

5.5.1 One-Stage Production Processes

Let X_0, Y_0 , Z_0 and Y_1, Y_1 denote finite dimensional commodity spaces.

We consider a one-stage production process involving technological matrices of the form

$$B_0W_0A_0x_0 + C_1x_1 \in M$$

where $M \subset Y_1$ is a closed convex set.

Here $x_0 \in X_0$ is regarded as a consumption or an output, $x_1 \in X_1$ as a production or an input, $A_0 \in \mathcal{L}(X_0, Y_0)$, $B_0 \in \mathcal{L}(Z_0, X_1)$ and $C_1 \in \mathcal{L}(X_1, Y_1)$ are given linear operators and $W_0 \in \mathcal{L}(Y_0, Z_0)$ is regarded as a "technological matrix" to be chosen for the above constraints to be satisfied.

Therefore, in such a situation, the state variables are the triples (x_0, x_1, W_0) made of pairs of vectors and of matrices. The above constraints are defined by inverse images under a bilinear map.

$$\left\{\begin{array}{ccc} & A_0 & \\ x_0 \in X_0 & \mapsto & Y_0 \\ & W_0 & \downarrow \\ & & Z_0 \\ & B_0 & \downarrow \\ x_1 \in X_1 & \mapsto & Y_1 \supset M \\ & C_1 & \end{array}\right.$$

Example Taking $Y_0 = Z_0 = Y_1 := \mathbf{R}$, $A_0 \in X_0^\star$ and $C_1 \in X_1^\star$ and $B_0 = 1$, the constraints become

$$w_0\langle A_0, x_0\rangle - \langle C_1, x_1\rangle \leq 0$$

For instance, $\langle C_1, x_1\rangle$ can be regarded as the number of labor units needed for the production of x_1, $\langle A_0, x_0\rangle$ as the "labor value" of the consumer-worker associated with the consumption of x_0 and w_0 the number of labor-value units needed to produce the output. □

But the linearity of the operators A_0, B_0 & C_1 does not bring any simplification in the viability analysis. This is the reason why we study this example in the more general form described in the following way:

1. "consumption" (or outputs) $x_0 \in L_0 \subset X_0$,
2. "production" (or input) $x_1 \in L_1 \subset X_1$,
3. continuously differentiable maps $l_0 : X_0 \mapsto Y_0$ and $h : Z_0 \times X_1 \mapsto Y_1$,
4. "technological matrix" $W_0 \in \mathcal{W}_0 \subset \mathcal{L}(Y_0, Z_0)$

5. "scarcity constraint" $h(W_0 l(x_0), x_1) \in M \subset Y_1$

The constrained set is defined by

$$K := \{(x_0, x_1, W_0) \in L_0 \times L_1 \times \mathcal{W}_0 \mid h(W_0 l(x_0), x_1) \in M\} \quad (5.5.1)$$

The "bilinear" example with which we started was obtained with $l_0 := A_0 \in \mathcal{L}(X_0, Y_0)$ and with $h(z_0, x_1) := B_0 z_0 + C_1 x_1$.

By the way, we reintroduce such (state dependent) linear operators denoted by

$$A_0 := l'(x_0), \; B_0 := h'_{z_0}(W_0 l(x_0), x_1) \; \& \; C_1 := h'_{x_1}(W_0 l(x_0), x_1)$$

Hence, the directional derivative is equal to

$$h'(W_0 l(x_0), x_1)(u_0, u_1, V_0) = C_1 u_1 + B_0 W_0 A_0 u_0 + B_0 V_0 l(x_0)$$

and can be written in the form

$$B_0 W_0 A_0 u_0 + C_1 u_1 + (l(x_0) \otimes B_0) V_0$$

using the notations of tensor product of liner operators (see Definition 10.2.1 in the Tool-Box). The transpose of this operator maps $q \in Y_1^\star$ to $X_0^\star \times X_1^\star \times \mathcal{L}(Y_0^\star, Z_0^\star)$ in the following way:

$$h'(W_0 l(x_0), x_1)^\star(q) := (A_0^\star W_0^\star B_0^\star q \;,\; C_1^\star q \;,\; l(x_0) \otimes B_0^\star q)$$

We posit a sufficient condition guaranteeing the transversality conditions:

$$\forall\, (x_0, x_1, W_0), \;\; h'_{x_1}(W_0 l(x_0), x_1) \text{ is surjective} \quad (5.5.2)$$

Let us assume now that the dynamical behavior of consumers and producers is described by (simple) dynamics of the form

$$\begin{cases} x_i'(t) = g_i(x_i(t)) - p_i(t) \;\; (i = 0, 1) \\ W_0'(t) = e(W_0(t)) - y_0(t) \otimes q(t) \end{cases}$$

where $y_0(t) \in Y_0$ and $q(t) \in Z_0^\star$.

We then deduce from Corollary 5.2.3 that *the constraints (5.5.1) involving technological matrices are viable under dynamical economies of the form*

$$\begin{cases} x_0'(t) = g_0(x_0(t)) - A_0^\star W_0^\star(t) B_0^\star p(t) \\ x_1'(t) = g_1(x_1(t)) - C_1^\star p(t) \\ W_0'(t) := e(W_0(t)) - l(x_0(t)) \otimes B_0^\star p(t) \end{cases}$$

The latter differential equation can be written explicitly in the following form:

$$\forall\,(i,j),\;\; W'_{0_{i,j}}(t) \;:=\; e(W_0(t))_{i,j} - l(x_0(t))_i \otimes (B_0^\star p(t))_j$$

The regulation map can be written in the following way: p belongs to $\Pi_K(x_0, x_1.W_0)$ if and only if p is a solution to the equation

$$(C_1C_1^\star + B_0W_0A_0A_0^\star W_0^\star B_0^\star)\,p + \langle p, B_0l(x_0)\rangle B_0l(x_0) \;=\; z$$

where $r_0(x_0, x_1, W_0)$ belongs to $T_{M_0}(h(W_0l(x_0), x_1))$ and where

$$z \;:=\; C_1g(x_1) + B_0W_0A_0g_0(x_0) + B_0e_0(W_0)l(x_0) - r_0(x_0, x_1, W_0)$$

The solution to this equation can be written

$$p \;=\; (C_1C_1^\star + B_0W_0A_0A_0^\star W_0^\star B_0^\star)^{-1}\left(z - B_0l(x_0)\frac{((z, B_0l(x_0)))}{1+\|B_0l(x_0)\|^2}\right)$$

Indeed, if J is invertible, one can check easily that the solution q to the equation $Jq + \langle q, y\rangle y = z$ is equal to

$$q \;=\; J^{-1}\left(z - \frac{\langle J^{-1}z, y\rangle}{1+\langle J^{-1}y, y\rangle}y\right)$$

5.5.2 Multi-Stage Production Processes

Let $X_j, Y_j\,, Z_j\; (j = 0, \ldots m-1)$ and $Y_j\; (j = i, \ldots m)$ denote finite dimensional commodity spaces.

Consider linear operators $A_j \in \mathcal{L}(X_j, Y_j)$, $B_j \in \mathcal{L}(Z_j, X_{j+1})$ and $C_j \in \mathcal{L}(X_j, Y_j)$ and "technological matrices" $W_j \in \mathcal{L}(Y_j, Z_j)$.

The multi-stage production process takes the form

$$\forall\, j = 1, \ldots m,\;\; B_{j-1}W_{j-1}A_{j-1}x_{j-1} + C_jx_j \;\in\; M_j$$

where $M_j \subset Y_j$ are closed convex subsets.

Here $x_0 \in X_0$ is regarded as an output, $x_m \in X_m$ as an input, whereas $x_j \in X_j$ is regarded either as an intermediate output or input.

Example Observe that taking $M_j := \{0\}$ for $j = 1, \ldots, m-1$ and $C_j := \mathbf{1}$ for $j = 1, \ldots, m$, the above constraints read:

$$B_{m-1}W_{m-1}A_{m-1}\cdots B_{j-1}W_{j-1}A_{j-1}\cdots B_0W_0A_0x_0 \;\in\; M_m$$

It encompasses the case when the technological operators are products of partial technological operators. □

Therefore, in a multi-stage process, the state variables are the sequences $(x_j)_{j=0,\ldots,m}$, $(W_j)_{j=0,\ldots,m-1}$ made of vectors and of matrices. The above constraints are defined by inverse images under a multilinear map.

$$\left\{\begin{array}{ccc} & A_{j-1} & \\ x_{j-1} \in X_{j-1} & \mapsto & Y_{j-1} \\ & W_{j-1} & \downarrow \\ & & Z_{j-1} \\ & B_{j-1} & \downarrow \\ x_j \in X_j & \mapsto & Y_j \supset M \\ & C_j & \end{array}\right.$$

As in the one-stage process, one can as well consider a more general framework described by

1. $x_j \in L_j \subset X_j$, $(j = 1, \ldots m)$,
2. continuously differentiable maps $l_j : X_j \mapsto Y_j$ and $h_j : Z_{j-1} \times X_j \mapsto Y_j$,
3. "technological matrices" $W_j \in \mathcal{W}_j \subset \mathcal{L}(Y_j, Z_j)$
4. "scarcity constraint" $h_j(W_{j-1}l_{j-1}(x_{j-1}), x_j) \in M_j \subset Y_j$

The constrained set is defined by

$$\begin{array}{l} K := \{((x_j)_{j=0,\ldots,m}, (W_j)_{j=0,\ldots,m-1}) \in \prod_{j=0}^m L_j \times \prod_{j=0}^{m-1} \mathcal{W}_j \\ \textit{such that} \\ h_j(W_{j-1}l_{j-1}(x_{j-1}), x_j) \in M_j\} \end{array} \tag{5.5.3}$$

We set

$$A_j := l'_j(x_j), \quad B_{j-1} := h'_{j_{x_{j-1}}}(W_{j-1}l_{j-1}(x_{j-1}), x_j)$$

and

$$C_j := h'_{j_{x_j}}(W_{j-1}l_{j-1}(x_{j-1}), x_j)$$

Hence, the directional derivative is equal to

$$\left\{\begin{array}{l} h'_j(W_{j-1}l_{j-1}(x_{j-1}), x_j)(u_{j-1}, u_j, V_{j-1}) \\ = C_j u_j + B_{j-1}W_{j-1}A_{j-1}u_{j-1} + B_{j-1}V_{j-1}l_{j-1}(x_{j-1}) \end{array}\right.$$

and can be written in the form

$$B_{j-1}W_{-1}A_{j-1}u_{j-1} + C_j u_j + (l_{j-1}(x_{j-1}) \otimes B_{j-1})V_{j-1}$$

The transpose of this operator maps $(q_j) \in \prod_{j=1}^m Y_j^\star$ to

$$(X_0 \times \mathcal{L}(Y_0^\star, Z_0^\star)) \times \left(\prod_{j=1}^{m-1} (X_j^\star \times \mathcal{L}(Y_j^\star, Z_j^\star)\right) \times X_m^\star$$

in the following way:

$$\begin{cases} (A_0^\star W_0^\star B_0^\star q_1 \,,\; l_0(x_0) \otimes B_0^\star q_1) \\ (A_j^\star W_j^\star B_j^\star q_{j+1} + C_j^\star q_j \,,\; l_j(x_j) \otimes B_j^\star q_{j+1}) \;\; (j = 1, \ldots, m-1) \\ C_m^\star q_m \end{cases}$$

We posit a sufficient condition guaranteeing the transversality conditions:

$$\forall\, ((x_j)_{j=0,\ldots,m},\; (W_j)_{j=0,\ldots,m-1}),\;\; h'_{j_{x_j}}(W_{j-1}l(x_{j-1}), x_j) \text{ are surjective} \quad (5.5.4)$$

Assume now that the dynamical behaviors are described by (simple) dynamics of the form

$$\begin{cases} x_j'(t) &=& g_j(x_j(t)) - p_j(t) \;\; (j = 0, \ldots m) \\ W_j'(t) &=& e(W_j(t)) - y_j(t) \otimes q_j(t) \;\; (j = 0, \ldots m-1) \end{cases}$$

where $y_j(t) \in Y_j$ and $q_j(t) \in Z_j^\star$.

We then deduce from Corollary 5.2.3 that *the constraints (5.5.3) involving technological matrices are viable under dynamical economies of the form*

$$\begin{cases} x_0'(t) &=& g_0(x_0(t)) - A_0^\star W_0^\star(t) B_0^\star p_1(t) \\ x_j'(t) &=& g_j(x_j(t)) - C_j^\star p_j(t) - A_j^\star W_j^\star(t) B_j^\star p_{j+1}(t) \;\; (j = 1, \ldots, m-1) \\ x_m'(t) &=& g_m(x_m(t)) - C_m^\star p_m(t) \\ W_j'(t) &:=& e(W_j(t)) - l_j(x_j(t)) \otimes B_j^\star p_{j+1}(t) \;\; (j = 0, \ldots, m-1) \end{cases}$$

The regulation map can be written in the form in the following way: $(p_j)_{j=1,\ldots,m}$ belongs to $\Pi_K((x_j)_{j=0,\ldots,m},\; (W_j)_{j=0,\ldots,m-1})$ if and only if p is a solution to the system of equations

$$\begin{cases} (C_1C_1^\star + B_0W_0A_0A_0^\star W_0^\star B_0^\star)\, p + \langle p, B_0 l_0(x_0)\rangle B_0 l_0(x_0) + C_1 A_1^\star W_1^\star B_1^\star p_2 \;=\; z_1 \\[2ex] \left(C_jC_j^\star + B_{j-1}W_{j-1}A_{j-1}A_{j-1}^\star W_{j-1}^\star B_{j-1}^\star\right) p_j \\ +\langle p_j, B_{j-1}l_{j-1}(x_{j-1})\rangle B_{j-1}l_{j-1}(x_{j-1}) + C_j A_j^\star W_j^\star B_j^\star p_{j+1} \\ +B_{j-1}W_{j-1}A_{j-1}C_{j-1}^\star p_{j-1} \;=\; z_j \;\; (j = 2, \ldots, m-1) \\[2ex] \left(C_mC_m^\star + B_{m-1}W_{m-1}A_{m-1}A_{m-1}^\star W_{m-1}^\star B_{m-1}^\star\right) p_m \\ +\langle p_m, B_{m-1}l_{m-1}(x_{m-1})\rangle B_{m-1}l_{m-1}(x_{m-1}) + B_{m-1}W_{m-1}A_{m-1}C_{m-1}^\star p_{m-1} \;=\; z_m \end{cases}$$

where where r_j belongs to $T_{M_j}(h_j(W_{j-1}l(x_{j-1}), x_j))$ and where

$$z_j \;:=\; C_j g(x_j) + B_{j-1}W_{j-1}A_{j-1}g_{j-1}(x_{j-1}) + B_{j-1}e_{j-1}(W_{j-1})l(x_{j-1}) - r_j$$

In other words, this solution is obtained by inverting a tri-diagonal block matrix:

$$\begin{pmatrix} \mathcal{J}_1 & \mathcal{M}_1 & 0 & \cdots & \cdots & \cdots & \cdots & \cdots & \cdots & \cdots & \cdots & \cdots & 0 \\ \mathcal{M}_1^\star & \mathcal{J}_2 & \mathcal{M}_2 & 0 & \cdots & \cdots & \cdots & \cdots & \cdots & \cdots & \cdots & \cdots & 0 \\ \vdots & \vdots & \vdots & \vdots & \vdots & \vdots & \vdots & \vdots & \vdots & \vdots & \vdots & \vdots & \vdots \\ 0 & \cdots & \cdots & \cdots & 0 & \mathcal{M}_{j-1}^\star & \mathcal{J}_j & \mathcal{M}_j & 0 & \cdots & \cdots & \cdots & 0 \\ \vdots & \vdots & \vdots & \vdots & \vdots & \vdots & \vdots & \vdots & \vdots & \vdots & \vdots & \vdots & \vdots \\ 0 & \vdots & \vdots & \vdots & \vdots & \vdots & \vdots & \vdots & \vdots & \vdots & \vdots\mathcal{M}_{m-2}^\star & \mathcal{J}_{m-1} & \mathcal{M}_{m-1} \\ 0 & 0 & 0 & \cdots & \cdots & \cdots & \cdots & \cdots & \cdots & \cdots & \cdots & \mathcal{M}_{m-1}^\star & \mathcal{J}_m \end{pmatrix}$$

where we set

$$\begin{cases} \mathcal{J}_j p_j \\ := \left(C_j C_j^\star + B_{j-1} W_{j-1} A_{j-1} A_{j-1}^\star W_{j-1}^\star B_{j-1}^\star\right) p_j + \langle p_j, B_{j-1} l_{j-1}(x_{j-1})\rangle B_{j-1} l_{j-1}(x_{j-1}) \end{cases}$$

and

$$\mathcal{M}_j := C_j A_j^\star W_j^\star B_j^\star$$

Chapter 6

Viability Kernels and Capture Basins

Introduction

When a closed subset K is not viable under a dynamical economy, then two questions arise naturally: find solutions starting from K which **remain viable in K as long as possible**, and starting outside of K, find solutions which **return to K as soon as possible** to restore viability, as it is advocated in [177, Doyen & Saint-Pierre].

In order to achieve this purpose as well as answering other related questions, we have to cover briefly several preliminary tasks that we shall now describe.

1. We could have observed long ago that solutions to a dynamical economy are solutions to the differential inclusion $x'(t) \in F(x(t))$ where, for each commodity x, $F(x) := c(x, P(x))$ is the subset of feasible velocities[1] (which are no longer univoquely related to the state as in deterministic systems). It is in the general framework of **differential inclusions** that the Viability Theorem and the Equilibrium Theorem are stated in this chapter and proved in the Tool-Box. It will be easier to handle differential inclusions instead of dynamical economies whenever the explicit use of either prices or perturbations is not required in order to save notations and to concentrate on the novel issues.

   Only in a stationary environment does the concept of equilibrium make sense. Still, the concept of equilibrium — regarded as a minimal viability domain — deserves another visit in the framework of viability domains. When K is convex and compact, the celebrated Brouwer fixed point theorem can be reformulated by stating that "viability implies stationarity". This is also true when the image $F(K)$ is convex,

[1] Conversely, a differential inclusion is an example of an economy in which the prices are the velocities ($c(x,p) =: p$ & $P(x) := F(x)$.)

or when there exists a viable solution of "average finite length". When no equilibrium does exist, we have to be satisfied with the existence of minimal viability domains. Otherwise, for nonconvex subsets K, algebraic topology should be requisitioned, but this is outside the scope of this book.

The use of Marchaud set-valued maps for which most of the viability theorems hold true, where convexity of the images compensates the minimal requirement of continuity (closedness of the graph instead of Lipschitzianity), may be regarded as overwhelming. First, we shall provide in Chapter 7 a way of disregarding this requirement by using smaller regulation maps for guaranteeing the viability of a given subset. But the lack of convexity can also be partly accepted when the evolution of the commodity is governed by the average of a large finite number of actions:

$$x'(t) \in \frac{1}{J}\sum_{j=1}^{J} F_j(x(t))$$

where the values of the set-valued maps F_j do not have necessarily convex values. However, thanks to the Shapley-Folkman Theorem[2], we shall show that there exists a solution starting at x_0 to the perturbed differential inclusion

$$x'(t) \in \frac{1}{J}\sum_{j=1}^{J} F_j(x(t)) + \varepsilon_J(t)B$$

(where B is the unit ball of X) in which the error $\varepsilon_J(t)$ converges to 0 when $J \to +\infty$.

2. After having characterized the subsets K viable under dynamical economies or differential inclusions, one can ask what happens when a given constrained set K is not viable under a differential inclusion $x' \in F(x)$? Or, regarding K as an objective or a target, can it be reached in prescribed finite time, in finite time or in infinite time, in minimal time ? Such problems are of utmost importance in mathematical planning, for instance.

   First, we introduce the **reachable tube** which associates with any compact subset K and any time t the values at time t of the solutions to the differential inclusion starting from K. We shall prove that its inverse associates with any t the subset of elements which reach K at the prescribed time t.

   Next, we shall answer two basic and natural questions:

[2] which is sophisticated version of the Carathéodory Convexity Theorem. In control theory, convexification procedures of the set of velocities are known under the name of "relaxation".

(a) For any time $T \geq 0$, what is the set of initial states $x \in K$ such that at least one solution starting from x is viable in K in the time interval $[0, T[$? We shall denote this set by $\text{Viab}_F(K, T)$ and call it the T-**viability kernel**[3] of K under F.

When x does belong to such a viability kernel, we shall also look for solutions which are viable in K as long as possible.

When $T := +\infty$, we define the **viability kernel** of K under F as the subset of initial states $x \in K$ such that at least one solution starting from x is viable in K forever.

(b) For any time $T \geq 0$, what is the set of initial states $x \notin K$ such that a least one solution starting from x reaches K before time T? We shall denote this set by $\text{Capt}_F(K, T)$ and call it the T-**capture basin**[4] of K because it assigns to its elements its "capture" time needed to reach K. In this perspective, K may be regarded as a target, so that the capture basin of K under F describes a "potential victory domain" of the problem of hitting the target in prescribed or finite time.

When x does belong to such a capture basin, we shall also look for solutions which hit K as soon as possible.

In retrospect, we should have called the T-capture basin the "recovery" basin, since it assigns to its elements the minimal time needed to recover the viability of the system.

When $T := +\infty$, the **capture basin** of K under F is defined as the subset of initial states from which starts at least a solution "hitting" the boundary in finite time.

The "tubes" $T \rightsquigarrow \text{Viab}_F(K, T)$ and $T \rightsquigarrow \text{Capt}_F(K, T)$ provide a precise description of the dynamics relatively to the closed subset K. If $x \in \text{Capt}_F(K, T)$, then it reaches K before T. If it reaches it at a point of the boundary of K which belongs to $\text{Viab}_F(K, S)$, then it remains in K during an interval $[T, S]$, and so on. They provide the dynamical portrait of the boundary of K, by describing the subsets of the elements of the boundary from which either all or at least one solution leave or enter K, forever, for a finite interval of time or instantaneously, and thus, describing the elements from which solutions "fluctuate" around — in and out — the boundary of the set K.

---

[3] In VIABILITY THEORY, [28, Aubin] this set was called the **exit set** and was denoted by $\text{Exit}_F(K, T)$. The recent proliferation of sets of this nature due to investigations of dynamical games led to rationalize further the terminology around few key concepts.

[4] In VIABILITY THEORY, [28, Aubin] was introduced rather the **hitting set** defined by

$$\forall\, T \geq 0, \ \text{Hit}_F(K, T) := \text{Capt}_F(\overline{X \backslash K}, T)$$

of initial states from which starts at least a solution "hitting" the boundary before T.

Furthermore, we shall prove that the **viability kernel** of the constrained set K (under the differential inclusion) is **the largest closed subset of K viable under this differential inclusion.**

We shall use this key concept over and over in many instances. The very first application of the viability kernel is the construction of the regulation map of dynamical economies with **bounded inflation** and the concept of **viability niches** which we shall investigate in the next chapter.

As it was seen in Part I, the examples of viability kernels and viability niches were obtained thanks to an algorithm — the viability kernel algorithm — devised by Patrick Saint-Pierre, and adapted by Luc Doyen, Philippe Lacoude, Katharina Müllers, Nicolas Seube and Patrick Saint-Pierre himself to many other examples.

These questions are related to the concepts of Lyapunov stability, of attractor and of absorbing sets. When time t goes to ∞, the upper limit of the reachable tube, also called the **limit set** (or ω-limit set) of the tube is a viability domain. The **attractor** is defined as the closure of the limit sets of all compact subsets. It is a viability domain which contains the asymptotic features of the differential inclusion. We shall prove that any bounded invariant absorbing set contains the attractor.

It also happened that the concept of viability kernel allows us to treat the "target problem" and related issues as **fluctuations** around the target or the **permanence** property introduced by biologists.

In summary, not only the concept of viability kernel is naturally important in the framework of economic models having motivated viability theory in the first place, but it happened that the notions of equilibria, of absorbing sets, of basins of absorption, of attractors, of "permanence", of "fluctuation", of Lyapunov stability", of optimal Lyapunov functions, of value function of an intertemporal optimization problem as well as other dynamical features can be studied by using the concept of the **viability kernel** as a mathematical tool, as we shall see in Chapters 7 and 8.

3. We next relate T-viability kernels and T-capture basins to exit time and hitting time functions. The **upper exit time function** is the function which assigns to every $x \in K$ the longest time for which at least one solution starting from x remains viable in K. When F is Marchaud, it is an upper semicontinuous function, the upper level sets of which are the finite time viability kernels. The **lower hitting time function** (or minimal time function) associates with every $x \notin K$ the shortest time at which at least one solution hits K. When F is Marchaud, it is a lower semicontinuous function, the lower level sets of which are the

finite time capture basins. We shall characterize the hypograph and the epigraph of these functions as viability kernels and as solutions to Hamilton-Jacobi functions in Chapter 8.

On the boundary of K, we define the **egress set** as the set of elements from which all solutions leave K instantaneously, and we prove that it is equal to the subset on which the upper exit time vanishes. When the interior of K is not empty, we define the **ingress set** as the set of elements from which at least one solution enters the interior of K instantaneously, and we prove that it is the subset of the boundary on which the lower hitting function vanishes. Together with the viability kernels of K and of the closure of its complement, these subsets draw a dynamical portrait of the boundary of K, where the viability scenarios of the differential inclusion or the dynamical economy leave their imprints.

4. The concept of **chaos à la Saari** involves the concept of capture basins and lower hitting time functions. The idea is the following: The subset K is covered by a finite number of subsets (called "cells", such as qualitative cells in qualitative economics). This covering is "chaotic à la Saari" when for every prescribed order of visit, there exists at least one solution visiting successively these cells in the prescribed order. This very important concept was used by Donald Saari for explaining many paradoxes in different fields, and in particular, in the field of social choice and voting.

6.1 Toward Differential Inclusions

We can simplify further the formulation of dynamical economies and games by introducing the set-valued map $F : K \rightsquigarrow X$ defined by

$$F(x) := c(x, P(x)) = \{c(x,p)\}_{p \in P(x)} \tag{6.1.1}$$

associating with any commodity the set of available velocities.

6.1.1 Differential Inclusions

Therefore, the solutions of the dynamical economy

$$\begin{cases} i) & x'(t) = c(x(t), p(t)) \\ ii) & p(t) \in P(x(t)) \end{cases}$$

are solutions to the differential inclusion[5]

$$x'(t) \in F(x(t)) \tag{6.1.2}$$

[5]Conversely, one can regard a differential inclusion as a dynamical economy where the velocities $c(x,p) = p$ are prices and where the pricing map $P := F$ is the right-hand side of the differential inclusion.

and there exists a viable equilibrium $(\overline{x}, \overline{p})$ of the dynamical economy if and only if $\overline{x}$ is an equilibrium of F in the sense that

$$\overline{x} \in K \;\&\; 0 \in F(\overline{x}) \tag{6.1.3}$$

Hence, we have to appeal to existence of viable solutions to differential inclusions (the Viability Theorem) and to existence of viable equilibria of set-valued maps (the Equilibrium Theorem) to prove the theorems stated in the preceding chapters for dynamical economies.

Therefore, dynamical economies are instances of differential inclusions. As long as we can use theorems on differential inclusions which do not use the specific features of dynamical economies, we shall state them in this generality, despite the potential danger of losing for some time some economic intuition. However, we shall need these results for other purposes on one hand, and on the other hand, this more abstract approach simplifies the exposition by using only the necessary properties for solving a given problem.

Therefore, we shall consider initial value problems (or Cauchy problems) associated with the differential inclusion (6.1.2).

Definition 6.1.1 (Viability and Invariance Properties) *Let K be a subset of the domain of F (usually defined by constraints on the state). We shall say that K is* viable under F on $[0,T]$ *if from any initial state x_0 in K starts* at least *one solution to differential inclusion (6.1.2) which is* viable *in K. When $T := +\infty$, we simply say that K is viable under F.*

The subset K is said to be invariant under F *on $[0,T]$ if starting from any initial state x_0 of K,* all *solutions to differential inclusion (6.1.2) are* viable in *K on the interval $[0,T]$.*

We shall characterize the viability property and provide sufficient conditions for the existence of an equilibrium in terms of "tangential conditions" on the boundary of the constrained subset K.

Definition 6.1.2 *Let $F : X \rightsquigarrow X$ be a nontrivial set-valued map. We shall say that a subset $K \subset \mathrm{Dom}(F)$ is a* viability domain *of F if*

$$\forall\, x \in K, \quad F(x) \cap T_K(x) \neq \emptyset$$

and an invariance domain *of F if*

$$\forall\, x \in K, \quad F(x) \subset T_K(x)$$

This means that for any point $x \in K$, there exists at least one direction $v \in F(x)$ which is contingent to K at x for viability domains and that all directions $v \in F(x)$ are contingent to K at x for invariance domains.

We now recall a very important characterization of viability domains (see Theorem 3.2.4 of VIABILITY THEORY, [28, Aubin]) which allows us to extend to the nonconvex case the dual characterization of the tangential condition involved in the definition of the viability domains:

Theorem 6.1.3 *Assume that the set-valued map $F : K \rightsquigarrow X$ is upper semicontinuous with convex compact values. Then the three following properties are equivalent:*

$$\begin{cases} i) & \forall\, x \in K, \;\; F(x) \cap T_K(x) \;\neq\; \emptyset \\ ii) & \forall\, x \in K, \;\; F(x) \cap \overline{co}\,(T_K(x)) \;\neq\; \emptyset \\ iii) & \forall\, x \in K, \;\; \forall\, p \in N_K(x), \;\; \sigma^{\flat}(F(x),p) \;:=\; \inf_{v \in F(x)} \langle p, v \rangle \;\leq\; 0 \end{cases} \tag{6.1.4}$$

(See the proof due to Hélène Frankowska of Theorem 3.2.4 in Chapter 3 of [28, Aubin]).

In order to deduce the viability and equilibrium theorems for dynamical economies, we have to prove that Marchaud dynamical economies yield differential inclusions the right-hand side of which are Marchaud maps:

Lemma 6.1.4 *If the dynamical economy (P, c) is Marchaud, then the set-valued map F defined by (6.1.1) is also Marchaud.*

Proof — First, since the map c is continuous and the graph of P is closed by assumption, then the graph of F is also closed.

Second, since c is affine with respect to p and since the images $P(x)$ are convex and compact, we note that the images $F(x)$ are convex and compact.

Third, since the growth of c and P is linear, the growth of F is also linear. □

Viability properties require differential inclusions governed by Marchaud maps. Invariance properties require the set-valued maps to be Lipschitz:

Definition 6.1.5 *When X and Y are normed spaces, we shall say that $F : X \rightsquigarrow Y$ is* Lipschitz *if there exist a positive constant λ such that*

$$\forall\, x_1, x_2 \in X, \;\; F(x_1) \;\subset\; F(x_2) + \lambda \|x_1 - x_2\| B_Y$$

When the images of F are closed, we denote by $\|F\|_\Lambda$ the smallest constant λ.

We now provide a characterization of invariance domains which allows to extend to the nonconvex case the dual characterization of the tangential condition involved in the definition of the viability domains:

Theorem 6.1.6 *Assume that the set-valued map $F : K \rightsquigarrow X$ is Lipschitz. Then the three following properties are equivalent:*

$$\begin{cases} i) & \forall\, x \in K, \;\; F(x) \subset T_K(x) \\ ii) & \forall\, x \in K, \;\; F(x) \subset \overline{co}(T_K(x)) \neq \emptyset \\ iii) & \forall\, x \in K, \;\; \forall\, p \in N_K(x), \;\; \sigma^\flat(F(x),p) \; := \; \sup_{v \in F(x)} \langle p, v \rangle \; \leq \; 0 \end{cases}$$

In order to deduce the invariance theorems for dynamical economies, we have to define Lipschitz dynamical economies and show that they yield differential inclusions the right-hand side of which are Lipschitz maps:

Definition 6.1.7 *We shall say that* dynamical economy (P, c) *is Lipschitz if the set-valued map P is Lipschitz with compact values and if the single-valued map c is Lipschitz.*

We thus observe that

Lemma 6.1.8 *If the dynamical economy (P, c) is Lipschitz, then the set-valued map F defined by (6.1.1) is also Lipschitz and the Lipschitz constants of the map $c(\cdot, P(\cdot))$ is estimated by $\|c\|_\Lambda(1 + \|P\|_\Lambda)$.*

6.1.2 The Viability Theorem

We now state the general statements from which we derived the above theorems on viability of dynamical economies and games and on the existence of equilibria.

Theorem 6.1.9 (Viability Theorem) *Consider a* Marchaud map $F : X \rightsquigarrow X$ *and a* closed *subset $K \subset \mathrm{Dom}(F)$ of a finite dimensional vector space X.*

Then K is viable under F if and only if K is a viability domain *of F.*

Furthermore, if we set

$$c_K \; := \; \sup_{x \in K} \frac{\|F(x)\|}{\|x\| + 1}$$

then every solution $x(\cdot)$ starting at x_0 satisfies the estimates

$$\begin{cases} \forall\, t \geq 0, \;\; \|x(t)\| \; \leq \; (\|x_0\| + 1)e^{c_K t} \\ \text{and} \\ \text{for almost all } t \geq 0, \;\; \|x'(t)\| \; \leq \; c_K(\|x_0\| + 1)e^{c_K t} \end{cases}$$

Finally, let $\mathcal{S}_F : X \rightsquigarrow \mathcal{C}(0, \infty, X)$ denote the solution map *associating with any initial state x the set $\mathcal{S}_F(x)$ of solutions to the differential inclusion (6.1.2) starting at x. Then the graph of $\mathcal{S}_F$ is closed (and the graph of its restriction to any compact subset is compact in $X \times \mathcal{C}(0, \infty; X)$).*

6.1.3 Monotone Solutions along a Preorder

Not only we wished to characterize the viability of constraints under a dynamical economy or a differential inclusion, but we also desired to characterize the fact that starting from any state, at least one solution increases along a preorder $\succeq$.

This means that from any initial state, there exists a solution to the differential inclusion

$$x'(t) \in F(x(t))$$

which **does not decrease** in the sense that

$$\forall t \geq s \geq 0, \quad x(t) \succeq x(s)$$

Such solutions are said to be **satisfycing** (or **monotone**) with respect to the preorder. A preorder is said to be **satisfycing** if from any initial state, there exists at least one satisfycing solution to the differential inclusion.

Proposition 6.1.10 *Let F be a Marchaud map and P be a preorder with closed graph whose domain is contained in the domain of F.*

The following statements are equivalent:

$$\left\{\begin{array}{ll} i) & \forall\, x \in \mathrm{Dom}(P),\ F(x) \cap T_{P(x)}(x) \neq \emptyset \\ ii) & \forall\, (x,y) \in \mathrm{Graph}(P),\ \ F(y) \cap DP(x,y)(0) \neq \emptyset \\ iii) & \forall\, x_0 \in \mathrm{Dom}(P),\ \exists\, x(\cdot) \in \mathcal{S}(x_0) \ \ \textit{such that} \\ & \forall\, t \geq s \geq 0,\ \ x(t) \succeq x(s) \end{array}\right. \tag{6.1.5}$$

Proof — Since any monotone solution starting at x remains in $P(x)$, condition (6.1.5)i) is obviously necessary.

It is sufficient, since it implies that the graph of P is a closed viability domain of the set-valued map $(x,y) \rightsquigarrow \{0\} \times F(y)$. Indeed, for any $y \in P(x)$, there exists $v \in F(y) \cap T_{P(y)}(y)$, i.e., such that $y + h_n v_n \in P(y) \subset P(x)$ for some sequences $h_n \to 0+$ and $v_n \to v$. Hence, the pair $(x + h_n 0, y + h_n v_n)$ belongs to the graph of P, i.e., $(0,v) \in T_{\mathrm{Graph}(P)}(x,y)$.

We infer that for any $(x_0, x_0) \in \mathrm{Graph}(P)$, there exists a solution $(x(\cdot), y(\cdot))$ to the system of differential inclusions $x' = 0$ and $y' \in F(y)$ which is viable in $\mathrm{Graph}(P)$, i.e., a solution $y(\cdot) \in \mathcal{S}(x_0)$ such that

$$\forall\, t \geq 0, \quad y(t) \in P(x_0) \tag{6.1.6}$$

We associate now with $h \to 0+$ the grid jh, $(j = 1, \ldots)$ and we build a solution $x_h(\cdot) \in \mathcal{S}(x_0)$ to differential inclusion (6.1.2) iteratively: for $j = 0$, we take $x_h(\cdot) = y(\cdot)$ on the interval $[0,h]$ satisfying (6.1.6), then we take $x_h(\cdot)$ on $[h, 2h]$ to be a solution starting at $x_h(h)$ and satisfying $x_h(t) \in P(x_h(h))$, etc.

Since the image $\mathcal{S}(x_0)$ is compact, a subsequence (again denoted by) x_h converges to some solution $x(\cdot) \in \mathcal{S}(x_0)$ in the space $\mathcal{C}(0,\infty)$. Let $t > s$ be approximated by $j_h h \geq k_h h$ so that

$$x_h(jh) \in P(x_h(kh)) \text{ or } (x_h(kh), x_h(jh) \in \text{Graph}(P))$$

The graph of P being closed, we infer that $(x(s), x(t)) \in \text{Graph}(P)$, i.e., that $x(t) \in P(x(s))$. □

When the preorder is defined though a given loss function (or a utility function), the monotonicity property is closely related to the study of Lyapunov functions which we shall present in Chapter 8. In this case, if the preorder defined by a loss function V is not consistent with a differential inclusion, we shall be able to build the smallest lower semicontinuous loss function larger than V which does not increase along at least one solution to the differential inclusion. We shall be able to reveal such a loss function concealed in the dynamical system by using the concept of viability kernel.

6.1.4 Averaging a Large Number of Actions

Let X be a finite dimensional vector space, representing for instance a commodity space. We assume that the evolution of the commodity is governed by the average of a large finite number of actions:

$$x'(t) \in \frac{1}{|J|} \sum_{j\in J} F_j(x(t))$$

where J is a finite set of actions and $|J|$ its cardinal.

Here, we shall not assume that the values of the set-valued maps F_j have convex values.

Theorem 6.1.11 *Assume that the set-valued maps $F_j : X \rightsquigarrow X$ have closed graph and linear growth : There exists a constant $c > 0$ such that*

$$\forall\, x \in K, \quad \sup_{j\in J} \|F_j(x)\| \leq c(\|x\| + 1)$$

If

$$\forall\, j \in J, \quad F_j(x) \cap T_K(x) \neq \emptyset \tag{6.1.7}$$

then, for any initial state $x_0 \in K$, there exists a solution to the differential inclusion

$$x'(t) \in \frac{1}{|J|} \sum_{j\in J} F_j(x(t)) + \frac{2c\dim(X)(\|x_0\| + 1)e^{ct}}{|J|} B$$

viable in K (where B is the unit ball of X).

Hence the error due to the lack of convexity of the images of the F_j decreases to 0 when the number $|J|$ of actions increases to ∞.

We shall prove this theorem thanks to the Shapley-Folkman Theorem, which is sophisticated version of the Carathéodory Theorem:

Theorem 6.1.12 (Shapley-Folkman) *Let us consider a finite family* $\{F_j\}_{j\in J}$ *of subsets* $F_j \subset X$. *Then, for any* $x \in \mathrm{co}\left(\sum_{j\in J} F_j\right)$[6], *there exists a subset* $J(x) \subset J$ *of at most* $\dim(X)$ *indexes such that*

$$x \in \left(\sum_{j\notin J(x)} F_j\right) + \sum_{j\in J(x)} \mathrm{co}(F_j)$$

If $\|K\| := \sup_{x\in K} < +\infty$ is bounded, we observe that

$$\mathrm{co}(K) \subset K + 2\|K\|B$$

As a consequence, we deduce from the Shapley-Folkman Theorem the following consequence using this concept of convexity gap:

Theorem 6.1.13 *Let us consider a finite family* $\{F_j\}_{j\in J}$ *of bounded subsets* $F_j \subset X$. *Then*

$$\mathrm{co}\left(\sum_{j\in J} F_j\right) \subset \left(\sum_{j\in J} F_j\right) + 2\max_{j\in J}\|F_j\|\dim(X)B$$

Proof of Theorem 6.1.11 — Let us set $F(x) := \frac{1}{|J|}\sum_{j\in J} F_j(x)$. First, we check that

$$\forall\, x \in K,\ \overline{\mathrm{co}}(F(x)) \cap T_K(x) \neq \emptyset \tag{6.1.8}$$

Indeed, the tangential condition (6.1.7) implies that

$$\forall\, x \in K,\ \left(\frac{1}{|J|}\sum_{j\in J} F_j(x)\right) \cap \overline{\mathrm{co}}(T_K(x)) \neq \emptyset$$

which is equivalent to (6.1.8) by Theorem 6.1.3.

By the Viability Theorem, we know that there exists one viable solution to the differential inclusion

$$x'(t) \in \overline{\mathrm{co}}(F(x(t)))$$

starting from x_0 and satisfying:

$$\|x(t)\| \leq (\|x_0\| + 1)e^{ct} - 1 \tag{6.1.9}$$

[6]co(K) denotes the convex hull of K.

Since the growth of the set-valued maps F_j is linear, we deduce that their convexity gaps are bounded by

$$\|F_j(x(t))\|B \subset c(\|x_0\|+1)e^{ct}B$$

By Theorem 6.1.13, we deduce that

$$\mathrm{co}(F(x(t))) \subset F(x(t)) + \frac{2c\dim(X)(\|x_0\|+1)e^{ct}}{|J|}B$$

Then, our result ensues. □

6.1.5 The Invariance Theorem

The invariance Theorem requires dynamics governed by Lipschitz maps without assuming that their images are necessarily convex:

Theorem 6.1.14 *Let us assume that $F : X \rightsquigarrow X$ is Lipschitz with closed values.*

Then a closed subset $K \subset X$ is locally invariant under F if and only if K is an invariance domain.

Furthermore, every solution $x(\cdot)$ to the differential inclusion $x' \in F(x)$ satisfies inequalities

$$\|x(t)-x(0)\| \le \|F(x(0))\|\frac{e^{\|F\|_\Lambda t}-1}{\|F\|_\Lambda}$$

and

$$x(t)-x(s) \in (t-s)F(x(s)) + \|F(x(0))\|e^{\|F\|_\Lambda s}\frac{e^{\|F\|_\Lambda (t-s)}-\|F\|_\Lambda(t-s)-1}{\|F\|_\Lambda}$$

Finally, the solution map $S_F : X \rightsquigarrow \mathcal{C}(0,\infty,X)$ is lower semicontinuous.

6.1.6 Viability and Equilibria

We now state one theorem on the existence of an equilibrium of a differential inclusion.

Theorem 6.1.15 (Equilibrium Theorem # 1) *Assume that X is a finite dimensional vector-space and that $F : X \rightsquigarrow X$ is an upper hemicontinuous set-valued map with closed convex images.*

If $K \subset X$ is a convex compact viability domain of F, then it contains an equilibrium of F.

Therefore, the Brouwer Fixed Point Theorem can be reformulated by saying that *any convex compact viability domain of a Marchaud map contains one equilibrium.*

More general statements are provided in the Tool-Box.

We shall derive from the Viability Theorem a comparable statement, where the assumption of the convexity of K is replaced by the convexity of the image $F(K)$ of K by F:

Theorem 6.1.16 (Equilibrium Theorem # 2) *Assume that X is a finite dimensional vector-space and that $F : X \rightsquigarrow X$ is Marchaud.*

If $K \subset X$ is a compact subset such that $F(K)$ is convex and if there exists at least one viable solution to the differential inclusion $x' \in F(x)$, then there exists a viable equilibrium of F in K.

Naturally, the Viability Theorem above implies the existence of such a viable solution whenever K is a viability domain of F.

Proof — Assume that there is no equilibrium. Hence, this means that 0 does not belong to the closed convex subset $F(K)$, so that the Separation Theorem implies the existence of some $p \in X^\star$ and $\varepsilon > 0$ such that

$$\sup_{x\in K, v\in F(x)} \langle v, -p\rangle = \sigma(F(K), -p) < -\varepsilon$$

Hence, let us take any viable solution $x(\cdot)$ to differential inclusion

$$\text{for almost all } t \geq 0, \ \ x'(t) \in F(x(t))$$

which exists by assumption. We deduce that

$$\forall\, t \geq 0, \ \ \langle -p, x'(t)\rangle \leq -\varepsilon$$

so that, integrating from 0 to t, we infer that

$$\varepsilon\, t \leq \langle p, x(t) - x(0)\rangle$$

But K being bounded, we thus derive a contradiction. □

We can even relax the assumption of the convexity of $F(K)$, as the following "ergodic" Theorem shows:

Theorem 6.1.17 (Equilibrium Theorem # 3) *Let us assume that F is Marchaud and that K is compact. If there exists a solution $x(\cdot)$ to the differential inclusion $x' \in F(x)$ viable in K such that*

$$\inf_{t>0} \frac{1}{t} \int_0^t \|x'(\tau)\| d\tau = 0$$

then there exists a viable equilibrium $\bar{x}$, i.e., a state $\bar{x} \in K$ solution to the inclusion $0 \in F(\bar{x})$.

Observe that it is sufficient that the lim inf of the Cesaro means $\frac{1}{t_n}\int_0^{t_n}\|x'(\tau)\|d\tau$ of the velocities of a viable solution is equal to 0 to derive the existence of an equilibrium.

Proof — Let us assume that there is no viable equilibrium, i.e., that for any $x \in K$, 0 does not belong to $F(x)$. Since the images of F are closed and convex, the Separation Theorem implies that there exists $p \in \Sigma$, the unit sphere, and $\varepsilon_p > 0$ such that $\sigma(F(x), -p) < -\varepsilon_p$. In other words, we can cover the compact subset K by the subsets

$$\mathcal{V}_p := \{ x \in K \mid \sigma(F(x), -p) < -\varepsilon_p \}$$

when p ranges over Σ. They are open thanks to the upper hemicontinuity of F, so that the compact subset K can be covered by q open subsets $\mathcal{V}_{p_j}$. Set $\varepsilon := \min_{i=1,\dots,q} \varepsilon_{p_i} > 0$.

Consider now a viable solution to the differential inclusion $x' \in F(x)$, which exists by assumption. Hence, for any $t \geq 0$, $x(t)$ belongs to some $\mathcal{V}_{p_j}$, so that

$$-\|x'(t)\| \leq \langle -p_j, x'(t)\rangle \leq \sigma(F(x(t)), -p_j) < -\varepsilon$$

and thus, by integrating from 0 to t, we have proved that there exists $\varepsilon > 0$ such that, for all $t > 0$,

$$\varepsilon < \frac{1}{t}\int_0^t \|x'(\tau)\|d\tau$$

a contradiction of the assumption of the theorem. □

6.1.7 Minimal Viability Domains

An equilibrium being a viability domain reduced to a singleton, we may investigate whether a compact (nonconvex) viability domain contains minimal non empty viability domain:

Proposition 6.1.18 *Let K be a nonempty compact viability domain of a Marchaud map F. Then nonempty minimal viability domains M do exist and are made of ω-limit sets of viable solutions. Actually, they enjoy the following property:*

$$\forall\, x \in M,\ \exists\, x(\cdot) \in \mathcal{S}_F(x) \mid M = \omega(x(\cdot))$$

Proof— We apply Zorn's lemma for the inclusion order on the family of nonempty closed viability domains of F. For that purpose, consider any decreasing family of closed viability domains M_i and its intersection $M_\star := \bigcap M_i$. It is a closed viability domain thanks to the Stability Theorem 12.1.5, and nonempty because K is compact. Therefore there exists a minimal element for this preorder.

Let $M \subset K$ be a minimal closed viability domain. Since it is a closed viability domain, we can associate with any $x \in M$ a viable solution $x(\cdot) \in \mathcal{S}_F(x)$ starting at x. Hence its limit set $\omega(x(\cdot))$ is contained in M. But limit sets being closed viability domains and M being minimal, it is equal to $\omega(x(\cdot))$, so that $x \in \omega(x(\cdot))$. □

6.2 Viability Kernels and Capture Basins

We shall answer in this section questions such as:

- starting from K, is it possible to remain viable in K as long as possible,
- starting outside from K, to reach as fast as possible the subset K being regarded in this case as a target.

These two very natural questions lead to the introduction of the following concepts.

6.2.1 Reachable, Viability and Capture Tubes

Definition 6.2.1 *Let $F : X \rightsquigarrow X$ be a set-valued map and $C \subset X$ be a subset. The* reachable map *$\vartheta_F(\cdot, x)$ is defined by*

$$\forall\, x \in X, \quad \vartheta_F(t,x) \;:=\; \{x(t)\}_{x(\cdot)\in\mathcal{S}_F(x)}$$

We associate with it the reachable tube *$t \rightsquigarrow \vartheta_F(t, C)$ defined by*

$$\vartheta_F(t,C) \;:=\; \{x(t)\}_{x(\cdot)\in\mathcal{S}_F(C)}$$

We derive the following properties:

Proposition 6.2.2 *The reachable map $t \rightsquigarrow \vartheta_F(t,x)$ enjoys the semi-group property: $\forall t, s \geq 0,\;\; \vartheta_F(t+s,x) = \vartheta_F(t, \vartheta_F(s,x))$.*

Furthermore,

$$(\vartheta_F(t,\cdot))^{-1} \;:=\; \vartheta_{-F}(t,\cdot)$$

Therefore, the subset $\vartheta_{-F}(t,C)$ is the subset of elements $x \in X$ which reach the subset at the prescribed time t.

If F is Marchaud and $K \subset X$ is compact, the graph of the reachable map $(x,t) \rightsquigarrow \vartheta_F(t,K)$ is closed.

Proof — The semi-group property is obvious. Let us prove the second one: If $y \in \vartheta_F(t,x)$, there exists a solution $x(\cdot)$ to the differential inclusion $x' \in F(x)$ starting at x such that $y = x(t)$. We set $y(s) := x(t-s)$ if $s \in [0,t]$ and we choose any solution $y(\cdot)$ to the differential inclusion $y' \in -F(y)$ starting at x at time t for $s \geq t$. Then such a function $y(\cdot)$ is a solution to the differential inclusion $y' \in -F(y)$ starting at y and satisfying $y(t) = x$. This shows that $x \in \vartheta_{-F}(t,y)$.

The last statement is a consequence of Theorem 6.1.9 on the upper semicontinuity of the solution map. □

We state here an extension of a theorem of Hélène Frankowska on the contingent derivative of the reachable map:

Theorem 6.2.3 *Assume that F is a Lipschitz set-valued map from X to X with nonempty convex compact values. Let K be a compact subset.*

$$\forall x \in \vartheta_F(t,K), \;\; D\vartheta_F(t,K,x)(1) \;=\; F(x) + T_{\vartheta_F(t,K)}(x)$$

Taking $K := \{x\}$ and $t = 0$, we obtain Frankowska's original result:

$$F(x) \;=\; \mathrm{Limsup}_{h\to 0+} \frac{\vartheta_F(h,x) - x}{h} \;=:\; D\vartheta_F(0,x,x)(1)$$

which states that a Lipschitz-Marchaud set-valued map F is the infinitesimal generator of the set-valued semigroup $t \rightsquigarrow \vartheta_F(t,K)$.

Theorem 6.2.3 follows from Theorem 12.5.2 of Chapter 12 of the Tool-Box with $K(t) := \vartheta_F(t,K)$.

Definition 6.2.4 *Let $F : X \rightsquigarrow X$ be a set-valued map and $C \subset \mathrm{Dom}(F)$ be any subset.*

Existential Concepts

1. *The subset* $\mathrm{Viab}_F(C,T)$ *of initial states $x_0 \in C$ such that one solution $x(\cdot)$ to differential inclusion $x' \in F(x)$ starting at x_0 is viable in C for all $t \in [0,T]$ is called the T-*viability kernel *and the subset*

$$\mathrm{Viab}_F(C) \;:=\; \bigcap_{T>0} \mathrm{Viab}_F(C,T)$$

is called the viability kernel *of C under F. A subset C is a repeller if its viability kernel is empty.*

2. *The subset* $\mathrm{Capt}_F(C,T)$ *of initial states $x_0 \in X$ such that C is reached before T by one solution $x(\cdot)$ to differential inclusion $x' \in F(x)$ starting at x_0 is called the T-*capture basin *and*

$$\mathrm{Capt}_F(C) \;:=\; \bigcup_{T>0} \mathrm{Capt}_F(C,T)$$

is said to be the capture basin *of C.*

Universal Concepts

1. *The set* $\mathrm{Inv}_F(C,T)$ *of initial states $x_0 \in C$ such that all solutions $x(\cdot)$ to differential inclusion $x' \in F(x)$ starting at x_0 remains in C for all $t \in [0,T]$ is said to be the T-*invariance kernel *and*

$$\mathrm{Inv}_F(C) \;:=\; \bigcap_{T\geq 0} \mathrm{Inv}_F(C,T)$$

is called the invariance kernel[7] *of* C *under* F.

2. *The subset* $\mathrm{Abs}_F(C,T)$ *of initial states* $x_0 \in X$ *such that* C *is reached before* T *by all solutions* $x(\cdot)$ *to differential inclusion* $x' \in F(x)$ *starting at* x_0 *is called the* T-absorption basin *and*

$$\mathrm{Abs}_F(C) := \bigcup_{T>0} \mathrm{Abs}_F(C,T)$$

is said to be the basin of absorption *of* C.

We say that an subset C *is* absorbing *if its absorption basin* $\mathrm{Abs}_F(C) = X$ *is the whole space.*

Remarks — We point out the obvious relations

$$\begin{cases} X\backslash\mathrm{Viab}_F(C,T) = \mathrm{Abs}_F(X\backslash C,T) \\ \\ X\backslash\mathrm{Capt}_F(C,T) = \mathrm{Inv}_F(X\backslash C,T) \end{cases} \quad (6.2.1)$$

In particular, a subset is absorbing if and only if and only if its complement is repeller.

We observe that if $T_1 \le T_2$,

$$\begin{cases} \begin{array}{ccccc} \mathrm{Inv}_F(C) & \subset & \mathrm{Inv}_F(C,T_2) & \subset & \mathrm{Inv}_F(C,T_1) \\ \cap & & \cap & & \cap \\ \mathrm{Viab}_F(C) & \subset & \mathrm{Viab}_F(C,T_2) & \subset & \mathrm{Viab}_F(C,T_1) \\ \cap & & \cap & & \cap \\ \mathrm{Capt}_F(C,0) & = & C & = & \mathrm{Viab}_F(C,0) \\ \cap & & \cap & & \cap \\ \mathrm{Abs}_F(C) & \supset & \mathrm{Abs}_F(C,T_2) & \supset & \mathrm{Abs}_F(C,T_1) \\ \cap & & \cap & & \cap \\ \mathrm{Capt}_F(C) & \supset & \mathrm{Capt}_F(C,T_2) & \supset & \mathrm{Capt}_F(C,T_1) \end{array} \end{cases}$$

One can write

$$\mathrm{Capt}_F(C,T) = \bigcup_{t\in[0,T]} \vartheta_{-F}(t,C)$$

and observe that the viability kernels satisfy

$$\forall\, S \in [0,T],\ \ \mathrm{Viab}_F(C,T) = \mathrm{Viab}_F(\mathrm{Viab}_F(C,S),T-S)$$

Let $0 < T_1 < T_2$ and $x \in \mathrm{Viab}_F(C,T_1)\backslash\mathrm{Viab}_F(C,T_2)$. Then all solutions $x(\cdot) \in \mathcal{S}_F(x)$ viable in C on $[0,T_1]$ remain in the complement of the exit tube:

$$\forall\, t \in [0,T_1],\ \ x(t) \in \mathrm{Viab}_F(C,T_1-t)\backslash\mathrm{Viab}_F(C,T_2-t)$$

[7] Viability and invariance kernels where defined in VIABILITY THEORY, [28, Aubin] for closed subsets K only as the largest closed subsets of K respectively viable and invariant under F. These definitions are equivalent thanks to Theorem 6.2.7 below. It appeared since then that in the framework of game theory and mathematical morphology it was much more useful to introduce these concepts for any type of subset C instead of closed subsets K.

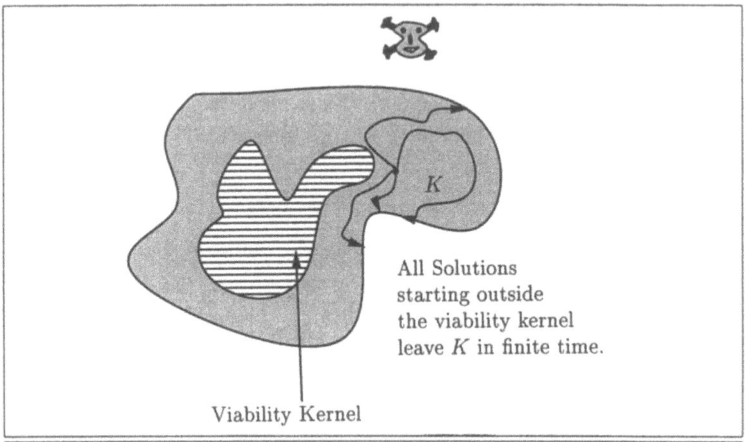

The viability kernel of a closed subset K is the subset of initials states from which starts one viable solution, and is also the largest closed viability domain contained in K. It complement is the set of initial points which violate the constraints in finite time.

Figure 6.1: Solutions starting from the viability kernel

In particular, for any $x \in \text{Viab}_F(C,T)\backslash\text{Viab}(C)$, all solutions $x(\cdot) \in \mathcal{S}_F(x)$ satisfy

$$\forall\, t \in [0,T], \;\; x(t) \;\in\; \text{Viab}_F(C, T-t)\backslash\text{Viab}(C)$$

We point out the following obvious properties:

Lemma 6.2.5 *The capture basin of any union of subsets C_i $(i \in I)$ is the union of the capture basins of the C_i and thus, that the invariance kernel of any intersection of subsets C_i $(i \in I)$ is the intersection of the invariance kernels of the C_i.*

Proposition 6.2.6 *The viability kernel* $\text{Viab}_F(C)$ *(respectively, the invariance kernel* $\text{Inv}_F(C)$*) of C under F is the largest subset of C viable (respectively, invariant) under F.*

Proof — Every subset $L \subset C$ viable (respectively invariant) under F is obviously contained in the viability kernel $\text{Viab}_F(C)$ (respectively the invariant kernel $\text{Inv}_F(C)$) of C under F.

On the other hand, if $x(\cdot)$ is a solution to the differential inclusion $x' \in F(x)$ viable in C, then for all $t > 0$, the function $y(\cdot)$ defined by $y(\tau) := x(t+\tau)$ is also a solution to the differential inclusion, starting at $x(t)$, viable in C.

Therefore, for any element $x_0 \in \mathrm{Viab}_F(C)$, there exists a viable solution $x(\cdot)$ to the differential inclusion starting from x_0, and thus, for all $t \geq 0$, $x(t) \in \mathrm{Viab}_F(C)$, so that it is viable under F.

In the same way, if for any element $x_0 \in \mathrm{Inv}_F(C)$, all solutions $x(\cdot)$ to the differential inclusion starting from x_0 being viable in C, then $x(t) \in \mathrm{Inv}_F(C)$ for every $t \geq 0$, so that it is invariant under F. □

6.2.2 Topological Properties

Theorem 6.2.7 *If $F : X \rightsquigarrow X$ is Marchaud, then the T-viability kernels of a closed subset $K \subset X$ and the T-capture basins of K under F are closed. The viability kernel of K under F is the largest closed subset of K viable under F.*

If $F : X \rightsquigarrow X$ is Lipschitz, then the T-invariance kernels of a closed subset $K \subset X$ and the T-absorption basins of K under F are closed. The invariance kernel of K under F is the largest closed subset of K invariant under F.

Proof — Let $T < +\infty$ be fixed. We set

$$v_T(C) := \{x(\cdot) \in \mathcal{C}(0,\infty;X) \mid \forall\, t \in [0,T],\ \ x(t) \in C\} \tag{6.2.2}$$

and

$$c_T(C) := \{x(\cdot) \in \mathcal{C}(0,\infty;X) \mid \exists\, t \in [0,T] \text{ such that } x(t) \in C\} \tag{6.2.3}$$

and we observe that

$$X \backslash v_T(C) = c_T(X \backslash C)$$

We note that $v_T(C)$and $c_T(C)$ are closed whenever C is closed and open whenever C is open.

Let us consider the **solution map** $\mathcal{S}_F(x)$ mapping any initial state $x \in X$ to the (possibly empty) set of solutions to differential inclusion $x' \in F(x)$. Then, we observe that[8]

$$\begin{cases} \mathrm{Viab}_F(C,T) = \mathcal{S}_F^{-1}(v_T(C)) & \& \quad \mathrm{Capt}_F(C,T) = \mathcal{S}_F^{-1}(c_T(C)) \\ \mathrm{Inv}_F(C,T) = \mathcal{S}_F^{+1}(v_T(C)) & \& \quad \mathrm{Abs}_F(C,T) = \mathcal{S}_F^{+1}(c_T(C)) \end{cases}$$

When $C := K$ is closed, the subsets $v_T(K)$ and $c_T(K)$ being closed, their inverse images under $\mathcal{S}_F$ are closed when $\mathcal{S}_F$ us upper semicontinuous (which is the case when F is Marchaud thanks to Theorem 6.1.9) and their core

[8]We recall that the core $F^{+1}(M)$ of a subset M under a set-valued map $F : X \rightsquigarrow Y$ is the set

$$F^{+1}(M) := \{x \in X \mid F(x) \subset M\}$$

under $\mathcal{S}_F$ are closed when $\mathcal{S}_F$ is lower semicontinuous (which is the case when F is Lipschitz thanks to Theorem 6.1.14). Therefore, the four above subsets are closed whenever $C := K$ is closed. □

We deduce at once the following consequence:

Proposition 6.2.8 *Let $F : X \rightsquigarrow X$ be a strict Marchaud map and $K \subset X$ be a closed subset. If $M \subset \mathrm{Abs}_F(X \backslash K) = X \backslash \mathrm{Viab}_F(K)$ is compact, there exists $T \geq 0$ such that, for every $x \in M$ and every solution $x(\cdot) \in \mathcal{S}_F(x)$, there exists $t \in [0, T]$ such that $x(t) \notin K$.*

Proof — Indeed, by (6.2.1), M is contained in the union of the complements

$$X \backslash \mathrm{Viab}_F(K, T) \;=\; \mathrm{Abs}_F(X \backslash K, T)$$

which form an increasing sequence of open subsets. Since M is compact, it is contained in one of the open $\mathrm{Abs}_F(X \backslash K, T)$. Hence, from every $x \in \mathrm{Abs}_F(X \backslash K, T)$ and any solution $x(\cdot) \in \mathcal{S}_F(x)$, there exists $t \in [0, T]$ such that $x(t) \in X \backslash K$. □

Proposition 6.2.9 *Let us assume that K is a compact and that $F : X \rightsquigarrow X$ is Marchaud.*

Then either the viability kernel of K is not empty or K is a repeller, , and in this case, there exists $\overline{T} \in]0, +\infty[$ such that

$$\mathrm{Viab}_F(K, \overline{T}) \neq \emptyset \;\;\&\;\; \forall\, T > \overline{T},\;\; \mathrm{Viab}_F(K, T) \;=\; \emptyset$$

so that all solutions leave K after $\overline{T}$.

Proof — When K is a repeller, Proposition 6.2.8 implies that for some $T > 0$, the T-viability kernel is empty. Define $\overline{T}$ as the supremum of $T \geq 0$ such that the T-viability kernel is not empty. Then, for every $\varepsilon > 0$, there exists $x_\varepsilon \in \mathrm{Viab}_F(K, \overline{T} - \varepsilon)$, and thus, a solution $x_\varepsilon(\cdot) \in \mathcal{S}_F(\varepsilon)$ which is viable in K on the interval $[0, \overline{T} - \varepsilon]$. By Theorem 6.1.9, a subsequence (again denoted by) $x_\varepsilon(\cdot)$ converges to a solution $\overline{x}(\cdot) \in \mathcal{S}_F(\overline{x})$ which is viable in K on the interval $[0, \overline{T}]$. This means that $\overline{x}$ belongs to $\mathrm{Viab}_F(K, \overline{T})$. □

In other words, when K is a compact repeller, there exists a smallest nonempty T-viability kernel of K, the "viability core", so to speak, because it is the subset of initial states from which one solution which enjoys the longest "life expectation" $\overline{T}$ in K.

Marc Quincampoix has proved that the boundary of the viability kernel enjoys the barrier property: Every solution starting from the boundary of the viability kernel can either remain in the boundary or leave the viability kernel, or equivalently, no solution starting from outside of the viability kernel can cross its boundary: such solutions can only remain on the boundary of the viability kernel, or leave it. This is a semi-permeability property of the viability kernel, which is very important in terms of interpretation. Viability is indeed a very fragile property, which cannot be reestablished from the outside. In other words, love it or leave it ...:

Theorem 6.2.10 (Quincampoix) *Let $F : X \rightsquigarrow X$ be a set-valued map both Marchaud and Lipschitz and $K \subset X$ be a closed subset. Assume that* $\mathrm{Viab}_F(K,T)$ *is contained in the interior of K. Then, if $x \in \partial(\mathrm{Viab}_F(K,T))$, every solution $x(\cdot) \in \mathcal{S}_F(x)$ viable in K on $[0,T]$ remains on the boundary of the viability kernel tube:*

$$\forall\, t \in [0,T], \quad x(t) \;\in\; \partial\left(\mathrm{Viab}_F(K, T-t)\right)$$

In particular, for $T := +\infty$,

$$\forall\, x \in \partial\mathrm{Viab}_F(K), \forall\, x(\cdot) \in \mathcal{S}_F(x), \quad \forall\, t \geq 0, \quad x(t) \in \partial\left(\mathrm{Viab}_F(K)\right)$$

(See [291,293, Quincampoix] or Theorem 5.5.2 of VIABILITY THEORY, [28, Aubin] for a proof).

6.2.3 The Viability Kernel Algorithm

The concept of viability kernel is not only an attractive concept in the framework of viability theory, but also an efficient mathematical tool we shall use very often.

This is the reason why an algorithm designed by Patrick Saint-Pierre and his collaborators to compute the viability kernel happens to be very useful. For the time being, it works only for low dimensional systems because it requires an amount of information which increases exponentially with the dimension of the state space.

We could interpret the definition of the viability kernel as the intersection of the T-viability kernels as an algorithm, but this is not operational because we do not know how to compute the reachable maps. However, the necessary condition of the Viability Theorem states that

$$\forall\, T \geq 0, \quad \mathrm{Viab}_F(K,T) \;\subset\; K_1 \;:=\; \{x \in K \mid F(x) \cap T_K(x) \neq \emptyset\}$$

Since the subsets K_j defined by induction by

$$K_j \;:=\; \{x \in K_{j-1} \mid F(x) \cap T_{K_{j-1}}(x) \neq \emptyset\}$$

can be "computed" more easily than the T-viability kernel, the question arises whether their intersection is the viability kernel. Unfortunately, this is not true in general, since the subsets K_j are not even necessarily closed[9].

[9] Hélène Frankowska and Marc Quincampoix proved that a "sharper" algorithm using "smaller" subsets than K_j do converge to the viability kernel. See [211, Frankowska & Quincampoix] or section 4.5 of chapter 4 of VIABILITY THEORY, [28, Aubin] and its bibliography for more informations on this very important algorithm.

However, if we approximate the contingent cone $T_K(x)$ by $\frac{K-x}{\rho}$, then the algorithm reads

$$K_j := \{x \in K_{j-1} \mid (x + \rho F(x)) \cap K_{j-1} \neq \emptyset\}$$

which brings to light the finite-difference approximation of F by $\frac{F-\mathbf{1}}{\rho} + \alpha\rho B$ in the case of the explicit scheme (or $\frac{(F+\mathbf{1})^{-1}}{\rho} + \alpha\rho B$ in the case of the implicit finite-difference scheme).

Patrick Saint-Pierre proved that the finite-difference version of this algorithm does actually approximate the viability kernel.

Case of the Explicit Scheme The viability kernel algorithm for the explicit scheme

$$x_{n+1} - x_n \in \rho F(x_n)$$

is defined by induction by

$$K_{j+1} := \{x \in K_j^\rho \mid (x + \rho F(x)) \cap K_j^\rho \neq \emptyset\}$$

and thus, in the form

$$K_{j+1}^\rho := K_j^\rho \cap (\mathbf{1} + \rho F)^{-1}(K_j^\rho)$$

Therefore, we need at each iteration to invert $(\mathbf{1} + \rho F)$.

Case of the Implicit Scheme The viability kernel algorithm for the implicit scheme

$$x_{n+1} - x_n \in \rho F(x_{n+1})$$

can be written

$$K_{j+1}^\rho := \{x \in K_j^\rho \mid (\mathbf{1} - \rho F)^{-1}(x) \cap K_j^\rho \neq \emptyset\}$$

and thus, in the form

$$K_{j+1}^\rho := K_j^\rho \cap (\mathbf{1} - \rho F)(K_j^\rho)$$

Therefore, we need at each iteration to compute the direct image of K_j^ρ by $(\mathbf{1} - \rho F)$, which is, so to speak, an "approximation" of the inverse of $(\mathbf{1} + \rho F)$. However, computing solutions to the implicit scheme is much more difficult than computing solutions to the explicit scheme since we have to inverse $\mathbf{1} - \rho F$ at each iteration.

Patrick Saint-Pierre proved in [319, Saint-Pierre] that when F is Marchaud, the upper limit of the subsets $K_{[\frac{T}{\rho}]}^\rho$ is contained in the T-viability kernel $\mathrm{Viab}_F(K,T)$ and, together with Marc Quincampoix in [298, Quincampoix & Saint-Pierre], that they are equal when F is also Lipschitz and the approximation is slightly modified. With Pierre Cardaliaguet, they generalized this algorithm for approximating invariant domain, capture basins and discriminating kernels for dynamical games (see [104, Cardaliaguet, Quincampoix & Saint-Pierre]).

6.2.4 Set of Equilibria

We mention the relationship between the set of equilibria of a dynamical economy and the viability kernel of an auxiliary differential equation due to Patrick Saint-Pierre:

Proposition 6.2.11 *Let us consider a dynamical economy (P, c). We associate with it the auxiliary dynamical system*

$$x'(t) = 0, \; p'(t) = 0 \; \& \; y'(t) = c(x(t), p(t)) \tag{6.2.4}$$

Then the set of equilibria of the dynamical economy is the viability kernel of $\mathrm{Graph}(P) \times cB$ *under the foregoing system (6.2.4).*

Proof— Indeed (x_0, p_0, y_0) belongs to the viability kernel of $\mathrm{Graph}(P) \times cB$ under the system (6.2.4) if and only if the solution $(x_0, p_0, tc(x_0, p_0))$ remains in $\mathrm{Graph}(P) \times cB$ for every $t \geq 0$. Since the ball cB is bounded, this is only possible when $c(x_0, p_0) = 0$, i.e., when $(x_0, p_0) \in \mathrm{Graph}(P)$ is an equilibrium. □

6.2.5 Lyapunov Stability

Let C be a subset. We shall use the concept of viability kernel to reformulate several concepts of Lyapunov stability.

Lyapunov stability of the a viability domain C means that for any $\varepsilon > 0$, there exists a neighborhood $\mathcal{U}$ of C such that for every initial state $x \in \mathcal{U}$, there exists at least one solution $x(\cdot) \in \mathcal{S}_F(x)$ which remains in the ball $B(C, \varepsilon)$ and *universal Lyapunov stability of the a viability domain* C means that this property holds true for all solutions.

Therefore, this Lyapunov stability property can be translated in the following way:

Proposition 6.2.12 *Let $F : X \rightsquigarrow X$ be a strict Marchaud map and C be a compact viability domain. It is* stable *in the sense of Lyapunov if and only if*

$$C \subset \bigcap_{\varepsilon>0} \mathrm{Int}(\mathrm{Viab}_F(B(C, \varepsilon)))$$

and universally stable if

$$C \subset \bigcap_{\varepsilon>0} \mathrm{Int}(\mathrm{Inv}_F(B(C, \varepsilon)))$$

6.2.6 Limit Sets

Limit sets of a solution $x(\cdot)$ to a differential inclusion is the set of cluster points of $t \mapsto x(t)$ when t goes to $+\infty$:

Definition 6.2.13 *Let F be a set-valued map and K be a closed subset. We denote by*

$$\omega_F(K) := \text{Limsup}_{t\to+\infty}\vartheta_F(t,K)$$

the ω-limit set or simply limit set *of the subset K under F.*

Equilibria, trajectories of periodic solutions and more generally, limit sets of solutions and compact sets provide examples of closed viability domains of F:

Proposition 6.2.14 *Assume that F is Marchaud and K is compact. Then the ω-limit set $\omega_F(K)$ is viable under F.*

It is not empty whenever there exists $T \geq 0$ such that $\bigcup_{t\geq T}\vartheta_F(t,K)$ is bounded.

Proof — Indeed, let ξ belong to $\omega_F(K)$. Then $\xi = \lim \xi_n$ where $\xi_n \in \vartheta_F(t_n,K)$. We associate with the solutions $x_n(\cdot)$ to the differential inclusion

$$x_n'(t) \in F(x_n(t)), \;\; x_n(t_n) = \xi_n$$

the functions $y_n(\cdot)$ defined by $y_n(t) := x_n(t+t_n)$ which are solutions to

$$y_n'(t) \in F(y_n(t)), \;\; y_n(0) = \xi_n$$

Theorem 6.1.9 implies that these solutions remain in a compact subset of $\mathcal{C}(0,\infty;X)$. Therefore, a subsequence (again denoted by) $y_n(\cdot)$ converges to $y(\cdot)$, which is a solution to

$$y'(t) \in F(y(t)), \;\; y(0) = \xi$$

Furthermore, this solution is viable in $\omega_F(K)$ since for all $t \geq 0$, $y(t)$ is the limit of a subsequence of $y_n(t) = x_n(t+t_n) \in \vartheta_F(t+t_n,K)$, and thus belongs to $\omega_F(K)$. □

6.2.7 Attractor

Definition 6.2.15 *The* **attractor** of F *is the closed subset*

$$\text{Attr}_F := \overline{\bigcup_{K \text{ compact}} \omega_F(K)}$$

We thus deduce an important relation between absorbing sets and the attractor, allowing to determine the attractor whenever one knows a bounded invariant absorbing set.

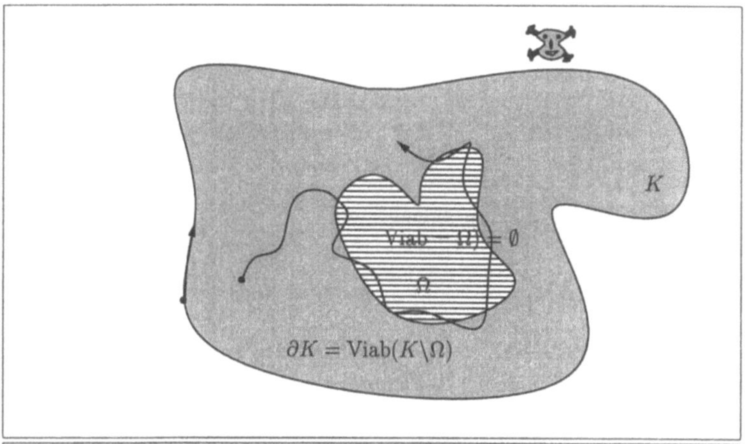

When $\partial K = \text{Viab}(K\backslash\Omega)$, the solutions starting from the interior of K successively enter and leave Ω in finite time.

Figure 6.2: Fluctuat nec mergitur

Theorem 6.2.16 *Assume that F is Marchaud. If a bounded open subset Ω is absorbing and if it is invariant under F, then the attractor exists and is equal to the limit set of $\overline{\Omega}$:*

$$\text{Attr}_F \;=\; \omega_F(\overline{\Omega})$$

Proof — First, since $\overline{\Omega}$ is compact, we observe that

$$\omega_F(\overline{\Omega}) \;\subset\; \text{Attr}_F$$

For proving the converse statement, we deduce that Ω being absorbing, then $\text{Abs}_F(\Omega) = X$. Therefore Corollary 6.2.8 implies that for any compact $K \subset X$, there exists $T_K \geq 0$ such that any solution starting from K reaches Ω before T_K, and remains in Ω afterwards since Ω is assumed to be invariant. Therefore for every $t \geq T_K$, we deduce that $\vartheta_F(t, K) \subset \overline{\Omega}$, so that $\omega_F(K) \subset \omega_F(\overline{\Omega})$, from which we infer that $\text{Attr}_F \subset \omega_F(\overline{\Omega})$. □

6.2.8 Fluctuations around a Target

Assume now that K is a viability domain of a set-valued map F with nonempty interior and that it contains an open target $\Omega \subset K$. We shall provide a situation in which all solutions starting from the interior of the viability domain K reach the open target Ω in finite time, then leaves the target in finite time, comes back to the target, and so on, **fluctuating** around the target. This may serve as a metaphor for some economic cycles that have been observed.

Theorem 6.2.17 *Let $F : X \rightsquigarrow X$ be a Marchaud map and K be a closed subset of its domain with a nonempty interior.*

Assume that there exists a nonempty open subset Ω such that $\overline{\Omega} \subset \mathrm{Int}(K)$ and

$$\partial K \ = \ \mathrm{Viab}_F(K\backslash\Omega)$$

Then

1. *the boundary ∂K of K is* semipermeable *under F in the sense that for every $x_0 \in \partial K$,* any solution $x(\cdot) \in \mathcal{S}_F(x_0)$ viable in $K\backslash\Omega$ is also viable in ∂K,
2. *the interior* $\mathrm{Int}(K)$ *of K is invariant under F,*
3. *Starting from $x_0 \in \mathrm{Int}(K)\backslash\Omega$, we can associate with any solution $x(\cdot) \in \mathcal{S}_F(x_0)$ an increasing sequence of instants t_n satisfying*

$$x(t_{2k+1}) \in \Omega \ \ \& \ \ x(t_{2k}) \in \mathrm{Int}(K)\backslash\Omega$$

such that $t_{2k+1} > t_{2k}$ is finite whenever t_{2k} is finite.

As a consequence,

1. *If Ω is invariant, all solutions starting from the interior of K reach Ω in finite time and remain there, and $\omega_F(\overline{\Omega})$ is the attractor. This is the* permanence property *introduced by Joseph Hofbauer and Karl Sigmund in biomathematics and extensively studied by these authors.*
2. *If Ω is a repeller, all the solutions starting from the interior of K "fluctuate" around Ω in the sense that they alternatively reach and leave Ω in finite time. It is the* fluctuation property *introduced by Vlastimil Krivan in the ecological framework of carrying capacity.*

Proof — See the proof of Theorem 4.1.9 of VIABILITY THEORY, [28, Aubin]. □

6.3 Exit and Hitting Times

6.3.1 Exit Time Function and Viability Kernels

We shall characterize the T-viability kernels and T-capture basins as level sets of upper exit and lower hitting functions that we are about to define:

Definition 6.3.1 *Let $C \subset X$ be a subset. We denote by*

$$\varpi_C : \mathcal{C}(0,\infty;X) \mapsto \mathbf{R}_+ \cup \{+\infty\}$$

the hitting functional *(or* minimal time functional*) associating with $x(\cdot)$ its* hitting time *$\varpi_C(x(\cdot))$ defined by*

$$\varpi_C(x(\cdot)) \ := \ \inf\{t \in [0,+\infty[\ | \ x(t) \in C\}$$

The functional $\tau_C : \mathcal{C}(0,\infty;X) \mapsto \mathbf{R}_+ \cup \{+\infty\}$ *associating with* $x(\cdot)$ *its* exit time $\tau_C(x(\cdot))$ *defined by*

$$\tau_C(x(\cdot)) := \inf\{t \in [0,\infty[\mid x(t) \notin C\} := \varpi_{X\setminus C}(x(\cdot))$$

is called the exit functional.

The crisis functional $\gamma_C(x(\cdot)) : \mathcal{C}(0,\infty;X) \mapsto \mathbf{R}_+ \cup \{+\infty\}$ *associates with* $x(\cdot)$ *its* crisis time *defined by*

$$\gamma_C(x(\cdot)) := \int_0^{+\infty} \chi_{X\setminus C}(x(\tau))d\tau = \operatorname{meas}\{t \geq 0 \mid x(t) \notin C\}$$

introduced by Luc Doyen & Patrick Saint-Pierre in [177, Doyen & Saint-Pierre], which measures the time spent by the trajectory of $x(\cdot)$ *outside* C.

We use the convention $\inf\{\emptyset\} := +\infty$ and we observe that

$$\forall\, s \in [0, \varpi_C(x(\cdot))],\ \ \varpi_C(x(\cdot+s)) = \varpi_C(x(\cdot)) - s \qquad (6.3.1)$$

Lemma 6.3.2 *The functional* ϖ_C *is lower semicontinuous when* C *is closed and when* $\mathcal{C}(0,\infty;X)$ *is supplied with the compact convergence topology.*

It is upper semicontinuous when C *is open and when* $\mathcal{C}(0,\infty;X)$ *is supplied with the pointwise convergence topology.*

(See Lemma 4.2.3 of VIABILITY THEORY, [28, Aubin] for a proof.)

Consider now a set-valued map $F : X \rightsquigarrow X$ and denote by $\mathcal{S}_F(x)$ the set of solutions $x(\cdot)$ to differential inclusion $x' \in F(x)$ starting at the initial state x.

Definition 6.3.3 *Let* $C \subset X$ *be a subset. The functions* $\varpi_C^\flat : C \mapsto \mathbf{R}_+ \cup \{+\infty\}$ *and* $\varpi_C^\sharp : C \mapsto \mathbf{R}_+ \cup \{+\infty\}$ *defined by*

$$\varpi_C^\flat(x) := \inf_{x(\cdot)\in\mathcal{S}_F(x)} \varpi_C(x(\cdot)) \ \ \&\ \ \varpi_C^\sharp(x) := \sup_{x(\cdot)\in\mathcal{S}_F(x)} \varpi_C(x(\cdot))$$

are called the lower and upper hitting function *and the functions* $\tau_C^\flat : C \mapsto \mathbf{R}_+ \cup \{+\infty\}$ *and* $\tau_C^\sharp : C \mapsto \mathbf{R}_+ \cup \{+\infty\}$ *defined by*

$$\tau_C^\flat(x) := \inf_{x(\cdot)\in\mathcal{S}_F(x)} \tau_C(x(\cdot)) \ \ \&\ \ \tau_C^\sharp(x) := \sup_{x(\cdot)\in\mathcal{S}_F(x)} \tau_C(x(\cdot))$$

the lower and upper exit function.

The function

$$\gamma_C^\flat(x) := \inf_{x(\cdot)\in\mathcal{S}_F(x)} \gamma_C(x(\cdot))$$

is called the crisis function *and its level sets*

$$\operatorname{Neigh}_F(C,T) := \{x \in X \mid \gamma_C^\flat(x) \leq T\}$$

their viability neighborhoods.

We observe that the crisis function coincides with the lower hitting function when C is viable under F.

Lemma 6.3.2 and Theorem 6.1.9 on the upper semicontinuity of the solution map $x \in K \rightsquigarrow \mathcal{S}_F(x) \subset \mathcal{C}(0,\infty;X)$ (supplied with the compact convergence topology) imply

Proposition 6.3.4 *Let $F : X \rightsquigarrow X$ be a strict Marchaud map and $K \subset X$ be a closed subset. Then the hitting function $\varpi_K^\flat$ is lower semicontinuous and the exit function $\tau_K^\sharp$ is upper semicontinuous. Furthermore, for any $x \in \mathrm{Dom}(\varpi_K^\flat)$, there exists one solution $x^\flat(\cdot) \in \mathcal{S}_F(x)$ which hits K as soon as possible*

$$\varpi_K^\flat(x) \;=\; \varpi_K(x^\flat(\cdot))$$

and for any $x \in \mathrm{Dom}(\tau_K^\sharp)$, there exists one solution $x^\sharp(\cdot) \in \mathcal{S}_F(x)$ which remains viable in K as long as possible:

$$\tau_K^\sharp(x) \;=\; \tau_K(x^\sharp(\cdot))$$

We deduce from (6.3.1) that

$$\begin{cases} \forall\, s \in [0, \varpi_K^\flat(x)], & \varpi_K(x^\flat(s)) \;=\; \varpi_K^\flat(x) - s \\ \forall\, s \in [0, \tau_K^\sharp(x)], & \tau_K(x^\sharp(s)) \;=\; \tau_K^\sharp(x) - s \end{cases}$$

When the solution map $\mathcal{S}_F$ is lower semicontinuous — which is the case when F is Lipschitz — we can deduce from Proposition 6.3.2 and the Maximum Theorem 11.2.5 imply that:

Proposition 6.3.5 *Let $F : X \rightsquigarrow X$ be a strict Lipschitz map with closed values and $K \subset X$ be a closed subset. Then the upper hitting function $\varpi_K^\sharp$ is upper semicontinuous and the lower exit function $\tau_K^\flat$ is lower semicontinuous.*

We shall prove in Chapter 8 that when F is Marchaud and K is closed, the *crisis function $\gamma_K^\flat$ is lower semicontinuous, vanishes on the viability kernel of K and that its domain is* $\mathrm{Capt}_F(K)$. We shall also characterize in Chapter 8 the hypograph of the upper exit function and the epigraphs of the lower hitting function and of the crisis function as viability kernels of auxiliary systems. This allows us to approximate them by using the viability kernel algorithm. We shall likewise characterize the upper exit function, the lower hitting function and the crisis function as generalized solutions (called contingent solutions) to Hamilton-Jacobi-Bellman partial differential equations.

For the time, we relate the exit and hitting functions with the viability kernels and capture basins:

Theorem 6.3.6 *Let $K \subset X$ be a closed subset. If F is Marchaud, then for any $T \geq 0$, the viability kernels and the capture basins are characterized by:*

$$\begin{cases} \mathrm{Viab}_F(K,T) &= \left\{x \in K \mid \tau_K^\sharp(x) \geq T\right\} \\ \\ \mathrm{Capt}_F(K,T) &= \left\{x \in X \mid \varpi_K^\flat(x) \leq T\right\} \end{cases} \quad (6.3.2)$$

If F is Lipschitz, then for any $T \geq 0$, the invariance kernels and the absorption basins are characterized by:

$$\begin{cases} \mathrm{Inv}_F(K,T) &= \left\{x \in K \mid \tau_K^\flat(x) \geq T\right\} \\ \\ \mathrm{Abs}_F(K,T) &= \left\{x \in X \mid \varpi_K^\sharp(x) \leq T\right\} \end{cases} \quad (6.3.3)$$

Proof — Let $T \geq 0$ be finite or infinite. We observe that if $x \in \mathrm{Viab}_F(K,T)$, then a solution $x(\cdot)$ to differential inclusion $x' \in F(x)$ starts at x and remains in K for all $t \in [0,T[$, so that satisfies $\tau_K^\sharp(x) \geq T$.

Conversely, we have to prove that for any x satisfying this property, there exists $x(\cdot) \in \mathcal{S}_F(x)$ which remains in K whenever $t \in [0,T]$. By definition of the exit function, we can associate with any sequence $t_n < T$ converging to T a solution $x_n(\cdot) \in \mathcal{S}_F(x)$ satisfying $t_n \leq \tau_K(x_n(\cdot))$. By Theorem 12.1.1, a subsequence of solutions $x_n(\cdot)$ converges to a solution $x(\cdot) \in \mathcal{S}_F(x)$. Let $t \in [0,T[$ be given and choose n such that $t < t_n < T$. Observing that $x_n(t) \in K$ converges to $x(t)$, we infer that $x(t) \in K$ for any $t < T$. Since K is closed and $x(\cdot)$ is continuous, it is also true for $t = T$.

In the same way, if $T < +\infty$ and if $x \in \mathrm{Capt}_F(K,T)$, then a solution $x(\cdot)$ to differential inclusion $x' \in F(x)$ starts at x and hits K at some $t \in [0,T[$, so that satisfies $\varpi_K^\flat(x) \geq T$. Conversely, if $\varpi_K^\flat(x) \geq T$, we can associate with every n a solution $x_n(\cdot) \in \mathcal{S}_F(x)$ such that

$$\varpi_K(x_n(\cdot)) < T + \frac{1}{n}$$

and thus, a time $t_n \in [0, T+\frac{1}{n}]$ such that $x_n(t_n) \in K$ by the very definition of the hitting function. A subsequence (again denoted by) t_n converges to some $t \in [0,T]$ and, by Theorem 12.1.1, a subsequence (again denoted by) $x_n(\cdot)$ converges to a solution $x(\cdot) \in \mathcal{S}_F(x)$. Hence $x(t)$ being the limit of elements $x_n(t_n) \in K$ belongs to K, so that x belongs to $\mathrm{Capt}_F(K,T)$.

When F is Lipschitz, the proofs are analogous. □

6.3.2 Dynamical Portrait of the Boundary

The knowledge of the exit and hitting functions provides a portrait of the dynamical behavior of the solutions under constraints.

Let us denote by Ω the complement of a closed subset K. We know that outside the viability kernel of K, all solutions leave K in finite time, and that they never enter the viability kernel before leaving K.

Starting from $x_0 \notin K$, any solution $x(\cdot)$ travels along the T-viability kernels:

$$\forall\, t \geq 0,\ \ \tau_K(x(t)) \ \geq\ \tau_K^\sharp(x_0) - t \ =\ \tau_K(x^\sharp(t))$$

Let us introduce the **egress subset** of the boundary of K defined by

$$\text{Egress}_F(K) \ :=\ \{x \in \partial K \mid \tau_K^\sharp(x) = 0\}$$

of initial states of the boundary from which all solutions leave K instantaneously: If $\tau_K^\sharp(x_0) = 0$, then for all solutions $x(\cdot) \in \mathcal{S}_F(x)$, there exists a sequence $t_n > 0$ converging to 0 such that

$$\forall\, n \geq 0,\ \ x(t_n) \ \in\ \Omega$$

One can then introduce the set-valued map $\Gamma_K^\sharp$ defined by

$$\Gamma_K(x) \ :=\ \{x^\sharp(\tau_K^\sharp(x))\}_{\tau_K(x^\sharp(\cdot))=\tau_K^\sharp(x)}$$

from $K\backslash\text{Viab}_F(K)$ to the set $\text{Egress}_F(K)$ through which all solutions leave K instantaneously.

When the interior $\overset{\circ}{K}$ of K is not empty, we can introduce the **ingress subset** of the boundary of K defined by

$$\text{Ingress}_F(K) \ :=\ \{x \in \partial K \mid \varpi^\flat_{\overset{\circ}{K}}(K) = 0\}$$

from which at least one solution at least enter the interior of K: If $\varpi^\flat_{\overset{\circ}{K}}(K) = 0$, then there exist at least one solution $x(\cdot) \in \mathcal{S}_F(x)$ and a sequence $t_n > 0$ converging to 0 such that

$$\forall\, n \geq 0,\ \ x(t_n) \ \in \overset{\circ}{K} \ \square$$

The set-valued map $\Gamma_K^\flat$ defined by

$$\Gamma_K(x) \ :=\ \{x^\flat(\varpi_K^\flat(x))\}_{\varpi_K(x^\flat(\cdot))=\varpi_K^\flat(x)}$$

from $\text{Capt}_F(K)$ to the boundary ∂K, which is a kind of projector onto K (the usual projector is associated with the constant differential inclusion $x' \in B$ where B is the unit ball) that can be called the **chronector**[10] onto K associated with the set-valued map F. Anne Gorre has proved that its graph is closed.

The boundary ∂K is then covered by sets such as $\text{Viab}_F(K)$, $\text{Ingress}_F(K)$, $\text{Inv}_F(K)$, $\text{Egress}_F(K)$ as well as $\text{Viab}_F(\overline{\Omega})$, $\text{Ingress}_F(\overline{\Omega})$, $\text{Inv}_F(\overline{\Omega})$, $\text{Egress}_F(\overline{\Omega})$.

From the view point of sustaining the viability of K, for instance, these sets describe the dynamic behavior of initial states of the subset $\text{Egress}_F(K)$:

[10]because the functions $x^\flat(\cdot)$ are solutions to the classical problem of the calculus of variations known under the name of **brachistochrone** (short time in Greek) solved by Bernouilli.

Proposition 6.3.7 *Let F be a Marchaud map and $K \subset X$ be a closed subset.*

1. *From $x_0 \in \mathrm{Egress}_F(K) \cap \mathrm{Inv}_F(\overline{\Omega})$, all solutions $x(\cdot) \in \mathcal{S}_F(x_0)$ leave K forever,*

2. *From $x_0 \in \mathrm{Egress}_F(K) \cap \mathrm{Viab}_F(\overline{\Omega})$, at least one solution $x(\cdot) \in \mathcal{S}_F(x_0)$ leaves K forever,*

3. *From $x_0 \in \mathrm{Egress}_F(K) \backslash \mathrm{Viab}_F(\overline{\Omega})$, all solutions $x(\cdot) \in \mathcal{S}_F(x_0)$ leave K instantaneously, but return to K in finite time,*

4. *From $x_0 \in \mathrm{Egress}_F(K) \cap \mathrm{Ingress}_F(K)$, at least one solution $x(\cdot) \in \mathcal{S}_F(x_0)$ oscillates around the boundary of K in the following sense: there exist one solution and a decreasing sequence $t_n > 0$ converging to 0 such that*
$$\forall\, n \geq 0,\ \ x(t_{2n}) \in \overset{\circ}{K} \ \ \&\ \ x(t_{2n+1}) \in \Omega$$

6.4 Chaos à la Saari

Let us consider now the situation in which a closed subset K is covered by a finite family $\{K_a\}_{a \in \mathcal{A}}$ of arbitrary closed "qualitative cells" $K_a \subset K$ with nonempty interior:
$$K = \bigcup_{a \in \mathcal{A}} K_a$$
"Monotonic cells" defined by (4.2.2) are examples of what one can call qualitative cells. Theorem 6.3.6 implies that
$$K_a \cap \mathrm{Capt}(K_b, T) \tag{6.4.1}$$
is the subset of elements of the qualitative cell K_a from which starts at least one solution reaching the qualitative cell K_b before time T.

In the framework of qualitative dynamic economics, one can say that b is a successor of a if $K_a \cap \mathrm{Capt}(K_b, T)$ is not empty.

We associate with the sequence $a_0, a_1, \ldots$ the subsets $M_{a_0 a_1 \cdots a_n}$ defined by induction by $M_{a_n} := K_{a_n}$,
$$M_{a_{n-1}a_n} := K_{a_{n-1}} \cap \mathrm{Capt}(K_{a_n}, T)$$
which is the subset of $x \in K_{a_{n-1}}$ such that there exists $x(\cdot) \in \mathcal{S}(x)$ satisfying $x(\varpi^\flat_{M_{a_n}}(x)) \in K_{a_n}$. For $j = n-2, \ldots, 0$, we define recursively the cells:
$$M_{a_j a_{j+1} \cdots a_n} := K_{a_j} \cap \mathrm{Capt}(M_{a_{j+1} \cdots a_n}, T)$$
which can be written
$$\begin{cases} M_{a_j a_{j+1} \cdots a_n} = \{x \in K_{a_j} \mid \exists\, x(\cdot) \in \mathcal{S}(x) \\ \text{such that } x(\varpi^\flat_{M_{a_{j+1} \cdots a_n}}(x)) \in M_{a_{j+1} \cdots a_n}\} \end{cases}$$

In other words, the set $M_{a_0a_1\cdots a_n}$ is the set of initial states $x_0 \in K_{a_0}$ from which at least one solution will visit the cells K_{a_j} is the prescribed order $a_0, \ldots a_n$.

A situation in which all the subsets $M_{a_0a_1\cdots a_n}$ are nonempty should generate a kind of chaos, which was introduced by Donald Saari:

Definition 6.4.1 *We shall say that the covering of K by a family of closed subsets K_a $(a \in \mathcal{A})$ is* chaotic à la Saari *under F if for any sequence $a_0, a_1, \ldots$, there exists at least one solution $x(\cdot) \in \mathcal{S}(x)$ to differential inclusion $x' \in F(x)$ and a sequence of elements $t^j \geq 0$ such that $x(t^j) \in K_{a_j}$ for all $j \geq 0$.*

Theorem 6.4.2 (Chaotic Behavior) *Let us assume that a compact viability domain K of the Marchaud map F is covered by a family of closed subsets K_a $(a \in \mathcal{A})$ satisfying the following* controllability assumption*: There exists $T < \infty$ such that*[11]

$$K \subset \bigcap_{a \in \mathcal{A}} \bigcup_{t \in [0,T]} \vartheta_F(t, K_a)$$

Then, for any sequence $a_0, a_1, \ldots, a_n, \ldots$, there exists at least one solution $x(\cdot) \in \mathcal{S}(x)$ to differential inclusion $x' \in F(x)$ and a sequence of elements $t^j \geq 0$ such that $x(t^j) \in K_{a_j}$ for all $j \geq 0$.

Proof — The controllability assumption implies that subsets $M_{a_0a_1\cdots a_n}$ are nonempty. They are closed thanks to Theorem 6.2.7. Since the family of subsets $M_{a_0a_1\cdots a_n}$ form a nonincreasing family and since K is compact, the intersection $K_\infty := \bigcap_{n=0}^{\infty} M_{a_0a_1\cdots a_n}$ is nonempty.

Let us take an initial state x in K_∞ and fix n. Hence there exists $x_n(\cdot) \in \mathcal{S}(x)$ and a sequence of $t_n^j \in [0, jT]$ such that

$$\forall\, j = 1, \ldots, n, \;\; x_n(t_n^j) \in M_{a_j\cdots a_n} \subset K_{a_j}$$

Indeed, there exist $y_1(\cdot) \in \mathcal{S}(x)$ and $\varpi^\flat_{M_{a_1\cdots a_n}}(y_1(\cdot)) \in [0,T]$ such that $y_1(\varpi^\flat_{M_{a_1\cdots a_n}}(y_1(\cdot)))$ belongs to $M_{a_1\cdots a_n}$. We set $t_n^1 := \varpi^\flat_{M_{a_1\cdots a_n}}(y_1(\cdot))$, $x_n^1 = y_1(t_n^1)$ and $x_n(t) := y_1(t)$ on $[0, t_n^1]$.

Assume that we have built $x_n(\cdot)$ on the interval $[0, t_n^k]$ such that $x_n(t_n^j) \in M_{a_j\cdots a_n} \subset K_{a_j}$ for $j = 1, \ldots, k$. Since $x_n(t_n^k)$ belongs to $M_{a_k\cdots a_n}$, there exist $y_{k+1}(\cdot) \in \mathcal{S}(x_n(t_n^k))$ and $\varpi^\flat_{M_{a_{k+1}\cdots a_n}}(y_{k+1}(\cdot)) \in [0,T]$ such that

$$y_{k+1}(\varpi^\flat_{M_{a_{k+1}\cdots a_n}}(y_{k+1}(\cdot))) \in M_{a_{k+1}\cdots a_n}$$

[11] or, in other words, such that

$$\forall\, a \in \mathcal{A}, \forall\, y \in K, \;\; \exists\, x \in K_a, \; x(\cdot) \in \mathcal{S}(x) \;\&\; t \in [0, T[\text{ with } x(t) = y$$

We set

$$t_n^{k+1} := t_n^k + \varpi^b_{M_{a_{k+1}\cdots a_n}}(y_{k+1}(\cdot)) \ \& \quad \ldots \quad (t + \varpi^b_{M_{a_{k+1}\cdots a_n}}(y_{k+1}(\cdot)))$$

on $[t_n^k, t_n^{k+1}]$. When $k = n$, we exten … $\infty[$ by any solution to the differential inclusion starting at $x_n($ …

Since the sequence $x_n(\cdot) \in \mathcal{S}(x)$ i … ıe space $\mathcal{C}(0, \infty; X)$ thanks to Theorem 6.1.9), a subsequence (… $x_n(\cdot)$) converges to some solution $x(\cdot) \in \mathcal{S}(x)$ to the differen … By extracting successive converging subsequences of $t^1_{n_1}, \ldots$ … fer the existence of t_j's in $[0, jT]$ such that $x_{n_j}(t^j_{n_j})$ converge … $_{a_j}$, because the functions $x_n(\cdot)$ remain in an equicontinuous s …

Chapter 7

Bounded Inflation and Heavy Evolution

Introduction

Let us still consider the problem of regulating a dynamical economy (P, c)

$$(i) \quad \text{for almost all } t \geq 0, \ x'(t) = c(x(t), p(t)) \ \text{ where } \ p(t) \in P(x(t))$$

where the pricing map $P : K \rightsquigarrow \mathcal{M}$ associates with each commodity x the set $P(x)$ of feasible prices and $c : \mathrm{Graph}(P) \mapsto X$ is the change function.

For simplicity, we take for allocation set the domain[1] $K := \mathrm{Dom}(P)$ of P.

We have seen in the preceding chapter that viable price functions (which provide viable solutions $x(t) \in K := \mathrm{Dom}(P)$) are the ones obeying the regulation law

$$\forall\, t \geq 0, \ \ p(t) \in \Pi_K(t) \ \ (\text{or} \ \ (x(t), p(t)) \in \mathrm{Graph}(\Pi_K))$$

where

$$\forall\, x \in K, \ \Pi_K(x) = \{\, p \in P(x) \mid c(x,p) \in T_K(x) \,\}$$

We also mentioned that the prices regulating viable solutions were at best **measurable**. Such a situation does not exclude brutal behaviors of prices, such as **impulses**. We did observe these impulses when simple examples (one commodity, one price) were investigated. Starting with solutions evolving with constant price, we observed that the price may have to be switched **instantaneously** when the boundary of the set of allocations was reached.

This is a general situation, because, when the commodity $x(t)$ belongs to the interior of K, $\Pi_K(x(t)) = P(x(t))$, and when it reaches the boundary at time $\bar{t}$, the regulation map can "shrink" abruptly: $\Pi_K(x(\bar{t})) = P(x(\bar{t}))$, ruling out a continuous change of price.

[1] or we replace P by its restriction to K, assigning empty values to elements $x \notin K$.

We may accept such regulation laws allowing **impulses**, i.e., infinite price velocities for switching instantaneously prices to sustain viability. This is what we observe in the actual management of firms by bankruptcy, and "real economic systems" are not immune to periods of crisis (devaluation, etc.) when the regulatory parameters are kept constant too long.

But we also may investigate how to avoid the use of such **impulses**. For that purpose, we **impose a bound on the velocity of the prices**, i.e., a constraint of the form

$$(ii) \text{ for almost all } t \geq 0, \quad \|p'(t)\| \leq \varphi(x(t), p(t))$$

and start anew the study of the evolution of the dynamical economy under this new constraint, which leads us to the **metasystem**

$$\begin{cases} x'(t) &= c(x(t), p(t)) \\ p'(t) &\in \varphi(x(t), p(t))B \end{cases}$$

whose metastate is the commodity-price pair and the metaregulee is the velocity of the price (inflation).

1. We assume here that this bound is chosen *a priori* and exogenously, as a bifurcation parameter. However, we can — in theory — assume that this bound $\varphi(\cdot, \cdot)$ is a solution to a dynamical system, but this is another problem which is not the topic of this chapter.

   Therefore, in this chapter, we are looking for systems of differential equations or of differential inclusions governing **the evolution of both viable commodities and prices**, so that we can **derive "supply and demand laws"** governing the evolution of prices and thus, of commodities which are viable in the sense that $p(t) \in P(x(t))$ for every t.

   In particular, we shall look for

   - **punctuated equilibria**, i.e., evolutions in which the price $\bar{p}$ remains constant whereas the commodity may evolve in the associated **viability niche**, which is the viability kernel of $x \mapsto c(x, \bar{p})$,
   - **heavy solutions**, which are evolutions where the prices evolve with minimal velocity

   The idea which allows us to achieve these aims is quite simple: **we differentiate the regulation law.**

   This is possible since we know how to differentiate set-valued maps (we already used the contingent derivative of tubes, and we refer to chapter 11 of the Tool-Box and to chapter 5 of SET-VALUED ANALYSIS for further information on the differential calculus of set-valued maps.)

We denote by $D\Pi_K(x,p)$ the contingent derivative of the regulation map $\Pi_K : X \rightsquigarrow \mathcal{M}$ at a point (x,p) of its graph[2]. We thus differentiate the regulation law and deduce that

$$(iii) \quad \text{for almost all } t \geq 0, \;\; p'(t) \in D\Pi_K(x(t),p(t))(c(x(t),p(t)))$$

whenever the viable price $p(\cdot)$ is absolutely continuous.

This is the second half of the system of differential inclusions, i.e., the largest multivalued **supply and demand law**, we were looking for.

Observe that this new differential inclusion has a meaning whenever the commodity-price pair $(x(\cdot),p(\cdot))$ remains viable in the graph of Π_K.

Fortunately, by the very definition of the contingent derivative, the graph of Π_K is a viability domain of the new system (i), (ii).

Unfortunately, as soon as viability constraints involve inequalities, there is no hope for the graph of the contingent cone, and thus, for the graph of the regulation map, to be closed, so that, the Viability Theorem cannot apply directly.

But, by imposing a bound on inflation, we can look for **graphs of closed set-valued maps** Π contained in Graph(P) which are viable under the system of differential inclusions. We already illustrated that in the simple economic example of Chapter 7.

Such set-valued maps Π are solutions to the **system of first-order partial differential inclusions**

$$\forall\, x \in K, \;\; 0 \in D\Pi(x,p)(c(x,p)) - \varphi(x,p)B$$

satisfying the constraint

$$\forall\, (x,p) \in \text{Graph}(\Pi), \;\; \Pi(x) \subset P(x)$$

For example, any continuously differentiable **single-valued** $\varpi : K \mapsto \mathcal{M}$ solution to this **partial differential inclusion** can be written

$$\forall\, j = 1,\ldots,m, \;\; \sum_{i=1}^{n} \frac{d\varpi_j(x)}{dx_i} c_i(x,\varpi(x)) + [-\varphi(x,\varpi(x)), +\varphi(x,\varpi(x))]$$

Since we shall show that such closed set-valued maps Π are all contained in the regulation map Π_K, we call them **subregulation maps** associated with the economy with bounded inflation $i), ii)$. In particular, there exists a largest subregulation map denoted Π^{φ} (the graph of which is

[2]We recall that the graph of the contingent derivative at (x,p) is the contingent cone to the graph of Π_K at (x,p).

— incidentally — the viability kernel of the graph of P for the system $(i), (ii)$).

In this case, **we do not have to assume anymore that the dynamical economy is Marchaud, but just to assume that the change map is continuous and the graph of the pricing map is closed.** We do not need anymore the convexity assumptions and the assumption that the change map is affine with respect to the price p.

But the price to pay for relaxing these assumptions is that — for the time at least — **we do not have much informations about the new subregulation maps Π**, which are set-valued solutions to the above system of first-order partial differential inclusions.

In particular, any **single-valued**[3] $\varpi : K \mapsto \mathcal{M}$ with closed graph which is a solution to the **partial differential inclusion**

$$\forall\, x \in K, \quad 0 \in D\varpi(x)(c(x, \varpi(x))) - \varphi(x, \varpi(x))B$$

satisfying the constraint

$$\forall\, x \in K, \quad \varpi(x) \; \in \; P(x)$$

provides "feedback prices" regulating smooth allocations of the dynamical economy.

Set-valued and single-valued solutions to these partial differential inclusions are studied in section 6 of chapter 8 of VIABILITY THEORY, [28, Aubin].

2. Let us consider such a subregulation map Π. Viability Theorem 6.1.9 implies that **whenever the initial commodity x_0 is chosen in** $\mathrm{Dom}(\Pi)$ **and the initial price p_0 in $\Pi(x_0)$, there exists a solution to the system of differential inclusions (i), (ii) viable in** $\mathrm{Graph}(\Pi)$. The regulation law for the viable commodity-prices becomes

$$(iv) \quad p'(t) \in D\Pi(x(t), p(t))(c(x(t), p(t))) \cap \varphi(x(t), p(t))B$$

This is the very supply and demand law we are looking for, regulating the evolution of prices in such a way that the viability property is satisfied. We also call it the **metaregulation law** associated with the subregulation map Π.

This is how we can obtain **viable commodity-price solutions under bounded inflation** to our pricing problem by solving the system of differential inclusions (i), (iv).

[3]We set $D\varpi(x) := D\varpi(x, \varpi(x))$ whenever ϖ is single-valued. When ϖ is Fréchet differentiable at x, then $D\varpi(x)(v) = \varpi'(x)v$ is reduced to the usual derivative.

We devote the next section to selection procedures of **adjustment laws**, which are selections $g(\cdot,\cdot)$ of the **metaregulation map**

$$(x,p) \rightsquigarrow G_\Pi(x,p) := D\Pi(x,p)(c(x,p)) \cap \varphi(x,p)B$$

They can be obtained through **selection procedures** $S_{G_\Pi}(x,p)$ "slicing out" the piece we desire, where the map $S_{G_\Pi}(\cdot,\cdot)$ has "good" properties that the selection does not have[4].

Naturally, under adequate assumptions, we shall check that Michael's Theorem implies the existence of a continuous adjustment law. But under the same assumptions, we can take as adjustment law the minimal selection $g^\circ(\cdot,\cdot)$ defined by $\|g^\circ(x,p)\| = \min_{v\in D\Pi(x,p)(c(x,p))} \|v\|$, which, in general, is not continuous.

3. However, we shall prove that this minimal adjustment law still yields smooth viable price-commodity solutions to the system of differential equations

$$x'(t) = c(x(t),p(t)) \quad \& \quad p'(t) = g^\circ(x(t),p(t))$$

called **heavy viable solutions**, heavy in the sense of "**heavy trends**". They are the ones for which **the price evolves with minimal velocity.**

Heavy viable solutions obey the **inertia principle**: "**keep the prices constant as long as they provide viable allocations**".

Indeed, if the zero velocity belongs to

$$D\Pi(x(T),p(T))(c(x(T),p(T)))$$

at some time T, then the price will remain equal to $p(T)$ as long as for $t \geq T$, a solution $x(\cdot)$ to the differential equation $x'(t) = c(x(t),p(T))$ satisfies the condition

$$0 \in D\Pi(x(t),p(T))(c(x(t),p(T)))$$

Definition 7.0.3 *When the bound φ is taken to be equal to 0, the viability kernel of the graph of Π for the system*

$$\begin{cases} i) & x'(t) = c(x(t),p(t)) \\ ii) & p'(t) = 0 \end{cases}$$

is called the **inertial set** $\mathcal{N}$. *It defines* **viability niches** *by*

$$N(p) := \{x \in K \mid (x,p) \in \mathcal{N}\}$$

We say that a price p is a **punctuated equilibrium** *if its viability niche is not empty.*

[4]The idea of introducing and using these selection procedures is due to Hélène Frankowska.

If at some time t_f, $p(t_f)$ is a "punctuated equilibrium", then heavy viable solution enters its viability niche and remains in this viability niche forever[5] whereas the price remains equal to this punctuated equilibrium. In other words, **punctuated equilibria are "locked-in" by heavy solutions.**

Heavy solutions are obtained through one of the many **selection procedures** of the metaregulation map that one can design. We shall provide other examples of selection procedures.

4. We provide an explicit formula of the adjustment law driving heavy solutions for dynamical economies of the form

$$x'(t) = g(x(t)) - p(t)$$

However, as it was explained in Chapter 1 and 4, one can regulate dynamical economies by connection matrices instead of prices:

$$x'(t) = W(t)g(x(t))$$

We now define a **dynamical index of (connectionist) complexity** of a time-dependent connection matrix $W(t)$, which is simply the norm of its velocity $\|W'(t)\|$, that we shall minimize over the metaregulation map for obtaining heavy evolution of connection matrices.

5. We next consider a dynamical economy for which the regulatory control which was constituted only by a price vector is now made of two components, the price and another regulee, as the frustration parameters as we did when we introduced dynamical games in the preceding chapter:

$$\begin{cases} i) & x'(t) = c(x(t), p(t), u(t)) \\ ii) & u(t) \in Q(x(t), p(t)) \\ iii) & p(t) \in P(x(t), u(t)) \end{cases}$$

By proceeding as above, we can associate with any subregulation map Π a metaregulation map G_Π with which the solutions to the system of differential inclusions

$$\begin{cases} i) & x'(t) = c(x(t), p(t), u(t)) \\ ii) & (p'(t), u'(t)) \in G_\Pi(x(t), p(t), u(t)) \end{cases}$$

yields viable solution.

Instead of selecting heavy solutions, which amounts here to select the pair of velocities $(p', u') \in G_\Pi(x, p, u)$ with minimal norm $\sqrt{\|p'\|^2 + \|u'\|^2}$,

[5] as long as the set of available resources does not change for external reasons which are not taken into account here.

we shall investigate the nonsymmetric case in which the two parameters play a different role according to the following hierarchy: we select first the velocity p' of the price with minimal norm among the set

$$G_{\Pi_1}(x,p,u) := \{p' \mid \exists\, u' \ \text{ such that } (p',u') \in G_\Pi(x,p,u)\}$$

and second, the velocity u' of the control with minimal norm in the set

$$G_{\Pi_2}(x,p,u,p') := \{u' \mid (p',u') \in G_\Pi(x,p,u)\}$$

The solutions evolving according to this rule are called **hierarchical heavy solutions.**

Such an evolution is "hierarchical" in the sense that the velocity of the price is chosen **independently** of the velocity of the control, whereas the velocity of the control (the decision) depends upon the velocity of the price (the inflation).

Hierarchical heavy solutions then satisfy the **hierarchical inertia principle** which states that **price and the control remain constant as long as the viability is satisfied and that when the viability is at stakes, then**

(a) **the control does evolve while the price is maintained constant as long as the viability is still maintained,**

(b) **both price and control evolve when the above behavior is no longer possible.**

As before, we can define the inertial set $\mathcal{N}_\Pi$ and the viability niches $N(p,u)$, which are the sets of initial states $x \in K$ from which starts a viable solution to the differential equation $x'(t) = c(x(t),p,u)$. But we can also define the **first order inertial set** $\mathcal{N}_\Pi^{(1)}$ and **first order viability niches** $N(p)$, which are the subsets of initial states $x \in K$ from which starts one viable solution to

$$\begin{cases} i) & x'(t) = c(x(t),p,u(t)) \\ ii) & u'(t) \in \psi(x(t),p,u(t))B \end{cases}$$

We thus can check that first order viability niches locks-in hierarchical heavy solutions by blocking first the price when it reaches it and letting both the commodity and the control evolve. Then, a viability niche locks-in the hierarchical heavy solution by blocking the control when it hits it, so that the commodity evolves in this viability niche.

For such hierarchical heavy evolutions, **the velocity of the control depends upon the velocity of the price, but the velocity of the price is independent of the velocity of the control.** Hierarchical selection principles by "cascades"[6] provide general mechanisms having this hierarchical property.

[6] introduced and studied by Katharina Müllers.

7.1 Bounded Inflation

7.1.1 Subregulation and Metaregulation Maps

Let us consider a dynamical economy (P, c) defined by a set-valued pricing map $P : X \leadsto \mathcal{M}$ and a single-valued change map $c : \text{Graph}(P) \mapsto X$, where X is regarded as the commodity space, $\mathcal{M}$ the price space. The evolution of a commodity-price solution $(x(\cdot), p(\cdot))$ viable in Graph(P) is governed by

$$x'(t) = c(x(t), p(t)), \quad p(t) \in P(x(t)) \tag{7.1.1}$$

For a Marchaud dynamical economy, there exist viable solutions in $K := \text{Dom}(P)$ whenever the regulation map

$$\forall\, x \in K, \quad \Pi_K(x) = \{\, p \in P(x) \mid c(x, p) \in T_K(x) \,\}$$

is strict.

In this case, viable evolutions are governed by measurable prices according to the regulation law

$$p(t) \in \Pi_K(x(t))$$

Since the time dependence of the price is only measurable, it does not bar impulses. We shall here answer the question whether it is possible to obtain smooth solutions: this is possible by preventing infinite inflation.

We shall now look for viable solutions in $K := \text{Dom}(P)$ which are smooth in the following sense:

Definition 7.1.1 *We say that the pair $(x(\cdot), p(\cdot))$ is* smooth *if both $x(\cdot)$ and $p(\cdot)$ are absolutely continuous.*

It is said to be bounded under inflation if in addition for almost all $t \geq 0$, $\|p'(t)\| \leq \varphi(x(t), p(t))$, where $\varphi : X \times \mathcal{M} \mapsto \mathbf{R}_+$ is a given bound on inflation.

We obtain smooth viable solutions by bounding inflation, i.e., setting a bound to the growth to the evolution of prices, as we did when we studied simple economic examples.

For that purpose, we associate with this dynamical economy and with any nonnegative continuous function $p \to \varphi(x, p)$ with linear growth[7] the metasystem of differential inclusions

$$\begin{cases} i) & x'(t) = c(x(t), p(t)) \\ ii) & p'(t) \in \varphi(x(t), p(t))B \end{cases} \tag{7.1.2}$$

and look for solutions $(x(\cdot), p(\cdot))$ which remain viable in the graph of P:

$$\forall\, t \geq 0, \quad (x(t), p(t)) \in \text{Graph}(P)$$

[7] which can be a constant ρ, or the function $(x,p) \to \rho\|p\|$, or the function $(x,p) \to \rho(\|p\| + \|x\| + 1)$. One could also take other dynamics $p' \in \Phi(x,p)$ where Φ is a Marchaud map. The point here is to stress the simplest situation.

Observe that any solution $(x(\cdot), p(\cdot))$ to (7.1.2) viable in Graph(P) is a solution bounded under inflation to the dynamical economy (7.1.1).

Therefore, we are looking for closed set-valued feedback maps Π contained in P whose graphs are viable under the system (7.1.2).

We thus deduce from the Viability Theorem 6.1.9 applied to the system (7.1.2) on the graph of P the following

Theorem 7.1.2 *Let us assume that the dynamical economy (7.1.1) satisfies*

$$\begin{cases} i) & \text{Graph}(P) \text{ is closed} \\ ii) & c \text{ is continuous and has linear growth} \end{cases} \tag{7.1.3}$$

Let $(x,p) \to \varphi(x,p)$ be a nonnegative continuous function with linear growth and $\Pi : \mathcal{M} \rightsquigarrow X$ a closed set-valued map contained in P. Then the two following conditions are equivalent:

a) — Π regulates viable solutions under bounded inflation in the sense that from any initial commodity $x_0 \in \text{Dom}(\Pi)$ and any initial price $p_0 \in \Pi(x_0)$ starts a commodity-price solution $(x(\cdot), p(\cdot))$ under bounded inflation to the dynamical economy (7.1.1) viable in the graph of Π_k (and thus, of P).

b) — Π is a solution to the partial differential inclusion

$$\forall (x,p) \in \text{Graph}(\Pi), \quad 0 \in D\Pi(x,p)(c(x,p)) - \varphi(x,p)B \tag{7.1.4}$$

satisfying the constraint: $\forall x \in K, \ \Pi(x) \subset P(x)$.

In this case, such a map Π is contained in the regulation map Π_K defined by

$$\forall x \in K, \ \Pi_K(x) = \{ p \in P(x) \mid c(x,p) \in T_K(x) \}$$

*and is thus called a φ-*subregulation map of P *or simply a* subregulation map.

There exists a largest φ-subregulation map denoted Π^φ contained in P.

The supply and demand law *or the* metaregulation law *regulating the evolution of commodity-price solutions viable in the graph of Π takes the form of the system of differential inclusions*

$$\begin{cases} i) & x'(t) = c(x(t), p(t)) \\ ii) & p'(t) \in G_\Pi(x(t), p(t)) \end{cases} \tag{7.1.5}$$

where the set-valued map G_Π defined by

$$G_\Pi(x,p) := D\Pi(x,p)(c(x,p)) \cap \varphi(x,p)B$$

is called the metaregulation map *associated with Π.*

Proof— Indeed, to say that Π is a regulation map regulating φ- smooth solutions amounts to saying that its graph is viable under the system (7.1.2).

In this case, we deduce that for any $(x_0, p_0) \in \text{Graph}(\Pi)$, there exists a solution $(x(\cdot), p(\cdot))$ viable in the graph of P, so that $x(\cdot)$ is in particular viable in K. Since $x'(t) = c(x(t), p(t))$ is absolutely continuous, we infer that $c(x_0, p_0)$ is contingent to K at x_0, i.e., that p_0 belongs to $\Pi_K(x_0)$.

The regulation map for the system (7.1.2) associates with any $(x,p) \in \text{Graph}(\Pi)$ the set of pairs $(x', p') \in \{c(x,p)\} \times \varphi(x,p)B$ such that (x', p') belongs to the contingent cone to the graph of Π at (x,p), i.e., such that

$$p' \in D\Pi(x,p)(c(x,p)) \cap \varphi(x,p)B =: G_\Pi(x,p)$$

The graph of the largest φ-subregulation map[8] Π^φ contained in P is the viability kernel of Graph(P) for the system of differential inclusions (7.1.2). □

We can be particularly interested in **single-valued regulation maps** $\varpi : K \mapsto \mathcal{M}$, which are **feedback prices regulating viable solutions under bounded inflation** and then be interpreted as **planning mechanisms**:

Proposition 7.1.3 *A closed single-valued continuous map ϖ is a feedback price regulating viable solutions under bounded inflation to the price problem if and only if ϖ is a single-valued solution to the* inclusion

$$\forall\, x \in K, \;\; 0 \in D\varpi(x)(c(x, \varpi(x))) - \varphi(x, \varpi(x))B$$

satisfying the constraint

$$\forall\, x \in K, \;\; \varpi(x) \in P(x)$$

Then for any $x_0 \in K$, there exists a solution to the differential equation $x'(t) = c(x(t), \varpi(x(t)))$ starting at x_0 such that

$$\forall\, t \geq 0, \; p(t) := \varpi(x(t)) \in P(x(t))$$

and

$$\text{for almost all } t \geq 0, \; \|p'(t)\| \leq \varphi(x(t), \varpi(x(t)))$$

Remark — We observe that any φ-subregulation map remains a ψ-subregulation map for $\psi \geq \varphi$. □

[8]Which is not always easy to characterize. This is the reason why we so not always choose it and why we introduce smaller regulation maps easier to "compute".

7.1.2 Example: Equality Constraints

Consider the case when $h : X \mapsto Y$ is a twice continuously differentiable map and when the viability domain is $K := h^{-1}(0)$.

Since $T_K(x) = \ker h'(x)$ when $h'(x)$ is surjective, we deduce that the regulation map is equal to

$$\Pi_K(x) = \{p \in P(x) \mid h'(x)c(x,p) = 0\}$$

Proposition 7.1.4 *Assume that $h'(x) \in \mathcal{L}(X,Y)$ is surjective whenever $h(x) = 0$, that the graph of P is sleek and that for any $y \in Y$ and $v \in X$, the subsets*

$$DP(x,p)(v) \cap (h'(x)c'_p(x,p))^{-1}(y - h''(x)(c(x,p),v) - h'(x)c'_x(x,p)v)$$

are not empty. Then the contingent derivative $D\Pi_K(x,p)(v)$ of the regulation map is equal to

$$DP(x,p)(v) \cap -(h'(x)c'_p(x,p))^{-1}(h''(x)(c(x,p),v) - h'(x)c'_x(x,p)v)$$

when $h'(x)v = 0$ and $D\Pi_K(x,v) = \emptyset$ if not. In particular, if $P(x) \equiv \mathcal{M}$, then it is sufficient to assume that $h'(x)c'_p(x,p)$ is surjective and we have in this case

$$D\Pi_K(x,p)(v) = -(h'(x)c'_p(x,p))^{-1}(h''(x)(c(x,p),v) - h'(x)c'_x(x,p)v)$$

when $h'(x)v = 0$ and $D\Pi_K(x,v) = \emptyset$ if not.

Proof — The graph of Π_K can be written as the subset of pairs $(x,p) \in \text{Graph}(P)$ such that $C(x,p) := (h(x), h'(x)c(x,p)) = 0$. Since the graph of P is closed and sleek, we know that the transversality condition

$$C'(x,p)T_{\text{Graph}(P)}(x,p) = C'(x,p)\text{Graph}(DP(x,p)) = Y \times Y$$

implies that the contingent cone to the graph of P is the set of elements $(v,w) \in \text{Graph}(DP(x,p))$ such that

$$\begin{cases} C'(x,p)(v,w) = \\ (h'(x)v, h'(x)c'_p(x,p)w + h'(x)c'_x(x,p)v + h''(x)(c(x,p),v)) = 0 \end{cases}$$

But the surjectivity of $h'(x)$ and the nonemptiness of the intersection imply this transversality condition. □

Therefore, the right-hand side of the metaregulation rule is equal to

$$\begin{cases} -(h'(x)c'_p(x,p))^{-1}(h''(x)(c(x,p),c(x,p)) - h'(x)c'_x(x,p)c(x,p)) \\ \cap DP(x,p)(c(x,p)) \cap \varphi(x,p)B \end{cases}$$

7.1.3 Example: Inequality Constraints

Consider the case when

$$K := \{x \in X : \forall i = 1, \ldots, p, \ g_i(x) \geq 0\}$$

is defined by inequality constraints (for simplicity, we do not include equality constraints.)

We denote by $I(x) := \{i = 1, \ldots, p \mid g_i(x) = 0\}$ the subset of active constraints and we assume once and for all that for every $x \in K$,

$$\exists\, v_0 \in C_L(x) \ \text{ such that } \forall\, i \in I(x), \ < g_i'(x), v_0 >> 0$$

so that, by Theorem 11.2.14,

$$\Pi_K(x) := \{p \in P(x) \mid \forall\, i \in I(x), \langle g_i'(x), c(x,p)\rangle \geq 0\}$$

We set $g(x) := (g_1(x), \ldots, g_p(x))$.

We have seen that the graph of the set-valued map $x \rightsquigarrow \Pi_K(x)$ is not necessarily closed. However, we can find explicit subregulation maps by using Theorem 11.2.15. We thus introduce the set-valued map $\Pi_K^\diamond : X \rightsquigarrow \mathcal{M}$ defined by

$$\Pi_K^\diamond(x) := \{p \in P(x) \mid g(x) + g'(x)c(x,p) \geq 0\} \subset \Pi_K(x)$$

We can regulate solutions viable in K by smooth open-loop prices by looking for solutions to the system of differential inclusions (7.1.2) which are viable in the graph of $\Pi_K^\diamond$.

We thus need to compute the derivative of $\Pi_K^\diamond$ in order to characterize the associated metaregulation map:

Proposition 7.1.5 *Assume that the stronger viability condition*[9]

$$\forall\, x \in K, \ \Pi_K^\diamond(x) \neq \emptyset$$

is satisfied. We set

$$I(x,p) := \{i = 1, \ldots, p \mid g_i(x) + \langle g_i'(x), c(x,p)\rangle = 0\}$$

Assume that P is sleek and closed and that for every $(x,p) \in \mathrm{Graph}(\Pi_K^\diamond)$, there exists $p_0' \in DP(x,p)(x_0')$ satisfying

$$\forall\, i \in I(x,p), \ \langle g_i'(x), x_0' + c_x'(x,p)x_0' + c_p'(x,p)p_0'\rangle + g_i''(x)(c(x,p), x_0') \geq 0$$

[9] which holds true whenever K is a viability domain for the dynamical economy and

$$\forall\, x \in K, \ \exists\, p \in P(x) \ \text{ such that } \|c(x,p)\| \leq \gamma_K(x)$$

where the function γ_K is defined by (11.2.5). See Theorem 11.2.15.

Then the contingent derivative $D\Pi_K^\circ(x,p)(v)$ *of the subregulation map* Π_K° *is defined by:* $p' \in D\Pi_K^\circ(x,p)(x')$ *if and only if* $p' \in DP(x,p)(x')$ *and*

$$\forall\, i \in I(x,p),\ \langle g_i'(x), x' + c_x'(x,p)x' + c_p'(x,p)p'\rangle + g_i''(x)(c(x,p),x') \geq 0$$

If $P(x) \equiv \mathcal{M}$, *then it is sufficient to assume that* $g'(x)c_p'(x,p)$ *is surjective. We then have in this particular case*

$$\begin{cases} D\Pi_K^\circ(x,p)(x') := \{p' \in \mathcal{M} \mid \forall\, i \in I(x,p), \\ \langle g_i'(x), c_p'(x,p)p'\rangle \ \geq\ -\langle g_i'(x), x' + c_x'(x,p)x'\rangle - g_i''(x)(c(x,p),x')\} \end{cases}$$

Proof — By Theorem 11.2.14 applied to $L :=$ Graph(P) and to the constraints defined by $\tilde{g}_i(x,p) := g_i(x) + \langle g_i'(x), c(x,p)\rangle$, we deduce that $p' \in D\Pi_K^\circ(x,p)(x')$ if and only if $p' \in DP(x,p)(x')$ and

$$\forall\, i \in I(x,p),\ \langle g_i'(x), x' + c_x'(x,p)x' + c_p'(x,p)p'\rangle + g_i''(x)(c(x,p),x') \geq 0 \ \square$$

We then deduce from the above Proposition and the Regularity Theorem the following consequence:

Proposition 7.1.6 *We posit the assumptions of Proposition 7.1.5. If for any* $(x,p) \in$ Graph(Π_K°), *there exists* p' *such that* $\|p'\| \leq \varphi(x,p)$, *then for any initial commodity* x_0 *and any* $p_0 \in \Pi_K^\circ(x_0)$, *there exists a solution* $(x(\cdot),p(\cdot))$ *to the dynamical economy (7.1.2) such that* $x(\cdot)$ *is viable in the set* K *defined by inequality constraints. The metaregulation law can then be written*

$$\begin{cases} i) & x'(t) = c(x(t),p(t)) \\ \\ ii) & p'(t) \in G(x(t),p(t)) \end{cases} \tag{7.1.6}$$

where the metaregulation map G *associated to* Π_K°

$$G(x,p) \ :=\ D\Pi_K^\circ(x,p)(c(x,p)) \cap \varphi(x,p)B$$

defined by:

$w \in G(x,p)$ *if and only if* $w \in DP(x,p)(c(x,p)) \cap \varphi(x,p)B$ *and*

$$\begin{cases} \forall\, i \in I(x,p),\ \langle g_i'(x), c_p'(x,p)p'\rangle \\ \geq\ -\langle g_i'(x), c(x,p) + c_x'(x,p)c(x,p)\rangle - g_i''(x)(c(x,p),c(x,p)) \end{cases}$$

Naturally, the graph of the metaregulation map G is not necessarily closed. However, we can still use Theorem 11.2.15 to obtain a "submetaregulation map" of this system of differential inclusions. We introduce the set-valued map G° defined by: $p' \in G^\circ(x,p)$ if and only if $\|p'\| \leq \varphi(x,p)$ and

$$\begin{cases} \forall\, i = 1,\ldots,p,\ \langle g_i'(x), c_p'(x,p)p'\rangle \\ \geq\ -g_i(x) - \langle g_i'(x), 2c(x,p) + c_x'(x,p)c(x,p)\rangle - g_i''(x)(c(x,p),c(x,p)) \end{cases}$$

Hence the system of differential inclusions

$$\begin{cases} i) & x'(t) = c(x(t),p(t)) \\ \\ ii) & p'(t) \in G^\circ(x(t),p(t)) \cap \varphi(x(t),p(t))B \end{cases} \tag{7.1.7}$$

regulates solutions under bounded inflation which are viable in K.

7.1.4 Other Methods for Bounding Inflation

Naturally, we can consider other evolution laws of prices of a dynamical economy (P, c) which provide smooth evolution of viable prices which are more sophisticated than requiring the price to belong to a ball of radius $\varphi(x,p)$.

We may wish, for instance, to have the price obey a second order differential inclusion, or that the balls $\varphi(x,p)B$ be replaced by subsets $\Phi(x,p)$. The situation is just identical: For instance, if $A \in \mathcal{L}(\mathcal{M},\mathcal{M})$ and $\Phi : X \times \mathcal{M} \rightsquigarrow \mathcal{M}$ is a Marchaud map, we can replace the dynamical economy (7.1.2) by the system

$$\begin{cases} i) & x'(t) = c(x(t),p(t)) \\ ii) & p'(t) \in Ap(t) + \Phi(x(t),p(t)) \end{cases} \quad (7.1.8)$$

(With an adequate choice of A, we are able to study the evolution of m time differentiable prices.)

Then Theorem 7.1.2 can be restated in this more general case in the following form:

Theorem 7.1.7 *Let us assume that the dynamical economy (7.1.1) satisfies*

$$\begin{cases} i) & \text{Graph}(P) \text{ is closed} \\ ii) & c \text{ is continuous and has linear growth} \end{cases}$$

Let $(x,p) \to \Phi(x,p)$ be a Marchaud and $\Pi : \mathcal{M} \rightsquigarrow X$ a closed set-valued map contained in P. Then the two following conditions are equivalent:

a) — Π regulates viable solutions in the sense that from any initial commodity $x_0 \in \text{Dom}(\Pi)$ and any initial price $p_0 \in \Pi(x_0)$ starts a commodity-price solution $(x(\cdot),p(\cdot))$ to the system (7.1.8) viable in the graph of Π (so that $x(\cdot)$ is still a viable solution to the dynamical economy (7.1.1)).

b) — Π is a solution to the partial differential inclusion

$$\forall (x,p) \in \text{Graph}(\Pi), \quad Ap \in D\Pi(x,p)(c(x,p)) - \Phi(x,p) \quad (7.1.9)$$

satisfying the constraint: $\forall x \in K$, $\Pi(x) \subset P(x)$. Such a set-valued map is called a Φ-subregulation map of P.

There exists a largest Φ-subregulation map denoted Π^Φ contained in P.

The metaregulation law *regulating the evolution of commodity-price solutions viable in the graph of Π takes the form of the system of differential inclusions*

$$\begin{cases} i) & x'(t) = c(x(t),p(t)) \\ ii) & p'(t) \in Ap(t) + G_\Pi(x(t),p(t)) \end{cases} \quad (7.1.10)$$

where the set-valued map G_Π defined by

$$G_\Pi(x,p) := D\Pi(x,p)(c(x,p)) \cap (Ap + \Phi(x,p))$$

is called the metaregulation map *associated with Π and Φ.*

7.2 Designing Adjustment Laws

Let us consider a dynamical economy (P, c), a regulation map $\Pi \subset P$ which is a solution to the partial differential inclusion (7.1.4) and the metaregulation map

$$(x,p) \rightsquigarrow G_\Pi(x,p) \ := \ D\Pi(x,p)(c(x,p)) \cap \varphi(x,p)B$$

regulating smooth commodity-price solutions viable in the graph of Π through the system (7.1.5) of differential inclusions.

The question arises as to whether we can construct selection procedures of the price component of this system of differential inclusions. It is convenient for this purpose to introduce the following definition.

Definition 7.2.1 *Let Π be a φ-growth subregulation map of P. We shall say that a selection g of the metaregulation map G_Π associated with Π mapping every $(x,p) \in \mathrm{Graph}(\Pi)$ to*

$$g(x,p) \ \in \ G_\Pi(x,p) \ := \ D\Pi(x,p)(c(x,p)) \cap \varphi(x,p)B \tag{7.2.1}$$

is a dynamical closed loop *of Π.*

The system of differential equations

$$\begin{cases} i) \quad x'(t) \ = \ c(x(t),p(t)) \\ ii) \quad p'(t) \ = \ g(x(t),p(t)) \end{cases} \tag{7.2.2}$$

is called the associated feedback differential system.

Clearly every solution to (7.2.2) is also a solution to (7.1.5). Therefore, an adjustment law being given, solutions to the system of ordinary differential equations (7.2.2) (if any) are smooth commodity-price solutions of the initial price problem (7.1.1).

In order to obtain adjustment laws, we shall use selection procedures of the metaregulation map $G_\Pi(x,p)$ (see Definition 4.4.1 of the Tool-Box) which provide ways to design adjustment laws of the metaregulation map G_Π:

Theorem 7.2.2 *Let us assume that the dynamical economy (7.1.1) satisfies*

$$\begin{cases} i) \quad \mathrm{Graph}(P) \text{ is closed} \\ ii) \quad c \text{ is continuous and has linear growth} \end{cases} \tag{7.2.3}$$

Let $(x,p) \rightarrow \varphi(x,p)$ be a nonnegative continuous function with linear growth and $\Pi : X \rightsquigarrow \mathcal{M}$ a closed sub regulation map contained in P.

Let $S_{G_\Pi} : \mathrm{Graph}(\Pi) \rightsquigarrow X$ be a selection procedure with convex values of the metaregulation map G_Π. Then, for any initial commodity-price $(x_0, p_0) \in \mathrm{Graph}(\Pi)$, there exists a commodity-price solution to the associated feedback system

$$x' = c(x,p), \quad p' \in G_\Pi(x,p) \cap S_{G_\Pi}(x,p) \tag{7.2.4}$$

defined on $[0,\infty[$ *and starting at* (x_0,p_0). *In particular, if for any* $(x,p) \in \text{Graph}(\Pi)$, *the intersection*

$$G_\Pi(x,p) \cap S_{G_\Pi}(x,p) = \{s(G_\Pi(x,p))\}$$

is a singleton, then there exists a commodity-price solution defined on $[0,\infty[$ *and starting at* (x_0,p_0) *to the associated feedback system*

$$x'(t) = c(x(t),p(t)), \quad p'(t) = s(G_\Pi(x(t),p(t)))$$

Proof — Consider the system of differential inclusions

$$x' = c(x,p), \quad p' \in S_{G_\Pi}(x,p) \cap \varphi(x,p)B \tag{7.2.5}$$

subject to the constraints

$$\forall\, t \geq 0, \quad (x(t),p(t)) \in \text{Graph}(\Pi)$$

Since the selection procedure S_{G_Π} has a closed graph and convex values, the right-hand side is an upper semicontinuous set-valued map with nonempty compact convex images and with linear growth.

On the other hand Graph(Π) is a viability domain of the map $\{c(x,p)\} \times (S_{G_\Pi}(x,p) \cap \varphi(x,p)B)$.

Therefore, the Viability Theorem can be applied: For any initial commodity-price $(x_0,p_0) \in \text{Graph}(\Pi)$, there exists a solution $(x(\cdot),p(\cdot))$ to (7.2.5) which is viable in Graph(Π). Consequently, for almost all $t \geq 0$, the pair $(x'(t),p'(t))$ belongs to the contingent cone to the graph of Π at $(x(t),p(t))$, which is the graph of the contingent derivative $D\Pi(x(t),p(t))$. In other words, for almost all $t \geq 0$, $p'(t) \in D\Pi(x(t),p(t))(c(x(t),p(t)))$. Since $\|p'(t)\| \leq \varphi(x(t),p(t))$, we deduce that $p'(t) \in G_\Pi(x(t),p(t))$ for almost all $t \geq 0$. Hence, the commodity-price pair $(x(\cdot),p(\cdot))$ is a solution to (7.2.4). □

7.3 Heavy Viable Evolutions

7.3.1 Punctuated Equilibria and their Viability Niches

The case when the inflation bound φ is equal to 0 is particularly interesting, because the inverse N^0 of the 0-growth regulation map Π^0 determines the areas $N^0(p)$ regulated by constant price p.

Definition 7.3.1 *We shall say that the graph* $\mathcal{N}_\Pi$ *of the* 0*-growth regulation map* Π^0 *is the* inertial set *of the dynamical economy and denote by*

$$N^0(p) := \{x \in K \mid (x,p) \in \mathcal{N}_\Pi\}$$

the viability niche *of* p. *A price* p *is called a* punctuated equilibrium *if and only if its viability niche is not empty.*

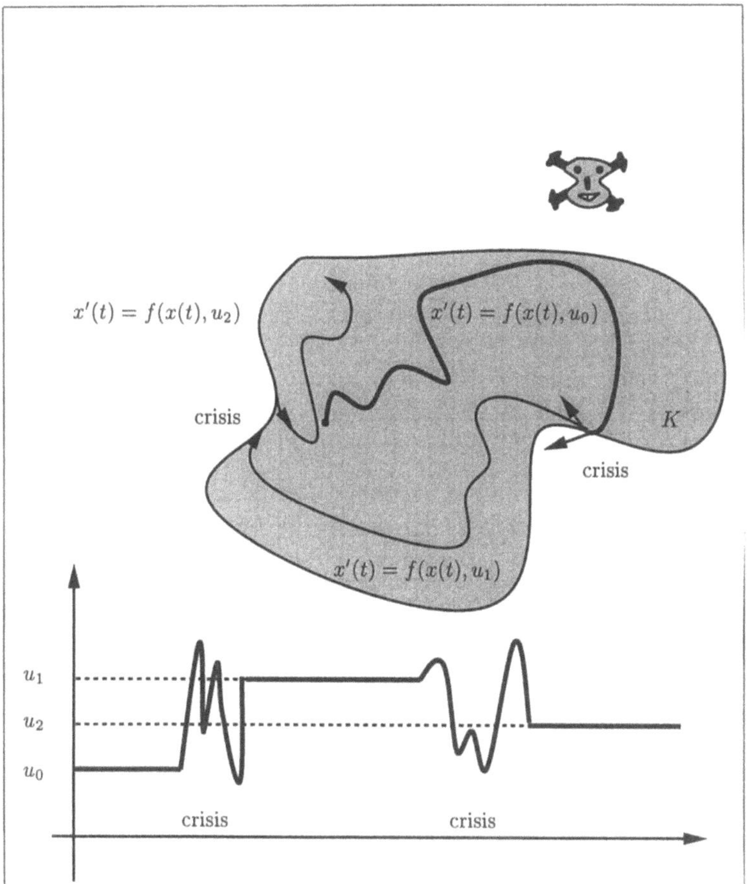

Starting from x_0 with the constant price p_0, the solution evolves in K until time t_1, (first *punctuated equilibrium phase*) when the velocity points outwards K. Then a *crisis* happens during which the price velocity also evolves (as slowly as possible) to sustain viability, until time $\bar{t}_1$ where the velocities are contingent to K during the *second punctuated equilibrium phase*.

Figure 7.1: Heavy Viable Solutions

Naturally, when the viability niche of a punctuated equilibrium is reduced to a point, this point is an equilibrium.

So, punctuated equilibria are constant prices which regulate the dynamical economies (in its viability niche):

Proposition 7.3.2 *The viability niche of a price p is the viability kernel of $P^{-1}(p)$ for the differential equation $x'(t) = c(x(t), p)$ parametrized by the constant price p.*

Proof — Indeed, viability niches describe the regions of $\mathrm{Dom}(P)$ which are regulated by the constant price p because for any initial commodity x_0 given in $N^0(p)$, there exists a viable solution $x(\cdot)$ to the differential inclusion

$$\begin{cases} i) & x'(t) = c(x(t), p(t)) \\ ii) & p'(t) = 0 \end{cases}$$

starting at (x_0, p), i.e., of the differential equation $x'(t) = c(x(t), p)$ which is viable in the viability niche $N^0(p)$ since $p \in \Pi^0(x(t))$ for every $t \geq 0$. □

Viability niches enjoy the lock-in property: When a commodity enters the viability niche of a punctuated equilibrium, it may remain there forever while the price stays equal to this punctuated equilibrium.

7.3.2 Heavy Solutions and the Inertia Principle

The simplest example of adjustment law is therefore the minimal selection of the metaregulation map G_Π, which in this case is equal to the map g_Π° associating with each commodity-price pair (x, p) the element $g_\Pi^\circ(x, p)$ of minimal norm of $D\Pi(x, p)(c(x, p))$ because for all (x, p), $\|g_\Pi^\circ(x, p)\| \leq \varphi(x, p)$ whenever $G_\Pi(x, p) \neq \emptyset$.

Definition 7.3.3 (Heavy Viable Solutions) *Denote by $g_\Pi^\circ(x, p)$ the element of minimal norm of $D\Pi(x, p)(c(x, p))$. We shall say that the solutions to the associated feedback differential system*

$$\begin{cases} i) & x'(t) = c(x(t), p(t)) \\ ii) & p'(t) = g_\Pi^\circ(x(t), p(t)) \end{cases}$$

are heavy viable solutions *to the dynamical economy (P, c) associated with Π.*

Proposition 7.3.4 *Any heavy viable solution $(x(\cdot), p(\cdot))$ to the dynamical economy (7.1.1) satisfies the inertia principle.*

Furthermore, a punctuated equilibrium is locked-in by any heavy viable solution.

Proof — Indeed, we observe that if for some t_1, the solution enters the subset $C_\Pi(p(t_1))$ where we set

$$C_\Pi(p) := \{x \in K \mid 0 \in D\Pi(x,p)(c(x,p))\}$$

the price $p(t)$ remains equal to $p(t_1)$ as long as $x(t)$ remains in $C_\Pi(p(t_1))$. Since such a subset is not necessarily a viability domain, the solution may leave it.

Furthermore, if for some $t_f > 0$, $p(t_f)$ is a punctuated equilibrium, then $p(t) = p_{t_f}$ for all $t \geq t_f$ and thus, $x(t)$ remains in the viability cell $N_1^0(p(t_f))$ for all $t \geq t_f$. □

For heavy viable solutions to exist, we need to assume that the metaregulation map G_Π associated with Π is lower semicontinuous with closed convex images. This happens whenever Π is lower semicontinuously differentiable:

Definition 7.3.5 *We shall say that a set-valued map Π is lower semicontinuously differentiable if*

$$(x,y,u) \in \text{Graph}(\Pi) \times X \rightsquigarrow D\Pi(x,y)(u) \text{ is lower semicontinuous} \quad (7.3.1)$$

Indeed, since φ is continuous, we infer from Proposition 11.2.11 that the metaregulation map G_Π is also lower semicontinuous.

Theorem 7.3.6 (Heavy Viable Solutions) *Let us assume that P is closed and that c, φ are continuous and have linear growth. Let $\Pi(\cdot) \subset P(\cdot)$ be a lower semicontinuously differentiable φ-growth subregulation map. Then for any initial commodity-price pair (x_0, p_0) in* Graph(Π), *there exists a heavy viable solution to the dynamical economy (7.1.1).*

Proof — The reason why this theorem holds true is that the minimal selection is obtained through the selection procedure defined by

$$S^\circ_{G_\Pi}(x,p) := \|g^\circ_\Pi(x,p)\| \, B \quad (7.3.2)$$

By the Maximum Theorem 11.2.5 the map $(x,p) \mapsto \|g^\circ_\Pi(x,p)\|$ is upper semicontinuous. It has a linear growth on Graph(Π). Thus the set-valued map $(x,p) \rightsquigarrow \|g^\circ_\Pi(x,p)\| \, B$ is a selection procedure satisfying the assumptions of Theorem 7.2.2. □

Since we know many examples of selection procedures, it is possible to multiply examples of adjustment laws as we did for usual feedbacks.

7.3.3 Heavy Solutions under Equality Constraints

Consider the case when $h : X \mapsto Y$ is a twice continuously differentiable map, when the viability domain is $K := h^{-1}(0)$ and when there are no constraints on the prices ($P(x) = \mathcal{M}$ for all $x \in K$). We derive from Proposition 7.1.4 the following explicit formulas for the adjustment law yielding heavy solutions.

Proposition 7.3.7 *We posit assumptions of Theorem 7.1.4. Assume further that* $P(x) \equiv \mathcal{M}$, *that the regulation map*

$$\Pi(x) := \{p \in \mathcal{M} \mid h'(x)c(x,p) = 0\}$$

has nonempty values, that $h(x)$ *is surjective whenever* $x \in K$ *and that* $h'(x)c'_p(x,p) \in \mathcal{L}(\mathcal{M}, Y)$ *is surjective whenever* $p \in \Pi(x)$.

Then there exist heavy solutions viable in K, *which are the solutions to the system of differential equations*

$$\begin{cases} i) & x' = c(x,p) \\ ii) & p' = -c'_p(x,p)^\star h'(x)^\star k(x,p) \text{ where} \\ & k(x,p) := (h'(x)c'_p(x,p)c'_p(x,p)^\star h'(x)^\star)^{-1} h'(x)c'_x(x,p)c(x,p) \end{cases}$$

Proof — The element $g(x,p) \in G(x,p)$ of minimal norm, being a solution to the quadratic minimization problem with equality constraints

$$h'(x)c'_p(x,p)w = -h'(x)c'_x(x,p)c(x,p) - h''(x)(c(x,p), c(x,p))$$

is equal to

$$\begin{array}{l} g(x,p) = -c'_p(x,p)^\star h'(x)^\star (h'(x)c'_p(x,p)c'_p(x,p)^\star h'(x)^\star)^{-1} \\ (h'(x)c'_x(x,p)c(x,p) + h''(x)(c(x,p), c(x,p))) \end{array}$$

because the linear operator $B := h'(x)c'_p(x,p) \in \mathcal{L}(\mathcal{M}, Y)$ is surjective.

Heavy solutions are obtained when $g(x,p)$ is the element of minimal norm in $G_\Pi(x,p)$.

Example: Heavy solutions viable in affine spaces. Consider the case when $K := \{x \in X \mid Lx = y\}$ where $L \in \mathcal{L}(X,Y)$ is surjective.

Let us assume that

$$\begin{cases} i) & \forall\, x \in K,\ \Pi(x) := \{p \in \mathcal{M} \ \text{ such that } \ Lc(x,p) = 0\} \neq \emptyset \\ ii) & \forall\, x \in K,\ \forall\, p \in \Pi(x),\ \ Lc'_p(x,p) \ \text{is surjective} \end{cases}$$

Then, for any initial commodity $x_0 \in K$ and initial velocity p_0 satisfying $Lc(x_0, p_0) = 0$, there exists a heavy viable solution given by the system of differential equations

$$\begin{cases} i) & x' = c(x,p) \\ ii) & p' = -c'_p(x,p)^\star L^\star (Lc'_p(x,p)c'_p(x,p)^\star L^\star)^{-1} Lc'_x(x,p)c(x,p) \end{cases}$$

When $Y := \mathbf{R}$ and $K := \{x \in X \mid \ < p,x >= y\}$ is an hyperplane, the above assumption becomes

$$\begin{cases} i) & \forall\, x \in K,\ \Pi(x) := \{p \in \mathcal{M} \ \text{ such that } \ < p, c(x,p) >= 0\} \neq \emptyset \\ ii) & \forall\, x \in K,\ \forall\, p \in \Pi(x),\ \ c'_p(x,p)^\star p \neq 0 \end{cases}$$

and heavy viable solutions are solutions to the system of differential equations

$$\begin{cases} i) & x' = c(x,p) \\ ii) & p' = -\dfrac{< p, c'_x(x,p)c(x,p) >}{\|c'_p(x,p)^\star p\|^2} c'_p(x,p)^\star p \end{cases}$$

7.4 Heavy Price and Connective Regulation

Let X, Y be finite dimensional vector spaces, $g : X \mapsto X$ describe the priceless dynamics of the consumers, $h : X \mapsto Y$ represent the resource map and $M \subset Y$ the set of available resources.

7.4.1 Heavy Price Regulation

We consider the simple dynamical economy

$$x'(t) = g(x(t)) - p(t)$$

subject to viability constraints of the form

$$\forall t \geq 0, \quad h(x(t)) \in M$$

which is regulated by

$$p(t) \in \Pi_M(x(t))$$

where

$$\Pi_M(x) := \{p \in X^\star \mid h'(x(t))p(t) \in h'(x(t))g(x(t)) - T_M(x)\}$$

We now set a bound on the velocity of the prices:

$$\|p'(t)\| \leq \varphi(x(t), p(t))$$

where B denotes the unit ball of the space $X^\star$.

If the graph of the contingent cone is not closed, we introduce a closed graph selection $T_M^\square(\cdot)$ of $T_M(\cdot)$, such as the Maderner selection $T_M^\diamond(\cdot)$ in the case of inequality constraints, or the set-valued map whose graph is the viability kernel of the graph of $T_M(\cdot)$ under the system $(x', v') \in \{v\} \times cB$.

Therefore, we solve the system of differential inclusions which governs the initial system together with the price which controls it:

$$\begin{cases} i) & x'(t) = g(x(t)) - p(t) \\ ii) & \|p'(t)\| \leq \varphi(x(t), p(t)) \end{cases} \tag{7.4.1}$$

subject to the new viability constraints

$$\mathcal{K} := \left\{(x, p) \mid h'(x)p \in h'(x)g(x) - T_M^\square(h(x))\right\}$$

When g is continuously differentiable and h twice continuously differentiable, we infer that (v, q) belongs to the contingent cone $T_{\mathcal{K}}(x, p)$ if and only if

$$\begin{cases} h'(x)q \in h''(x)(v, g(x) - p) + h'(x)g'(x)v \\ -DT_M^\square(h(x), h'(x)(g(x) - p))(h'(x)v) \end{cases}$$

Hence, when $v := g(x) - p$, we set

$$\begin{cases} H_M(x,p) := h''(x)(g(x)-p, g(x)-p) + h'(x)g'(x)(g(x)-p) \\ -DT_M^{\square}(h(x), h'(x)(g(x)-p))(h'(x)g(x)-p) \end{cases}$$

We deduce from the Viability Theorem the

Proposition 7.4.1 *Let us assume that g is continuously differentiable and h twice continuously differentiable, that*

$$\forall\, x \in K,\quad h'(x) \text{ is surjective } \& \ \sup_{x \in K} \|h'(x)^+\| \ < \ +\infty \tag{7.4.2}$$

and that $(x,p) \rightsquigarrow H_M(x,p)$ is lower semicontinuous.

Then for any initial state $x_0 \in K$ and p_0, there exist a solution $x(\cdot)$ to the dynamical economy (7.4.1) starting at x_0 and a solution $p(\cdot)$ to the implicit differential inclusion

$$h'(x(t))p'(t) \ \in \ H_M(x(t), p(t))$$

starting at p_0 sustaining the viability of the system in the sense that the viability constraints:

$$\forall\, t \geq 0,\quad h'(x(t))(g(x(t)) - p(t)) \ \in \ T_M^{\square}(h(x(t)))$$

(implying the initial scarcity constraints) are always met.

Recall that when $B \in \mathcal{L}(X,Y)$ is surjective, we can supply Y with the final norm μ^B defined by $\mu^B(z) := \|B^+ z\|$ and that we denote by π_K^B the projector of best approximation onto the closed subset K for this final norm.

We also recall that when M is sleek, then its tangent cones $T_M(y)$ are convex.

Theorem 7.4.2 *We posit the assumptions of Proposition 7.4.1.*

Then the differential equation governing the evolution of the heavy evolution of the connecting matrix is given by the system

$$\begin{cases} i) \quad x' \ = \ g(x) - p \\ ii) \quad p' \ = \ h'(x)^+ v(x,p) \\ \quad\quad \text{where} \\ \quad\quad v(x,p) \ := \ \left(\mathbf{1} - \pi^{h'(x)}_{DT_M^{\square}(h(x),h'(x)(g(x)-p))(h'(x)g(x)-p)}\right) \\ \quad\quad (h''(x)(g(x)-p, g(x)-p) + h'(x)g'(x)(g(x)-p) \end{cases}$$

Then for any initial state $x_0 \in K$ and p_0, there exist a solution $(x(\cdot), p(\cdot))$ to the above system of differential equations starting at (x_0, p_0) sustaining the viability of the system in the sense that the viability constraints

$$\forall\, t \geq 0,\quad h'(x(t))(g(x(t)) - p(t)) \ \in \ T_M^{\square}(h(x(t)))$$

are always met.

7.4.2 Heavy Evolution of Connection Matrices

We still consider the initial — priceless — dynamical economy

$$x'(t) = g(x(t))$$

subjected to scarcity constraints of the form

$$\forall\, t \geq 0, \;\; h(x(t)) \in M$$

When the necessary and sufficient viability condition $h'(x)g(x) \in T_M(h(x))$ is not met by every $x \in K$, we already suggested in Chapter 4 to "correct" the priceless dynamical economy by introducing both prices and connection matrices

$$x'(t) = W(t)g(x(t)) - p(t)$$

The **disconnected system** is obtained when the connection matrix is equal to the identity matrix $\mathbf{1}$.

The prices and the connection matrices regulating viable solutions obey the regulation law

$$(W(t), p(t)) \in R_M(x(t))$$

where

$$R_M(x) := \{(W,p) \in \mathcal{L}(X,X) \times X^\star \mid h'(x)(Wg(x) - p) \in T_M(h(x))\} \tag{7.4.3}$$

We still introduce a closed graph selection $T_M^\square(\cdot)$ of $T_M(\cdot)$ whenever the graph of the contingent cone is not closed, such as the Maderner selection. We denote by

$$\mathcal{K} := \left\{(x,W,p) \mid h'(x)Wg(x) - h'(x)p \in T_M^\square(h(x))\right\}$$

the graph of his regulation map, regarded as a viability set of the augmented system obtained by setting bonds on the velocity of the connection matrices and the prices:

$$\begin{cases} i) & x'(t) = W(t)g(x(t)) - p(t) \\ ii) & \|W'(t)\| \leq \varphi(x(t), W(t), p(t)) \\ iii) & \|p'(t)\| \leq \psi(x(t), W(t), p(t)) \end{cases} \tag{7.4.4}$$

When $g(x) \neq 0$ and $h'(x)$ is surjective, when g is continuously differentiable and h twice continuously differentiable, we infer that (v, V, q) belongs to the contingent cone $T_{\mathcal{K}}(x, W, p)$ if and only if

$$\begin{cases} h'(x)Vg(x) - h'(x)q + h'(x)Wg'(x)v + h''(x)(v, Wg(x) - p) \\ \in DT_M^\square(h(x), h'(x)(Wg(x) - p))(h'(x)v) \end{cases}$$

From now on, we set $v := Wg(x) - p$ and

$$G_M(x, W, p) := DT_M^\square(h(x), h'(x)v)(h'(x)v) - h'(x)Wg'(x)v - h''(x)(v,v)$$

so that (v, V, q) belongs to $T_{\mathcal{K}}(x, W, p)$ if and only if

$$h'(x)Vg(x) - h'(x)q \in G_M(x, W, p)$$

Hence, we deduce from the Viability Theorem the following

Proposition 7.4.3 *Let us assume that φ and ψ are continuous with linear growth, that g is continuously differentiable and that h is twice continuously differentiable.*

Then from any initial $(x_0, W_0, p_0) \in \mathcal{K}$ starts one solution $(x_\cdot, W_\cdot, p_\cdot)$ to the connected system (7.4.4) satisfying the implicit differential inclusion

$$h'(x(t))(W'(t)g(x(t)) - p'(t)) \in G_M(x(t), W(t), p(t))$$

sustaining the viability of the system in the sense that the viability constraints

$$\forall\, t \geq 0,\;\; h'(x(t))(W(t)g(x(t)) - p(t)) \in T_M^{\square}(h(x(t)))$$

are always met.

In particular, for any selection $r(x, W, p)$ of $h'(x)^{-1}(G_M(x, W, p))$ and any real-valued function $\lambda(\cdot)$, the solutions to the system of differential equations

$$\left\{ \begin{array}{lll} i) & x' & = & Wg(x) - p \\ ii) & W' & = & \lambda(x)g(x) \otimes r(x, W, p) \\ iii) & p' & = & (\lambda(x)\|g(x)\|^2 - 1)r(x, W, p) \end{array} \right.$$

are viable.

Proof — Indeed, we observe that

$$W'g(x) - p' = \lambda(x)\|g(x)\|^2 r(x, W, p) - (\lambda(x)\|g(x)\|^2 - 1)r(x, W, p) = r(x, W, p)$$

so that

$$h'(x)(W'g(x) - p') = h'(x)r(x, W, p) \in G_M(x, W, p) \;\square$$

From this, we derive the differential equations governing the evolution of heavy solutions, the ones which choose velocities of the connection matrices and of the prices with minimal norm when we supply the space $\mathcal{L}(X, X) \times X$ with the norm $\nu_{\alpha,\beta}$ defined by

$$\nu_{\alpha,\beta}(W, p)^2 := \alpha\|W\|^2 + \beta\|p\|^2$$

Theorem 7.4.4 *We posit the assumptions of Proposition 7.4.3 and we assume that*

$$\forall\, x \in K,\;\; h'(x) \text{ is surjective } \& \sup_{x \in K} \frac{\|h'(x)^+\|}{\|g(x)\|} < +\infty \tag{7.4.5}$$

and that $(x,W) \rightsquigarrow G_M(x,W,p)$ *is lower semicontinuous. We set*

$$r(x,W,p) := h'(x)^+ \left(\pi^{h'(x)}_{DT^\square_M(h(x),h'(x)v)(h'(x)v)} - \mathbf{1}\right)(h'(x)Wg'(x)v+h''(x)(v,v))$$

where $v := Wg(x) - p$, *which is a map independent from* α *and* β. *Then the differential equation governing the evolution of the heavy evolution of the connecting matrix is given by the system*

$$\begin{cases} i) & x' = Wg(x) - p \\ ii) & W' = \dfrac{\beta g(x)}{\alpha + \beta\|g(x)\|^2} \otimes r(x,W,p) \\ iii) & p' = \dfrac{-\alpha}{\alpha + \beta\|g(x)\|^2} r(x,W,p) \end{cases}$$

Proof — Proposition 11.2.10 implies that the regulation map R_M defined by (7.4.3) is lower semicontinuous because the set-valued map $G_M(\cdot)$ is lower semicontinuous and that the surjectivity assumption (7.4.5) implies assumption implies assumption (11.2.1).

Heavy solutions are obtained by choosing the minimal velocity $(W'(t), p(t))$ in the set $G_M(x(t), W(t), p(t))$ for the norm $\nu_{\alpha,\beta}$. Setting $x := x(t)$, $W := W(t)$ and $p(t) = p$, then $(V,q) := (W'_{\alpha,\beta}(t), p'_{\alpha,\beta}(t))$ is the pair (V,q) solution to the minimization problem:

$$\inf_{h'(x)Vg(x) - h'(x)q \,\in\, G_M(x,W,p)} \nu_{\alpha,\beta}(V,q)$$

We infer from Theorem 10.3.7 that since the map $g(x)\otimes h'(x)$ is surjective from the space $\mathcal{L}(X,X)$ of connection matrices to the resource space Y, the solution is given by

$$\begin{cases} W' = \dfrac{\beta g(x)}{\alpha + \beta\|g(x)\|^2} \otimes h'(x)^+(\pi^{h'(x)}_{DT^\square_M(h(x),h'(x)v)(h'(x)v)} - \mathbf{1})q(x,W,p) \\ \text{where} \\ q(x,W,p) := h'(x)Wg'(x)v + h''(x)(v,v) \\ \\ p' = \dfrac{-\alpha}{\alpha + \beta\|g(x)\|^2} h'(x)^+(\pi^{h'(x)}_{DT^\square_M(h(x),h'(x)v)(h'(x)v)} - \mathbf{1})q(x,W,p) \end{cases}$$

The velocities $W'(t)$ and $p'(t)$ depend upon α and β. However, we observe that

$$W'(t)g(x(t)) - p'(t) = r(x(t), W(t), p(t))$$

is independent of the parameters α and β.

7.5 Hierarchical Games

We consider the case when we have a dynamical game $(\mathcal{P}, \mathcal{Q}, c)$ described by a finite dimensional vector space X of **states**, a space $\mathcal{M}$ of **prices** and a space $\mathcal{N}$ of other regulees — which could be the frustrations we introduced in Chapter 5, for instance. We introduce the pricing map $\mathcal{P} : X \times \mathcal{N} \rightsquigarrow \mathcal{M}$ associating with each commodity-regulee pair (x, u) a subset $\mathcal{P}(x, u)$ of potential prices and a map $\mathcal{Q} : X \times \mathcal{M} \rightsquigarrow \mathcal{N}$ associating with each commodity-price pair (x, p) a subset $\mathcal{Q}(x, p)$ of regulees.

Hence the evolution of the commodity, the regulee and the price is governed by

$$\begin{cases} i) & x'(t) = c(x(t), p(t), u(t)) \\ ii) & u(t) \in \mathcal{Q}(x(t), p(t)) \\ iii) & p(t) \in \mathcal{P}(x(t), u(t)) \end{cases} \tag{7.5.1}$$

The purpose of the game is to find both prices $p(t)$ and $u(t)$ in order to keep the commodity $x(t)$ inside a given closed subset $K \subset X$.

Let us point out that what we shall say for two regulees (the price p and the regulee u) could be extended to n different strategies, without any new difficulty other than the cumbersome notations which have to be used to write down all these parameters.

When these two parameters play the same role, we could regard this problem as a mere viability problem for the dynamical economy (7.5.1), and, as we did in the first section, set bounds $\varphi(x, p, u)$ and $\psi(x, p, u)$ on the velocities $p'(\cdot)$ and $u'(\cdot)$ of the price $p(\cdot)$ and the regulee $u(\cdot)$ (the velocity of a regulee is often interpreted as a **decision**). We then look for a subregulation map $\Pi : X \rightsquigarrow \mathcal{M} \times \mathcal{N}$ contained in $P := \hat{\mathcal{Q}} \cap \check{\mathcal{P}}$ regulating smooth solutions of the dynamical economy (7.5.1) in order to maintain the viability of its solutions.

Theorem 7.1.2 associates with such a subregulation map Π of this dynamical economy (7.5.1) a **metaregulation** map G_{Π} for which the solutions of the system

$$\begin{cases} i) & x'(t) = c(x(t), p(t), u(t)) \\ ii) & (p'(t), u'(t)) \in G_{\Pi}(x(t), p(t), u(t)) \end{cases} \tag{7.5.2}$$

satisfy at each instant the viability constraints

$$\begin{cases} i) & u(t) \in \mathcal{Q}(x(t), p(t)) \\ ii) & p(t) \in \mathcal{P}(x(t), u(t)) \end{cases} \tag{7.5.3}$$

This is the point where we now **distinguish** the behavior of the two parameters, by looking to **dynamical cascades** which are metasystems of the form

$$\begin{cases} i) & x'(t) = c(x(t), p(t), u(t)) \\ ii) & p'(t) \in H_1(x(t), p(t), u(t)) \\ ii) & u'(t)) \in H_2(x(t), p(t), u(t), p'(t)) \end{cases} \tag{7.5.4}$$

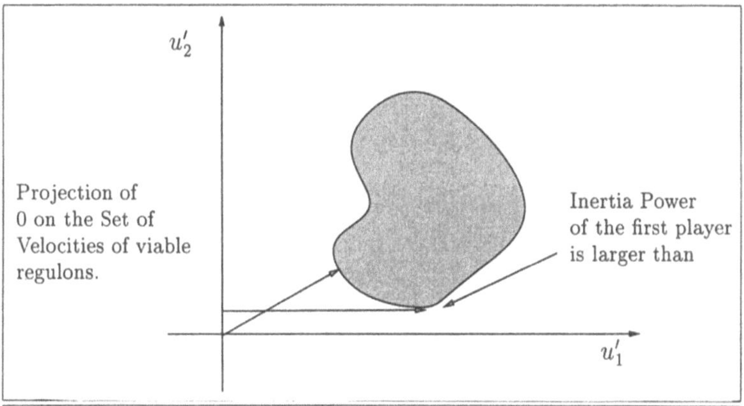

The axes represent the norms of the velocities of the two regulees. The choice (A) corresponds to the case when the first player has a smallest inertia (the smallest velocity of the regulee) and the choice (B) corresponds to the opposite scenario. The choice (V), is a virtual minimum. The choice (D) provides the heavy evolution, of a cooperative nature. The Nash equilibrium (N) can be anywhere. The choice C is the best compromise between the Nash equilibrium and the virtual minimum. The south-west part of the boundary is the set of "Pareto velocities".

Figure 7.2: Selection Mechanisms of Velocities

the solutions of which satisfy the viability constraints (7.5.3).

Such a system is "hierarchical" in the sense that the velocity of the price is chosen **independently of** the velocity of the regulee, whereas the velocity of the regulee (the decision) depends upon the velocity of the price (the inflation).

Therefore, this section is devoted to **nonsymmetric** selection procedures (S_1, S_2) called **cascades** which provide systems of differential inclusions of the foregoing form.

7.5.1 Cascades

Let us consider a set-valued map $G : X \rightsquigarrow \mathcal{M} \times \mathcal{N}$[10]. The **hierarchical decomposition** of this set-valued map G induces two set-valued maps $G_1 : X \rightsquigarrow \mathcal{M}$ and $G_2 : X \times \mathcal{M} \rightsquigarrow \mathcal{N}$ defined in the following way

$$\begin{cases} G_1(x) := \{p \in \mathcal{M} \mid \exists\, u \in \mathcal{N} \text{ such that } (p,u) \in G(x)\} \\ G_2(x,p) := \{u \in \mathcal{N} \mid (p,u) \in G(x)\} \end{cases}$$

[10]Here, X stands for $X \times \mathcal{M} \times \mathcal{N}$ in the situation described previously. What matters is the decomposition of the arrival space $\mathcal{M}$ as the product $\mathcal{M} := \mathcal{M} \times \mathcal{N}$.

We observe that

$$\begin{cases} \mathrm{Dom}(G_1) = \mathrm{Dom}(G), & \mathrm{Graph}(G_1) = \mathrm{Proj}_{X\times\mathcal{M}}(\mathrm{Graph}(G)) \\ \mathrm{Dom}(G_2) = \mathrm{Graph}(G_1), & \mathrm{Graph}(G_2) = \mathrm{Graph}(G) \end{cases}$$

Definition 7.5.1 *Let us consider a set-valued map $G : X \rightsquigarrow \mathcal{M} \times \mathcal{N}$. We shall say that a pair (S_1, S_2) of set-valued maps $S_1 : X \rightsquigarrow \mathcal{M}$ and $S_2 : X \times \mathcal{M} \rightsquigarrow \mathcal{N}$ constitute a* cascade *of G (or a hierarchical selection of G) if they are closed and if*

$$\begin{cases} \mathrm{Dom}(S_1) = \mathrm{Dom}(G), \quad \mathrm{Dom}(S_2) = \mathrm{Dom}(G_2) \\ \forall\, x \in \mathrm{Dom}(G), \quad S_1(x) \cap G_1(x) \neq \emptyset \\ \forall\, (x,p) \in \mathrm{Dom}(G_2), \quad S_2(x,p) \cap G_2(x,p) \neq \emptyset \end{cases}$$

We associate with the cascade (S_1, S_2) of G the set-valued map $S : X \rightsquigarrow \mathcal{M} \times \mathcal{N}$ defined by the formula

$$(p,u) \in S(x) \text{ if and only if } p \in S_1(x) \;\&\; u \in S_2(x,p)$$

The map $S : X \rightsquigarrow \mathcal{M} \times \mathcal{N}$ associated with a cascade (S_1, S_2) of G is a selection procedure of G (see Definition 4.4.1), since we observe that its graph is closed and that $S(x) \cap G(x) \neq \emptyset$.

We also observe that

Lemma 7.5.2 *A cascade S has convex values if and only if S_1 has convex values and the graphs of the maps $S_2(x,\cdot)$ are convex*[11]*. In this case, we shall say that S is a* convex cascade.

7.5.2 Dynamical Cascades

The Selection Theorem 7.2.2 can be adapted to cascades:

Theorem 7.5.3 *Let us assume that the dynamical game $(\mathcal{P}, \mathcal{Q}, c)$ satisfies*

$$\begin{cases} i) & \mathrm{Graph}(\mathcal{P}) \;\&\; \mathrm{Graph}(\mathcal{Q}) \text{ are closed} \\ ii) & c \text{ is continuous and has linear growth} \end{cases} \tag{7.5.5}$$

Let $(x,p) \to \varphi(x,p,u), \psi(x,p,u)$ be nonnegative continuous functions with linear growth and $\Pi : X \rightsquigarrow \mathcal{M} \times \mathcal{N}$ a subregulation map of (7.5.1).

Let $G_\Pi : X \times \mathcal{M} \times \mathcal{N} \rightsquigarrow \mathcal{M} \times \mathcal{N}$ be the metaregulation map associated with a subregulation map Π[12]*. Let us choose a convex cascade (S_1, S_2) of the*

[11] This means that the maps $S_2(x,\cdot)$ are convex (see Definition 9.1.2).

[12] which is defined by

$$G_\Pi(x,p,u) = D\Pi(x,p,u) \cap (\varphi(x,p,u)B \times \psi(x,p,u)B)$$

metaregulation map G_Π. *Then, from any initial triple* $(x_0, p_0, u_0) \in \mathrm{Graph}(\Pi)$ *starts a solution to*

$$\begin{cases} i) & x'(t) = c(x(t), p(t), u(t)) \\ ii) & p'(t) \in S_1(x(t), p(t), u(t)) \cap G_{\Pi_1}(x(t), p(t), u(t)) \\ iii) & u'(t) \in S_2(x(t), p(t), u(t), p'(t)) \cap G_{\Pi_2}(x(t), p(t), u(t), p'(t)) \end{cases} \tag{7.5.6}$$

defined on $[0, \infty[$ *and viable in* $\mathrm{Graph}(\Pi)$. *In particular, if for any* $(x, p, u) \in \mathrm{Graph}(\Pi)$, *the intersections*

$$\begin{cases} S_1(x, p, u) \cap G_1(x, p, u) := \{s_1(x, p, u)\} \\ S_2(x, p, u, p') \cap G_2(x, p, u, p') := \{s_2(x, p, u, p')\} \end{cases}$$

are singleta, then from any initial triple $(x_0, p_0) \in \mathrm{Graph}(\Pi)$ *starts one viable solution defined on* $[0, \infty[$ *to the system of differential equations*

$$\begin{cases} i) & x'(t) = c(x(t), p(t), u(t)) \\ ii) & p'(t) = s_1(x(t), p(t), u(t)) \\ iii) & u'(t) = s_2(x(t), p(t), u(t), p'(t)) \end{cases}$$

Proof — We associate with the cascade S of G_Π the set-valued map $F : X \times \mathcal{M} \times \mathcal{N} \leadsto X \times \mathcal{M} \times \mathcal{N}$ defined by

$$F(x, p, u) = \{c(x, p, u)\} \times (S(x, p, u) \cap (\varphi(x, p, u)B \times \psi(x, p, u)B))$$

This map F is Marchaud because the cascade S of G_Π has a closed graph and convex values. Furthermore, we claim that $\mathrm{Graph}(\Pi)$ is a viability domain of F, i.e., that

$$\forall\, (x, p, u) \in \mathrm{Graph}(\Pi), \quad F(x, p, u) \cap T_{\mathrm{Graph}(\Pi)}(x, p, u) \neq \emptyset$$

Since $G_\Pi(x, p, u) = D\Pi(x, p, u) \cap (\varphi(x, p, u)B \times \psi(x, p, u)B)$, since

$$T_{\mathrm{Graph}(\Pi)}(x, p, u) = \mathrm{Graph}(D\Pi(x, p, u))$$

since $F(x, p, u) \cap \mathrm{Graph}(D\Pi(x, p, u)) \neq \emptyset$ and since S is a cascade of G_Π, and thus, a selection procedure, these intersections are not empty.

Therefore, the Viability Theorem can be applied. For any initial $(x_0, p_0, u_0) \in \mathrm{Graph}(\Pi)$, there exists a solution $(x(\cdot), p(\cdot), u(\cdot))$ to the differential inclusion

$$(x'(t), u'(t), p'(t)) \in F(x(t), p(t), u(t))$$

which is viable in Graph(Π).

Consequently, for almost all $t \geq 0$, the triple $(x'(t), p'(t), u'(t))$ belongs to the contingent cone to the graph of Π at $(x(t), p(t), u(t))$, which is the graph of the contingent derivative $D\Pi(x(t), p(t), u(t))$.

In other words, for almost all $t \geq 0$,

$$(p'(t), u'(t)) \in D\Pi(x(t), p(t), u(t))(c(x(t), p(t), u(t)))$$

Since $\|p'(t)\| \leq \varphi(x(t), p(t), u(t))$ and $\|u'(t)\| \leq \psi(x(t), p(t), u(t))$, we deduce that $(p'(t), u'(t)) \in G_\Pi(x(t), p(t), u(t))$ for almost all $t \geq 0$.

7.5.3 Hierarchical Heavy Viable Solutions

Among the single values cascade of the metaregulation map G_Π, we would like to select the velocities with minimal norm of the hierarchical decomposition of the metaregulation map G_Π. They are denoted by

$$\begin{cases} G^\circ_{\Pi_1}(x,p,u) := \{p' \in G_{\Pi_1}(x,p,u) \mid \|p'\| = \min_{q \in G_{\Pi_1}(x,p,u)} \|q\|\} \\ \\ G^\circ_{\Pi_2}(x,p,u,p') := \{u' \in G_{\Pi_2}(x,p,u,p') \mid \|u'\| = \min_{v \in G_{\Pi_2}(x,p,u,p')} \|v\|\} \end{cases}$$

Definition 7.5.4 *We shall say that the solutions to the system of differential equations*

$$\begin{cases} i) & x'(t) = c(x(t), p(t), u(t)) \\ \\ ii) & p'(t) = G^\circ_{\Pi_1}(x(t), p(t), u(t)) \\ \\ iii) & u'(t) = G^\circ_{\Pi_2}(x(t), p(t), u(t), p'(t)) \end{cases} \tag{7.5.7}$$

are the hierarchical heavy viable solutions *to the dynamical game associated with the subregulation map* Π.

We then can prove

Theorem 7.5.5 *We posit the assumptions of Theorem 7.5.3 and we assume that the set-valued maps* (G_{Π_1}, G_{Π_2}) *of the hierarchical decomposition of the metaregulation map* G_Π *are lower semicontinuous with convex images. Then, from any initial triple* $(x_0, p_0, u_0) \in$ Graph(Π) *starts a* hierarchical heavy viable solution *of the dynamical game.*

Proof — The minimal hierarchical decomposition is obtained by the "minimal cascade" (S°_1, S°_2) defined by

$$\begin{cases} S^\circ_1(x,p,u) := \{p' \in \mathcal{M} \mid \|p'\| = \min_{q \in G_{\Pi_1}(x,p,u)} \|q\|\} \\ S^\circ_2(x,p,u,p') := \{u' \in \mathcal{N} \mid \|u'\| = \min_{v \in G_{\Pi_2}(x,p,u,p')} \|v\|\} \end{cases}$$

The lower semicontinuity of the both the maps G_{Π_1} and G_{Π_2} of the hierarchical decomposition of the metaregulation map G_Π implies that the minimal cascade has a closed graph and thus deserves the qualification of being a cascade. □

7.5.4 Hierarchical Viability Niches

Hierarchical heavy viable obey the **hierarchical inertia principle**, which states that price and the regulee remain constant as long as the viability is satisfied and that when the viability is at stakes, then

1. the regulee does evolve while the price is maintained constant as long as the viability is still maintained,
2. both price and regulee evolve when the above behavior is no longer possible.

Indeed, assume that starting from an initial triple $(x_0, p_0, u_0) \in \text{Graph}(\Pi)$, we have

$$G^{\circ}_{\Pi_1}(x_0, p_0, u_0) \;=\; 0 \;\&\; G^{\circ}_{\Pi_2}(x_0, p_0, u_0, 0) \;=\; 0$$

so that the commodity evolving according the differential equation

$$x'(t) \;=\; c(x(t), p_0, u_0)$$

as long as $x(t) \in K$. If at time t_1 the viability is at stakes, then the regulee will start to evolve and the system will evolve according to the dynamical economy

$$\begin{cases} x'(t) \;=\; c(x(t), p_0, u(t)) \\ \\ u'(t) \;=\; G^{\circ}_{\Pi_2}(x(t), p_0, u(t), 0) \end{cases}$$

as long as $x(t) \in K$. If at tome t_2, the viability is again threatened, then both the commodity, the regulee and the price move according to (7.5.7).

Furthermore, we can define hierarchical viability niches in the following way:

We associate with the dynamical game $(\mathcal{P}, \mathcal{Q}, c)$ the dynamical system:

$$\begin{cases} i) & x'(t) \;=\; c(x(t), p(t), u(t)) \\ ii) & p'(t) \;=\; 0 \\ iii) & u'(t) \;=\; 0 \end{cases}$$

A subregulation map Π being given, the **inertial set** of the associated dynamical economy is the viability kernel of $\text{Graph}(\Pi)$, which is the set $\mathcal{N}_\Pi$ of triples (x_0, p_0, u_0) such that there exists one solution to

$$x'(t) \;=\; c(x(t), p_0, u_0)$$

starting from x_0 and viable in the **viability niche**

$$N(p_0, u_0) \;:=\; \{x \in K \mid (x, u_0, v_0) \in \mathcal{N}\}$$

We next associate with the dynamical game $(\mathcal{P}, \mathcal{Q}, c)$ the dynamical system:

$$\begin{cases} i) & x'(t) \;=\; c(x(t), p(t), u(t)) \\ ii) & p'(t) \;=\; 0 \\ iii) & u'(t) \;=\; G_{\Pi_2}(x(t), u(t), p(t), 0) \end{cases}$$

We regard the viability kernel of Graph(Π) as the **first order inertial set**: It is the set $\mathcal{N}_{\Pi}^{(1)}$ of triples (x_0, p_0, u_0) such that there exists one solution to

$$\begin{cases} i) & x'(t) = c(x(t), p_0, u(t)) \\ ii) & u'(t) = G_{\Pi_2}(x(t), u(t), p_0, 0) \end{cases}$$

viable in K. It furnishes **first order viability** niches

$$N^{(1)}(p) := \{(x, u) \mid (x, p, u) \in \mathcal{N}_{\Pi}^{(1)}\}$$

Since $\mathcal{N}_{\Pi}$ is contained in $\mathcal{N}_{\Pi}^{(1)}$, we note that whenever $x \in \mathcal{N}(p, u)$ for some u, then $x \in \mathcal{N}^{(1)}(p)$

Hierarchical heavy viable solutions enjoys a **hierarchical lock-in property of hierarchical viability niches**: As soon as at some time T a hierarchical heavy solution enters a first order viability niche $\mathcal{N}^{(1)}(p(T))$, then the price remains constant and the commodity-regulee evolve according to the system of differential equations

$$\forall\, t \geq T, \begin{cases} i) & x'(t) = c(x(t), p(T), u(t)) \\ ii) & u'(t) = G_{\Pi_2}^{\circ}(x(t), u(t), p(T), 0) \end{cases}$$

Then, as soon as at some time $T^{(1)} \geq T$ the solution enters the viability niche $\mathcal{N}(p(T), u(T^{(1)}))$, then the solution evolve according the differential equation

$$\forall\, t \geq T^{(1)}, \ x'(t) = c(x(t), p(T), u(T^{(1)}))$$

In other words, first order viability niches locks-in hierarchical heavy solutions by blocking first the price and letting both the commodity and the regulee evolve. Then, a viability niche locks-in the heavy solution by blocking the regulee, so that the commodity evolves in this viability niche.

7.5.5 Designing Cascades

We could naturally use and combine the other examples of selection procedures for designing cascades. The ones based on selection mechanisms through optimization or the use of Ky Fan inequalities require the lower semicontinuity of both the maps G_{Π_1} and G_{Π_2} of the hierarchical decomposition of the metaregulation map G_{Π}.

If it is easy to deduce the lower semicontinuity of G_{Π_1} from the lower semicontinuity of G_{Π}, we need extra assumptions to derive the lower semicontinuous of G_{Π_2}.

Proposition 7.5.6 *Let us consider the hierarchical decomposition* (G_1, G_2) *of a set-valued map* $G : X \rightsquigarrow \mathcal{M} \times \mathcal{N}$.

1. *If* G *is upper semicontinuous with compact images, then the graph of* G_1 *is closed.*

2. If G is lower semicontinuous, then

- *G_1 is lower semicontinuous,*
- *G_2 is lower semicontinuous provided that for every $x \in \text{Dom}(G)$ and every $\varepsilon > 0$, there exist $\delta > 0$ such that, whenever $\|x-y\| \leq \delta$, $p, q \in G_1(y)$ satisfy $\|p-q\| \leq \delta$, then, $\mathbf{d}$ denoting the Hausdorff distance*

$$\mathbf{d}(G_2(y,p_1), G_2(y,p_2)) \leq \varepsilon$$

- *The above condition is necessary whenever we assume that G is upper semicontinuous with compact images.*

Proof

- Assume that G is upper semicontinuous with compact images. Let $(x_n, p_n) \in \text{Graph}(G_1)$ converge to (x,p). There exists u_n such that $(p_n, u_n) \in G(x_n)$. Since G is upper semicontinuous with compact values, we deduce that a subsequence (again denoted by) u_n converges to some u and that (p,u) belongs to $G(x)$. Hence $p \in G_1(x)$, so that the graph of G_1 is closed.

- Assume that G is lower semicontinuous. Let x_n converge to x, choose $p \in G_1(x)$ and fix some $u \in G_2(x,p)$. Since $(p,u) \in G(x)$, there exists a sequence $(p_n, u_n) \in G(x_n)$ converging to (p,u). This means that $p_n \in G_1(x_n)$ converges to p, so that G_1 is lower semicontinuous.

Let x_n converge to x, $p_n \in G_1(x_n)$ converge to p and choose $u \in G_2(x,p)$. Since $(p,u) \in G(x)$, there exists a sequence $(u_n, q_n) \in G(x_n)$ converging to (p,u). Hence u_n converges to u and both p_n and q_n converge to the same p. Therefore, associating with any $\varepsilon = \frac{1}{k}$ a positive number δ_k according to the assumption, we can build recursively an increasing sequence $n_{1<n_2} < \ldots n_k < \ldots$ such that

$$\forall\, n \geq n_k, \;\; \|x - x_n\| \leq \delta_k, \;\; \|p - p_n\| \leq \frac{\delta_k}{2} \;\; \& \;\; \|p - q_n\| \leq \frac{\delta_k}{2}$$

so that

$$\mathbf{d}(G_2(x_n, p_n), G_2(x_n, q_n)) \leq \frac{1}{k}$$

Setting $k_n := \max_{n_k \leq n} k \to +\infty$, we deduce the existence of $v_n \in G_2(x_n, p_n)$ such that $\|v_n - u_n\| \leq \frac{1}{k_n}$ converges to 0, and thus that v_n converges to u. Hence G_2 is lower semicontinuous.

In order to prove the converse statement, assume that G is upper semicontinuous with compact images and that this condition is false: There exists $\varepsilon > 0$ and sequences x_n converging to x, $p_n, q_n \in G_1(x_n)$ such that $\|p_n - q_n\|$ converges to 0 and $v_n \in G_2(x_n, p_n)$ satisfying

$$\inf_{v \in G_2(x_n, q_n)} \|v_n - v\| = d(G_2(x_n, p_n), G_2(x_n, q_n)) \geq \varepsilon$$

Since G is upper semicontinuous with compact images, a subsequence (again denoted by) (p_n, v_n) converges to some $(p, u) \in G(x)$. If G_2 were lower semicontinuous, there would exist $w_n \in G_2(x_n, q_n)$ converging to u because (x_n, q_n) converges to (x, p). This is a contradiction of the foregoing inequality which implies that $\|v_n - w_n\| \geq \varepsilon$. □

Chapter 8

Dynamical Inequalities

Introduction

This chapter is devoted to specific viability constraints — called, dynamical inequalities — which can be written in the form

$$\forall\, t \in [0,T], \quad V(x(t)) \;\leq\; w(t)$$

where $V : X \mapsto \mathbf{R}_+ \cup \{+\infty\}$ is a given nontrivial nonnegative extended function.

Here, the evolution of $t \mapsto x(t)$ is governed by a dynamical economy (P, c) and the evolution of $t \mapsto w(t)$ is governed by some differential equation

$$w'(t) \;=\; -\varphi(x(t), w(t))$$

In this case, V can be regarded as loss function (or $U := -V$ as a utility function) which decreases along at least a solution $x(\cdot)$ of the dynamical economy faster that $w(\cdot)$. They are called "satisfycing" evolutions in the economic literature after Herbert Simon. For instance, taking $\varphi(x,w) = aw$, we would obtain solutions along which the loss function decreases exponentially:

$$\forall\, t \geq 0, \quad V(x(t)) \;\leq\; V(x_0)e^{-at}$$

where $a \geq 0$. Such functions can legitimately be called exponential Lyapunov functions.

If a given loss function does not satisfy such a requirement, the question then arises whether there exists another loss function larger than or equal to V satisfying the above property. We may even ask the question whether the smallest among the functions satisfying this property still enjoys it. In some sense, this smallest loss function larger than or equal to the original one "reveals" the loss function consistent with the dynamical economy.

Before raising other issues and reviewing more motivating examples, let us mention right away the link between dynamical inequalities and viability constraints: Denoting by

$$\mathcal{E}p(V) \;:=\; \{(x,w) \in X \times \mathbf{R} \mid V(x) \;\leq\; w\}$$

the **epigraph** of the function V, the above dynamical inequality constraints can be written

$$\forall\, t \in [0,T], \;\; (x(t), w(t)) \;\in\; \mathcal{E}p(V)$$

Hence we are now in familiar grounds whenever the epigraph of the extended function V is closed[1].

1. Using the results obtained so far, we shall be able to characterize the functions V along which solutions to a dynamical economy satisfies such dynamical inequalities in terms of generalized solutions[2] (called **contingent solutions**) to Hamilton-Jacobi equations

$$\inf_{q \in P(x)} \langle V'(x), c(x,q)\rangle + \varphi(x, V(x)) \;=\; 0$$

In this case, the solutions $x(t)$ to the dynamical economy satisfying such a dynamical inequality are regulated by price $p(t)$ obeying the regulation law: for almost all ≥ 0,

$$p(t) \;\in\; \Pi_V(x) \;:=\; \{q \in P(x) \mid \langle V'(x), c(x,q)\rangle + \varphi(x, V(x)) \;\leq\; 0\}$$

However, if a given function V does not satisfy such a property, we shall be able to construct such a (lower semicontinuous) solution $V_\propto$ smaller than or equal to any (lower semicontinuous) function larger than V: we shall prove that *the epigraph of such a function $V_\propto$ is the viability kernel of the epigraph of V* under an auxiliary system. In this case, we are able to use the viability kernel algorithm for approximating these functions by finite-difference schemes and prove their convergence (even in the presence of viability constraints).

---

[1]Extended functions the epigraph of which are closed are called **lower semicontinuous** functions. This property can be used as a handy definition of upper semicontinuous instead of the one proposed by René Baire at the beginning of the century.

[2]Actually, since we require only that V is lower semicontinuous — which is mandatory whenever constraints are involved — we have to relax the classical definition of the gradient of the function V. Several competing ways have be devised to achieve this purpose. However, using the links between dynamical inequalities and the Viability Theorem, it is natural to introduce the contingent cone to the epigraph of the function V and to observe that it is the epigraph

$$T_{\mathcal{E}p(V)}(x, V(x)) \;=\; \mathcal{E}p(D_\uparrow V(x))$$

of the function $D_\uparrow V(x)$ defined by:

$$\forall\, u \in X, \;\; D_\uparrow V(x)(u) \;=\; \liminf_{h \to 0+, u' \to u} (V(x+hu') - V(x))/h$$

It is called the **contingent epiderivative** of V at x and is a generalized directional derivative because $D_\uparrow V(x)(u) = \langle V'(x), u\rangle$ whenever V is differentiable. We refer to [208, Frankowska] and to the monograph CONTROL OF NONLINEAR SYSTEMS AND DIFFERENTIAL INCLUSIONS, [209, Frankowska], for an exhaustive exposition of generalized solutions (both contingent and viscosity) to Hamilton-Jacobi-Bellman equations as well as other issues in control theory.

Doing so, we are closing a loop. Hamilton-Jacobi equations, adapted by Richard Bellman to optimal control problems and by Rufus Isaacs for differential games — sometime called **quantitative problems** — were used for solving problems with state constraints (such target problems) — sometimes called **qualitative problems**. Here, we shall derive results about quantitative problems such as intertemporal optimization from viability theorems obtained for solving qualitative problems. The link can be made by characterizing a given subset $K \subset X$ by its indicator function $\psi_K : X \mapsto \mathbf{R} \cup \{+\infty\}$ defined by

$$\psi_K(x) := \begin{cases} 0 & \text{if} \quad x \in K \\ +\infty & \text{if} \quad x \notin K \end{cases}$$

because whenever $w(t)$ is finite,

$$x(t) \in K \text{ if and only if } \psi_K(x(t)) < +\infty$$

Although not differentiable, the indicator ψ_K of a closed subset K viable under a dynamical economy is a contingent solution to the Hamilton-Jacobi equation $\inf_{q \in P(x)} \langle V'(x), c(x,q) \rangle = 0$.

2. Before going further, let us review other motivating examples. A loss function is a (exponential) Lyapunov function under a dynamical economy — or a differential inclusion — if it decreases (exponentially) along at least one solution of this dynamical economy:

$$\forall\, t \geq 0, \quad V(x(t)) \leq V(x_0)e^{-at}$$

In mathematics, Lyapunov functions have been used since Alexander Lyapunov proposed his "second method" in 1892 for studying asymptotic properties (asymptotic stability for instance) of dynamical systems. Hence studying asymptotic stability of equilibria and their basin of attraction requires the knowledge of a Lyapunov function. This amounted to find for each class of problems a "natural" Lyapunov function. But it is now possible thanks to the concept of viability kernel to build Lyapunov functions — actually, "optimal" Lyapunov functions.

3. Using hitting and exit time functions for describing T-viability kernels and capture basins also leads to constraints of the above form because saying that x belongs to $\text{Viab}_F(K,T)$ amounts to saying that there exists a solution to the differential inclusion $x' \in F(x)$ such that

$$\forall\, t \in [0,T], \quad \tau_K^\sharp(x(t)) \geq \tau^\sharp(x) - t$$

and that x belongs to $\mathrm{Capt}_F(K,T)$ amounts to saying there exists a solution to the differential inclusion $x' \in F(x)$ such that

$$\forall\, t \in [0,T], \quad \varpi_K^\flat(x(t)) \;\leq\; \varpi^\flat(x) - t$$

We shall present a theorem due to Pierre Cardaliaguet, Marc Quincampoix and Patrick Saint-Pierre characterizing the hypograph of the upper exit time function and the epigraph of the lower hitting time function as viability kernels of auxiliary systems. Following Luc Doyen and Patrick Saint-Pierre, we shall also characterize the epigraph of the crisis function as the viability kernel of an auxiliary system, since the crisis function is the value function of an intertemporal optimization problem studied in the next section.

These characterizations of the epigraphs and hypographs of those functions allows us to approximate them thanks to the viability kernel algorithm.

4. Lyapunov functions are actually closely linked with the characterization of the value function of an infinite horizon discounted optimization problem.

   Indeed, let us introduce a nonnegative lower semicontinuous loss function

$$W : (x,q) \in \mathrm{Graph}(P) \mapsto W(x,q) \in \mathbf{R}_+$$

   (called the Lagrangian) assumed to be convex with respect to q and to have linear growth. We shall study the infinite horizon intertemporal minimization of the discounted cost

$$V_\propto(x_0) \;:=\; \inf_{x(\cdot)\in\mathcal{S}_{(P,c)}(x_0)} \int_0^\infty e^{a\tau} W(x(\tau),p(\tau))d\tau \;\in\; [0,+\infty]$$

   over the solutions of a dynamical economy (P,c) starting at a given initial state x_0.

   The initial state dependent function $V_\propto$ introduced above is called the **value function** of the optimization problem. It plays an important role because its knowledge provides the regulation map regulating the prices governing the evolution of optimal solutions $x(\cdot)$. Indeed, after Hamilton, Jacobi, Carathéodory and Bellman among many other mathematicians, the value function is a (contingent) solution to the Hamilton-Jacobi-Bellman equation

$$\inf_{q\in P(x)}(\langle V'(x), c(x,q)\rangle + W(x,q)) + aV(x) \;=\; 0$$

   We thus define the regulation map

$$\Pi_{V_\propto}(x) \;:=\; \{q \in P(x) \mid \langle V_\propto'(x), c(x,q)\rangle + W(x,q) + aV_\propto(x) \;\leq\; 0\}$$

We shall prove that optimal solutions are regulated by prices satisfying

$$\text{for almost all } t \geq 0, \;\; p(t) \;\in\; \Pi_{V_\alpha}(x)$$

This justifies the importance of finding the value function for solving the optimization problem (we leave aside the problem of studying the Pontryagin principle, which is the adaptation to this problem of the Fermat rule providing necessary conditions for optimization). We shall prove in essence that the value function V_α of the intertemporal optimization problem **is the smallest of the lower semicontinuous nonnegative extended functions V satisfying the following dynamical inequalities: From any initial state $x_0 \in \mathrm{Dom}(V)$ starts a solution to the dynamical economy satisfying**

$$\forall\, t \geq 0, \;\; V(x(t)) \;\leq\; w(t) \;:=\; e^{-at}V(x_0) + \int_0^t e^{a(\tau - t)}W(x(\tau), p(\tau))d\tau \tag{8.0.8}$$

Actually, we shall characterize the epigraph of the value function as the **viability kernel** of an auxiliary differential inclusion.

5. Next, we minimize

$$J_\alpha(x,p) \;:=\; \inf_{(x(\cdot),p(\cdot))\in\mathcal{S}(x,p)} \int_0^\infty e^{a\tau}W(x(\tau),p(\tau))d\tau$$

over the solutions to the dynamical economy satisfying both the state constraints, the price constraints and **bounds on the inflation rates of prices**, as in Chapter 7.

For simplicity of the exposition, we assume in this introduction that there are no constraints on the commodities and the prices and we concentrate on the presence of bounds on the velocities of the prices (bounded inflation).

We shall prove that there exists a commodity-price dependent function denoted $(x,p) \mapsto J_\alpha(x,p)$ (called the metavalue function) which we can approximate and compute. We then derive from this metavalue function both the value function V_α of the optimal dynamical economy subjected to inflation bounds and the feedback set-valued map Π_{V_α} providing the optimal prices:

$$\begin{cases} i) & V_\alpha(x) \;=\; \inf_{p\in\mathcal{M}} J_\alpha(x,p) \\ ii) & \Pi_{V_\alpha}(x) \;=\; \{p \in \mathcal{M} \mid V_\alpha(x) \;=\; J_\alpha(x,p)\} \end{cases} \tag{8.0.9}$$

so that, optimal solutions subjected to inflation bounds are solutions to the dynamical economy

$$\begin{cases} i) & x'(t) \;=\; c(x(t),p(t)) \\ ii) & p(t) \;\in\; \Pi_{V_\alpha}(x(t)) \end{cases} \tag{8.0.10}$$

In particular, any selection $\varpi(x) \in \Pi_{V_\alpha}(x)$ provides feedback prices driving solutions to the system $x'(t) = c(x(t), \varpi(x(t)))$. Furthermore, we shall provide also **dynamical feedback controllers** $(x,p) \mapsto g(x,p)$ (driving the inflation) such that the commodity-price which are solutions to the system of differential equations

$$\begin{cases} i) \quad x'(t) = c(x(t), p(t)) \\ ii) \quad p'(t) = g(x(t), p(t)) \end{cases} \tag{8.0.11}$$

We observe also that imposing a bound on the inflation of the prices actually allows us to relax the assumptions usually made on the dynamical economy, and in particular, to drop the assumption of linearity of the dynamics and of convexity of the Lagrangian W with respect to the prices. The price to pay is to find the metavalue function which is defined on an augmented space the dimension of which being the sum of the dimensions of the state space and the price space.

6. Having characterized the epigraph of the value function as the **viability kernel** of an auxiliary differential inclusion., we can apply the **viability kernel algorithm** to this epigraph, which provides the value function of the discretized problem. In this way, we obtain an algorithm for computing the discrete value function. This (approximated) discrete value function is then used for approximating an optimal solution.

8.1 Epigraphs

8.1.1 Extended Functions and their Epigraphs

A function $V : X \mapsto \mathbf{R} \cup \{+\infty\}$ is called an *extended (real-valued) function*. Its *domain* is the set of points at which V is finite:

$$\text{Dom}(V) := \{x \in X \mid V(x) < +\infty\}$$

A function is said to be *nontrivial*[3] if its domain is not empty. Any function V defined on a subset $K \subset X$ can be regarded as the extended function V_K equal to V on K and to $+\infty$ outside of K, whose domain is K.

Since the order relation on the real numbers is involved in the definition of the Lyapunov property (as well as in minimization problems and other dynamical inequalities), we no longer characterize a real-valued function by its graph, but rather by its *epigraph*[4]

$$\mathcal{E}p(V) := \{(x,\lambda) \in X \times \mathbf{R} \mid V(x) \leq \lambda\}$$

[3]Such a function is said to be *proper* in convex and non smooth analysis. We chose this terminology for avoiding confusion with proper maps.

[4]The *hypograph* of a function $V : X \mapsto \mathbf{R} \cup \{-\infty\}$ is defined in a symmetric way by

$$\mathcal{H}p(V) := \{(x,\lambda) \in X \times \mathbf{R} \mid V(x) \geq \lambda\} = -\mathcal{E}p(-V)$$

We also remark that some properties of a function are actually properties of their epigraphs. For instance, *an extended function V is convex (resp. positively homogeneous) if and only if its epigraph is convex (resp. a cone).* The epigraph of V is closed if and only if V is lower semicontinuous:

$$\forall\, x \in X,\;\; V(x) \;=\; \liminf_{y \to x} V(y)$$

We also observe that any positively homogeneous extended function is nontrivial (never takes the value $-\infty$) whenever $V(0) \neq -\infty$. In this case, $V(0) = 0$. For more details, we refer to Chapter 6 of [54, Aubin & Frankowska].

Indicators ψ_K of subsets K are cost functions defined by

$$\psi_K(x) := 0 \;\text{ if }\; x \in K \;\text{ and }\; +\infty \;\text{ if not}$$

which characterize subsets (as *characteristic functions* do for other purposes) provide important examples of extended functions. It can be regarded as a **membership cost**[5] to K: it costs nothing to belong to K, and $+\infty$ to step outside of K.

Since

$$\mathcal{E}p(\psi K) \;=\; K \times \mathbf{R}_+$$

we deduce that the indicator ψ_K is lower semicontinuous if and only if K is closed and that ψ_K is convex if and only if K is convex. One can regard the sum $V + \psi_K$ as the restriction of V to K.

We recall the convention $\inf(\emptyset) := +\infty$.

Epigraphs of extended functions are subsets of $\mathcal{E} \subset X \times \mathbf{R}$ satisfying

$$\mathcal{E}p(V) \;=\; \mathcal{E}p(V) + \{0\} \times \mathbf{R}_+$$

Conversely:

Lemma 8.1.1 *Let $\mathcal{M} \subset X \times \mathbf{R}_+$ be a closed subset. Then*

$$\mathcal{M} + \{0\} \times \mathbf{R}_+ \;=\; \mathcal{E}p(V_{\mathcal{M}})$$

is the epigraph of the lower semicontinuous function $V_{\mathcal{M}} : X \mapsto \mathbf{R}_+ \cup \{+\infty\}$ defined by

$$V_{\mathcal{M}}(x) \;:=\; \inf_{(x,w) \in \mathcal{M}} w$$

Proof — The only fact to check is that $V_{\mathcal{M}}$ is lower semicontinuous whenever $\mathcal{M}$ is closed. Let (x_n, λ_n) be a sequence of $\mathcal{E}p(V_{\mathcal{M}})$ converging to (x, λ). We know that there exists $\alpha_n \leq \lambda_n$ such that (x_n, α_n) belongs to $\mathcal{M}$. Since $0 \leq \alpha_n \leq \lambda_n$, a subsequence (again denoted by) α_n converges to some $\alpha \leq \lambda$, and thus, (x, α) belongs to $\mathcal{M}$ and consequently, (x, λ) belongs to $\mathcal{E}p(V_{\mathcal{M}})$. □

The graph of a real-valued (finite) function is then the intersection of its epigraph and its hypograph.

[5]Functions $V : X \mapsto [0, +\infty]$ can be regarded as some kind of fuzzy sets, called toll sets (see Chapter 5). For "fuzzy differential inclusions" and "fuzzy viability", we refer to chapter 10, section 3 of [28, Aubin].

8.1.2 Contingent Epiderivatives

Let $V : X \mapsto \mathbf{R} \cup \{\pm\infty\}$ be a nontrivial extended function and x belong to its domain. We recall from Proposition 11.4.2 of the Tool-Box that the contingent epiderivative $D_\uparrow V(x)$ satisfies

$$\forall\, u \in X, \;\; D_\uparrow V(x)(u) \;=\; \liminf_{h\to 0+, u'\to u} \frac{V(x+hu') - V(x)}{h}$$

and that the epigraph of the contingent epiderivative $D_\uparrow V(\cdot)$ is equal to the contingent cone to the epigraph of V at $(x, V(x))$ is

$$\mathcal{E}p(D_\uparrow V(x)) \;=\; T_{\mathcal{E}p(V)}(x, V(x))$$

Formula (8.1.2) allows us to "translate" the viability theorems to constraints defined by dynamical inequalities. The link is made through the indicators: Let ψ_K be the *indicator* of a subset K. Then it is easy to check that the contingent epiderivative of the indicator is the indicator of the contingent cone:

$$D_\uparrow \psi_K(x) = \psi_{T_K(x)}$$

We shall need the following technical Lemma:

Lemma 8.1.2 *Let $\mathcal{M} \subset X \times \mathbf{R}_+$ be a closed subset and $V_{\mathcal{M}} : X \mapsto \mathbf{R}_+ \cup \{+\infty\}$ be its associated extended function.*

If for some $(x, \alpha) \in \mathcal{M}$, the pair (u, λ) belongs to $T_{\mathcal{M}}(x, \alpha)$, then for every $\mu \in \mathbf{R}$ and every $w > \alpha$, the pair (u, μ) belongs to $T_{\mathcal{E}p(V)_{\mathcal{M}}}(x, w)$.

In particular, if $V : X \mapsto \mathbf{R}_+ \cup \{+\infty\}$ is an extended function and if $D_\uparrow V(x)(u)$ is finite, then, for every $w > V(x)$ and every $\mu \in \mathbf{R}$, the pair (u, μ) belongs to the contingent cone $T_{\mathcal{E}p(V)}(x, w)$ to the epigraph of V at (x, w).

Proof— Let (u, λ) belong to $T_{\mathcal{M}}(x, \alpha)$. Then we know that there exists sequences $h_n > 0$ converging to 0, u_n converging to u and λ_n converging to λ such that $(x + h_n u_n, \alpha + h_n \lambda_n)$ belongs to $\mathcal{M} \subset \mathcal{E}p(V_{\mathcal{M}})$. Therefore, for $w > \alpha$ and $\mu \in \mathbf{R}$ and h_n is small enough,

$$(x + h_n u_n, w + h_n \mu) \;=\; (x + h_n u_n, \alpha + h_n \lambda_n) + (0, w - \alpha + h_n(\mu - \lambda_n))$$

belongs to the epigraph of $V_{\mathcal{M}}$.

The second statement follows with $\mathcal{M} := \mathcal{E}p(V)$ and $\alpha := V(x)$. □

8.2 The Quantitative Viability Theorem

We consider a differential inclusion

$$\text{for almost all } t \geq 0, \;\; x'(t) \;\in\; F(x(t)) \tag{8.2.1}$$

and a time-dependent function $w(\cdot)$ defined as a solution to the differential equation

$$w'(t) \;=\; -\varphi(x(t), w(t)) \tag{8.2.2}$$

where $\varphi : X \times \mathbf{R}_+ \to \mathbf{R}$ is a given continuous function with linear growth.

Theorem 8.2.1 *Let V be a nonnegative contingently epidifferentiable lower semicontinuous extended function and $F : X \rightsquigarrow X$, be a Marchaud map and $\varphi : X \times \mathbf{R}_+ \to \mathbf{R}$ be continuous with linear growth.*

Then the two following conditions are equivalent:

1. *for any initial state $x_0 \in \mathrm{Dom}(V)$, there exist solutions $x(\cdot)$ to (8.2.1) and $w(\cdot)$ to (8.2.2) satisfying property*

$$\forall\, t \geq 0,\;\; V(x(t)) \;\leq\; w(t), \quad w(0) = V(x(0)) \tag{8.2.3}$$

2. *V is a contingent solution to the* Hamilton-Jacobi inequalities

$$\forall\, x \in \mathrm{Dom}(V), \;\; \inf_{v \in F(x)} D_\uparrow V(x)(v) + \varphi(x, V(x)) \;\leq\; 0 \tag{8.2.4}$$

If V does not satisfy either one of these properties, there exists a smallest nonnegative lower semicontinuous solution $V_\propto^\varphi : X \mapsto \mathbf{R} \cup \{+\infty\}$ to the contingent Hamilton-Jacobi inequalities (8.2.4) larger than or equal to V (which can be the constant $+\infty$), which thus enjoys the property:

$$\begin{array}{c} \forall\, x_0 \in \mathrm{Dom}(V_\propto^\varphi), \textit{ there exist solutions to (8.2.1) and (8.2.2)} \\ \text{satisfying } \forall\, t \geq 0, \;\; V(x(t)) \;\leq\; V_\propto^\varphi(x(t)) \;\leq\; w(t) \end{array}$$

Consequently, if $V(x_0) < V_\propto^\varphi(x_0)$, all solutions $x(\cdot)$ to differential inclusion (8.2.1) and all solutions $w(\cdot)$ to differential equation (8.2.2) starting at $(x_0, V(x_0))$ satisfy

$$\begin{cases} i) & \forall\, t \geq 0,\; V_\propto^\varphi(x(t)) \;>\; w(t) \text{ as long as } V(x(t)) \;\leq\; w(t) \\ & \\ ii) & \exists\, T > 0 \;\; \textit{such that} \;\; V(x(T)) \;>\; w(T) \end{cases}$$

This happens for any solution $w(\cdot)$ whenever the initial state x_0 does not belong to the domain of $V_\propto^\varphi$.

Proof — We set $G(x, w) := F(x) \times \{-\varphi(x, w)\}$. Obviously, the system (8.2.1), (8.2.2) has a solution satisfying (8.2.3) if and only if the system of differential inclusions

$$(x'(t), w'(t)) \in G(x(t), w(t)) \tag{8.2.5}$$

has a solution starting at $(x_0, V(x_0))$ viable in $\mathcal{E}p(V)$. Since G is Marchaud, this is equivalent to

$$\forall\, (x, w) \in \mathcal{E}p(V), \;\; \exists\, u \in F(x) \;\; \text{such that} \;\; (u, -\varphi(x, w)) \in T_{\mathcal{E}p(V)}(x, w)$$

1. This condition implies (8.2.4) because by taking $w = V((x))$, we infer that
$$(v, -\varphi(x, V(x))) \in T_{\mathcal{E}p(V)}(x, V(x)) = \mathcal{E}p(D_\uparrow V(x))$$
for some $v \in F(x)$. Hence V is a contingent solution to the Hamilton-Jacobi inequality (8.2.4).

   Conversely, since $F(x)$ is compact and $v \mapsto D_\uparrow V(x)(v)$ is lower semicontinuous, (8.2.4) implies that there exists $v \in F(x)$ such that the pair $(v, -\varphi(x, V(x)))$ belongs to $T_{\mathcal{E}p(V)}(x, V(x))$. If $w > V(x)$, Lemma 8.1.2 implies that for every $\mu \in \mathbf{R}$, (v, μ) belongs to $T_{\mathcal{E}p(V)}(x, w)$, and in particular, for $\mu := -\varphi(x, w)$.

2. We know that the viability kernel $\text{Viab}_G(\mathcal{E}p(V))$ is the largest closed viability domain of $\mathcal{E}p(V)$ of the set-valued map $(x, w) \rightsquigarrow G(x, w) := F(x) \times \{-\varphi(x, w)\}$. If it is empty, it is the epigraph of the constant function equal to $+\infty$.

   If not, it is the epigraph of the nonnegative lower semicontinuous function $V_\propto^\varphi$ defined by
$$V_\propto^\varphi(x) \;:=\; \inf_{(x,\lambda)\in \text{Viab}_G(\mathcal{E}p(V))} \lambda$$
we are looking for.

   Indeed, since $\text{Viab}_G(\mathcal{E}p(V))$ is viable under G, we deduce that for any $(x, \alpha) \in \text{Viab}_G(\mathcal{E}p(V))$, the pair $(u, -\varphi(x, w))$ belongs to the contingent cone to $\text{Viab}_G(\mathcal{E}p(V))$ at (x, α). Taking $\alpha := V_\propto^\varphi(x)$, we infer that
$$(u, -\varphi(x, w)) \;\in\; T_{\mathcal{E}p(V_\propto^\varphi)}(x, V_\propto^\varphi(x))$$
Since
$$\mathcal{E}p(V_\propto^\varphi) \;=\; \text{Viab}_G(\mathcal{E}p(V)) + \{0\} \times \mathbf{R}_+$$
Lemma 8.1.2 implies that if $w > V_\propto^\varphi(x)$,
$$(u, -\varphi(x, w)) \;\in\; T_{\mathcal{E}p(V_\propto^\varphi)}(x, w)$$
Therefore $\mathcal{E}p(V_\propto^\varphi)$ is a closed viability domain of $\mathcal{E}_p(V)$ under G, and therefore, contained in its viability kernel:
$$\mathcal{E}p(V_\propto^\varphi) \;\subset\; \text{Viab}_G(\mathcal{E}p(V))$$
and thus, equal to it.

   Hence $V_\propto^\varphi$ is a contingent solution to the Hamilton-Jacobi inequalities, larger than or equal to V because the epigraph of $V_\propto$ is contained in the epigraph of V. Furthermore, the epigraph of any contingent solution to the dynamical inequalities (8.2.4) being a closed viability domain of the set-valued map G is contained in the epigraph of $V_\propto^\varphi$. This means that $V_\propto^\varphi$ is smaller than any lower semicontinuous solution to (8.2.4) larger than V. □

Remark — We can reformulate the viability theorem in the following way:

Corollary 8.2.2 *Let $F : X \rightsquigarrow X$ be a Marchaud map. A closed subset K enjoys the viability property if and only if its indicator ψ_K is a contingent solution to the Hamilton-Jacobi equation*

$$\inf_{v \in F(x)} D_{\uparrow}\psi_K(x)(v) = 0$$

8.3 Lyapunov Functions

Consider a differential inclusion $x' \in F(x)$ and a nontrivial lower semicontinuous extended function $V : X \mapsto \mathbf{R} \cup \{+\infty\}$ assumed to be bounded from below: we set

$$v_0 := \inf_{x \in X} V(x)$$

The function $V : X \mapsto \mathbf{R}_+ \cup \{+\infty\}$ is said *to enjoy the a-Lyapunov property* if and only if for any initial state x_0, there exists a solution to the differential inclusion satisfying

$$\forall t \geq 0, \;\; V(x(t)) \;\leq\; w(t) := (V(x_0) - v_0)e^{-at} + v_0 \qquad (8.3.1)$$

along at least one solution $x(\cdot)$ to differential inclusion (8.2.1).

Such inequalities allow us to deduce many properties on the asymptotic behavior of V along the solutions to the differential inclusion.

If $\mathcal{S}_F(x)$ denotes the set of solutions to the differential inclusion starting from x, we associate with the function V

$$\rho_{V,a}(x) \;:=\; \inf_{x(\cdot) \in \mathcal{S}_F(x)} \sup_{t \geq 0} \frac{V(x(t)) - v_0}{e^{-at}}$$

and we define the a-basin of attraction of V as

$$\text{Bas}_a(V) \;:=\; \{x \in X \mid \rho_{V,a}(x) \;<\; +\infty\}$$

In other words, a state x_0 belongs to the a-basin of attraction if and only if there exist a solution $x(\cdot) \in \mathcal{S}(x_0)$ and a constant c such that

$$\forall t \geq 0, \;\; V(x(t)) \;\leq\; ce^{-at} + v_0$$

This may be quite useful when V is the distance function $d_M(\cdot)$ to a subset. The domain of this Lyapunov function $d_{M\propto}$ provides the a-basin of attraction of M, which is the set of states from which a solution $x(\cdot)$ to the differential inclusion converges exponentially to M:

$$\forall\, x_0 \in \text{Dom}(D_{M\propto}), \;\; d_M(x(t)) \leq d_{M\propto}(x(t)) \leq d_{M\propto}(x_0)e^{-at}$$

The main question we face is *how to construct Lyapunov functions*. Ever since Lyapunov proposed a century ago his second method for studying the behavior of a solution around an equilibrium, finding Lyapunov functions for such and such differential equation (or differential inclusion) has been a source of numerous problems requiring most often many clever tricks.

8.3.1 Construction of Lyapunov Functions

We deduce from Theorem 8.2.1 with $\varphi(x,w) := a(w - v_0)$ the following characterization of Lyapunov functions:

Theorem 8.3.1 *Let V be a nonnegative contingently epidifferentiable lower semicontinuous extended function and $F : X \rightsquigarrow X$ be a Marchaud map. Then the two following properties are equivalent:*

1. *for any initial state $x_0 \in \mathrm{Dom}(V)$,* there exists one solution $x(\cdot)$ to (8.2.1) satisfying property (8.3.1)

2. *V is a contingent solution to the* Hamilton-Jacobi inequalities

$$\forall\, x \in \mathrm{Dom}(V), \quad \inf_{v \in F(x)} D_\uparrow V(x)(v) + a(V(x) - v_0) \;\leq\; 0 \qquad (8.3.2)$$

We call them Lyapunov functions[6] (with respect to a).

If V is not a such a Lyapunov function, there exists a smallest nonnegative lower semicontinuous solution $V_\propto^a : X \mapsto \mathbf{R} \cup \{+\infty\}$ to the contingent Hamilton-Jacobi inequalities (8.3.2) larger than or equal to V (which can be the constant $+\infty$), which thus enjoys the property: for all $x_0 \in \mathrm{Dom}(V_\propto^a)$, there exist solutions to (8.2.1) satisfying

$$\forall\, t \geq 0, \;\; V(x(t)) \;\leq\; V_\propto^a(x(t)) \;\leq\; (V_\propto(x_0) - v_0)e^{-at} + v_0$$

The domain of $V_\propto^a : \mathrm{Dom}(F) \mapsto \mathbf{R} \cup \{+\infty\}$ is equal to the a-basin of attraction of the function V.

Proof — The first statements following from Theorem 8.2.1, it remains to prove that the domain of $V_\propto^a$ is equal to the a-basin of attraction of the function V.

Since

$$\rho_{V,a}(x_0) \;\leq\; \sup_{t \geq 0} \frac{V(x(t)) - v_0}{e^{-at}} \;\leq\; V_\propto(x_0)$$

we see at once that the domain of $V_\propto$ is contained in the a-basin of V.

Conversely, assume that $V_\propto(x_0) = +\infty$. We shall prove that x_0 does not belong to the a-basin of V. Indeed, for any $w \geq V(x_0)$, the pair (x_0, w) belongs to the epigraph of V, but does not belong to its viability kernel, which is the epigraph of $V_\propto$. This implies that for all solution $x(\cdot) \in \mathcal{S}(x_0)$, all solutions $(x(\cdot), (w - v_0)e^{-a\cdot} + v_0)$ to the system of differential inclusions $(x'(t), w'(t)) \in F(x(t)) \times \{av_0 - aw(t)\}$ leave the epigraph of V in finite time.

[6] because, when V is differentiable and $F \equiv f$ is single-valued, we recognize the classical definition of a Lyapunov function, solution to

$$\langle V'(x), f(x)\rangle + a(V(x) - v_0) \;\leq\; 0$$

Therefore, we can associate with any $w \geq V(x_0)$ a time T_w such that $V(x(T_w)) = we^{-aT_w}$, so that $\sup_{t\geq 0} \frac{V(x(t))-v_0}{e^{-at}} = +\infty$. Since this is true for all solutions $x(\cdot) \in \mathcal{S}(x_0)$, we deduce that x_0 does not belong to the a-basin of attraction of V. □

Corollary 8.3.2 *We posit the assumptions of Theorem 8.3.1. For all $a \geq 0$, there exists a smallest lower semicontinuous function $d_{M\propto} : X \to \mathbf{R} \cup \{+\infty\}$ larger than or equal to d_M such that*

$$\forall\, x_0 \in \mathrm{Dom}(d_{M\propto}),\ \text{there exists a solution } x(\cdot) \text{ to (8.2.1) such that} \\ d_M(x(t)) \leq d_{M\propto}(x_0)e^{-at}$$

the domain of which coincides with the a-basin of attraction of M.

We refer to Chapter 6 of [39, Aubin & Cellina] and Chapter 9 of [28, Aubin] for more informations on asymptotic properties of solutions to differential inclusions.

8.3.2 The Montagnes Russes Algorithm for Global Optimization

Let us consider the minimization problem

$$v_0 := \inf_{x\in X} V(x)$$

where $V : X \mapsto \mathbf{R} \cup \{+\infty\}$ is a nontrivial lower semicontinuous extended function assumed to be bounded from below.

A way to introduce the "Gradient Method" is to use the simple differential inclusion

$$\forall\, t \geq 0,\ x'(t) \in B$$

where B denotes the unit ball of the finite dimensional vector space X, leaving open the direction to be chosen by the algorithm, as in the methods of simulated annealing. But instead of choosing the velocities at random and being satisfied by convergence in probability, we shall ask wheth V can be an exponential Lyapunov function for the differential inclusion $x' \in B$.

Theorem 8.3.1 implies that

$$\forall\, x \in \mathrm{Dom}(V),\ \inf_{u\in B} D_\uparrow V(x)(u) + a(V(x) - v_0) \leq 0 \tag{8.3.3}$$

is necessary and sufficient for the existence of one solution $x(\cdot)$ to $x'(t) \in B$ satisfying (8.3.1) starting from any given initial state $x_0 \in \mathrm{Dom}(V)$.

Observe also that the Fermat rule states that at every local minimum x of V, $0 \leq \inf_{u\in B} D_\uparrow V(x)(u)$. Therefore, an exponential Lyapunov function satisfying

$$\forall\, x \ \text{ such that } V(x) > v_0,\ \inf_{u\in B} D_\uparrow V(x)(u) \leq -a(V(x) - v_0) < 0$$

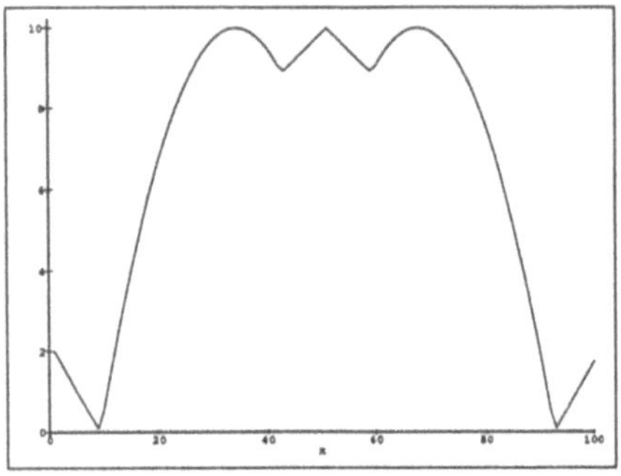

(a) Original function

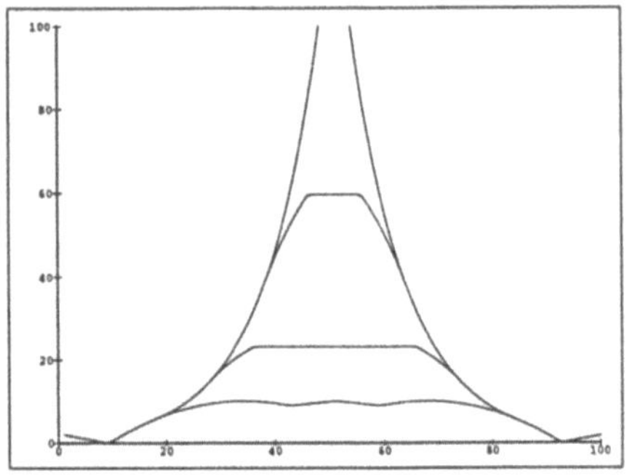

(b) Lyapunov function at a different scale

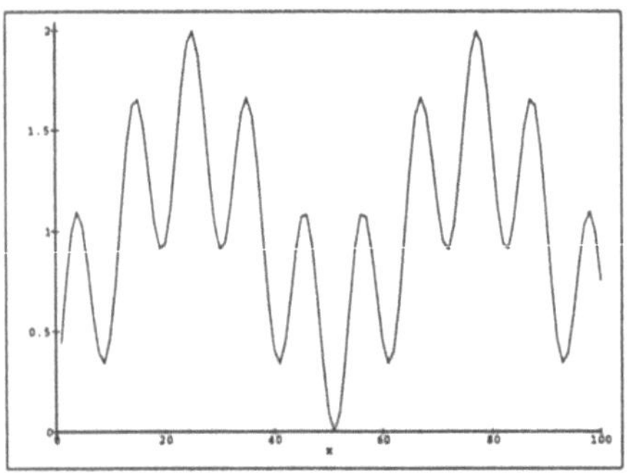

(c) Original function

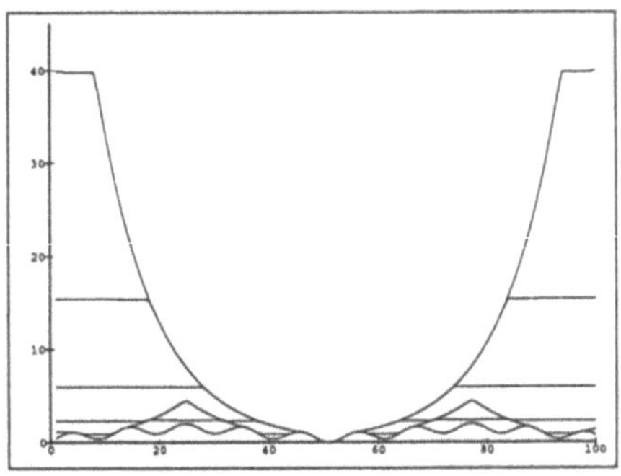

(d) Lyapunov function at a different scale

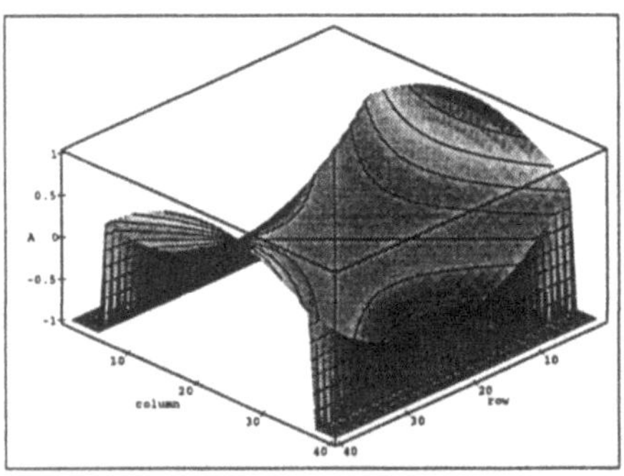

(e) Original function

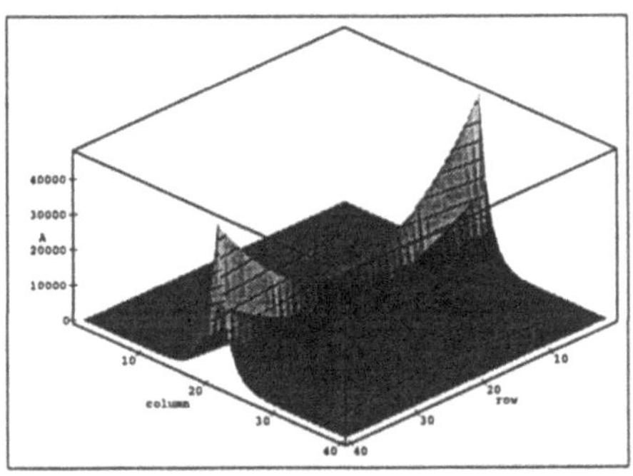

(f) Lyapunov function at a different scale

Various examples of optimal Lyapunov functions. Intermediate steps of computation are shown.

Figure 8.1: Optimal Lyapunov Functions

does not have local minima when $V(x) > v_0$.

Unfortunately, even when V is differentiable, we observe that condition (8.3.3) amounts to saying that V is still an exponential Lyapunov function of the gradient equation.

However, if this is not the case, Theorem 8.3.1 implies the existence of the optimal exponential Lyapunov function $V_\propto$, which is the smallest lower semicontinuous exponential Lyapunov function $V_\propto$ larger than or equal to V. Its minima are global and coincide with the global minima of the original function V. This exponential Lyapunov function can take infinite values.

Consequently, for any initial state $x_0 \in \mathrm{Dom}(V_\propto)$, there exists at least one solution to the differential inclusion $x' \in B$ starting from x_0 satisfying

$$V(x(t)) \leq V_\propto(x(t)) \leq (V_\propto(x_0) - v_0)e^{-at} + v_0$$

Although we know that $V(x(t))$ converges to v_0 when $t \to +\infty$, the function $t \mapsto V(x(t))$ is not necessarily decreasing. Along such a solution, the function V jumps above local maxima, leaves local minima, play "Montagnes Russes" (called "American Mountains" in Russian and "Big Dipper" in American !), but, ultimately, converges to its infimum. Furthermore, all solutions to the "gradient method" or "steepest descent method" of the exponential Lyapunov function $V_\propto$

$$x'(t) \in \Pi_{V_\propto}(x) := \{u \in B \mid D_\uparrow V_\propto(x)(u) = \inf_{v \in B} D_\uparrow V_\propto(x)(v)\}$$

converge to some global minimum of V.

8.4 Construction of Hitting, Exit and Crisis Functions

Theorem 6.3.6 states that the T-viability kernels and the T-capture basins are respectively the upper level sets and lower level sets of the upper exit and lower hitting functions defined by

$$\tau_K^\sharp(x) := \sup_{x(\cdot)\in\mathcal{S}_F(x)} \tau_K(x(\cdot)) \quad \& \quad \varpi_K^\flat(x) := \inf_{x(\cdot)\in\mathcal{S}_F(x)} \varpi_K(x(\cdot))$$

Actually, following P. Cardaliaguet, M. Quincampoix & P. Saint-Pierre, one can regard the hypograph of the exit function and the epigraph of the hitting function as viability kernels of auxiliary systems.

We associate now with the set-valued map F the set-valued map $\Psi_F : K \times \mathbf{R}_+ \rightsquigarrow X \times \mathbf{R}$ defined by

$$\Psi_F(x,w) := \begin{cases} F(x) \times \{-1\} & \text{if } \quad w > 0 \\ \overline{co}(F(x) \cup \{0\}) \times [-1, 0] & \text{if } \quad w = 0 \end{cases}$$

which is a Marchaud map whenever F is a Marchaud map.

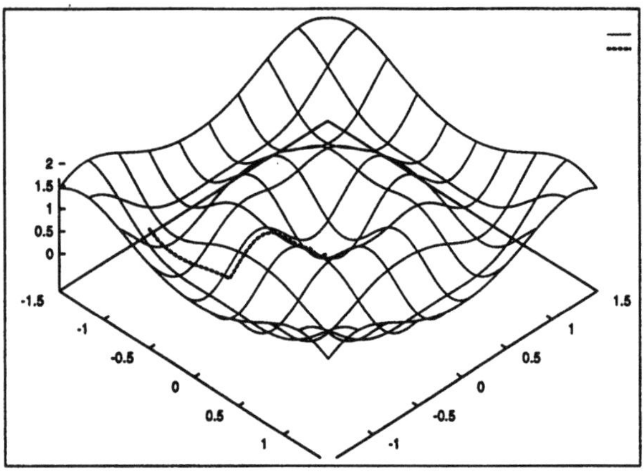

(a) Trajectory towards minimum

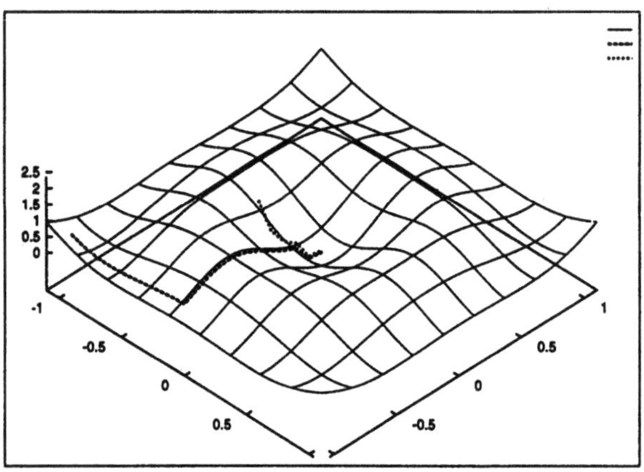

(b) Value of Lyapunov function along trajectory

Example of a the trajectory of solution converging to the global minimum of the function

$$V(x_1, x_2) = 1 - \cos(2\|x\|)\cos(3\|x\|)$$

Figure 8.2: Montagnes Russes

Theorem 8.4.1 *Assume that F is Marchaud and that $K \subset \mathrm{Dom}(F)$ is closed. Then the hypograph of the upper exit function $\tau_K^\sharp$ is the viability kernel of $K \times \mathbf{R}_+$ under the set-valued map Ψ_F:*

$$\mathcal{H}yp(\tau_K^\sharp) = \mathrm{Viab}_{\Psi_F}(K \times \mathbf{R}_+)$$

The upper exit function is the largest nonnegative upper semicontinuous contingent solution to the Hamilton-Jacobi equation

$$\sup_{v \in F(x)} D_\downarrow \tau_K^\sharp(x)(v) \geq -1$$

Proof — First, let us check that whenever (x, T) belongs to the viability kernel of $K \times \mathbf{R}_+$ under Ψ_F, we deduce that $T \leq \tau_K^\sharp(x)$, i.e., that the pair (x, T) belongs to the hypograph of the exit function. Let $(x(\cdot), w(\cdot))$ be a solution to the differential inclusion $(x', w') \in \Psi_F(x, w)$ starting at (x, T) and viable in $K \times \mathbf{R}_+$.

When $T > 0$ and when $t \in [0, T[$, we know that $w(t) = T - t > 0$ so that $x(\cdot)$ is a solution to $x' \in F(x)$ which remains in K whenever $t \in [0, T[$. Therefore $T \leq \tau_K^\sharp(x)$.

When $T = 0$, we know that $w(t) = 0$ and that $x(\cdot)$ is a solution to the differential inclusion $x'(t) \in \lambda(t) F(x(t))$ which remains in K and stops when $\lambda(t) = 0$.

Second, let us check that when (x, T) belongs to the hypograph of the exit function, then the pair (x, T) belongs to the viability kernel of $K \times \mathbf{R}_+$ under Ψ_F. For that purpose, we have to construct a solution $(x(\cdot), w(\cdot))$ to the differential inclusion $(x', w') \in \Psi_F(x, w)$ starting at (x, T) viable in $K \times \mathbf{R}_+$.

Let us consider one solution $x^\sharp(\cdot) \in \mathcal{S}_F(x)$ to $x' \in F(x)$ satisfying

$$\tau_K^\sharp(x) = \tau_K(x^\sharp(\cdot))$$

which exists according to Theorem 6.3.4.

We set

$$(x(t), w(t)) := \begin{cases} (x^\sharp(t), T - t) & \text{if } t \in [0, T] \\ (x^\sharp(t), 0) & \text{if } t \in [T, \tau_K^\sharp(x)] \\ (x^\sharp(\tau_K^\sharp(x)), 0) & \text{if } t \in [\tau_K^\sharp(x), +\infty[\end{cases}$$

which is the solution to the differential inclusion $(x', w') \in \Psi_F(x, w)$ viable in $K \times \mathbf{R}_+$ we were looking for. □

For the lower hitting function of a closed subset, we associate with the set-valued map $F : X \rightsquigarrow X$ the set-valued map $\Phi_F : X \times \mathbf{R}_+ \rightsquigarrow X \times \mathbf{R}$ defined by

$$\Phi_F(x, w) := \begin{cases} F(x) \times \{-1\} & \text{if } x \notin K \\ \overline{\mathrm{co}}((F(x) \times \{-1\}) \cup (0, 0)) & \text{if } x \in K \end{cases}$$

Theorem 8.4.2 *Assume that F is Marchaud and that $K \subset X$ is closed and satisfies $K \cap \mathrm{Dom}(F) \neq \emptyset$. The epigraph of the lower hitting function $\varpi_K^\flat$ is the viability kernel of $X \times \mathbf{R}_+$ under the set-valued map Φ_F:*

$$\mathcal{E}p(\varpi_K) = \mathrm{Viab}_{\Phi_F}(X \times \mathbf{R}_+)$$

The lower hitting function is the smallest nonnegative lower semicontinuous contingent solution to the Hamilton-Jacobi equation

$$\inf_{v \in F(x)} D_\uparrow \varpi_K^\flat(x)(v) \leq -1$$

Proof — First, let us check that $w \geq \varpi_K^\flat(x)$ whenever (x,w) belongs to $\mathrm{Viab}_{\Phi_F}(X \times \mathbf{R}_+)$. Indeed, from (x,w) starts a solution $(x(\cdot), w(\cdot))$ to the differential inclusion $(x', w') \in \Phi_F(x,w)$ viable in $X \times \mathbf{R}_+$. If $x \notin K$, then $x(\cdot)$ is a solution to the differential inclusion $x' \in F(x)$ viable in X and $w(t) = w - t$ as long as $x(t) \notin K$. Since $w(t) \geq 0$, we infer that $\varpi_K(x(\cdot)) \leq w$, i.e., that (x,w) belongs to the epigraph of $\varpi_K^\flat(x)$.

Conversely, assume that (x,w) belongs to the epigraph of $\varpi_K^\flat(x)$: we have to construct a solution $(\widetilde{x}(\cdot), \widetilde{w}(\cdot))$ to the differential inclusion $(x', w') \in \Phi_F(x,w)$ viable in $K \times \mathbf{R}_+$ starting form (x,w). Le us choose a solution $x^\flat(\cdot) \in \mathcal{S}_F^K(x)$ with minimal time, i.e., satisfying

$$\varpi_K(x^\flat(\cdot)) = \varpi_K^\flat(x) := \inf_{x(\cdot) \in \mathcal{S}_F^K} \varpi_K(x(\cdot))$$

We thus set

$$(\widetilde{x}(t), \widetilde{w}(t)) := \begin{cases} (x^\flat(t), w - t) & \text{if } t \in [0, \varpi_K^\flat(x)] \\ (x^\flat(\varpi_K^\flat(x)), w - \varpi_K^\flat(x)) & \text{if } t \geq \varpi_K^\flat(x) \end{cases}$$

which is obviously the viable solution we were looking for. □

Luc Doyen & Patrick Saint-Pierre in [177, Doyen & Saint-Pierre] has proved that the crisis function is lower semicontinuous and characterized its epigraph as the viability kernel of an auxiliary system.

We introduce for this purpose, we associate with the set-valued map $F : X \leadsto X$ the set-valued map $G_F : X \times \mathbf{R}_+ \leadsto X \times \mathbf{R}$ defined by

$$G_F(x,w) := F(x) \times [-1, -\chi_{X \setminus K}(x)]$$

Theorem 8.4.3 *Assume that F is Marchaud and that $K \subset X$ is closed and satisfies $K \cap \mathrm{Dom}(F) \neq \emptyset$. The epigraph of the crisis function $\gamma_K^\flat$ is the viability kernel of $X \times \mathbf{R}_+$ under the set-valued map G_F:*

$$\mathcal{E}p(\gamma_K) = \mathrm{Viab}_{G_F}(X \times \mathbf{R}_+)$$

The crisis function is the smallest nonnegative lower semicontinuous contingent solution to the Hamilton-Jacobi equation

$$\inf_{v \in F(x)} D_\uparrow \varpi_K^\flat(x)(v) \leq -\chi_{X \setminus K}(x)$$

Proof — When K is closed, the function $W := \chi_{X\setminus K}$ is lower semicontinuous. Hence the crisis function $\gamma_K^\flat$ is the value function of the infinite horizon intertemporal optimization problem

$$\gamma_K^\flat(x) := \inf_{x(\cdot)\in\mathcal{S}_F(x)} \int_0^{+\infty} W(x(\tau))d\tau$$

Therefore, Theorem 8.4.3 is a corollary of Theorems 8.5.1 and 8.5.5 on value functions of the next section. □

8.5 Optimal Evolution under Infinite Horizon

We shall study infinite horizon intertemporal discounted optimization problem.

The evolution of a dynamical economy (P, c) with a priori pricing map is governed by

$$\begin{cases} i) & x'(t) = c(x(t), p(t)) \\ ii) & p(t) \in P(x(t)) \end{cases} \tag{8.5.1}$$

We admit that scarcity constraints are implicitly taken into account in the definition of the domain of the pricing map P by setting $K := \mathrm{Dom}(P)$.

We denote by $L^1(0,\infty;X,e^{at}dt)$ the Lebesgue space of classes of measurable functions from $[0,\infty[$ to X integrable for the weighted measure $e^{at}dt$ and by $W_a^{1,1}(0,\infty;X)$ the space of functions $x(\cdot) \in L^1(0,\infty;X,e^{at}dt)$ such that their distributional derivative belongs to $L^1(0,\infty;X)$.

Denote by $\mathcal{S}(x_0)$ the set commodity-price pairs

$$(x(\cdot),p(\cdot)) \in W_a^{1,1}(0,\infty;X) \times L^1(0,\infty;\mathcal{M},e^{at}dt)$$

solutions to the dynamical economy (8.5.1) starting from x_0 at time 0, i.e., such that $x(0) = x_0$.

Let us introduce now a nonnegative lower semicontinuous "Lagrangian"

$$W : (x,q) \in \mathrm{Graph}(P) \mapsto W(x,q) \in \mathbf{R}_+$$

assumed to be convex with respect to q and to have linear growth.

Therefore, the discounted cost

$$V_\propto(x) := \int_0^\infty e^{a\tau} W(x(\tau),p(\tau))d\tau \in [0,+\infty]$$

is well defined over the solutions of the dynamical economy.

In this chapter, we shall prove that the value function $V_\propto$ of the intertemporal optimization problem is the smallest of the lower semicontinuous nonnegative extended functions V satisfying the following monotonicity property:

From any initial state $x_0 \in \mathrm{Dom}(V)$ starts a solution to the dynamical economy satisfying

$$\forall\, t \geq 0,\;\; e^{at}V(x(t)) - V(x_0) + \int_0^t e^{a\tau}W(x(\tau),p(\tau))d\tau \;\leq\; 0 \qquad (8.5.2)$$

which is equivalent to

$$\inf_{q\in P(x)} (D_\uparrow V(x)(c(x,q)) + W(x,q)) + aV(x) \;\leq\; 0$$

We thus define the regulation map

$$\Pi_{V_\propto}(x) \;:=\; \{p \in P(x) \mid \langle V_\propto'(x), c(x,p)\rangle + W(x,p) + aV_\propto(x) \;\leq\; 0\}$$

We shall prove that optimal solutions are regulated by prices satisfying

$$\text{for almost all } t \geq 0,\;\; p(t) \;\in\; \Pi_{V_\propto}(x)$$

Actually, we shall characterize the epigraph of the value function as the **viability kernel** of an auxiliary differential inclusion. This being done, we may apply the **viability kernel algorithm** to this epigraph, which provides the value function of the discretized problem. In this way, we obtain an algorithm for computing the discrete value function.

8.5.1 Monotone Solutions

Theorem 8.5.1 *Let (P,c) be a Marchaud dynamical economy,*

$$W : (x,q) \in \mathrm{Graph}(P) \mapsto W(x,q) \in \mathbf{R}_+$$

a nonnegative lower semicontinuous function convex with respect to q and $V : X \mapsto \mathbf{R}_+ \cup \{+\infty\}$ a nonnegative extended lower semicontinuous function (regarded as a loss function).

We assume that there exists a positive constant c such that

$$\left\{\begin{array}{ll} i) & \forall\,(x,q) \in \mathrm{Graph}(P),\;\; W(x,q) \;\leq\; c(\|x\|+1) \\ ii) & \forall\, x \in \mathrm{Dom}(V),\;\; \inf_{q\in P(x)} D_\uparrow V(x)(c(x,q)) \;\geq\; -c(\|x\|+1) \end{array}\right. \qquad (8.5.3)$$

Then the two following properties are equivalent:

1. *For any initial state $x_0 \in \mathrm{Dom}(V)$, there exists a solution $(x(\cdot),p(\cdot)) \in S(x_0)$ to the dynamical economy (8.5.1) monotone in the sense that:*

$$\forall\, t \geq 0,\;\; e^{at}V(x(t)) - V(x_0) + \int_0^t e^{a\tau}W(x(\tau),p(\tau))d\tau \;\leq\; 0 \qquad (8.5.4)$$

2. V is a contingent solution[7] *to the Hamilton-Jacobi-Bellman inequality*

$$\forall\, x \in \mathrm{Dom}(V),\quad \inf_{q\in P(x)} (D_\uparrow V(x)(c(x,q)) + W(x,q)) + aV(x) \;\leq\; 0 \tag{8.5.5}$$

Furthermore, if we denote by

$$\Pi_V(x) \;:=\; \{p \in P(x) \mid D_\uparrow V(x)(c(x,p)) + W(x,p) + aV(x) \;\leq\; 0\}$$

then the monotone solutions are governed by the optimal regulation law:

$$\text{for almost all } t \geq 0,\;\; p(t) \;\in\; \Pi_V(x(t)) \tag{8.5.6}$$

If V does not satisfy either one of the above properties, then there exists a smallest nonnegative lower semicontinuous solution $V_\propto : X \mapsto \mathbf{R}_+ \cup \{+\infty\}$ to the contingent Hamilton-Jacobi inequalities (8.5.5) larger than or equal to V (which can be the constant $+\infty$), which thus enjoys the monotonicity property: $\forall\, x_0 \in \mathrm{Dom}(V_\propto)$, there exists one solution to (8.5.1) starting from x_0 at time 0 and satisfying

$$\forall\, t \geq 0,\;\; V(x(t)) \;\leq\; V_\propto(x(t)) \;\leq\; e^{-at}V_\propto(x_0) - \int_0^t e^{-a(t-\tau)}W(x(\tau),p(\tau))d\tau \tag{8.5.7}$$

Remark — Recall that Theorem 4.1.8 implies that if we assume also that there exists a constant $\rho > 0$ such that

$$\forall\, (x,p) \in \mathrm{Graph}(P),\;\; W(x,p) + aV(x) \;\geq\; \rho\|c(x,p)\|$$

then there exists an equilibrium of the dynamical economy. □

Proof of Theorem 8.5.1 — We introduce the set-valued map $G : X \times \mathbf{R} \rightsquigarrow X \times \mathbf{R}$ defined by

$$G(x,w) := \{(c(x,q),\lambda) \mid q \in P(x)\;\&\;\lambda + aw \in [-c(\|x\|+1), -W(x,q)]\} \tag{8.5.8}$$

1. It is clear that the graph of G is closed and its values are convex and nonempty by assumption (8.5.3). Its growth is linear by construction. Furthermore, the epigraph of V is a closed viability domain of G: take

[7] When the function V is continuous instead of lower semicontinuous, so that we can define "viscosity solutions" to Hamilton-Jacobi equation, Hélène Frankowska has shown that the contingent solution coincides with the (upper) viscosity solution introduced by M. Crandall & P.-L. Lions. But lower semicontinuity is compulsory whenever we deal with intertemporal optimization problems under constraints, which is often the case in economics.

$q \in P(x)$ achieving the minimum of the lower semicontinuous function $D_{\uparrow}V(x)(c(x,\cdot)) + W(x,\cdot)$ on the compact subset $P(x)$. It satisfies

$$D_{\uparrow}V(x)(c(x,q)) + W(x,q) + aV(x) \leq 0$$

by assumption (8.5.5), so that the pair $(c(x,q), -aV(x) - W(x,q))$ belongs to the contingent cone to the epigraph of V at $(x, V(x))$. By Lemma 8.1.2, it also belongs to the contingent cone to the epigraph of V at (x, w) for every $w > V(x)$.

Hence $\mathcal{E}p(V)$ being a closed viability domain of $G(\cdot,\cdot)$, there exists a solution $(x(\cdot), w(\cdot))$ to differential inclusion

$$\text{for almost all } t \geq 0, \quad (x'(t), w'(t)) \in G(x(t), w(t))$$

starting from $(x_0, V(x_0))$ at time 0 and viable in the epigraph of V. Inequalities

$$w'(\tau) + aw(\tau) \leq -W(x(\tau), p(\tau)) \ \& \ V(x(t)) \leq w(t), \ \ V(x_0) = w(0)$$

imply that

$$V(x(t)) \leq w(t) \leq e^{-at}V(x_0) + \int_0^t e^{-a(t-\tau)} W(x(\tau), p(\tau)) d\tau$$

2. Conversely, let us consider a solution $(x(\cdot), p(\cdot))$ to the dynamical economy

$$x'(t) = c(x(t), p(t))$$

which is monotone:

$$\forall\, t \geq 0, \ \ e^{at}V(x(t)) - V(x_0) + \int_0^t e^{a\tau} W(x(\tau), p(\tau)) d\tau \leq 0$$

We shall prove that there exists $p_0 \in P(x_0)$ such that

$$D_{\uparrow}V(x_0)(c(x_0, p_0)) + W(x_0, p_0) + aV(x_0) \leq 0 \tag{8.5.9}$$

The above monotonicity condition means that

$$\left(x(h), e^{-ah}V(x_0) - e^{-ah} \int_0^h e^{a\tau} W(x(\tau), p(\tau)) d\tau\right) \in \mathcal{E}p(V)$$

Setting

$$u_h := \frac{1}{h} \int_0^h x'(\tau) d\tau$$

and

$$\lambda_h = \frac{e^{-ah} - 1}{h} V(x_0) - \frac{e^{-ah}}{h} \int_0^h e^{a\tau} W(x(\tau), p(\tau)) d\tau$$

it can be rewritten in the form

$$(x_0 + hu_h, V(x_0) + h\lambda_h) \in \mathcal{E}p(V)$$

Since P is upper semicontinuous with compact values, we can associate with any $\varepsilon > 0$ an $\eta_0 \in]0, \varepsilon]$ such that $P(x) \subset B(P(x_0), \varepsilon)$ whenever $d(x, x_0) \leq \eta_0$. Since c is continuous, we deduce that

$$x'(\tau) = c(x(\tau), p(\tau)) \in c(x_0, P(x_0)) + \varepsilon B$$

for τ small enough. Hence, since c is affine with respect to p,

$$u_h \in c\left(x_0, \frac{1}{h}\int_0^h p(\tau)d\tau\right) + \varepsilon B \tag{8.5.10}$$

The proof then follows from the following

Lemma 8.5.2 *We posit the assumptions of Theorem 8.5.1. If $(x(\cdot), p(\cdot))$ is a monotone solution, then $\frac{1}{h}\int_0^h p(\tau)d\tau$ converges to some $p_0 \in P(x_0)$ satisfying (8.5.9).*

Proof — Since W is lower semicontinuous, we can associate with any p_i an $\eta_i \in]0, \varepsilon]$ such that

$$\text{if } d(x, x_0) \leq \eta_i \;\&\; \|p - p_i\| \leq \eta_i, \;\text{ then } W(x_0, p_i) \leq W(x, p) + \varepsilon$$

But $B(P(x_0), \varepsilon)$ being compact, it can be covered by a finite number r of such balls. Let us set $\eta := \min_{i=0,\ldots r} \eta_i$. Therefore, for any $p \in P(x_0)$, we can find a price p_i such that $\|p - p_i\| \leq \eta_i$, so that

$$\forall\, x \in B(x_0, \eta),\; (p_i, W(x, p) + \varepsilon) \in \mathcal{E}p(W(x_0, \cdot))$$

Hence, for every $\varepsilon > 0$, there exists η such that for every $x \in B(x_0, \eta)$ and every $p \in P(x)$,

$$(p, W(x, p) + \varepsilon) \in \mathcal{E}p(W(x_0, \cdot)) + \varepsilon B \times \{0\}$$

because

$$(p, W(x, p) + \varepsilon) = (p_i, W(x, p) + \varepsilon) + (p - p_i, 0)$$

Let us consider now $p(\tau) \in P(x(\tau))$ and α such that $d(x_0, x(\tau)) \leq \eta$ whenever $\tau \leq \alpha$. We deduce that

$$(p(\tau), e^{a\tau} W(x(\tau), p(\tau)) + \varepsilon) \in \mathcal{E}p(W(x_0, \cdot)) + 2\varepsilon B \times \{0\}$$

Since the epigraph of $W(x_0,\cdot)$ is closed and convex, this implies that

$$\left(\frac{1}{h}\int_0^h p(\tau)d\tau, \frac{1}{h}\int_0^h e^{a\tau}W(x(\tau),p(\tau))d\tau + 2\varepsilon\right) \in \mathcal{E}p(W(x_0,\cdot))+2\varepsilon B\times\{0\}$$

Let $h_n > 0$ be such that

$$\lim_{n\to\infty}\frac{1}{h_n}\int_0^{h_n} e^{a\tau}W(x(\tau),p(\tau))d\tau \;=\; \liminf_{h\to 0+}\frac{1}{h}\int_0^h e^{a\tau}W(x(\tau),p(\tau))d\tau$$

Since $p(x(\tau)) \in P(x(\tau)) \subset B(P(x_0),\varepsilon)$ for τ small enough, and since the values of P are convex and compact, we infer that

$$p_n \;:=\; \frac{1}{h_n}\int_0^{h_n} p(\tau)d\tau \;\in\; B(P(x_0),\varepsilon)$$

This set being compact, a subsequence (again denoted by) p_n converges to some $p_0 \in P(x_0)$.

Therefore, by taking the limit when $n \to \infty$, we deduce that

$$\left(p_0, \liminf_{h\to 0+}\frac{1}{h}\int_0^h e^{a\tau}W(x(\tau),p(\tau))d\tau + \varepsilon\right) \in \mathcal{E}p(W(x_0,\cdot)) + \varepsilon B \times \{0\}$$

Since this is true for every ε, we infer that

$$\left(p_0, \liminf_{h\to 0+}\frac{1}{h}\int_0^h e^{a\tau}W(x(\tau),p(\tau))d\tau\right) \in \mathcal{E}p(W(x_0,\cdot))$$

Therefore

$$W(x_0,p_0) \;\leq\; \liminf_{h\to 0+}\frac{1}{h}\int_0^h e^{a\tau}W(x(\tau),p(\tau))d\tau$$

We thus conclude from (8.5.10) that u_{h_n} converges to $c(x_0,p_0)$ and that λ_{h_n} converges to some λ satisfying

$$\lambda \;\leq\; -aV(x_0) - W(x_0,p_0)$$

This implies (8.5.9).

3. The fact that for almost all $t \geq 0$, the derivative of the viable solution $(x'(t),w'(t))$ belongs to the contingent cone of the epigraph of V implies the regulation law (reglawinteroptim).

4. We know that the viability kernel $\mathrm{Viab}_G(\mathcal{E}p(V))$ is the largest closed viability domain $\mathcal{E}p(V)$ of the set-valued map $(x,w) \rightsquigarrow G(x,w)$ defined by (8.5.8). If it is empty, it is the epigraph of the constant function equal to $+\infty$.

   If not, Lemma 8.1.2 implies that it is the epigraph of the nonnegative lower semicontinuous function $V_\propto$ defined by

   $$V_\propto(x) \;:=\; \inf_{(x,\lambda)\in\mathcal{K}} \lambda$$

   we are looking for as in the proof of Theorem 8.2.1. □

Remark — We refer to Proposition 11.4.4 for conditions under which the optimal regulation map is lower semicontinuous, so that it can have continuous selections, minimal selections and other types of selection procedures. □

As a consequence, we deduce the following

Theorem 8.5.3 *We add to the hypothesis of Theorem 8.5.1 the assumption that the loss function V is continuous on its domain. Then the two following properties are equivalent:*

1. *For any initial state $x_s \in \mathrm{Dom}(V)$, there exists a solution to the dynamical economy (8.5.1) starting from x_s at time s and satisfying property:*

$$\forall\, t \geq s, \quad e^{at}V(x(t)) - e^{as}V(x(s)) + \int_s^t e^{a\tau} W(x(\tau), p(\tau))d\tau \;\leq\; 0 \tag{8.5.11}$$

2. *V is a contingent solution to the Hamilton-Jacobi inequality (8.5.5).*

Proof — We associate with $h \to 0+$ the grid jh, $(j = 1, \ldots)$ and we build a solution $x_h(\cdot) \in \mathcal{S}(x_0)$ to differential inclusion (8.5.1) by using Theorem 8.5.1 iteratively: for $j = 0$, we take $x_h(\cdot)$ on the interval $[0, h]$ satisfying (8.5.4). For $j > 0$, we consider the solution starting at $x_h(jh)$ and satisfying

$$e^{at}V(y_j(t)) - V(x_h(jh)) + \int_0^t e^{a\tau} W(y_j(\tau), q_j(\tau))d\tau$$

Setting $x_h(t) := y_j(t - jh)$ and $p_h(t) := q_j(t - jh)$ on the interval $[jh, (j+1)h]$, we see that the solution satisfies

$$\forall\, t \in [jh, (j+1)h], \;\; e^{at}V(x_h(t)) - e^{ajh}V(x_h(jh)) + \int_{jh}^{t-jh} e^{a\sigma} W(x_h(\sigma), p_h(\sigma))d\sigma \;\leq\; 0$$

Let $t > s$ be fixed. Since the Convergence Theorem implies that the image $\mathcal{S}(x_0)$ is compact in $\mathcal{C}(0, \infty; X) \times L^1(0, \infty, \mathcal{M}, e^{at}dt)$ when the first space is supplied with the compact topology and the second with the weak topology, a subsequence (again denoted) $(x_h(\cdot), p_h(\cdot))$ converges to some solution $(x(\cdot), p(\cdot)) \in \mathcal{S}(x_0)$ in $\mathcal{C}(s, t; X) \times L^1((s, t); \mathcal{M})$.

An adaptation of the proof of the Convergence Theorem implies the following property:

Lemma 8.5.4 *Assume that $W : (x, p) \in \mathrm{Graph}(P) \mapsto W(x, p) \in \mathbf{R}_+$ is a lower semicontinuous function convex with respect to p. Then the functional*

$$x(\cdot), p(\cdot) \mapsto \int_s^t e^{a\tau} W(x(\tau), p(\tau))d\tau$$

from $\mathcal{C}(s, t; X) \times L^1(s, t, X)$ to $\mathbf{R} \cup \{+\infty\}$ is lower semicontinuous when $\mathcal{C}(s, t; X)$ is supplied with the uniform convergence and $L^1(s, t, X)$ with the weak convergence.

(We refer to Proposition 6.3.4 of DIFFERENTIAL INCLUSIONS, [39, Aubin & Cellina], for the proof of this Lemma.) Hence

$$\int_s^t e^{a\tau} W(x(\tau), p(\tau))d\tau \leq \liminf_{h\to 0+} \int_s^t e^{a\tau} W(x_h(\tau), p_h(\tau))d\tau$$

Let $t > s$ be approximated by $j_h h \geq k_h h$ so that

$$e^{aj_h h} V(x_h(j_h h)) - e^{ak_h h} V(x_h(k_h h)) + \int_{k_h h}^{j_h h} e^{a\tau} W(x_h(\tau), p_h(\tau))d\tau \leq 0$$

The function V being continuous on its domain, inequality (8.5.4) ensues. □

8.5.2 The Value Function

Denote by $\mathcal{S}_t(x)$ the set commodity-price pairs of $(x(\cdot), p(\cdot))$ solutions to the dynamical economy (8.5.1) starting from x at time t.

This smallest lower semicontinuous loss function $V_\propto$ is closely related to the value function

$$U(t,x) := \inf_{(x(\cdot),p(\cdot))\in\mathcal{S}_t(x)} \int_t^\infty e^{a\tau} W(x(\tau), p(\tau))d\tau \in [0, +\infty[\quad (8.5.12)$$

of the intertemporal discounted optimization problem over the solutions $(x(\cdot), p(\cdot))$ to the dynamical economy starting at time t from x of the discounted functional

$$\int_t^\infty e^{a\tau} W(x(\tau), p(\tau))d\tau$$

Theorem 8.5.5 *Let (P, c) be a Marchaud dynamical economy and*

$$W : (x,q) \in \mathrm{Graph}(P) \mapsto W(x,q) \in \mathbf{R}_+$$

a nonnegative lower semicontinuous function convex with respect to q and with linear growth. We assume that the domain $K := \mathrm{Dom}(P)$ is closed. We posit the assumptions of Theorem 8.5.3. Then the value function $U(t,x)$ and the smallest lower semicontinuous function $V_\propto$ satisfying (8.5.4) larger than or equal to the indicator function ψ_K of K are related by the formula:

$$U(t,x) = e^{at} V_\propto(x) \quad (8.5.13)$$

Furthermore, a solution $(\widehat{x}(\cdot), \widehat{p}(\cdot)) \in \mathcal{S}(x_0)$ to (8.5.1) satisfies inequality (8.5.4) for $V_\propto$ if and only if it is an optimal solution to the intertemporal discounted optimization problem:

$$\int_0^\infty e^{a\tau} W(\widehat{x}(\tau), \widehat{p}(\tau))d\tau = \inf_{(x(\cdot),p(\cdot))\in\mathcal{S}(x_0)} \int_0^\infty e^{a\tau} W(x(\tau), p(\tau))d\tau$$

In this case, it obeys the "optimality principle":

$$\forall\, t \geq 0,\; e^{at} V_\propto(\widehat{x}(t)) = \int_t^\infty e^{a\tau} W(\widehat{x}(\tau), \widehat{p}(\tau))d\tau \quad (8.5.14)$$

and satisfies actually the equation

$$e^{at}V_{\propto}(\widehat{x}(t)) - e^{as}V_{\propto}(\widehat{x}(s)) + \int_s^t e^{a\tau}W(\widehat{x}(\tau),\widehat{p}(\tau))d\tau \; = \; 0 \qquad (8.5.15)$$

If we denote by

$$\Pi_{V_{\propto}}(x) \; := \; \{p \in P(x) \mid D_{\uparrow}V_{\propto}(x)(c(x,p) + W(x,p)) + aV_{\propto}(x) \; \leq \; 0\}$$

then the optimal solutions to the intertemporal optimization problem are governed by the optimal regulation law:

$$\text{for almost all } t \geq 0, \;\; p(t) \; \in \; \Pi_{V_{\propto}}(x(t))$$

Proof

1. Let $(x(\cdot), p(\cdot)$ be a monotone solution with respect to $V_{\propto}$ starting from x_0. We deduce in particular that, V being nonnegative,

$$\forall \, t \geq 0, \;\; \int_0^t e^{a\tau}W(x(\tau),p(\tau))d\tau \; \leq \; V_{\propto}(x_0)$$

Since W is nonnegative, the sequence $\int_0^t e^{a\tau}W(x(\tau),p(\tau))d\tau$ is nondecreasing, so that, taking the limit when $t \to +\infty$, we obtain

$$U(0,x_0) \; := \; \inf_{(x(\cdot),p(\cdot)) \in \mathcal{S}(x_0)} \int_0^\infty e^{a\tau}W(x(\tau),p(\tau))d\tau \; \leq \; V_{\propto}(x_0) \quad (8.5.16)$$

2. We observe that any solution $(\bar{x}(\cdot), \bar{p}(\cdot))$ monotone with respect to the value function in the sense that

$$U(s,\bar{x}(s)) - U(t,\bar{x}(t)) + \int_t^s e^{a\tau}W(\bar{x}(\tau),\bar{p}(\tau))d\tau \; \leq \; 0$$

is an optimal solution to the intertemporal optimization problem because

$$\begin{cases} \int_t^\infty e^{a\tau}W(\bar{x}(\tau),\bar{p}(\tau))d\tau \\ \leq \; U(t,\bar{x}(t)) \; := \; \inf_{(x(\cdot),p(\cdot)) \in \mathcal{S}_t(\bar{x}(t))} \int_t^\infty e^{a\tau}W(x(\tau),p(\tau))d\tau \end{cases}$$

Furthermore, property

$$\forall \, t \geq 0, \;\; U(t,\bar{x}(t)) \; = \; \int_t^\infty e^{a\tau}W(\bar{x}(\tau),\bar{p}(\tau))d\tau \qquad (8.5.17)$$

holds true.

Conversely, if a solution $(\widehat{x}(\cdot), \widehat{p}(\cdot))$ is optimal, it is monotone with respect to the value function, because

$$U(t,\widehat{x}(t)) + \int_0^t e^{a\tau}W(\widehat{x}(\tau),\widehat{p}(\tau))d\tau \leq \int_0^\infty e^{a\tau}W(\widehat{x}(\tau),\widehat{p}(\tau))d\tau = U(0,x_0)$$

3. We also observe that along optimal solutions, we have

$$U(t,\widehat{x}(t)) \;=\; e^{at}U(0,\widehat{x}(t))$$

Indeed, setting $\widehat{y}(\tau) := \widehat{x}(\tau+t)$ and $\widehat{q}(\tau) := \widehat{p}(\tau+t)$, we note that the pair $(\widehat{y}(\cdot),\widehat{q}(\cdot))$ is a solution to the dynamical economy starting at $\widehat{x}(t)$ and that by a change of variables that

$$U(t,\widehat{x}(t)) \;=\; \int_t^{\infty} e^{a\tau}W(\widehat{x}(\tau),\widehat{p}(\tau))d\tau = \; e^{at}\int_0^{\infty} e^{a\tau}W(\widehat{y}(\tau),\widehat{q}(\tau))d\tau$$

The same change of variables shows that the pair $(\widehat{y}(\cdot),\widehat{q}(\cdot))$ is an optimal solution starting from $\widehat{x}(t)$ at time 0.

4. Therefore, the function $x \mapsto U(0,x)$ satisfies property

$$e^{at}U(0,\widehat{x}(t)) - U(0,x_0) + \int_0^t e^{a\tau}W(\widehat{x}(\tau),\widehat{p}(\tau))d\tau \;\leq\; 0$$

Since the solution map $\mathcal{S}(\cdot)$ is upper semicontinuous with compact values from X to $\mathcal{C}(0,\infty;X) \times L^1(0,\infty,X)$ when $\mathcal{C}(0,\infty;X)$ is supplied with the compact convergence and $L^1(0,\infty,X)$ with the weak convergence by Theorem 3.5.2 of [28, Aubin], the Maximum Theorem and Lemma 8.5.4 imply that the value function $x \mapsto U(0,x)$ is lower semicontinuous.

Therefore, by Theorem 8.5.1, $U(0,\cdot)$ is a lower semicontinuous solution to the Hamilton-Jacobi inequality (8.5.5) and thus, larger than or equal to V_∞. By virtue of (8.5.16), it is then equal to V_∞. □

Remark — Monotone solutions to a dynamical economy enjoy asymptotic properties we just mention without proof.

Theorem 8.5.6 *We posit the assumptions of Theorem 8.5.5 and we assume that $K := \mathrm{Dom}(P)$ is compact. Then the optimal solution has "almost cluster points" $(\widehat{x}_\star,\widehat{p}_\star)$ satisfying*

$$c(\widehat{x}_\star,\widehat{p}_\star) \;=\; 0,\;\; \widehat{p}_\star \in P(\widehat{c}_\star) \;\;\&\;\; W(\widehat{x}_\star,\widehat{p}_\star) \;=\; 0$$

We refer to [15, Aubin] or to chapter 6 of DIFFERENTIAL INCLUSIONS, [39, Aubin & Cellina] for the precise definition of "almost convergence" and further asymptotic properties of monotone solutions. But we refer above all to the monograph CONTROL OF NONLINEAR SYSTEMS AND DIFFERENTIAL INCLUSIONS, [209, Frankowska], for an exhaustive exposition of the consequences of such an inequality and of generalized solutions (both contingent and viscosity) to Hamilton-Jacobi-Bellman equations as well as other issues in control theory. □

Example: Solutions with Minimal Length
Solutions with minimal length are obtain in the particular case when $W(x,q) = \|c(x,q)\|$ since

$$\int_0^\infty W(x(\tau),p(\tau))d\tau = \int_0^\infty \|x'(\tau)\|d\tau$$

measures the length of the solution to the differential equation $x'(t) = c(x(t),p(t))$. Then the optimal solutions are solutions with minimal length and the domain of the value function is the subset of initial states from which there exists at least one solution to the dynamical economy with finite length. Furthermore, any solution with minimal length converges when $t \to \infty$ to an equilibrium $x_\star$ of the system. i.e., a solution to

$$c(x_\star,p_\star) = 0 \ \& \ p_\star \in P(x_\star)$$

(see chapter 6 of [39, Aubin & Cellina] for a proof).
Remark: The Crisis Function — Taking $W(x,v) := \chi_{X\backslash K}(x)$. we deduce the Doyen & Saint-Pierre Theorem 8.4.3 on the crisis function. □

8.6 Infinite Horizon Optimization Problems under Bounded Inflation

The objective of this section is to find optimal solutions among solutions with bounded inflation, i.e., satisfying a constraint on the velocity of the prices. These solutions are driven by absolutely continuous open loop prices (instead of measurable open loop prices). Next, we shall provide the pricing maps regulating optimal solutions with bounded inflation.

For that purpose, we associate to the dynamical economy and to any nonnegative continuous function $p \to \varphi(x,p)$ with linear growth the system of differential inclusions

$$\begin{cases} i) & x'(t) = c(x(t),p(t)) \\ ii) & \|p'(t)\| \leq \varphi(x(t),p(t)) \\ iii) & p(t) \in P(x(t)) \end{cases} \qquad (8.6.1)$$

Let us denote by $\mathcal{S}^\varphi(x,p)$ the set of viable solutions to the new dynamical economy (8.6.1) starting at (x,p). We shall call it the **metasystem**, in which the commodity-price pairs of the initial dynamical economy are the states of the metasystem — subjected to viability constraints — and the "regulees" of the metasystem are the velocities of the prices of the original system. One observes that $\mathcal{S}^{\varphi}(x,p)$ is the set of solutions to the system of differential inclusions (8.6.1)i) & ii) **which are viable in the graph of** P (condition (8.6.1)iii)).

Any absolutely continuous function $p(\cdot)$ satisfying $\|p'(t)\| \leq \varphi(x(t),p(t))$ can be written

$$p(t) \; = \; p(0) + \int_0^t p'(\tau)d\tau$$

i.e., as a solution to (8.6.1)ii) starting at $p(0)$. Therefore, the value function V_{α}^{φ} of the **smooth optimal dynamical economy** can be written in the form

$$V_{\alpha}^{\varphi}(x) := \inf_{p\in\mathcal{M}} \inf_{(x(\cdot),p(\cdot))\in\mathcal{S}^{\varphi}(x,p)} \int_0^{\infty} e^{a\tau}W(x(\tau),p(\tau))d\tau \; \in \; [0,+\infty] \quad (8.6.2)$$

Setting

$$J_{\alpha}^{\varphi}(x,p) := \inf_{(x(\cdot),p(\cdot))\in\mathcal{S}^{\varphi}(x,p)} \int_0^{\infty} e^{a\tau}W(x(\tau),p(\tau))d\tau \; \in \; [0,+\infty] \quad (8.6.3)$$

(which takes infinite value outside the graph of P, and thus, subsumes the meta constraints) we can write

$$V_{\alpha}^{\varphi}(x) := \inf_{p\in\mathcal{M}} J_{\alpha}^{\varphi}(x,p)$$

so that properties of the value function V_{α}^{φ} can be derived from the value function J_{α}^{φ} of the metasystem, which we call the **metavalue function**.

This metasystem both has a simpler structure in the sense that the Lagrangian $W(x,p)$ depends only on the state of the metasystem (the commodity-price pair of the original system) and a ... larger commodity space (the dimension of which is the sum of the dimensions of the state space and the price space of the original systems), which make the algorithms more costly to implement. Commodity-dependent constraints on the prices of the original system become metastate constraints only of the metasystem, and easier to handle.

Theorem 8.6.1 *Assume that the graph of P is closed and that c, φ and W are continuous. Let J :* Graph$(P) \mapsto \mathbf{R}_+ \cup \{+\infty\}$ *a nonnegative extended lower semicontinuous function. If J is a contingent solution to the Hamilton-Jacobi inequality*

$$\forall \, (x,p) \in \text{Graph}(P), \quad \inf_{q\in\varphi(x,p)B} D_{\uparrow}J(x,p)(c(x,p),q) + W(x,p) + aJ(x) \; \leq \; 0 \quad (8.6.4)$$

then, for any initial commodity-price $(x_0,p_0) \in \text{Dom}(J)$, *there exists a solution to the metasystem (8.5.1) satisfying property:*

$$\forall \, t \geq 0, \;\; J(x(t),p(t)) - J(x_0,p_0) + \int_0^t e^{a\tau}W(x(\tau),p(\tau))d\tau \; \leq \; 0 \quad (8.6.5)$$

The value function J_α is the smallest lower semicontinuous nonnegative function larger that or equal to $\psi_{\text{Graph}(P)}$ satisfying (8.6.4), which means that it is a contingent solution to Hamilton-Jacobi-Bellman inequality

$$\frac{d}{dx}J(x,p)\cdot c(x,p)+W(x,p)+aJ(x)\ \le\ \varphi(x,p)\left\|\frac{d}{dp}J(x,p)\right\| \tag{8.6.6}$$

Proof — The epigraph of the function J_α^φ is the viability kernel of the subset $\text{Graph}(P)\times\mathbf{R}_+$ under the auxiliary system

$$\begin{cases} i) & x'(t) \;=\; c(x(t),p(t)) \\ ii) & p'(t) \;\in\; \varphi(x(t,p(t)))B \\ iii) & w'(t) \;=\; -aw(t)-W(x(t),p(t)) \end{cases} \tag{8.6.7}$$

and thus, at least a lower semicontinuous function. □

Knowing this function, we derive that the set-valued feedback map regulating all smooth optimal solutions is given by

$$\Pi^\varphi(x) \;=\; \{p\in X \mid J_\alpha^\varphi(x,p)=V_\alpha^\varphi(x)\}\ \subset\ P(x) \tag{8.6.8}$$

Therefore, smooth optimal prices are regulated by the feedback law

$$\forall\, t\ge 0,\ \ p(t)\ \in\ \Pi^\varphi(x(t))$$

Hence, we derive that *the graph of Π^φ is closed whenever J_α^φ is continuous* on $\text{Graph}(P)$.

We observe that this is consistent with the fact that the value function $V_\alpha^\varphi(x) = J_\alpha^\varphi(x,\bar p(x))$ where $\bar p(x)\in\Pi^\varphi(x)$ is a solution to the Hamilton-Jacobi-Bellman equation

$$\inf_{p\in\mathcal{M}} D_\uparrow(V(x)(c(x,p))+W(x,p))+aV(x)\ \le\ 0 \tag{8.6.9}$$

since the Fermat rule implies that when $\bar p(x)\in\Pi^\varphi(x)$,

$$D_\uparrow V_\alpha^\varphi(x)(c(x,\bar p(x)))\ \le\ \inf_w D_\uparrow J_\alpha^\varphi(x,\bar p(x))(c(x,\bar p(x)),w)$$

We denote by G^φ the set-valued map defined by

$$\begin{cases} G^\varphi(x,p) \;:=\; \{v\in\varphi(x,p)B \mid \\ D_\uparrow J_\alpha^\varphi(x,p)(c(x,p),q) \;=\; \inf_{q'\in\varphi(x,p)B} D_\uparrow J_\alpha^\varphi(x,p)(c(x,p),q')\} \end{cases} \tag{8.6.10}$$

When the function J_α^φ is differentiable, $G^\varphi(x,p)$ provides the descent directions of the function $p\mapsto J_\alpha^\varphi(x,p)$. This is natural because the optimal prices minimize J_α^φ with respect to p.

The system of differential inclusions regulating smooth optimal solutions is then

$$\begin{cases} i) & x'(t) \;=\; c(x(t),p(t)) \\ ii) & p'(t) \;\in\; G^\varphi(x(t),p(t)) \end{cases} \tag{8.6.11}$$

Consequently, any selection $g(x,p) \in G^{\varphi}(x,p)$ for which the system of differential equations

$$\begin{cases} i) & x'(t) = c(x(t),p(t)) \\ ii) & p'(t) = g(x(t),p(t)) \end{cases}$$

has solutions is a **dynamical closed-loop or dynamical controller** regulating smooth optimal solutions.

In particular, **heavy** smooth optimal solutions to the optimal dynamical economy, i.e., the ones which minimize the inflation of the price, are driven by the system of differential equations

$$\begin{cases} i) & x'(t) = c(x(t),p(t)) \\ ii) & p'(t) = g^{\varphi^\circ}(x(t),p(t)) \end{cases} \tag{8.6.12}$$

where $g^{\varphi^\circ}(x,p)$ is the element of minimal norm of the set $G^{\varphi}(x,p)$.

8.7 The Viability Kernel Algorithm

Let the discretization step $\rho \in]0, \frac{1}{a}[$ be fixed. The dynamical economy is replaced by the **discrete dynamical economy**

$$x^{\rho}_{s+1} \in x^{\rho}_{s} + \rho c(x^{\rho}_{s}, p^{\rho}_{s}) \text{ where } p^{\rho}_{s} \in P(x^{\rho}_{s}) \tag{8.7.1}$$

and we denote by $\mathcal{S}^{\rho}(x)$ the set of solutions $(x^{\rho},p^{\rho}) = ((x^{\rho}_{s},p^{\rho}_{s}))_{s\geq 0}$ of the discrete system (8.7.1) starting from an initial state $x \in \mathrm{Dom}(P)$.

We define the discrete value function of the discrete intertemporal optimization problem by

$$U_{\rho}(0,x) := \inf_{(x^{\rho},p^{\rho})\in\mathcal{S}^{\rho}(x)} \rho \sum_{r=0}^{\infty}(1-a\rho)^{-r-1}W(x^{\rho}_{r},p^{\rho}_{r})$$

As in the continuous case, we can characterize the discrete value function as the smallest lower semicontinuous nonnegative function V^{ρ} which is not increasing along at least one solution to the discrete dynamical economy in the following sense: from any $x_0 \in \mathrm{Dom}(P)$, starts one solution $(x^{\rho},p^{\rho}) \in \mathcal{S}^{\rho}(x_0)$ to the discrete dynamical economy (8.7.1) satisfying

$$(1-a\rho)^{-t}V^{\rho}(x^{\rho}_{t}) - V^{\rho}(x_0) + \rho\sum_{r=0}^{t-1}(1-a\rho)^{-r-1}W(x^{\rho}_{r},p^{\rho}_{r}) \leq 0 \tag{8.7.2}$$

We shall associate the synthesis map Π^{ρ}_{V} defined by

$$\begin{cases} \Pi^{\rho}(x) := \{p \in P(x) \quad \text{such that} \\ V^{\rho}_{\propto}(x+\rho c(x,p)) + \rho W(x,p) \\ = \inf_{q\in P(x)}(V^{\rho}_{\propto}(x+\rho c(x,q)) + \rho W(x,q))\} \end{cases}$$

providing the solutions to the discrete dynamical economy (8.7.1) which are monotone with respect to V^ρ:

$$\bar{x}^\rho_{s+1} := \bar{x}^\rho_s + \rho c(\bar{x}^\rho_s, \bar{p}^\rho_s) \text{ where } \bar{p}^\rho_s \in \Pi^\rho_V(\bar{x}^\rho_s) \tag{8.7.3}$$

We set $V^\rho_0 := \psi_K$, the indicator of the domain K of P and we define recursively the non decreasing sequence of functions V^ρ_n by the "viability kernel algorithm"

$$V^\rho_n(x) := \max\left(V^\rho_{n-1}(x), \frac{1}{1-a\rho}\inf_{p\in P(x)}(V^\rho_{n-1}(x+\rho c(x,p)) + \rho W(x,p))\right) \tag{8.7.4}$$

Assume that P is upper semicontinuous with compact images, that c is continuous and that W is lower semicontinuous with respect to the price. Let $V^\rho \geq \psi_K$ be a lower semicontinuous extended function.

From any $x_0 \in \text{Dom}(P)$, starts one solution $(x^\rho, p^\rho) \in S^\rho(x_0)$ to the discrete dynamical economy (8.7.1) satisfying (8.7.2) if and only if V^ρ satisfies

$$\inf_{p\in P(x)}(V^\rho(x+\rho c(x,p)) + \rho W(x,p)) \leq (1-a\rho)V^\rho(x) \tag{8.7.5}$$

Furthermore, the discrete value function $x_0 \mapsto U^\rho(0, x_0)$ is the smallest lower semicontinuous function V^ρ_∞ larger than or equal to ψ_K satisfying (8.7.2) and can be obtained through:

$$V^\rho_\infty(x) := \sup_{n\geq 0} V^\rho_n(x) \tag{8.7.6}$$

where the functions V^ρ_n are defined recursively by the "viability kernel algorithm" (8.7.4).

The next question we may ask is the following: *Is the limit of a sequence of discrete functions V^ρ_∞ a lower semicontinuous function satisfying (8.5.5)?*

If we modify slightly the finite-difference scheme and if we assume that the dynamical economy is also Lipschitz, then the lower epilimit of the sequence of discrete smallest functions V^ρ_∞ satisfying (8.7.5) is the smallest lower semicontinuous function V_∞ satisfying (8.5.5) larger than or equal to ψ_K.

In the same way, we can discretize the optimization problem with bounded inflation.

The dynamical economy is replaced by the discrete metasystem

$$\begin{cases} i) & x^\rho_{j+1} \in x^\rho_j + \rho c(x^\rho_j, p^\rho_j) \\ ii) & p^\rho_{j+1} \in p^\rho_j + \rho\varphi^\rho(x^\rho_j, p^\rho_j)B \end{cases} \tag{8.7.7}$$

subjected to the constraints

$$(x^\rho_j, p^\rho_j) \in \text{Graph}(P) \tag{8.7.8}$$

and we denote by $\mathcal{S}^\rho(x,p)$ the set of commodity-price solutions $(x^\rho, p^\rho) = \left((x_j^\rho, p_j^\rho)\right)_{j\geq 0}$ of the discrete metasystem starting at an initial commodity-price pair $(x,p) \in \text{Graph}(P)$.

We define the discrete value function of the discrete optimal dynamical economy by

$$J_\infty^\rho(x,p) \;:=\; \inf_{(x^\rho,p^\rho)\in\mathcal{S}^\rho(x,p)} \rho\sum_{q=0}^{\infty}(1-a\rho)^{-q}W(x_q^\rho,p_q^\rho)$$

Assume that the graph of P is closed and that c, φ^ρ and W are continuous. Let us consider a nonnegative lower semicontinuous extended function J^ρ.

The discrete value function J_∞^ρ is the smallest lower semicontinuous nonnegative discrete monotone function. It is equal to

$$J_\infty^\rho(x,p) \;:=\; \sup_{n\geq 0} J_n^\rho(x,p) \tag{8.7.9}$$

where the functions J_n^ρ are defined recursively by the "viability kernel algorithm" described as follows. We set $J_0^\rho := \psi_{\text{Graph}(P)}$, the indicator of the graph of P and we define recursively the non decreasing sequence of functions J_n^ρ by

$$\begin{array}{l} J_n^\rho(x,p) \;:=\; \max\left(J_{n-1}^\rho(x,p),\right. \\ \left.\dfrac{1}{1-a\rho}\inf_{w\in\varphi^\rho(x,p)B}\left(J_{n-1}^\rho(x+\rho c(x,p), p+\rho q) + \rho q^\rho(x,p)\right)\right) \end{array} \tag{8.7.10}$$

We shall associate the synthesis map G^ρ defined by

$$\left\{\begin{array}{l} G^\rho(x,p) \;:=\; \{v \in \varphi^\rho(x,p)B \quad \text{such that} \\ J_\infty^\rho(x+\rho c(x,p), p+\rho q) \\ = \inf_{w\in\varphi^\rho(x,p)B} J_\infty^\rho(x+\rho c(x,p), p+\rho q)\} \end{array}\right.$$

providing the optimal solutions of the discrete metasystem (8.7.7):

$$\left\{\begin{array}{lll} i) & x_{j+1}^\rho & := \; x_j^\rho + \rho c(x_j^\rho, p_j^\rho) \\ ii) & p_{j+1}^\rho & \in \; p_j^\rho + \rho G^\rho(x_j^\rho, p_j^\rho) \end{array}\right. \tag{8.7.11}$$

As in the unbounded inflation case, the limit of a sequence of discrete value functions J_∞^ρ satisfy the monotonicity condition (8.6.5) under Lipschitzianity conditions.

Chapter 9

Dynamical Games

Introduction

This chapter is devoted to a short introduction to dynamical games

$$\begin{cases} i) & x'(t) = c(x(t), p(t), v(t))) \\ ii) & p(t) \in P(x(t)) \\ iii) & v(t) \in Q(x(t)) \end{cases}$$

which are, so to speak, dynamical economies regulated by two parameters, $p(t)$ and $v(t)$.

With two players, the number of questions and problems increases dramatically. We shall be concerned with a very small number of them in this chapter, concentrating on "**games against nature**", which appear for instance in financial or insurance markets. Other games with specific scenarios are investigated in other chapters.

Dynamical economies with frustrations presented in the preceding chapter are examples of **dynamical games**, where the first parameter is the price regulating so to speak collective scarcity constraints and the frustrations $v := (v_1, \ldots, v_n)$ the individual constraints. In this example, the parameters $p(\cdot)$ and $v(\cdot)$ are chosen to guarantee together the viability of the dynamical economy. The regulation map thus associates with any state x a set $\Pi_K(x)$ of pairs (p, v).

We may then appeal to static game theoretical concepts to select in each set $\Pi_K(x)$ an adequate pair (p, v) as we did in Chapter 4. We did also investigate the problem of selecting "cascades" in the case of **hierarchical games** where one player as a stronger "inertia power" than the other.

But, in games against nature we have in mind in this chapter, regulees like prices play the role of the first parameter p, whereas the second parameters $v(t)$ are uncontrolled perturbations or disturbances under the "control" of an incorruptible player, nature. In this framework, the prices have to be chosen independently of nature's purposes.

Therefore, two distinct assumptions on the "information structure" can be made:

1. The first player — the economy — "knows" at each moment not only the state $x(t)$, but also the perturbation $v(t)$ chosen by "nature", and thus, can choose the regulee $p(t)$ in terms of both the state and the disturbance.

2. The first player does not know at all the implemented perturbation. Then he has to choose the regulee as a function of the state alone, against all potential perturbations.

In two-person zero-sum static games, for instance, the opposite objectives of the two players are to minimize and maximize a loss function depending upon the variables chosen by each variable. The two above scenarios are translated into the familiar "sup-inf" and "inf-sup" inequalities involved in the theory, which actually describes **two combinations of quantifiers**, such as

- $\forall\ \exists$: *"for all ..., there exists ..."*
- $\exists\ \forall$: *"there exists ... such that for all ..."*

We shall face analogous situations here, but these two combinations of quantifiers are not necessarily translated into "sup-inf" and "inf-sup" inequalities.

In our viability framework, for instance, one player would like to maintain the viability of at least one solution $x(t)$ to the game in a given subset K. In a noncooperative context, the opponent's objective is then to insure that all solutions to the game leave K in finite time, or, equivalently, reach in finite time its complement $\Omega := X\backslash K$ regarded as a target.

Therefore, we have to

1. associate with the "viability concept" — remain forever in a subset C — a "capture concept" — reach a target C in finite time, as we did in Chapter 6,

2. investigate whether one of these two properties is achieved for at least one solution (existential concepts) or all solutions (universal concepts)

and combine these properties in various ways.

1. We begin the investigation of dynamical games from the viewpoint of one player. We shall relate with a dynamical game a **comprehensive dynamical economy** (P, C), for which the change map $(x,p) \rightsquigarrow C(x,p)$ is defined by

$$C(x,p) := c(x,p,Q(x))$$

Then the change map is set-valued, and the dynamics are described by

$$\begin{cases} i) & x'(t) \in C(x(t),p(t)) \\ ii) & p(t) \in P(x(t)) \end{cases}$$

Such a comprehensive dynamical economy is naturally a differential inclusion to which we can apply the Viability Theorem. We also introduce in this framework the concept of **feedback price** $x \mapsto \tilde{p}(x)$, among which continuous selections of the set-valued map P will play a major role later on.

For the time, we regard the viability kernel of a given subset K under the differential inclusion $x' \in C(x, \tilde{p}(x))$ as the **viability niche** of $\tilde{p}$: it specifies the subset of elements of K which can be regulated by this feedback price.

2. Feedback prices $\tilde{p}$ and perturbations $\tilde{v}$ are used for defining solution concepts of dynamical games. Such feedbacks are usually called "strategies" in the differential game literature[1].

   Indeed, we associate with each feedback $\tilde{p}$ not only the viability kernel of the differential inclusion $x' \in C(x, \tilde{p}(x))$, but also its invariance kernel, or its capture and absorption basins. For instance, the union over a class of feedback selections $\tilde{p}$ of the invariance kernels of the differential inclusions $x' \in C(x, \tilde{p}(x))$ provides *the set of elements of K from which there exists a feedback $\tilde{p}$ such that, for all perturbations $t \mapsto v(t)$, the solutions to*

$$x'(t) = c(x(t), \tilde{p}(x(t)), v(t))$$

   are viable in K. We shall call such a set the "**guaranteed viability kernel**" of K, prove that it is closed and relate it to the infimum over the feedback selections $\tilde{p}$ of the exit time functions of the differential inclusions $x' \in C(x, \tilde{p}(x))$.

   This is a *"there exists ... such that for all ..."* situation. The *"for all ..., there exists ..."* situation consists in associating with any feedback disturbance $\tilde{v}$, a continuous selection of the set-valued map Q, the viability kernels $\text{Viab}(K)$ of the differential inclusion $x' \in c(x, P(x), \tilde{v}(x))$.

   The intersection over the class of continuous selections $\tilde{v}$ of the viability kernels of the differential inclusion $x' \in c(x, P(x), \tilde{v}(x))$ furnishes *the set of elements of K such that, for all feedbacks $\tilde{v}$, there exists one solution to*

$$x'(t) = c(x(t), p(t), \tilde{v}(x(t)))$$

   viable in K. We shall call such a set the "**conditional viability kernel**" of K, prove that it is closed and relate it to the supremum over the feedback selections $\tilde{v}$ of the exit time functions of the differential inclusions $x' \in c(x, P(x), \tilde{v}(x))$.

[1] Actually, such feedbacks are the simplest of the many concepts of strategies introduced by specialists of differential games. But these topics go beyond the scope of this book. Our objective here is to provide the minimal information necessary to give a game theoretical solution to options pricing, for instance.

The conditional viability kernel does not enjoy the **discriminating property**: for every continuous selection $\tilde{v}$, the solution viable in K is not necessarily viable in the conditional viability kernel. The largest closed subset of K enjoying this property is called the **discriminating kernel**. Its existence and its characterizations are due to Pierre Cardaliaguet.

3. Finally, we conclude this introduction to dynamical games with the adaptation of the above problems to the original intertemporal differential games where one player minimizes a discounted infinite horizon cost function. As for optimal intertemporal optimization problems treated in Chapter 8, we shall be able to characterize the value function of the differential game both as a generalized (contingent) solution to the Hamilton-Jacobi-Isaacs partial differential equation and as intersections of viability kernels of auxiliary systems, allowing their approximation by using the viability kernel algorithm.

9.1 Comprehensive Dynamical Economies

We start to investigate two-player dynamical games, which are dynamical economies, so to speak, regulated by two parameters, each controlled by a given decision maker or player.

Definition 9.1.1 *A (P,Q,c) is defined by*

- *a "pricing" set-valued map $P : X \rightsquigarrow \mathcal{M}$*
- *a "perturbation" set-valued map $Q : X \rightsquigarrow \mathcal{N}$*
- *a single-valued map $c : X \times \mathcal{M} \times \mathcal{N} \mapsto X$ describing the evolution law of the dynamical game:*

$$\begin{cases} i) & x'(t) = c(x(t), p(t), v(t))) \\ ii) & p(t) \in P(x(t)) \\ iii) & v(t) \in Q(x(t)) \end{cases}$$

A dynamical game (P,Q,c) is said to be Marchaud if it satisfies the following conditions:

$$\begin{cases} i) & P \text{ and } Q \text{ are Marchaud} \\ ii) & c \text{ is continuous and affine with respect to } (p,v) \\ iii) & c \text{ has linear growth} \end{cases} \tag{9.1.1}$$

In this case, we shall often set

$$C(x,p) := c(x,p,Q(x))$$

The example of dynamical economies with frustrated consumers motivates also the introduction of dynamical economies with two regulees: the prices and the "frustrations". Both of them were introduced as regulation parameters, frustrations dealing with the individual constraints and prices with collective constraints.

We shall need the following definition:

Definition 9.1.2 *We say that a set-valued map F is* convex *if its graph is convex*[2].

It easy to check that a set-valued map F is convex if and only if for any x_i and $\lambda_i \geq 0$ summing up to 1, we have

$$\sum_{i=1}^{n} \lambda_i F(x_i) \subset F\left(\sum_{i=1}^{n} \lambda_i x_i\right)$$

Dynamical games can be regarded as examples of "comprehensive dynamic economies":

Definition 9.1.3 (Comprehensive Dynamic Economy) *A* comprehensive dynamic economy (P, C) *is defined by*

— *a "pricing" set-valued map $P : X \rightsquigarrow \mathcal{M}$*

— *a set-valued map C : $\mathrm{Graph}(P) \mapsto X$ describing the set-valued dynamics of the system.*

The evolution of the commodity and of the price is governed by the differential inclusion

$$\begin{cases} i) & \text{for almost all } t, \ x'(t) \in C(x(t), p(t)) \\ ii) & \text{where } p(t) \in P(x(t)) \end{cases} \tag{9.1.2}$$

We associate with any subset $K \subset \mathrm{Dom}(P)$ the regulation map Π_K : $K \rightsquigarrow \mathcal{M}$ *defined by*

$$\forall x \in K, \ \ \Pi_K(x) := \{p \in P(x) \mid C(x,p) \cap T_K(x) \neq \emptyset\}$$

A comprehensive dynamic economy (P, C) is a Marchaud comprehensive dynamic economy *if*

$$\begin{cases} i) & \mathrm{Graph}(P) \text{ is closed} \\ ii) & C \text{ is upper hemicontinuous with convex compact images} \\ iii) & C \text{ is convex with respect to } p \text{ and } P \text{ has convex values} \\ iv) & C \text{ and } P \text{ have linear growth} \end{cases} \tag{9.1.3}$$

[2]See chapter 2 of [54, Aubin & Frankowska] for an exhaustive presentation of closed convex maps and closed convex processes, which are the "set-valued analogues" of continuous linear operators.

We also obtain comprehensive dynamical economies whenever the change function c is derived from a utility function which is no longer continuously differentiable. In the case of a nondifferentiable utility function U, the change map $C(x,p) := \partial U(x) - p$ where $\partial U(x)$ denotes the generalized gradient of U at x is set-valued.

For comprehensive dynamic economies, we obtain:

Theorem 9.1.4 *Let us consider a Marchaud comprehensive dynamic economy (P,C). Then a closed subset $K \subset \text{Dom}(P)$ is viable*[3] *under F if and only if the regulation map $\Pi_K(\cdot)$ is strict, which is equivalent to the dual form:*

$$\sup_{q\in N_K(x)} \inf_{u\in C(x,p)} \langle q,u\rangle \leq 0$$

Furthermore, any price $p(\cdot)$ regulating a viable solution $x(\cdot)$ in the sense that

$$\text{for almost all } t,\ x'(t) \in C(x(t),p(t))$$

obeys the regulation law

$$\text{for almost all } t,\ p(t) \in \Pi_K(x(t)) \tag{9.1.4}$$

Furthermore, if K is convex and compact, then there exists a viable equilibrium $\bar{x} \in K$ of the dynamical economy.

Indeed, comprehensive dynamic economies lead to differential inclusions: the set-valued map F associating with any state x the set of available velocities is defined by

$$F(x) := \bigcup_{p\in P(x)} C(x,p) \tag{9.1.5}$$

9.1.1 Comprehensive Dynamic Economies as Differential Inclusions

For proving that Marchaud comprehensive dynamic economies yield differential inclusions with Marchaud right-hand sides, we need the following topological results. Without introducing right away the whole machinery concerning continuity issues of set-valued maps[4], we just introduce **hemicontinuity** which is sufficient for our immediate purpose.

Definition 9.1.5 (Upper Hemicontinuous Map) *We shall say that a set-valued map $F : X \leadsto Y$ is* **upper hemicontinuous** *at $x_0 \in \text{Dom}(F)$ if and only if for any $p \in Y^\star$, the function*

$$x \mapsto \sigma(F(x),p) := \sup_{v\in F(x)} \langle p,v\rangle$$

[3]This means that for any initial commodity $x_0 \in K$, there exists a solution on $[0,\infty[$ to the comprehensive dynamic economy (9.1.2) *viable in K*.

[4]for which we refer to SET-VALUED ANALYSIS, [54, Aubin & Frankowska].

is upper semicontinuous at x_0 *(or the function*

$$x \mapsto \sigma^{\flat}(F(x),p) \;:=\; \inf_{v\in F(x)} \langle p,v\rangle$$

is lower semicontinuous). It is said to be upper hemicontinuous if and only if it is upper hemicontinuous at every point of $\mathrm{Dom}(F)$.

Proposition 9.1.6 *The graph of an upper hemicontinuous set-valued map with closed convex values is closed.*

Conversely, if the graph of F *is closed and if the images of* F *remain in a compact set, then* F *is upper hemicontinuous.*

Proof

a) Consider a sequence of elements (x_n, y_n) of $\mathrm{Graph}(F)$ converging to a pair (x,y). Then, for every $p \in Y^\star$,

$$\langle p,y\rangle \;=\; \lim_{n\to\infty}\langle p,y_n\rangle \le \limsup_{n\to\infty}\sigma_{F(x_n)}(p) \;\le\; \sigma_{F(x)}(p)$$

by the upper semicontinuity of $x \mapsto \sigma_{F(x)}(p)$. This inequality implies that $y \in F(x)$ since these subsets are closed and convex, thanks to the Separation Theorem 3.3.2.

We thus have shown that (x,y) belongs to $\mathrm{Graph}(F)$.

b) To show that F is upper hemicontinuous, take any $\lambda \in \mathbf{R}$ and show that the upper sections

$$\left\{x \mid \lambda \;\le\; \sigma_{F(x)}(q)\right\}$$

are closed. Let a sequence of elements x_n of such an upper section converge to x. Since the values $F(x_n)$ are compact, there exist $y_n \in F(x_n)$ such that $\langle q, y_n\rangle = \sigma_{F(x_n)}(q) \ge \lambda$. But the sequence y_n residing in a compact set, a subsequence (again denoted by) y_n converges to some y, which belongs to $F(x)$ since the graph of F is closed. Hence $\lambda \le \langle q,y\rangle \le \sigma_{F(x)}(q)$, and the second statement of the Proposition ensues. □

We deduce from Proposition 9.1.6 the following result:

Corollary 9.1.7 *If* Y *is a finite dimensional vector-space, to say that a nontrivial set-valued map* F *is a Marchaud map amounts to saying that it is nontrivial, upper hemicontinuous, has compact convex images and linear growth.*

We are now able to prove

Lemma 9.1.8 *If the comprehensive dynamic economy* (P,C) *is Marchaud, then the set-valued map* F *defined by (9.1.5) is also Marchaud.*

Proof— First, since the set-valued map C is a closed convex set-valued map with respect to p and since the images $P(x)$ are convex and compact, we infer that the images

$$F(x) := \bigcup_{p \in P(x)} C(x,p)$$

are closed and compact. Indeed, if $v_i \in C(x,p_i)$ $(i = 1,2)$, then

$$\alpha_1 v_1 + \alpha_2 v_2 \in \alpha_1 C(x,p_1) + \alpha_2 C(x,p_2) \subset C(x, \alpha_1 p_1 + \alpha_2 p_2) \subset C(x, P(x))$$

Since

$$\|C(x,p)\| \leq \lambda_x \|p\|$$

we deduce that the values of F are closed: Indeed, if $v_n \in C(x,p_n)$ converges to some v and since $p_n \in P(x)$ remains in a compact set, a subsequence (again denoted by) p_n converges to some $p \in P(x)$. Hence $(x, p_n, v_n) \in \text{Graph}(P)$ converges to (x,p,v), and, since the graph of C is closed, we infer that $v \in C(x, P(x)) =: F(x)$.

Finally, we derive that F is upper hemicontinuous. Indeed,

$$\sigma(F(x),q) = \sup_{p \in P(x)} \sigma(C(x,p),q)$$

Since the map C is upper hemicontinuous, the graph of P is closed and the images $P(x)$ remain in the compact set, we infer that $x \mapsto \sigma(F(x),q)$ is upper semicontinuous.

Hence F is an upper hemicontinuous with nonempty closed convex images. □

9.1.2 Viability Niches

Let us consider a comprehensive dynamic economy (P, C), a closed subset K and the regulation map $\Pi_K(\cdot)$. We did provide in Chapter 4 sufficient conditions under which the regulation map does possess continuous selections.

However, the question arises to know what happens when an arbitrary continuous single-valued map — regarded as a feedback price — is used.

For that purpose, we associate with any single-valued map $\tilde{p} : K \mapsto \mathcal{M}$ the subset

$$P_K^{-1}(\tilde{p}) := \{x \in K \mid p(x) \in P(x)\}$$

of elements of K for which $\tilde{p}$ is a selection of P.

One can extend the definition 7.0.3 of a viability niche of a price p to viability niches of selections $\tilde{p}$ in the following way:

Definition 9.1.9 *The T-viability niche*[5] *of a single-valued map $\tilde{p} : X \mapsto \mathcal{M}$ is the T-viability kernel of $P_K^{-1}(\tilde{p})$ for the differential inclusion $x'(t) \in C(x(t), \tilde{p}(x))$, i.e., the subset*

$$\text{Niche}_{\tilde{p}}(K,T) := \text{Viab}_{C(\cdot,\tilde{p}(\cdot))}(P_K^{-1}(\tilde{p}), T)$$

[5]Naturally, the T-invariance kernel of $P_K^{-1}(\tilde{p})$ for the differential inclusion $x'(t) \in C(x(t), \tilde{p}(x))$ could also be called the T-invariance niche of $\tilde{p}$. They pop up in the definition 9.2.16 of the guaranteed viability kernel.

Therefore,

$$\bigcup_{\tilde{p}} \text{Niche}_{\tilde{p}}(K,T) \subset \text{Viab}_{(P,C)}(K,T)$$

A subset of single-valued maps $\tilde{p}$ is said to form a "viability basis" of the dynamical economy (P, C) on K if the union of the viability niches coincides with the viability kernel of the dynamical economy.

If, for instance, K is viable under a dynamical economy (P, C) and if the regulation map $\Pi_K(\cdot)$ has a continuous selection (or a minimal selection), such a selection is constitutes a viability basis because by construction, K is viable under the differential equation $x' = c(x, \tilde{p}(x))$.

We denote by $\mathcal{P}_\lambda$ the set of λ-Lipschitz selections of the set-valued map P. If P is Lipschitz with convex compact values, then $\mathcal{P}_\lambda \neq \emptyset$ when $\lambda \geq \|P\|_\Lambda$, because such maps possess Lipschitz selections thanks to Theorem 4.4.9.

We mention that the Ascoli Theorem implies that $\mathcal{P}_\lambda$ is compact in $\mathcal{C}(X, \mathcal{M})$ supplied with the compact convergence whenever P has a closed graph and linear growth.

Proposition 9.1.10 *Assume that the dynamical economy (P, c) is Marchaud. Then*

$$\bigcup_{\tilde{p}\in\mathcal{P}_\lambda} \text{Niche}_c(\tilde{p}, T) \text{ is closed}$$

Proof — Let us set

$$M := \bigcup_{\tilde{p}\in\mathcal{P}_\lambda} \text{Viab}_{c(\cdot,\tilde{p}(\cdot)))}(K,T)$$

Let $x_n \in M$ converge to some x. Then there exists a λ-Lipschitz selection $\tilde{p}_n$ of P such that x_n belongs to $\text{Viab}_{C(\cdot,\tilde{p}_n(\cdot))}(K,T)$. Therefore, there exists a solution $x_n(\cdot)$ to the differential inclusion $x' \in C(x, \tilde{p}_n(x))$ viable in K on the interval $[0, T]$. On the other hand, by the Ascoli Theorem, we know that the selections $\tilde{p}_n$ remain in a compact subset of $\mathcal{C}(X, \mathcal{M})$ supplied with the compact convergence topology, so that a subsequence (again denoted by) $\tilde{p}_n$ converges to $\tilde{p}$ uniformly on compact subsets. We observe that $\tilde{p}$ is still a selection of P.

Since the comprehensive dynamic economy is Marchaud, we deduce form the Convergence Theorem that a subsequence of the solutions (again denoted by) $x_n(\cdot)$ converges uniformly on compact intervals to a solution $x(\cdot)$ to the differential inclusion $x' \in C(x, \tilde{p}(x))$ viable in K on the interval $[0, T]$. □

9.2 Games Against Nature

Let us consider a dynamical game (P, Q, c) and, more generally, a (comprehensive) dynamical game (P, Q, C) defined in the following way

Definition 9.2.1 *A* (comprehensive) dynamical game (P,Q,C) *is defined by*

- *a "pricing" set-valued map* $P : X \rightsquigarrow \mathcal{M}$
- *a "perturbation" set-valued map* $Q : X \rightsquigarrow \mathcal{N}$
- *a set-valued map* $C : X \times \mathcal{M} \times \mathcal{N} \rightsquigarrow X$ *describing the dynamics of the dynamical game:*

$$\begin{cases} i) & x'(t) \in C(x(t), p(t), v(t))) \\ ii) & p(t) \in P(x(t)) \\ iii) & v(t) \in Q(x(t)) \end{cases}$$

A (comprehensive) dynamical game (P,Q,C) *is said to be Marchaud if it satisfies the following conditions:*

$$\begin{cases} i) & P \text{ and } Q \text{ are Marchaud} \\ ii) & C \text{ is upper hemicontinuous with convex compact images} \\ iii) & C \text{ is convex with respect to } p \text{ and } P \text{ has convex values} \\ iv) & C \text{ and } P \text{ have linear growth} \end{cases} \qquad (9.2.1)$$

For the time being, and for fixing the ideas, we regard the parameters v as "disturbances" or "perturbations". Hence the "economy", regarded as a "player", chooses the regulee p, a price, for instance, and the "nature", the adverse player, picks out disturbances. Therefore, in this section, we regard this dynamical game as "a game against nature".

The concept of viable solution does not make sense in this a-cooperative context, because of its "existential character": A viable solution is associated with the existence of two functions $y \mapsto p(t)$ and $t \mapsto v(t)$, assuming implicitly that the regulation process can choose both one price function and one perturbation function in a "cooperative" way, so to speak.

But in a game against nature, perturbations are imposed by "nature" in an independent way, for unknown purpose, i.e., in a "a-cooperative" way. In the framework of a viability problem, the question of sustaining the viability of an economic system is not "nature's" business!

In this framework, the prices have to be chosen independently of nature's purposes.

Therefore, two distinct assumptions on the "information structure" can be made:

1. The first player — the economy — "knows" at each moment not only the state $x(t)$, but also the perturbation $v(t)$ chosen by "nature", and thus, can choose the regulee $p(t)$ in terms of both the state and the disturbance. A "discriminating" strategy in this case is a "feedback map" depending on both the states and the perturbations.

2. The first player does not know at all the implemented perturbation. Then he has to choose the regulee as a function of the state alone, against all potential perturbations.

Let us consider a given problem, a viability problem, for instance. The purpose of the first player, the economy, is then to find a regulee such that the state $x(t)$ of the system remains viable

1. either knowing at each instant the perturbation implemented by "nature",

2. or not knowing at all the perturbation chosen by "nature".

9.2.1 Discriminating and Leadership Properties

We shall denote by $\mathcal{Q}$ the set of continuous selections $x \mapsto \tilde{v}(x) \in Q(x)$ of the set-valued map Q.

Definition 9.2.2 *We shall say that a subset L enjoys:*

1. *the* discriminating property *— or* conditional viability property *or* retorting *property — if for any continuous selection $\tilde{v}(x) \in Q(x)$ played by "nature" and any initial state $x \in L$, there exists a solution $x(\cdot)$ to the associated comprehensive dynamical economy*

$$x'(t) \in C(x(t), P(x(t)), \tilde{v}(x(t)))$$

starting at x and viable in L, i.e., if L is viable under $C(\cdot, P(\cdot), \tilde{v}(\cdot))$.

2. *the* leading property *— or* guaranteed viability property *or* blind property *— if there exists a feedback price $\tilde{p}(x) \in P(x)$ such that for any initial state $x \in L$, all solutions $x(\cdot)$ to the differential inclusion*

$$x'(t) \in C(x(t), \tilde{p}(x(t)), Q(x(t)))$$

starting at x are viable in L, i.e., if L is invariant under $C(\cdot, \tilde{p}(\cdot), Q(\cdot))$.

When the selections $\tilde{p}$ are λ-Lipschitz, we say that K enjoy the λ-leading property.

As for viability and invariance, we would like to characterize these two properties by tangential conditions which do not require the knowledge of actual solutions to the dynamical game. Further, if they are not satisfied, we would like to find the associated "kernels", i.e., in loose terms, the set of initial states for which these properties are satisfied.

9.2.2 Conditional Viability Kernel

Definition 9.2.3 *Denote by $Q|_L$ the restriction of the set-valued map Q to the subset L. We associate with L the* discriminating map $\Delta_L : \mathrm{Graph}(Q|_L) \rightsquigarrow \mathcal{M}$ *defined by*

$$\forall\,(x,v) \in \mathrm{Graph}(Q|_L),\;\; \Delta_L(x,v) \;:=\; \{p \in P(x) \mid C(x,p,v) \cap T_L(x) \neq \emptyset\}$$

Selections $\tilde{p} : (x,v) \mapsto \tilde{p}(x,v) \in \Delta_L(x,v)$ of the discriminating map are called discriminating strategies.

We shall say that L is a discriminating domain *— or a* conditional viability domain *— if*

$$\forall\, x \in L,\; \forall\, v \in Q(x),\; \Delta_L(x,v) \;\neq\; \emptyset \tag{9.2.2}$$

The discriminating map is a set-valued feedback map associating with both the state x and the perturbation v a set of regulees p. Using the discriminating map amounts to assuming that at each instant, the knowledge of both the state and the actual perturbation is known.

Theorem 9.2.4 *Let us assume that the dynamical game (P,Q,C) is Marchaud. If L is a discriminating domain, then it enjoys the discriminating property. The converse is true if we assume that Q is lower semicontinuous.*

Proof — Assume that L is a discriminating domain. Then for any continuous selection $\tilde{v} \in \mathcal{Q}$ of Q, the set-valued map $x \rightsquigarrow C(x,P(x),\tilde{v}(x))$ satisfies the assumptions of the Viability Theorem 6.1.9, so that from any initial state $x \in L$ starts one solution to the inclusion $x' \in C(x,P(x),\tilde{v}(x))$ viable in L.

Conversely, assume that L satisfies the discriminating property. Since Q is lower semicontinuous with convex values, the Michael Selection Theorem 4.4.7 implies that for all $x_0 \in L$ and $v_0 \in Q(x_0)$, there exists a continuous selection $\tilde{v}(\cdot)$ of Q such that $v(x_0) = v_0$. Then L enjoys the viability property under the set-valued map $C(x,P(x),\tilde{v}(x))$, and thus,

$$\forall\, x \in L,\;\; C(x,P(x),\tilde{v}(x)) \cap T_L(x) \;\neq\; \emptyset$$

In particular, for x_0, this implies that there exists $p_0 \in \Delta_L(x_0,v_0)$. □

Remark — When L is a discriminating domain, for any selection $\tilde{v}(\cdot) \in \mathcal{Q}$ of Q and from any $x \in L$ starts one solution to the differential inclusion

$$x'(t) \;\in\; C(x(t),p(t),\tilde{v}(x(t)))$$

satisfying

$$\text{for almost all } t \geq 0,\;\; p(t) \;\in\; \Delta_L(x(t),\tilde{v}(x(t)))$$

Hence, the time-dependent (open-loop) price $p(t)$ is a kind of "open-loop" selection of the discriminating map Δ_L.

This remark raises the question whether a discriminating domain can be characterized in terms of discriminating strategies $\tilde{p}(x,v) \in \Delta_L(x,v)$ instead of time-dependent prices. This question is answered by the Cardaliaguet Theorem 9.2.10 below. □

Definition 9.2.5 *Let (P, Q, C) be a dynamical game and $K \subset X$ be a closed subset. The closed subset*

$$[\mathrm{Inv}_v\mathrm{Viab}_p]_{(P,Q,C)}(K,T) := \bigcap_{\tilde{v}\in\mathcal{Q}} \mathrm{Viab}_{C(\cdot,P(\cdot),\tilde{v}(\cdot))}(K,T) \quad (9.2.3)$$

of initial states $x_0 \in K$ such that for any continuous selection $\tilde{v} \in \mathcal{Q}$ of Q starts one solution to the differential inclusion

$$x'(t) \in C(x(t), P(x(t)), \tilde{v}(x(t)))$$

*viable in K on the interval $[0,T]$ is called the T-*conditional viability kernel[6]. *When $T = +\infty$, we say that*

$$[\mathrm{Inv}_v\mathrm{Viab}_p]_{(P,Q,C)}(K) := \bigcap_{\tilde{v}\in\mathcal{Q}} \mathrm{Viab}_{C(\cdot,P(\cdot),\tilde{v}(\cdot))}(K)$$

the conditional viability kernel *of the dynamical game.*

The closed subset

$$[\mathrm{Abs}_v\mathrm{Capt}_p]_{(P,Q,C)}(K,T) := \bigcap_{\tilde{v}\in\mathcal{Q}} \mathrm{Capt}_{C(\cdot,P(\cdot),\tilde{v}(\cdot))}(K,T) \quad (9.2.4)$$

of initial states $x_0 \in X$ such that for any continuous selection $\tilde{v} \in \mathcal{Q}$ of Q starts one solution to the differential inclusion

$$x'(t) \in C(x(t), P(x(t)), \tilde{v}(x(t)))$$

*hitting K before T is called the T-*conditional capture basin. *When $T = +\infty$, we say that*

$$[\mathrm{Abs}_v\mathrm{Capt}_p]_{(P,Q,C)}(K) := \bigcap_{\tilde{v}\in\mathcal{Q}} \mathrm{Capt}_{C(\cdot,P(\cdot),\tilde{v}(\cdot))}(K)$$

the conditional viability basin *of the dynamical game.*

The conditional viability kernel $[\mathrm{Inv}_v\mathrm{Viab}_p]_{(P,Q,C)}(K)$ and conditional viability basin $[\mathrm{Abs}_v\mathrm{Capt}_p]_{(P,Q,C)}(K)$ are appealing concept because they provide the set of initial states from which, for every observed perturbation, the economy can regulate a viable solution or reestablish viability respectively.

Unfortunately, *the conditional viability kernel is not necessarily a discriminating domain.* We shall introduce later the concept of discriminating

[6] which is, then, the intersection of T-viability niches of the selections $\tilde{v}$.

kernel of K defined as the largest closed discriminating domain contained in K.

The discriminating domains enjoy the same type of stability properties than the viability domains:

Theorem 9.2.6 *Let us consider a sequence of discriminating domains K_n of Marchaud dynamical game (P, Q, C).*

Then the upper limit $K^\sharp$ of the K_n is still a discriminating domain of the dynamical game (P, Q, C).

Proof — Let $x := \lim_{n\to+\infty} x_n$ be the limit of a sequence of elements $x_n \in K_n$. Take any continuous selection $\tilde{v} \in \mathcal{Q}$ of Q. Since K_n is a discriminating domain of the dynamical game (P, Q, C), K_n is a viability domain of the set-valued map $C(\cdot, P(\cdot), \tilde{v}(\cdot))$. We thus deduce from the Stability Theorem 12.1.5 of viability domains that the upper limit $K^\sharp$ of the K_n is still a viability domain of the set-valued map $C(\cdot, P(\cdot), \tilde{v}(\cdot))$. This shows that $K^\sharp$ is a discriminating domain of the dynamical game (P, Q, C). □

Remark — Actually, we can extend this result to the case when the K_n are discriminating domains of dynamical games (P_n, Q, C_n) and a sequence of discriminating domains K_n of dynamical games.

Assume that the linear growth constants of P_n and C_n are bounded and that the dynamical games (P_n, Q, C_n) converge to a Marchaud dynamical game (P, Q, C) in the sense that for every continuous selection $\tilde{v} \in \mathcal{Q}$ of Q, the upper limits of the graphs of $c_n(\cdot, P_n(\cdot), \tilde{v}(\cdot))$ is contained in the graph of $C(\cdot, P(\cdot), \tilde{v}(\cdot))$.

Then the upper limit $K^\sharp$ of the K_n is still a discriminating domain of the dynamical game (P, Q, C). □

We now relate the concepts of conditional viability kernels and capture basins with exit and hitting time functions:

Theorem 9.2.7 *Let $K \subset X$ be a closed subset and (P, Q, C) a Marchaud dynamical game. We associate with any continuous selection $\tilde{v}$ of Q the $\tilde{v}$-upper* exit function *and lower* hitting function *defined by*

$$\tau^\sharp_{K,\tilde{v}}(x) := \sup_{x(\cdot)\in\mathcal{S}_{C(\cdot,P(\cdot),\tilde{v}(\cdot))}(x)} \tau_K(x(\cdot)) \ \ \& \ \ \varpi^\flat_{K,\tilde{v}}(x) := \inf_{x(\cdot)\in\mathcal{S}_{C(\cdot,P(\cdot),\tilde{v}(\cdot))}(x)} \varpi_K(x(\cdot))$$

Then for any $T \geq 0$, the conditional T-viability kernels and the conditional T-capture basins are characterized by

$$[\mathrm{Inv}_v\mathrm{Viab}_p]_{(P,Q,C)}(K, T) \ = \ \left\{x \in K \mid \inf_{\tilde{v}\in\mathcal{Q}} \tau^\sharp_{K,\tilde{v}}(x) \ \geq \ T\right\}$$

and

$$[\mathrm{Abs}_v\mathrm{Capt}_p]_{(P,Q,C)}(K, T) \ = \ \left\{x \in X \mid \sup_{\tilde{v}\in\mathcal{Q}} \varpi^\flat_{K,\tilde{v}}(x) \ \leq \ T\right\}$$

Proof — By Theorem 6.3.6, we know that when (P, Q, C) is Marchaud, then for any $T \geq 0$, the viability kernels and the capture basins are characterized by:

$$\begin{cases} \text{Viab}_{C(\cdot,P(\cdot),\tilde{v}(\cdot))}(K,T) & = \left\{x \in K \mid \tau^{\sharp}_{K,\tilde{v}}(x) \geq T\right\} \\ \text{Capt}_{C(\cdot,P(\cdot),\tilde{v}(\cdot))}(K,T) & = \left\{x \in X \mid \varpi^{\flat}_{K,\tilde{v}}(x) \leq T\right\} \end{cases} \quad (9.2.5)$$

Since

$$[\text{Inv}_v\text{Viab}_p]_{(P,Q,C)}(K,T) := \bigcap_{\tilde{v}\in\mathcal{Q}} \text{Viab}_{C(\cdot,P(\cdot),\tilde{v}(\cdot))}(K,T)$$

and since

$$[\text{Abs}_v\ \text{Capt}_p]_{(P,Q,C)}(K,T) := \bigcap_{\tilde{v}\in\mathcal{Q}} \text{Capt}_{C(\cdot,P(\cdot),\tilde{v}(\cdot))}(K,T)$$

we conclude proof of the Theorem because the lower level sections of the supremum is the intersection of the lower level sections. □

9.2.3 Discriminating Kernel

Definition 9.2.8 *Let (P, Q, C) be a dynamical game and $K \subset X$ be a closed subset. The largest closed discriminating domain* $\text{Disc}_{(P,Q,C)}(K)$ *contained in K is called the* discriminating kernel *of K under the dynamical game (P, Q, C).*

The discriminating kernel is contained in the conditional viability kernel, but may be different because the conditional viability kernel is not necessarily a discriminating domain. Actually, Pierre Cardaliaguet provided in [98, Cardaliaguet] an example of a game where these two subsets are different. He also proved a formula — actually, an algorithm — relating these two concepts and also, characterized the discriminating kernel in terms of discriminating strategies.

Theorem 9.2.9 (Cardaliaguet) *Let us assume that the dynamical game (P, Q, C) is Marchaud and that Q is lower semicontinuous. Setting*

$$K_0 := K \ \&\ \forall\, i \geq 0,\ K_i := [\text{Inv}_v\text{Viab}_p]_{(P,Q,C)}(K_{i-1})$$

then

$$\text{Disc}_{(P,Q,C)}(K) = \bigcap_{i=1}^{\infty} K_i$$

Proof — Let us denote by $K_\infty := \bigcap_{i\geq 0} K_i$ the intersection of the closed subsets K_i.

First, we check that K_∞ is a discriminating domain, i.e., thanks to Theorem 9.2.4, that for each $\tilde{v} \in \mathcal{Q}$, K_∞ is a viability domain for $C(\cdot, P(\cdot), \tilde{v}(\cdot))$. But, for a fixed $\tilde{v}$,

$$K_{i+1} \subset \text{Viab}_{C(\cdot,P(\cdot),\tilde{v}(\cdot))}(K_i) \subset K_i$$

So, K_∞ is the intersection of the decreasing sequence of the viability kernels $\mathrm{Viab}_{C(\cdot,P(\cdot),\tilde{v}(\cdot))}(K_i)$ viable under the set-valued map $C(\cdot,P(\cdot),\tilde{v}(\cdot))$, and so, is itself viable under this map. Therefore, K_i is viable under every set-valued maps $C(\cdot,P(\cdot),\tilde{v}(\cdot))$, and thus, enjoys the discriminating property. Thus, being a discriminating domain, it is contained in K_∞.

It remains to prove that the discriminating kernel is contained in K_∞, and, for that purpose, to show that it is contained in every K_i. This is true for $i = 1$, since it is contained in $K = K_0$. Assume that $\mathrm{Disc}_{(P,Q,C)}(K)$ is contained in K_i. Since the discriminating kernel is viable under every set-valued map $C(\cdot,P(\cdot),\tilde{v}(\cdot))$, then

$$\mathrm{Disc}_{(P,Q,C)}(K) \;=\; \mathrm{Viab}_{C(\cdot,P(\cdot),\tilde{v}(\cdot))}(\mathrm{Disc}_{(P,Q,C)}(K)) \;\subset\; \mathrm{Viab}_{C(\cdot,P(\cdot),\tilde{v}(\cdot))}(K_i)$$

so that, by taking the intersection over every continuous selection $\tilde{v}$, we conclude that

$$\mathrm{Disc}_{(P,Q,C)}(K) \;\subset\; \bigcap_{\tilde{v}\in\mathcal{Q}} \mathrm{Viab}_{C(\cdot,P(\cdot),\tilde{v}(\cdot))}(K_i) \;=:\; K_{i+1}$$

We deduce that $\mathrm{Disc}_{(P,Q,C)}(K) \subset K_\infty$ by taking the intersection over the integers $i \geq 0$. □

However, one would like to characterize the discriminating kernel as the subset if initial states of K satisfying some property. This is possible thanks to another characterization of discriminating domains involving the use of continuous discriminating strategies and a larger class of perturbations, the class of finite concatenations of continuous selections of the set-valued map Q.

We associate with a finite sequence $0 \leq t_0 \leq \ldots \leq t_p$ and a finite sequence of continuous selections $\tilde{v}_0, \ldots, \tilde{v}_p \in \mathcal{Q}$ of Q the **concatenated perturbation** $(t,x) \mapsto \tilde{\mathbf{v}}(t,x)$ defined by

$$\forall\, j = 0,\ldots,p-1, \;\; \forall\, t \in [t_j, t_{j+1}[, \;\; \tilde{\mathbf{v}}(t,x) \;:=\; \tilde{v}_j(x)$$

We thus can use these finite concatenations of continuous feedbacks as feedback controls of problems $x' \in F(x,u)$ by defining a solution to the differential inclusion

$$x'(t) \;\in\; F(x(t), \tilde{\mathbf{v}}(t, x(t)))$$

starting from x_0 in the following way: On the interval $[0,t_0]$, $x(\cdot)$ is a solution to the differential inclusion

$$\forall\, t \in [0,t_0], x'(t) \;\in\; F(x(t), \tilde{v}_0(x(t)))$$

Next, on the interval $[t_0,t_1]$, $x(\cdot)$ is a solution to the differential inclusion

$$\forall\, t \in [0,t_0], x'(t) \;\in\; F(x(t), \tilde{v}_1(x(t)))$$

starting at time t_0 from $x(t_0)$, and so on.

It will be convenient to "enlarge" the class $\mathcal{Q}$ of continuous selections $\tilde{v} \in \mathcal{Q}$ of Q to the class $\tilde{\mathcal{Q}}$ of finite concatenations of continuous selections $\tilde{\mathbf{v}} \in \mathcal{Q}$ of Q.

Proposition 9.2.10 *Let us assume that the dynamical game (P, Q, C) is Marchaud and that Q is lower semicontinuous.*

If there exists a continuous discriminating strategy $\tilde{p} : (x,u) \mapsto \tilde{p}(x,u) \in \Delta_L(x,u)$ such that all finite concatenations $\tilde{\mathbf{v}} \in \tilde{\mathcal{Q}}$ and from any initial state $x_0 \in L$ starts one solution to the differential inclusion

$$x'(t) \in C(x(t), \tilde{p}(x(t), \tilde{\mathbf{v}}(t, x(t))), \tilde{\mathbf{v}}(t, x(t)))$$

viable in L, then L is a discriminating domain.

The converse is true if we assume that the discriminating map Δ_L is lower semicontinuous with closed convex values[7].

Proof — The first statement follows from Theorem 9.2.4. Conversely, assume now that L is a discriminating domain. When the discriminating map Δ_L is lower semicontinuous, the Michael Selection Theorem implies the existence of a continuous discriminating strategy $(x,v) \in \text{Graph}(Q|_L) \rightsquigarrow \tilde{p}(x,v)$. Therefore, for every finite concatenation $\tilde{\mathbf{v}} \in \tilde{\mathcal{Q}}$, the closed subset L is a viability domain under the single-valued map $x \rightsquigarrow C(x, \tilde{p}(x, \tilde{\mathbf{v}}(t,x)), \tilde{\mathbf{v}}(t,x))$ since $\tilde{p}$ is a selection of the discriminating map Δ_L. Hence, from every $x_0 \in L$ starts one solution to differential inclusion $x' \in C(x, \tilde{p}(x, \tilde{\mathbf{v}}(t,x)), \tilde{\mathbf{v}}(t,x))$ viable in L. □

Theorem 9.2.11 (Cardaliaguet) *Assume that the discriminating map $\Delta_{\text{Disc}_{(P,Q,C)}(K)}$ of the discriminating kernel $\text{Disc}_{(P,Q,C)}(K)$ is lower semicontinuous. Then the discriminating kernel $\text{Disc}_{(P,Q,C)}(K)$ is equal to the set of initial states $x_0 \in K$ such that there exists a continuous discriminating strategy $\tilde{p}(x,v)$ such that, for any finite concatenation $\tilde{\mathbf{v}}$ of continuous feedback perturbations, there exists a solution to the differential inclusion*

$$x'(t) \in C(x(t), \tilde{p}(x(t), \tilde{\mathbf{v}}(t, x(t))), \tilde{\mathbf{v}}(t, x(t)))$$

starting from x_0 and viable in K.

Proof — By Proposition 9.2.10, the discriminating kernel being a discriminating domain, there exists a continuous discriminating strategy $\tilde{p}$: $(x,u) \mapsto \tilde{p}(x,u)$ such that for all finite concatenation $\tilde{v} \in \tilde{\mathcal{Q}}$ and from any

[7]This assumption is satisfied when L is sleek and when the game satisfies stronger tangential conditions. See the lower semicontinuity criteria of the Tool-Box (Propositions 11.2.10 and 11.2.11).

initial state $x_0 \in \mathrm{Disc}_{(P,Q,C)}(K)$ starts one solution to the differential inclusion

$$x'(t) \in C(x(t), \tilde{p}(x(t), \tilde{\mathbf{v}}(t, x(t))), \tilde{\mathbf{v}}(t, x(t)))$$

viable in $\mathrm{Disc}_{(P,Q,C)}(K)$, and thus, in K.

Conversely, assume that $x_0 \in K \backslash \mathrm{Disc}_{(P,Q,C)}(K)$. Let $(x,u) \mapsto \tilde{p}(x,u)$ be any discriminating strategy. We shall construct a finite concatenation $\tilde{\mathbf{v}}(\cdot)$ of continuous selections of Q such that one solution to the differential inclusion

$$x'(t) \in C(x(t), \tilde{p}(x(t), \tilde{\mathbf{v}}(t, x(t))), \tilde{\mathbf{v}}(t, x(t)))$$

starting from x_0 leaves K in finite time.

Indeed, by Proposition 9.2.9, there exists $i_0 \geq 0$ such that $x_0 \in K_{i_0} \backslash K_{i_0+1}$. Since

$$K_{i_0+1} := \bigcap_{\tilde{v} \in \mathcal{Q}} \mathrm{Viab}_{C(\cdot, P(\cdot), \tilde{v}(\cdot))}(K_{i_0})$$

this implies that there exists $\tilde{v}_0 \in \mathcal{Q}$ such that $x_0 \notin \mathrm{Viab}_{C(\cdot, P(\cdot), \tilde{v}_0(\cdot))}(K_{i_0})$. Then all solutions to $x' \in C(x, P(x), \tilde{v}(x))$ leave K_{i_0} in finite time. In particular, there exist $t_0 \geq 0$ and a solution $x_0(\cdot)$ to the differential inclusion $x'(t) \in C(x(t), \tilde{p}(x(t), \tilde{v}_0(x(t))), \tilde{v}_0(x(t)))$ such that $x(t_0) \notin K_{i_0}$. If $x_1 := x(t_0) \notin K$, the proof is completed. Otherwise, there exists $i_1 < i_0$ such that $x_1 \in K_{i_1} \backslash K_{i_1+1}$. By the same reasoning, there exists $\tilde{v}_1 \in \mathcal{Q}$ such that $x_1 \notin \mathrm{Viab}_{C(\cdot, P(\cdot), \tilde{v}_1(\cdot))}(K_{i_1})$, so that there exist $t_1 \geq t_0$ and a solution $x_1(\cdot)$ to the differential inclusion $x'(t) \in C(x(t), \tilde{p}(x(t), \tilde{v}_1(x(t))), \tilde{v}_1(x(t)))$ starting at time t_0 from x_1 such that $x(t_1) \notin K_{i_1}$. Continuing by induction at most i_0 times, we construct a finite number of times $0 \leq t_0 \leq \ldots \leq t_j$, j continuous feedback $\tilde{v}_j \in \mathcal{Q}$ so that their finite concatenation $\tilde{\mathbf{v}}$ provides a solution to the differential inclusion

$$x'(t) \in C(x(t), \tilde{p}(x(t), \tilde{\mathbf{v}}(t, x(t))), \tilde{\mathbf{v}}(t, x(t)))$$

starting at x_0 and leaving K in finite time. □

9.2.4 Guaranteed Viability Kernel

We would like now to "play against nature in a blind way", i.e., to find a feedback price $\tilde{p} : x \mapsto \tilde{p}(x) \in P(x)$ such that K is invariant under the associated differential inclusion

$$x'(t) \in C(x, \tilde{p}(x), Q(x))$$

Definition 9.2.12 *We associate with L the* guaranteed regulation map

$$\Gamma_L(x) := \{p \in P(x) \mid C(x, p, Q(x)) \subset T_L(x)\}$$

Selections $\tilde{p} : x \mapsto \tilde{p}(x) \in \Gamma_L(x)$ of the guaranteed regulation map are called leading or guaranteed strategies.

We shall say that L is a **guaranteed viability domain** *— or a* **leadership domain** *— if*

$$\forall\, x \in L,\ \Gamma_L(x) \neq \emptyset \tag{9.2.6}$$

The guaranteed regulation map is a set-valued feedback map associating with the state x a set of regulees p. *Using the guaranteed regulation map amounts to assuming that at each instant, the knowledge of the state without requiring the knowledge of the actual perturbation.*

Since invariance properties are involved in this concept of guaranteed viability, Lipschitz conditions will be involved:

Definition 9.2.13 *We shall say that* **dynamical game** (P, Q, C) *is Lipschitz if the set-valued maps P and Q are Lipschitz with compact values and if the set-valued map C is Lipschitz.*

We denote by $\mathcal{P}_\lambda$ the set of λ-Lipschitz selections of the set-valued map P. If P is Lipschitz with convex compact values, then $\mathcal{P}_\lambda \neq \emptyset$ when $\lambda \geq \|P\|_\Lambda$, because such maps possess Lipschitz selections thanks to Theorem 4.4.9.

We thus observe that the Lipschitz constants of maps $C(\cdot, \tilde{p}(\cdot), Q(\cdot))$ are estimated by $c_\lambda := \|C\|_\Lambda(1 + \lambda + \|Q\|_\Lambda)$. We mention that the Ascoli Theorem implies that $\mathcal{P}_\lambda$ is compact in $\mathcal{C}(X, \mathcal{M})$ supplied with the compact convergence whenever P has a closed graph and linear growth.

Theorem 9.2.14 *Let us assume that the dynamical game (P, Q, C) is Lipschitz. Then L enjoys the leading property whenever Γ_L possesses at least one Lipschitz selection $\tilde{p}$.*

Proof — The Invariance Theorem 12.3.4 states that if the set-valued map $x \rightsquigarrow C(x, \tilde{p}(x), Q(x))$ is Lipschitz, then a necessary and sufficient condition for the invariance of L under this map is that

$$\forall\, x \in L,\ \ C(x, \tilde{p}(x), Q(x)) \ \subset\ T_L(x)$$

i.e., if and only if $\tilde{p}(\cdot)$ is a selection of the guaranteed regulation map Γ_L.

Then L enjoys the leading property whenever Γ_L possesses at least one Lipschitz selection $\tilde{p}$. □

We begin by proving a stability property:

Proposition 9.2.15 *Let us assume that the dynamical game (P, Q, C) is Lipschitz. Then the upper limit of closed subsets K_n which enjoy the T-λ-leading property under the dynamical game (P, Q, C) still enjoys the T-λ-leading property.*

Proof — Let $x_n \in K_n$ converge to some $x \in K^\sharp$. Then there exists a λ-Lipschitz selection $\widetilde{p}_n$ of P such that x_n belongs to $\mathrm{Inv}_{C(\cdot,\widetilde{p}_n(\cdot),Q(\cdot))}(K_n,T)$. On the other hand, by the Ascoli Theorem, we know that the selections $\widetilde{p}_n$ remain in a compact subset of $\mathcal{C}(X,\mathcal{M})$ supplied with the compact convergence topology, so that a subsequence (again denoted by) $\widetilde{p}_n$ converges to $\widetilde{p}$ uniformly on compact subsets. We observe that $\widetilde{p}$ is still a selection of P.

Let $x(\cdot) \in \mathcal{S}_{C(\cdot,\widetilde{p}(\cdot),Q(x))}(x)$. We have to prove that it is viable in $K^\sharp$. By the Filippov Theorem 12.2.1, we know that there exists a solution $x_n(\cdot) \in \mathcal{S}_{C(\cdot,\widetilde{p}_n(\cdot),Q(\cdot))}(x_n)$ satisfying inequality

$$\begin{cases} \|x_n(t) - x(t)\| \\ \leq e^{c_\lambda t}\left(\|x - x_n\| + \int_0^t d(x'(s), C_n(x(s), \widetilde{p}_n(x(s)), Q(x_n(s))))e^{-c_\lambda s}ds\right) \end{cases}$$

Observing that

$$\begin{cases} x'(s) \in C(x(s),\widetilde{p}(s),Q(x(s))) \\ \subset C(x(s),\widetilde{p}_n(s),Q(x_s)) + c_\Lambda\|\widetilde{p}_n(x(s)) - \widetilde{p}(x(s))\|B \end{cases}$$

we deduce that $x_n(\cdot)$ converges to $x(\cdot)$ uniformly on compact subsets. The viability of the solutions $x_n(\cdot)$ in K_n on $[0,T]$ thus implies the viability of $x(\cdot)$ in $K^\sharp$. Consequently, $K^\sharp$ enjoys the -λ-leading property under (P,Q,C) on the interval $[0,T]$. □

When K does not enjoy the guaranteed viability property, one can always define the **guaranteed viability kernel** in the following way:

Definition 9.2.16 *Let (P,Q,C) be a dynamical game and $K \subset X$ be a closed subset. The closed subset*

$$[\mathrm{Viab}_p\mathrm{Inv}_v]_{(P,Q,C)}(K,T) := \bigcup_{\widetilde{p}\in\mathcal{P}_\lambda} \mathrm{Inv}_{C(\cdot,\widetilde{p}(\cdot),Q(\cdot))}(K,T) \quad (9.2.7)$$

where $\widetilde{p}$ ranges over $\mathcal{P}_\lambda$ is called the T-guaranteed viability kernel[8]*. It is the subset of initial states $x_0 \in K$ such that there exists a λ-Lipschitz feedback price $x \mapsto \widetilde{p}(x)$ such that all solutions to the differential inclusion $x' \in C(x,\widetilde{p}(x),Q(x))$ starting from x_0 remain viable in K on the interval $[0,T]$. When $T = +\infty$, we say that*

$$[\mathrm{Viab}_p\mathrm{Inv}_p]_{(P,Q,C)}(K) := \bigcup_{\widetilde{p}\in\mathcal{P}_\lambda} \mathrm{Inv}_{C(\cdot,\widetilde{p}(\cdot),Q(\cdot))}(K)$$

the **guaranteed viability kernel** *of the dynamical game.*

The closed subset

$$[\mathrm{Capt}_p\mathrm{Abs}_v]_{(P,Q,C)}(K,T) := \bigcup_{\widetilde{p}\in\mathcal{P}_\lambda} \mathrm{Abs}_{C(\cdot,\widetilde{p}(\cdot),Q(\cdot))}(K,T) \quad (9.2.8)$$

---

[8] which is, then, the union of T-invariance niches of the selections $\widetilde{p}$.

of initial states $x_0 \in X$ such that there exists a Lipschitz selection $\tilde{p} \in \mathcal{P}_\lambda$ of P from which all solutions to the differential inclusion

$$x'(t) \in C(x(t), \tilde{p}(x(t)), Q(x(t)))$$

*hit K before T is called the T-*guaranteed capture basin. *When $T = +\infty$, we say that*

$$[\mathrm{Capt}_p\mathrm{Abs}_v]_{(P,Q,C)}(K) := \bigcup_{\tilde{p}\in\mathcal{P}_\lambda} \mathrm{Abs}_{C(\cdot,\tilde{p}(\cdot),Q(\cdot))}(K)$$

the guaranteed capture basin *of the dynamical game.*

The guaranteed viability kernels and capture basins depend upon the choice of λ, but we shall not mention this in the notations.

Taking $K_n := [\mathrm{Viab}_p\mathrm{Inv}_v]_{(P,Q,C)}(K)$, Proposition 9.2.15 implies that the T-guaranteed viability kernel $[\mathrm{Viab}_p\mathrm{Inv}_v]_{(P,Q,C)}(K)$ are closed. In the same way, one can prove that the T-guaranteed capture basins $[\mathrm{Capt}_p\mathrm{Abs}_v]_{(P,Q,C)}(K)$ are closed:

Proposition 9.2.17 *Let us assume that the dynamical game (P, Q, C) is Lipschitz. Then the T-guaranteed viability kernel $[\mathrm{Viab}_p\mathrm{Inv}_v]_{(P,Q,C)}(K, T)$ and T-guaranteed capture basins $[\mathrm{Capt}_p\mathrm{Abs}_v]_{(P,Q,C)}(K, T)$ are closed.*

We thus deduce that

Theorem 9.2.18 *The guaranteed viability kernel $[\mathrm{Viab}_p\mathrm{Inv}_v]_{(P,Q,C)}(K)$ is the largest closed subset of K satisfying the λ-leading property.*

Proof — First, the guaranteed viability kernel enjoys the λ-leading property because whenever x belongs to it, there exists a Lipschitz selection $\tilde{p}$ of P such that x belongs to the invariance kernel of K under the Lipschitz set-valued map $x \rightsquigarrow C(x, \tilde{p}(x), Q(x))$. Therefore, every solution of the differential inclusion $x' \in C(x, \tilde{p}(x), Q(x))$ is viable in K, and actually, viable in $\mathrm{Inv}_{C(\cdot,\tilde{p}(\cdot),Q(\cdot))}(K) \subset [\mathrm{Viab}_p\mathrm{Inv}_v]_{(P,Q,C)}(K)$.

It contains obviously every closed subset enjoying the λ-leading property. □

We define the closure $\overline{\inf}_{i\in I} V_i$ of the infimum and the closure $\overline{\sup}_{i\in I} V_i$ of the supremum of a family of functions V_i by

$$\mathcal{E}p(\overline{\inf}_{i\in I} V_i) := \overline{\bigcup_{i\in I} \mathcal{E}p(V_i)} \ \ \& \ \ \mathcal{H}yp(\overline{\sup}_{i\in I} V_i) := \overline{\bigcup_{i\in I} \mathcal{H}yp(V_i)}$$

Theorem 9.2.19 *Let $K \subset X$ be a closed subset and (P, Q, C) a dynamical game. We associate with any continuous selection $\tilde{p}$ of P the $\tilde{p}$-lower* exit function *and upper* hitting function *defined by*

$$\tau^{\flat}_{K,\tilde{p}}(x) := \inf_{x(\cdot)\in\mathcal{S}_{C(\cdot,\tilde{p}(\cdot),Q(\cdot))}(x)} \tau_K(x(\cdot)) \ \ \& \ \ \varpi^{\sharp}_{K,\tilde{p}}(x) := \inf_{x(\cdot)\in\mathcal{S}_{C(\cdot,\tilde{p}(\cdot),Q(\cdot))}(x)} \varpi_K(x(\cdot))$$

If the game (P,Q,C) is Lipschitz, then for any $T \geq 0$, the guaranteed T-viability kernels are characterized by

$$[\mathrm{Viab}_p\mathrm{Inv}_v]_{(P,Q,C)}(K,T) \;=\; \left\{x \in K \mid \overline{\sup}_{\tilde{p}\in\mathcal{P}_\lambda}\tau^{\flat}_{K,\tilde{p}}(x) \;\geq\; T\right\}$$

and the guaranteed T-capture basins by

$$[\mathrm{Capt}_p\mathrm{Abs}_v]_{(P,Q,C)}(K,T) \;=\; \left\{x \in X \mid \overline{\inf}_{\tilde{p}\in\mathcal{P}_\lambda}\varpi^{\sharp}_{K,\tilde{p}}(x) \;\leq\; T\right\}$$

Proof — By Theorem 6.3.6, we know that when $C(\cdot,\tilde{p}(\cdot),Q(\cdot))$ is Lipschitz, then for any $T \geq 0$, the invariance kernels and the absorption basins are characterized by:

$$\left\{\begin{array}{lll}\mathrm{Inv}_{C(\cdot,\tilde{p}(\cdot),Q(\cdot))}(K,T) & = & \left\{x \in K \mid \tau^{\flat}_{K,\tilde{p}}(x) \;\geq\; T\right\} \\ \\ \mathrm{Abs}_{C(\cdot,\tilde{p}(\cdot),Q(\cdot))}(K,T) & = & \left\{x \in X \mid \varpi^{\sharp}_{K,\tilde{p}}(x) \;\leq\; T\right\}\end{array}\right. \tag{9.2.9}$$

Since

$$[\mathrm{Viab}_p\mathrm{Inv}_v]_{(P,Q,C)}(K,T) \;:=\; \bigcup_{\tilde{p}\in\mathcal{P}} \mathrm{Inv}_{C(\cdot,\tilde{p}(\cdot),Q(\cdot))}(K,T)$$

and since

$$[\mathrm{Capt}_v\mathrm{Abs}_p]_{(P,Q,C)}(K,T) \;:=\; \bigcup_{\tilde{p}\in\mathcal{P}} \mathrm{Capt}_{C(\cdot,\tilde{p}(\cdot),Q(\cdot))}(K,T)$$

we deduce (9.2.19) because the lower level sections of the closure $\overline{\inf}_{i\in I}V_i$ of the infimum of a family of functions Q_i is the closure $\bigcup_{i\in I}\{x \mid V_i(x) \leq \lambda\}$ of the unions of the lower level sections $\{x \mid V_i(x) \leq \lambda\}$. □

Remark — Let us denote by $\Omega := X\backslash K$ the complement of K, regarded as an open target.

The complement $K \setminus [\mathrm{Inv}_v\mathrm{Viab}_p]_{(P,Q,C)}(K)$ is the set of initial states x_0 for which there exists at least one continuous selection $\tilde{v} \in \mathcal{Q}$ of Q such that every solution to the differential inclusion

$$x'(t) \;\in\; C(x(t), P(x(t)), \tilde{v}(x(t)))$$

starting from x_0 reaches the target Ω in finite time.

In the same way, the complement $K \setminus [\mathrm{Viab}_p\mathrm{Inv}_v]_{(P,Q,C)}(K)$ of the guaranteed viability kernel is the set of initial states x_0 such that, for every λ-Lipschitz selection $\tilde{p}$ of P, there exist at least one solution $x(\cdot) \in \mathcal{S}_{C(\cdot,\tilde{p}(\cdot),Q(\cdot))}(x)$

which reaches the target Ω in finite time. In other words, outside the guaranteed viability kernel, "nature" may win in the sense that whatever the choice of the feedback strategy $\tilde{p}$, there exists one solution which reaches the target Ω in finite time.

These statements can be translated in the following relations

$$\begin{cases} X \setminus [\text{Viab}_p\text{Inv}_v]_{(P,Q,C)}(K) & = & [\text{Abs}_p\text{Capt}_v]_{(P,Q,C)}(\Omega) \\ X \setminus [\text{Inv}_v\text{Viab}_p]_{(P,Q,C)}(K) & = & [\text{Capt}_v\text{Abs}_p]_{(P,Q,C)}(\Omega) \end{cases}$$

9.3 Intertemporal Dynamical Games

Let us consider a dynamical game

$$\begin{cases} i) & x'(t) = c(x(t), p(t), v(t))) \\ ii) & p(t) \in P(x(t)) \\ iii) & v(t) \in Q(x(t)) \end{cases}$$

and a nonnegative lower semicontinuous Lagrangian function

$$W : (x, q, v) \in X \times \mathcal{M} \times \mathcal{N} \mapsto W(x, q, v) \in \mathbf{R}_+$$

with which we associate the infinite horizon intertemporal discounted cost

$$\int_0^\infty e^{a\tau} W(x(\tau), p(\tau), v(\tau)) d\tau \in [0, +\infty]$$

We shall only adapt to this problem the definition 9.2.2 of discriminating property defined for subsets, leaving the extension of the leading property concepts as an exercise.

Definition 9.3.1 *We shall say that a function V enjoys the* discriminating monotonicity *— or* conditional monotonicity *— if for any continuous selection $\tilde{v}(x) \in \mathcal{Q}$ of Q played by "nature" and any initial state x, there exists a solution $x(\cdot)$ to the associated dynamical economy*

$$\begin{cases} i) & x'(t) = c(x(t), p(t), \tilde{v}(x(t))) \\ ii) & p(t) \in P(x(t)) \end{cases}$$

starting at x and monotone in the sense that

$$\forall\, t \geq 0,\;\; e^{at} V(x(t)) - V(x_0) + \int_0^t e^{a\tau} W(x(\tau), p(\tau), \tilde{v}(x(\tau))) d\tau \leq 0 \quad (9.3.1)$$

We associate with V the discriminating map $\Delta_V : X \rightsquigarrow \mathcal{M}$ defined on $(x, v) \in \text{Graph}(Q)$ by

$$\Delta_V(x, v) := \{p \in P(x) \mid D_\uparrow V(x)(c(x, p, v)) + W(x, q, v) + aV(x) \leq 0\}$$

Selections $\tilde{p} : (x, u) \mapsto \tilde{p}(x, u) \in \Delta_V(x, v)$ of the discriminating map are called discriminating strategies.

We shall say that V is a discriminating function if

$$\forall\, x \in X,\; \forall\, v \in Q(x),\; \Delta_V(x, v) \neq \emptyset \quad (9.3.2)$$

Theorem 9.3.2 *Let us assume that the dynamical game (P,Q,c) is Marchaud and that the set-valued map Q is lower semicontinuous. Assume also that*

$$W : (x,q,v) \in X \times \mathcal{M} \times \mathcal{N} \mapsto W(x,q,v) \in \mathbf{R}_+$$

is a nonnegative lower semicontinuous function convex with respect to q, that $V : X \mapsto \mathbf{R}_+ \cup \{+\infty\}$ is a nonnegative extended lower semicontinuous function (regarded as a loss function) and that there exists a positive constant γ such that

$$\begin{cases} i) & \forall (x,q,v) \in X \times \mathcal{M} \times \mathcal{N}, \quad W(x,q,v) \ \le \ \gamma(\|x\|+1) \\ ii) & \forall\, x \in \mathrm{Dom}(V), \quad \inf_{q\in P(x), v\in Q(x)} D_\uparrow V(x)(c(x,q,v)) \ \ge \ -\gamma(\|x\|+1) \end{cases} \tag{9.3.3}$$

The function V is discriminating if and only if it enjoys the discriminating monotonicity property. In this case, V is a contingent solution to the Hamilton-Jacobi-Isaacs inequalities

$$\inf_{p\in P(x)} \sup_{v\in Q(x)} \left(D_\uparrow V(x)(c(x,p,v)) + W(x,q,v)\right) + aV(x) \ \le \ 0 \tag{9.3.4}$$

Proof — We introduce the set-valued map $G : X\times\mathcal{M}\times\mathcal{N}\times\mathbf{R} \rightsquigarrow X\times\mathbf{R}$ defined by

$$G(x,w,p,v) \ := \ \{c(x,p,v)\} \times [aw - \gamma(\|x\|+1), aw - W(x,p,v)] \tag{9.3.5}$$

It is clear that the graph of G is closed and the values $G(x,w,P(x),v)$ are convex and nonempty by assumption (9.3.3). Its growth is linear by construction.

We observe that V is a discriminating function of the game (P,Q,c) for the Lagrangian W if and only if V is a contingent solution to the Hamilton-Jacobi-Isaacs inequalities (9.3.4): take $p \in P(x)$ achieving the minimum of the lower semicontinuous function $\sup_{v\in Q(x)}(D_\uparrow V(x)(c(x,\cdot),v)+W(x,\cdot,c)$ on the compact subset $P(x)$. It satisfies

$$\forall\, v \in Q(x), \quad D_\uparrow V(x)(c(x,p,v)) + W(x,p,v) + aV(x) \ \le \ 0$$

so that

$$\Delta_V(x,v) \ \neq \ \emptyset$$

Next, we observe that

$$\Delta_V(x,v) \ = \ \Delta_{\mathcal{E}p(V)}(x,V(x),v)$$

because the pairs $(c(x,q,v), -aV(x) - W(x,q,v))$ belong to the contingent cone to the epigraph of V at $(x,V(x))$ for every $v \in Q(x)$. By Lemma 8.1.2, they also belong to the contingent cone to the epigraph of V at (x,w) for every $w > V(x)$:

$$\forall\, w \ge V(x), \quad \Delta_V(x,v) \ \subset \ \Delta_{\mathcal{E}p(V)}(x,w,v)$$

In other words, we have shown that V is a discriminating function of the game (P, Q, c) if and only if the epigraph $\mathcal{E}p(V)$ of V is a discriminating domain under the dynamical game (P, Q, G).

Therefore Theorem 9.3.2 follows from Theorem 9.2.4. □

In the same way, the conditional viability kernel

$$[\mathrm{Inv}_v \mathrm{Viab}_p]_{(P,Q,G)}(\mathcal{E}p(V)) := \bigcap_{\tilde{v}\in Q} \mathrm{Viab}_{G(\cdot,P(\cdot),\tilde{v}(\cdot))}(\mathcal{E}p(V))$$

can be written

$$[\mathrm{Inv}_v \mathrm{Viab}_p]_{(P,Q,G)}(\mathcal{E}p(V)) := \bigcap_{\tilde{v}\in Q} (\mathcal{E}p(V_{\tilde{v},\propto}))$$

where $V_{\tilde{v},\propto}$ is the smallest lower semicontinuous function larger than or equal to V satisfying the monotonicity property (9.3.1) associated with the continuous selection $\tilde{v}$. Hence

$$[\mathrm{Inv}_v \mathrm{Viab}_p]_{(P,Q,G)}(\mathcal{E}p(V)) := \mathcal{E}p\left(\sup_{\tilde{v}\in Q} V_{\tilde{v},\propto}\right)$$

is the epigraph of the "**conditional monotone function**" associated with V.

In particular, when $V = 0$, then

$$\begin{cases} [\mathrm{Inv}_v \mathrm{Viab}_p]_{(P,Q,G)}(X \times \mathbf{R}_+) \\ \mathcal{E}p\left(\sup_{\tilde{v}\in Q} \inf_{x(\cdot))\in S_{\tilde{v}}(x)} \int_0^\infty e^{a\tau} W(x(\tau), p(\tau), \tilde{v}(x(\tau)))d\tau\right) \end{cases}$$

is the supremum of the value functions associated with the continuous selections $\tilde{v}$.

Since the conditional viability kernel is not a discriminating domain, the function $\sup_{\tilde{v}\in Q} V_{\tilde{v},\propto}$ is not iminating. Cardaliaguet's Theorem 9.2.9 implies the following result:

Theorem 9.3.3 *We posit the assumptions of Theorem 9.3.2. Starting with a nontrivial lower semicontinuous function $V^0 : X \mapsto \mathbf{R}_+ \cup \{+\infty\}$, we define by induction the functions V^j by*

$$V^{j+1} := \sup_{\tilde{v}\in Q} V^j_{\tilde{v},\propto}$$

Therefore, the lower semicontinuous function

$$V^\infty := \sup_{j\geq 0} V^j$$

is the smallest lower semicontinuous function larger than or equal to V satisfying the discriminating monotonicity property or equivalently, which is a solution to the Hamilton-Jacobi-Isaacs inequalities (9.3.4).

III

The Mathematical Tool Box

Chapter 10

Pseudo-Inverses and Tensor Products

Introduction

1. The Projection Theorem allows us to construct the orthogonal right-inverse of a linear surjective operator A, associating with any data y the solution x to the equation $Ax = y$ with minimal norm.

   In the same way, it allows us to construct the orthogonal left-inverse of a linear injective operator A, associating with any data y the solution x to the equation $Ax = \bar{y}$ where $\bar{y}$ is the orthogonal projection of y onto the image of A.

   More generally, when A is any linear operator between finite dimensional Hilbert spaces, the **pseudo-inverse** of A associates with any data y the solution x with minimal norm to the equation $Ax = \bar{y}$ where $\bar{y}$ is the orthogonal projection of y onto the image of A.

   These definitions show how useful the concept of pseudo-inverse is in many questions, and here, to compute the slow and heavy solutions of dynamical economies. It is used explicitly or implicitly in many domains of statistics and data analysis. It is then quite natural that they play an important role in the study of evolution of connection matrices.

2. Since we are looking for connection matrices when we deal with connectionist problems, we have to study the structure of the space of linear operators, of its dual and of a **tensor products** of linear operators $A^\star \otimes B$ of linear operators, which are the linear operators of the form

   $$W \mapsto BWA$$

   naturally involved in matrix equations $BWA = U$. Actually, these tensor products appear naturally in many formulas (such as the gradients of functionals defined on matrices) and are needed to compute the regulation maps of dynamical economies involving connection matrices.

3. The main result which will is used over and over is that **the pseudo-inverse of a tensor product is the tensor product of the pseudo-inverses.**

10.1 Pseudo-Inverses of Linear Operators

We denote by $(e^j)_{j=1,\ldots,n}$ a basis of X and by $(f^i)_{i=1,\ldots,m}$ a basis of Y.

Let $X^\star$ and $Y^\star$ denote the dual spaces of X and Y respectively. The bilinear form $\langle p, x\rangle := p(x)$ on $X^\star \times X$ is called the duality product. We associate with the basis $(e^j)_j$ its dual basis $(e_j^\star)_{j=1,\ldots,n}$ of $X^\star$ defined by $\langle e_j^\star, e^l\rangle = 0$ when $j \neq l$ and $\langle e_j^\star, e^j\rangle = 1$. Hence the jth component x_j of x in the basis $(e^j)_j$ is given by $x_j = \langle e_j^\star, x\rangle$, the jth component p^j of $p \in X^\star$ in the dual basis $(e_j^\star)_j$ is given by $p^j = \langle p, e^j\rangle$ and $\langle p, x\rangle := \sum_{j=1}^n p^j x_j$ and $\langle q, y\rangle := \sum_{i=1}^m q^i y_i$.

We recall that the transpose $W^\star \in \mathcal{L}(Y^\star, X^\star)$ is defined by

$$\forall\, q \in Y^\star,\ \forall\, x \in X,\ \ \langle W^\star q, x\rangle \ = \ \langle q, Wx\rangle$$

Let p belong to $X^\star$. We shall identify the linear form $p \in \mathcal{L}(X, \mathbf{R})$: $x \in X \mapsto \langle p, x\rangle$ with its transpose $p \in \mathcal{L}(\mathbf{R}, X^\star)$: $\lambda \mapsto \lambda p \in X^\star$. In the same way, we shall identify $x \in X$ with both the maps $p \in X^\star \mapsto \langle p, x\rangle$ and $\lambda \in \mathbf{R} \mapsto \lambda x \in X$.

We shall begin by defining and characterize the orthogonal right-inverse of a surjective linear operator and then, the orthogonal left-inverse of an injective linear operator. These concepts depend respectively upon scalar products l and m defined on the finite dimensional vector-spaces X and Y, which are then given once and for all. We shall denote by $L \in \mathcal{L}(X, X^\star)$ and $M \in \mathcal{L}(Y, Y^\star)$ their **duality mappings** defined respectively by

$$\begin{cases} \forall\, x_1,\ x_2 \in X,\ \langle Lx_1, x_2\rangle := l(x_1, x_2) \\ \\ \forall\, y_1,\ y_2 \in Y,\ \langle My_1, y_2\rangle := m(y_1, y_2) \end{cases}$$

(The matrix $(l(e^i, e^j)_{i,j=1,\ldots,n})$ of the bilinear form l coincides with the matrix of the linear operator L).

We shall denote by

$$\lambda(x) := \sqrt{l(x,x)} \ \text{ and } \ \mu(y) := \sqrt{m(y,y)}$$

the norms associated with these scalar products.

The bilinear form $l_\star(p_1, p_2) := \langle p_1, L^{-1}p_2\rangle = l(L^{-1}p_1, L^{-1}p_2)$ is then a scalar product on the dual of $X^\star$, called the **dual scalar product.**

It is quite important to keep options open for the choice of a scalar product. The choice of the Euclidian scalar product is not always wise despite its simplicity. We shall see later that we can associate with a scalar product and a linear operator initial and final scalar product.

Another instance of a choice of a scalar product is to make a given basis orthonormal.

Lemma 10.1.1 *Assume that $\{e^1, \dots, e^n\}$ be a basis of a finite dimensional vector-space X and $B \in \mathcal{L}(\mathbf{R}^n, X)$ the associated operator defined by*

$$\forall x \in \mathbf{R}^n, \quad Bx = \sum_{i=1}^{n} x_i e^i$$

Then

$$l(x,y) := \langle B^{-1}x, B^{-1}y \rangle$$

is a scalar product for which the basis $\{e^1, \dots, e^n\}$ is orthonormal. The associated duality mapping is equal to $L = (BB^\star)^{-1}$.

Proof — Indeed, to say that the sequence $\{e^1, \dots, e^n\}$ is linearly independent amounts to saying that the associated linear operator B is bijective and maps the canonical basis of $\mathbf{R}^n$ onto the basis $\{e^1, \dots, e^n\}$. Hence the bilinear form l defined above is a scalar product satisfying

$$l(e^i, e^j) := \langle B^{-1}e^i, B^{-1}e^j \rangle = \delta_{i,j} \quad \square$$

Let us consider a surjective linear operator $A \in \mathcal{L}(X, Y)$. Then, for any $y \in Y$, the problem $Ax = y$ has at least a solution. We may select the solution $\bar{x}$ with minimal norm $\lambda(x)$, i.e., a solution to the minimization problem with linear equality constraints

$$A\bar{x} = y \quad \& \quad \lambda(\bar{x}) = \min_{Ax=y} \lambda(x)$$

The solution to this problem is given by the formula

$$\bar{x} = L^{-1}A^\star(AL^{-1}A^\star)^{-1}y$$

Indeed, for any $v \in \operatorname{Ker}(A)$ and h, we have $\lambda(\bar{x}) \leq \lambda(\bar{x} + hv)$ so that, taking the limit when $h \longrightarrow 0$ of $(\lambda(\bar{x} + hv)^2 - \lambda(\bar{x})^2)/2h$, we infer that $l(\bar{x}, v) = \langle L\bar{x}, v \rangle = 0$ for any $v \in \operatorname{Ker}(A)$. This means that $L\bar{x}$ belongs to the orthogonal of $\operatorname{Ker}(A)$, which is the image of $A^\star$. Thus there exists some $q \in Y^\star$ such that $L\bar{x} = A^\star q$ so that $A\bar{x} = AL^{-1}A^\star q = y$. But $AL^{-1}A^\star$ is positive-definite because $A^\star$ is injective (A being surjective): Indeed, for all $q \in Y^\star$,

$$\langle AL^{-1}A^\star q, q \rangle = \langle L^{-1}A^\star q, A^\star q \rangle = \lambda_\star(A^\star q)^2 = 0$$

is equivalent to

$$A^\star q = 0 \iff q = 0$$

We thus infer that $q = (AL^{-1}A^\star)^{-1}y$.

Definition 10.1.2 *If $A \in \mathcal{L}(X,Y)$ is surjective, we say that the linear operator $A^+ := L^{-1}A^\star(AL^{-1}A^\star)^{-1} \in \mathcal{L}(Y,X)$ is the* orthogonal right-inverse *of A (associated with the scalar product l on X).*

Indeed, A^+ is obviously an right-inverse of A because $AA^+y = y$ for any $y \in Y$. We observe that $1 - A^+A$ is the orthogonal projector onto the kernel of A.

10.1.1 Projections onto Inverse Images of Convex Sets

Hence, we can write explicitly the solution $\bar{x}$ to the quadratic optimization problem with linear equality constraints

$$A\bar{x} = y \ \ \& \ \ \frac{1}{2}\lambda(\bar{x} - u)^2 = \min_{Ax=y} \frac{1}{2}\lambda(x-u)^2$$

Proposition 10.1.3 *Let us assume that $A \in \mathcal{L}(X,Y)$ is surjective. Then the unique solution $\bar{x}$ to the above quadratic minimization problem is given by*

$$\begin{cases} \bar{x} \ = \ u - L^{-1}A^\star\bar{q} = u - A^+(Au - y) \\ \\ \text{where } \ \bar{q} := (AL^{-1}A^\star)^{-1}(Au - y) \ \text{ is the Lagrange multiplier} \end{cases}$$

i.e., a solution to the dual minimization problem

$$\min_{q \in Y^\star} \left(\frac{1}{2}\lambda_\star(A^\star q - Lu)^2 + \langle q, y\rangle \right)$$

We observe easily that $\xi(q) := u - L^{-1}A^\star(q)$ minimizes over X the function $x \mapsto \frac{1}{2}\lambda(x-u)^2 + \langle q, Ax\rangle$ and that the Lagrange multiplier minimizes $q \mapsto \frac{1}{2}\lambda_\star(\xi(q))^2 + \langle q, y\rangle$. Consider now a closed convex subset $M \subset Y$ and the minimization problem and elements $u \in X$ and $v \in Y$:

$$\inf_{Ax \in M+v} \lambda(x-u)$$

When $A \in \mathcal{L}(X,Y)$ is surjective, we can supply the space Y with the final scalar product

$$m^A(y_1, y_2) \ := \ l(A^+y_1, A^+y_2)$$

its associated final norm[1]

$$\mu^A(y) \ := \ \lambda(A^+y) \ = \ \inf_{Ax=y} \lambda(x)$$

[1]which may also be denoted by $\mu^A := A \cdot \mu$ by contrast with the initial norm $\lambda \cdot A$. We observe that $A \cdot \mu = \lambda \cdot A^+$.

Its duality mapping is equal to $(AL^{-1}A^\star)^{-1} \in \mathcal{L}(Y, Y^\star)$ and its dual norm is equal to

$$\mu_\star^A(p) = \lambda_\star(A^\star p)$$

In this case, we denote by π_M^A the projector of best approximation onto the closed convex subset M when Y is supplied with the norm μ^A and by $\pi_{M^-}^{A^\star}$ the projector of best approximation onto the polar cone $M^- := \{q \in Y^\star \mid \forall\, y \in M,\ \langle q, y\rangle \leq 0\}$ when $Y^\star$ is supplied with the norm $\mu_\star^A$.

Proposition 10.1.4 *Assume that $A \in \mathcal{L}(X,Y)$ is surjective and that $M \subset Y$ is closed and convex. Let $u \in X$ and $v \in Y$ be given. Then the unique solution $\bar{x}$ to the minimization problem*

$$\inf_{Ax \in M+v} \lambda(x-u)$$

is equal to

$$\bar{x} = u - A^+(\mathbf{1} - \pi_M^A)(Au - v)$$

When $M \subset Y$ is a closed convex cone, the solution can also be written in the form

$$\bar{x} = u - L^{-1}A^\star \pi_{M^-}^{A^\star}(AL^{-1}A^\star)^{-1}(Au - v)$$

Proof — Indeed, we can write

$$\begin{cases} \lambda(\bar{x}-u) = \inf_{Ax\in M+v} \lambda(x-u) \\ = \inf_{y\in M+v} \inf_{Ax=y} \lambda(x-u) \\ = \inf_{y\in M+v} \lambda(A^+(y - Au)) \\ \text{(thanks to Proposition 10.1.3)} \\ = \inf_{z\in M} \lambda(A^+(z - (Au - v))) \\ = \lambda(A^+(\pi_M^A(Au-v)) - (Au-v)) \end{cases}$$

Hence $\bar{x} = u - A^+(\mathbf{1} - \pi_M^A)(Au - v)$.

When M is a closed convex cone, the second statement follows from the following

Lemma 10.1.5 *Let $M \subset Y$ be a closed convex cone, $w \in Y$. If $A \in \mathcal{L}(X,Y)$ is surjective and if Y is supplied with the final norm m^A, then*

$$\left(\mathbf{1} - \pi_M^A\right) w = (AL^{-1}A^\star)\pi_{M^-}^{A^\star}((AL^{-1}A^\star)^{-1}w)$$

and thus

$$A^+(w - \pi_M^A w) = L^{-1}A^\star \pi_{M^-}^{A^\star}(AL^{-1}A^\star)^{-1}w$$

Proof — Let us denote by $K := (AL^{-1}A^\star)^{-1}$ the duality map associate with the final scalar product m^A. The projection $\pi_M^A w$ is characterized by

$$\forall\, y \in M,\ \ m^A(w - \pi_M^A w, y - \pi_M^A w) \ = \ m_\star^A(K(w - \pi_M^A w), Ky - K\pi_M^A w) \ \leq \ 0$$

When M is a closed convex cone, we can take successively $y = 0$ and $y = 2\pi_M^A w$ for deducing that this characterization is equivalent to

$$\begin{cases} m^A(w - \pi_M^A w, \pi_M^A w) \ = \ m_\star^A(K(w - \pi_M^A w), K\pi_M^A w) \ = \ 0 \\ \forall\, y \in M,\ \ m^A(w - \pi_M^A w, y) \ = \ \langle K(w - \pi_M^A w), y\rangle \ \leq \ 0 \end{cases}$$

The second statement means that

$$K(w - \pi_M^A w) \ \in \ M^-$$

so that the first statement means that $K(w - \pi_M^A w)$ is the projection of Kw onto M^- when the dual $Y^\star$ is supplied with the dual final scalar product $m_\star^A$ since the negative polar cone M^- is a closed convex cone. Hence

$$K(w - \pi_M^A w) \ = \ \pi_{M^-}^{A^\star} Kw$$

Consequently,

$$A^+(w - \pi_M^A w) = L^{-1}A^\star(AL^{-1}A^\star)^{-1}(w - \pi_{M^-}^A w) = L^{-1}A^\star\pi_{M^-}^{A^\star}(AL^{-1}A^\star)^{-1}w \ \square$$

10.1.2 Orthogonal Left-Inverses

Let us consider now an injective linear operator $B \in \mathcal{L}(X, Y)$. Since the problem $Bx = y$ may not have a solution (when y does not belong to the image of B), we shall project y onto this image and take for an approximated solution the inverse of this projection. In other words, the approximated solution to this problem is the element $\bar{x} \in X$ defined by

$$\mu(B\bar{x} - y) \ = \ \min_{x \in X} \mu(Bx - y)$$

Since the derivative of the convex function $\frac{1}{2}\mu(\cdot)^2$ at y is equal to $\langle My, \cdot\rangle$, the Fermat rule states that $\bar{x}$ is a solution to the equation

$$\forall\, v \in X,\ \ \langle M(B\bar{x} - y), Bv\rangle = \langle B^\star M(B\bar{x} - y), v\rangle = 0$$

The self-transposed operator $B^\star MB$ is positive-definite for B is injective (because $\langle B^\star MBx, x\rangle = \mu(Bx)^2 = 0$ if and only if $Bx = 0$, i.e., if and only if $x = 0$). Therefore $B^\star MB$ is invertible and we derive that $\bar{x}$ is equal to $(B^\star MB)^{-1}B^\star My$.

Definition 10.1.6 *If $B \in \mathcal{L}(X, Y)$ is injective, we say that the linear operator $B^- := (B^\star MB)^{-1}B^\star M \in \mathcal{L}(Y, X)$ is the* orthogonal left-inverse *of B.*

Indeed, B^- is obviously a left-inverse of B because $B^-Bx = x$ for any $x \in X$ and BB^- is the orthogonal projector onto the image of B.

Proposition 10.1.7 *Let us assume that $B \in \mathcal{L}(X,Y)$ is injective. Then*

$$\begin{cases} i) & (B^-)^\star = (B^\star)^+ \\ ii) & B = (B^-)^+ \\ iii) & (B^\star MB)^{-1} = B^-M^{-1}(B^-)^\star \\ iv) & \text{If } V \in \mathcal{L}(Z,X) \text{ is invertible, then } (BV)^- = V^{-1}B^- \end{cases}$$

If $A \in \mathcal{L}(X,Y)$ is surjective, then

$$\begin{cases} i) & (A^+)^\star = (A^\star)^- \\ ii) & A = (A^+)^- \\ iii) & (AL^{-1}A^\star)^{-1} = (A^+)^\star LA^+ \\ iv) & \text{If } W \in \mathcal{L}(Y,Z) \text{ is invertible, then} (WA)^+ = A^+W^{-1} \end{cases}$$

Example Orthogonal Projector

Let $P := [b^1, \ldots, b^n]$ be the vector space spanned by n linearly independent vectors b^l of a finite dimensional vector-space Y, supplied with the scalar product $m(y_1, y_2)$. If we denote by $B \in \mathcal{L}(\mathbf{R}^n, Y)$ the linear operator defined by $Bx := \sum_{i=1}^n x_i b^i$, which is injective because the vectors b^i are independent, we infer that BB^- is the orthogonal projector on the subspace $P = \operatorname{Im}(B)$.

The entries of the matrix of $B^\star MB$ are equal to $m(b^i, b^j)$ $(i,j = 1, \ldots, n)$. Let us denote by g_{ij} the entries of its inverse.

We infer that $B^-y = (\sum_{j=1}^n g_{ij} m(y, b^j))_{i=1,\ldots,n}$ and therefore, that

$$BB^-y = \sum_{i,j=1}^n g_{ij} m(y, b^j) b^i$$

When the vectors are m-orthogonal, the formula becomes

$$BB^-y = \sum_{i=1}^n \frac{m(b^i, y)}{\mu(b^i)^2} b^i \quad \square$$

In particular, we can regard an element $y \in Y$ different from 0 as the injective linear operator $y \in \mathcal{L}(\mathbf{R}, Y)$ associating with α the vector αy. It transpose $y^\star \in \mathcal{L}(Y^\star, \mathbf{R}) = Y$ is the map $p \mapsto \langle p, y\rangle$ and its left-inverse $y^- \in \mathcal{L}(Y, \mathbf{R}) = Y^\star$ is equal to

$$y^- = \frac{My}{\lambda(y)^2} \in Y^\star$$

10.1.3 Pseudo-inverses

Let us consider now any linear operator $C \in \mathcal{L}(X,Y)$. It can be split as the product $C = BA$ of a surjective linear operator $A \in \mathcal{L}(X,Z)$ and an injective linear operator $B \in \mathcal{L}(Z,Y)$: we can take for instance $Z := \operatorname{Im}(C)$, $A := C \in \mathcal{L}(X,Z)$ and $B := I \in \mathcal{L}(Z,Y)$.

Take now two decompositions $C = B_1A_1 = B_2A_2$ where $A_i \in \mathcal{L}(X,Z_i)$ is surjective and where $B_i \in \mathcal{L}(Z_i,Y)$ is injective $(i = 1,2)$. Then the images of the B_i's are equal to $\operatorname{Im}(C)$ and the kernels of the A_i's are also equal to a vector subspace $\ker(C) \subset X$. Let $\varphi \in \mathcal{L}(X, X/\ker(C))$ denote the canonical surjection from X to the factor space $X/\ker(C)$. Hence the surjective operators A_i split as $\tilde{A}_i\varphi$ where $\tilde{A}_i \in \mathcal{L}(X/\ker(C), Z_i)$ are invertible $(i = 1,2)$. Then $V := \tilde{A}_1\tilde{A}_2^{-1}$ is invertible and we have the relations

$$A_1 = VA_2 \ \& \ B_2 = B_1V$$

We thus deduce that $A_1^+B_1^- = A_2^+B_2^-$ does not depend upon the decomposition of C as a product of an injective operator and an surjective operator because $A_1^+B_1^- = A_2^+V^{-1}VB_2^- = A_2^+B_2^-$.

Definition 10.1.8 *Let $C = BA \in \mathcal{L}(X,Y)$ be any linear operator split as the product of an injective operator B and a surjective operator A. Then the operator $C^{\ominus 1} := A^+B^- \in \mathcal{L}(Y,X)$ is called the* pseudo-inverse *of C.*

If $y \in Y$ is given, we can regard $\bar{x} = C^{\ominus 1}y$ as the closest solution with minimal norm. Indeed, by taking $A = C$ and $B = I$, we see that $\bar{x} = A^+\bar{y}$ where $\bar{y} := I^-(y)$ is the projection of y onto the image of C and that $\bar{x}$ is the solution with minimal norm to the equation $Cx = \bar{y}$. □

Naturally, if C is surjective, the pseudo-inverse $C^{\ominus 1} = C^+$ coincides with the orthogonal right-inverse, if C is injective, the pseudo-inverse $C^{\ominus 1} = C^-$ coincides with the orthogonal left-inverse and if C is bijective, the pseudo-inverse $C^{\ominus 1} = C^{-1}$ coincides with the inverse.

We list below the (obvious) properties of the pseudo-inverses.

$$\begin{cases} i) & C^{\ominus 1}CC^{\ominus 1} = C^{\ominus 1} \\ ii) & CC^{\ominus 1}C = C \\ iii) & C^{\ominus 1}C \text{ is the orthogonal projector onto the orthogonal of } \operatorname{Ker}(C) \\ iii) & CC^{\ominus 1} \text{ is the orthogonal projector onto } \operatorname{Im}(C) \end{cases}$$

We also observe that

$$\begin{cases} i) & (C^{\ominus 1})^{\ominus 1} = C \\ ii) & (C^{\ominus 1})^{\star} = (C^{\star})^{\ominus 1} \\ iii) & (C_2C_1)^{\ominus 1} = C_1^{\ominus 1}C_2^{\ominus 1} \end{cases}$$

10.2 Tensor Products of Linear Operators

Let us introduce the following notations: We associate with any $p \in X^\star$ and $y \in Y$ the linear operator $p \otimes y \in \mathcal{L}(X,Y)$ defined by

$$p \otimes y \ : x \ \mapsto \ (p \otimes y)(x) \ := \ \langle p, x\rangle y$$

the matrix of which is

$$\left(p^i y_j\right)_{\substack{i=1,\dots,m \\ j=1,\dots,n}}$$

Its transpose $(p \otimes y)^\star \in \mathcal{L}(Y^\star, X^\star)$ maps $q \in Y^\star$ to

$$(p \otimes y)^\star(q) = \langle q, y\rangle p$$

because $\langle q, \langle p, x\rangle y\rangle = \langle \langle q, y\rangle p, x\rangle$. Hence, we shall identify from now on the transpose $(p \otimes y)^\star$ with $y \otimes p$.

In the same way, if $q \in Y^\star$ and $x \in X$ are given, $q \otimes x$ denotes the map $y \in Y \mapsto \langle q, y\rangle x \in X$ belonging to $\mathcal{L}(Y, X)$ and its transpose $p \in X^\star \mapsto \langle p, x\rangle q \in Y^\star$ belonging to $\mathcal{L}(X^\star, Y^\star)$ is identified with $x \otimes q$.

We observe that the map $(p, y) \in X^\star \times Y \mapsto p \otimes y \in \mathcal{L}(X, Y)$ is bilinear.

The operators $(e_j^\star \otimes f^i)_{\substack{i=1,\dots,m \\ j=1,\dots,n}}$ do form a basis of $\mathcal{L}(X, Y)$ because we can write $x = \sum_{j=1}^n \langle e_j^\star, x\rangle e^j$ and

$$\begin{aligned} Wx &= \textstyle\sum_{i=1}^m \langle f_i^\star, Wx\rangle f^i = \sum_{i=1}^m \sum_{j=1}^n \langle f_i^\star, W(e^j)\rangle \langle e_j^\star, x\rangle f^i \\ &= \left(\textstyle\sum_{i=1}^m \sum_{j=1}^n \langle f_i^\star, W(e^j)\rangle e_j^\star \otimes f^i\right)(x) \end{aligned}$$

where the entries

$$w_i^j := \langle f_i^\star, W(e^j)\rangle$$

of the matrix W are the components of W in this basis.

It is useful to set

$$X^\star \otimes Y \ := \ \mathcal{L}(X, Y)$$

to recall this structure.

Definition 10.2.1 *Let us consider two pairs (X, X_1) and (Y, Y_1) of finite dimensional vector-spaces. Let $A \in \mathcal{L}(X_1, X)$ and $B \in \mathcal{L}(Y, Y_1)$ be given. We denote by $A^\star \otimes B$ the linear operator from $\mathcal{L}(X, Y)$ to $\mathcal{L}(X_1, Y_1)$ defined by*

$$\forall\, W \in \mathcal{L}(X, Y), \ (A^\star \otimes B)(W) := BWA$$

and call it the **tensor product** *of $A^\star$ by B.*

We observe that when $W = p \otimes y$, we have

$$(A^\star \otimes B)(p \otimes y) = A^\star p \otimes By$$

In particular, $x \otimes B \in \mathcal{L}(\mathcal{L}(X, Y), Z)$ maps W to BWx.

We also observe that

$$(A^\star \otimes B)^\star \ = \ A \otimes B^\star \tag{10.2.1}$$

It is enough to check that

$$\begin{cases} \langle (A^\star \otimes B)^\star (x_1 \otimes q_1), p \otimes y\rangle \ = \ \langle (x_1 \otimes q_1), (A^\star \otimes B))p \otimes y)\rangle \\ = \ \langle (x_1 \otimes q_1), A^\star p \otimes By)\rangle \ = \ \langle A^\star p, x_1, \rangle \langle q_1, By\rangle \\ = \ \langle p, Ax_1, \rangle\langle B^\star q, x_1\rangle \ = \ \langle (A \otimes B^\star)(x_1 \otimes q_1), p \otimes y\rangle \end{cases}$$

For example, the transpose of $x \otimes B$ is equal to $(x \otimes B)^\star = x \otimes B^\star$. It maps any $q \in Z^\star$ to the linear operator $x \otimes B^\star q \in \mathcal{L}(X^\star, Y^\star)$.

Let $A_1 \in \mathcal{L}(X_2, X_1)$ and $B_1 \in \mathcal{L}(Y_1, Y_2)$. Then, it is easy to check that

$$(A_1^\star \otimes B_1)(A^\star \otimes B) \ = \ (A_1^\star A^\star) \otimes (B_1 B)$$

Then the following formulas are straightforward:

$$\begin{cases} i) & \text{If } A \text{ and } B \text{ are invertible, then} \\ & (A^\star \otimes B)^{-1} \ = \ (A^{-1})^\star \otimes B^{-1} \\ ii) & \text{If } A \text{ is left-invertible and } B \text{ is right-invertible,} \\ & \text{then } A^\star \otimes B \text{ is right-invertible} \\ iii & \text{If } A \text{ is right-invertible and } B \text{ is left-invertible,} \\ & \text{then } A^\star \otimes B \text{ is left-invertible} \\ iv) & \text{If } A \text{ and } B \text{ are projectors, so is } A^\star \otimes B \end{cases}$$

Example: Gradients of Functionals on Linear Operators

Let us consider three spaces X, Y and Z, an element $x \in X$, a differential map $g : Y \mapsto Z$.

Proposition 10.2.2 *The derivative of the map $H : W \mapsto H(W) := g(Wx)$ at $W \in \mathcal{L}(X,Y)$ is defined by*

$$H'(W) \;=\; x \otimes g'(Wx)$$

let E be a differentiable functional from Z to $\mathbf{R}$. *We set $\Psi(W) := E(g(Wx))$. The gradient of Ψ at $W \in \mathcal{L}(X,Y)$ is defined by*

$$\Psi'(W) \;=\; x \otimes g'(Wx)^{\star} E'(g(Wx)) \in \mathcal{L}(X^{\star}, Y^{\star})$$

Proof — Indeed,

$$\begin{cases} \lim_{\lambda \to 0} \dfrac{H(W+\lambda U) - H(W)}{\lambda} \\ = \lim_{\lambda \to 0} \dfrac{g(Wx + \lambda U x) - g(Wx)}{\lambda} \\ = g'(Wx)Ux \;=\; (x \otimes g'(Wx))(U) \;\square \end{cases}$$

The chain rule implies:

$$\Psi'(W)(U) \;=\; \langle E'(g(Wx)), g'(Wx))Ux\rangle \;=\; \langle g'(Wx))^{\star} E'(g(Wx)), Ux\rangle \;\square$$

10.3 Pseudo-Inverses of Tensor Products

10.3.1 Scalar Products on $\mathcal{L}(X,Y)$

If we supply the spaces X and Y with scalar products $l(\cdot,\cdot)$ and $m(\cdot,\cdot)$, we shall denote by $L \in \mathcal{L}(X, X^{\star})$ and $M \in \mathcal{L}(Y, Y^{\star})$ their duality mappings and by

$$\lambda(x) := \sqrt{l(x,x)} \text{ and } \mu(y) := \sqrt{m(y,y)}$$

the norms associated with these scalar products.

It is clear that if $(e^j)_{j=1,\dots,n}$ is a l- orthonormal basis of X (i.e., satisfying $l(e^j, e^l) = 0$ when $j \neq l$ and $l(e^j, e^j\rangle = 1)$, then the basis $(e_j^{\star} := Le^j)_j$ is the dual basis of $(e^j)_j$ because $\langle e_j^{\star}, e^l\rangle = l(e^j, e^l)$. The matrix of the duality mapping L in these basis is the identity matrix.

Let us consider now an l-orthonormal basis (e^j) of X and a m-orthonormal basis (f^i) of Y. Let U and V be two operators of $\mathcal{L}(X,Y)$. We observe that

$$\sum_{j=1}^{n} m(Ue^j, Ve^j) = \sum_{i=1}^{m} l_{\star}(U^{\star} f_i^{\star}, V^{\star} f_i^{\star}) = \sum_{i,j} \langle f_i^{\star}, Ue^j\rangle \langle f_i^{\star}, Ve^j\rangle$$

so that this expression is independent of the choice of the l- and m-orthonormal basis of X and Y.

We thus associate with the two scalar products l and m the scalar product $l_\star \otimes m$ on $\mathcal{L}(X,Y)$ defined by

$$\begin{aligned}(l_\star \otimes m)(U,V) &:= \sum_{i,j} \langle f_i^\star, Ue^j\rangle\langle f_i^\star, Ve^j\rangle \\ &= \sum_{j=1}^n m(Ue^j, Ve^j) = \sum_{i=1}^m l_\star(U^\star f_i^\star, V^\star f_i^\star)\end{aligned}$$

For matrices $p_i \otimes y_i$, this formula becomes

$$(l_\star \otimes m)(p_1 \otimes y_1, p_2 \otimes y_2) = l_\star(p_1,p_2)m(y_1,y_2)$$

We observe that the duality mapping from $\mathcal{L}(X,Y)$ to $\mathcal{L}(X^\star, Y^\star)$ of the scalar product $l_\star \otimes m$ is $L^{-1} \otimes M$ because

$$\begin{aligned}(l_\star \otimes m)(U,V) &= \sum_{i,j} \langle f_i^\star, Ue^j\rangle\langle f_i^\star, Ve^j\rangle \\ &= \sum_{i,j} \langle f^i, MUL^{-1}e_j^\star\rangle\langle f_i^\star, Ve^j\rangle \\ &= \langle MUL^{-1}, V\rangle = \langle (L^{-1} \otimes M)U, V\rangle\end{aligned}$$

If we consider the matrices (u_i^j) and (v_i^j) of the operators U and V for the orthonormal basis $(e^j)_j$ and $(f^i)_i$, then

$$(l_\star \otimes m)(U,V) = \sum_{i=1}^m \sum_{j=1}^n u_i^j v_i^j$$

We shall denote by

$$(\lambda_\star \otimes \mu)(U) := \sqrt{(l_\star \otimes m)(U,U)}$$

the norm associated with this scalar product.

10.3.2 Pseudo-Inverse of $A \otimes B$

We begin by characterizing the orthogonal right-inverse of a tensor product:

Proposition 10.3.1 *Let us assume that $A \in \mathcal{L}(X_1, X)$ is injective and that $B \in \mathcal{L}(Y, Y_1)$ is surjective. Then $A^\star \otimes B$ is surjective and its orthogonal right-inverse is given by the formula:*

$$(A^\star \otimes B)^+ = (A^\star)^+ \otimes B^+$$

Then $\overline{W} := B^+ W_1 A^-$ is the solution to the operator equation $BWA = W_1$ with minimal norm (for the norm $\lambda_\star \otimes \mu$ on $\mathcal{L}(X,Y)$).

Proof — In order to prove that $(A^\star \otimes B)^+ = (A^\star)^+ \otimes B^+$, we apply the formula for the orthogonal right-inverse: Let us set $\mathcal{A} := A^\star \otimes B$ and recall that the duality map $J \in \mathcal{L}(\mathcal{L}(X,Y), \mathcal{L}(X^\star, Y^\star))$ is equal to $L^{-1} \otimes M$, so that its inverse is equal to $J^{-1} = L \otimes M^{-1}$.

Hence

$$\mathcal{A}^+ = J^{-1}\mathcal{A}^\star \left(\mathcal{A}J^{-1}\mathcal{A}^\star\right)^{-1}$$

But we see that

$$\mathcal{A}J^{-1}\mathcal{A}^\star = A^\star LA \otimes BM^{-1}B^\star$$

so that

$$\mathcal{A}^+ = LA(A^\star LA)^{-1} \otimes M^{-1}B(BM^{-1}B^\star)^{-1}$$

It is enough to observe that $M^{-1}B(BM^{-1}B^\star)^{-1} = B^+$ and that

$$LA(A^\star LA)^{-1} = (A^\star)^+ = \left((A^\star LA)^{-1}A^\star L\right)^\star = (A^-)^\star$$

so that the formula holds true. □

As a corollary, we obtain the following result:

Corollary 10.3.2 *Let us assume that $A \in \mathcal{L}(X_1, X)$ is injective. Then $\overline{W} := W_1 A^-$ is the solution to the operator equation $WA = W_1$ with minimal norm (for the norm $\lambda_\star \otimes \mu$ on $\mathcal{L}(X,Y)$). This solution does not depend on the choice of the scalar product m on Y.*

In particular, we shall use quite often the following consequence when we take $X_1 := \mathbf{R}$ and $A := x \in \mathcal{L}(\mathbf{R}, X)$:

Corollary 10.3.3 *Let us assume $B \in \mathcal{L}(Y, Y_1)$ is surjective and take $x \neq 0$. Then $\overline{W} := Lx/\lambda(x)^2 \otimes B^+ y$ is the solution to the operator equation $BWx = y$ with minimal norm (for the norm $\lambda_\star \otimes \mu$ on $\mathcal{L}(X,Y)$).*

By transposition, we obtain a formula for the orthogonal left-inverse of a tensor product of linear operators:

Proposition 10.3.4 *Let us assume that $A \in \mathcal{L}(X_1, X)$ is surjective and that $B \in \mathcal{L}(Y, Y_1)$ is injective . Then $A^\star \otimes B$ is injective and its orthogonal left-inverse is given by the formula:*

$$(A^\star \otimes B)^- = (A^\star)^- \otimes B^-$$

Proof — Let us set $\mathcal{A} := A^\star \otimes B$, the transpose of which is

$$\mathcal{A}^\star = A \otimes B^\star$$

Since the transpose is surjective, we deduce from the preceding proposition that

$$(\mathcal{A}^\star)^+ = A^+ \otimes (B^\star)^+ = A^+ \otimes (B^-)^\star$$

Therefore, by taking the transpose and using the fact that the transpose of the orthogonal right-inverse is the orthogonal left-inverse of the transpose, we infer that

$$\mathcal{A}^{-} = (A^{+})^{\star} \otimes B^{-} = (A^{\star})^{-} \otimes B^{-} \quad \square$$

The definition and properties of pseudo-inverses imply the following formulas.

Theorem 10.3.5 *Let $C \in \mathcal{L}(X_1, X)$ and $D \in \mathcal{L}(Y, Y_1)$ be given. Then the pseudo-inverse of $C^\star \otimes D$ is equal to $(C^{\ominus 1})^\star \otimes D^{\ominus 1}$. In particular, if $W_1 \in \mathcal{L}(X_1, Y_1)$ is given, then the linear operator $\overline{W} := D^{\ominus 1} W_1 C^{\ominus 1} \in \mathcal{L}(X, Y)$ is the closest solution with minimal norm to the operator equation $DWC = W_1$.*

Proof— We split the operators $C = B_C A_C$ and $D = B_D A_D$ as products of injective and surjective linear operators, so that $C^\star \otimes D = (A_C^\star B_C^\star) \otimes B_D A_D$ is the product $(A_C^\star \otimes B_D)(B_C^\star \otimes A_D)$ of the injective operator $A_C^\star \otimes B_D$ and the surjective operator $B_C^\star \otimes A_D$. Hence the pseudo-inverse of $C^\star \otimes D$ is the product $(B_C^\star \otimes A_D)^+ (A_C^\star \otimes B_D)^-$, which is equal to

$$\begin{cases} (B_C^\star \otimes A_D)^+ (A_C^\star \otimes B_D)^- = ((B_C^-)^\star \otimes A_D^+)((A_C^+)^\star \otimes B_D^-) \\ = (B_C^{-\star} A_C^{+\star}) \otimes A_D^+ B_D^- = (A_C^+ B_C^-)^\star \otimes A_D^+ B_D^- \\ = (C^{\ominus 1})^\star \otimes (D^{\ominus 1}) \quad \square \end{cases}$$

10.3.3 Orthogonal Projection on Subsets of Matrices

Corollary 10.3.3 and Proposition 10.1.4 imply the following formula for the optimal solution to the quadratic minimization problem under constraints

$$(\lambda_\star \otimes \mu)(\overline{W} - U) \;=\; \inf_{BWx \in K+v} (\lambda_\star \otimes \mu)(W - U)$$

where $U \in \mathcal{L}(X, Y)$ is given and where $K \subset Z$ is a closed convex subset. Recall that when $B \in \mathcal{L}(Y, Z)$ is surjective, we can supply Z with the final norm μ^B defined by $\mu^B(z) := \mu(B^+ z)$ and that we denote by π_K^B the projector of best approximation onto the closed subset K for this final norm.

Theorem 10.3.6 *Let $B \in \mathcal{L}(Y, Z)$ be a surjective operator, $K \subset Z$ be a closed convex subset and $x \in X$ be different from 0. Then the matrix*

$$\overline{W} \;:=\; U - \frac{Lx}{\lambda(x)^2} \otimes B^+ \left(\mathbf{1} - \pi_K^B\right)(BUx - v)$$

is the best approximation of U in the subset of linear operators W satisfying the constraints

$$BWx \;\in\; K + v$$

If $K \subset Z$ is a closed convex cone, then the optimal matrix $\overline{W}$ can also be written in the form

$$\overline{W} \;:=\; U - \frac{Lx}{\lambda(x)^2} \otimes M^{-1} B^\star \pi_{K^-}^{B^\star} (BM^{-1}B^\star)^{-1} (BUx - v)$$

Proof — By Corollary 10.3.3, we know that the right inverse of the map $W \mapsto BWx$ is equal to the map $z \mapsto Lx/\lambda(x)^2 \otimes B^+z$. Therefore, the final norm on the vector space Z associated with this operator is equal to $\mu(B^+z)\lambda(x)$. Consequently, the orthogonal projector onto K associated with this norm is equal to the projector π_K^B associated with the final norm $\mu(B^+z)$.

Consequently, by Proposition 10.1.4, we infer that the solution to

$$(\lambda_\star \otimes \mu)(\overline{W} - U) = \inf_{BWx \in K+w} (\lambda_\star \otimes \mu)(W - U)$$

is equal to $U - (x \otimes B)^+(\mathbf{1} - \pi_K^B)(BUx - v)$. □

Theorem 10.3.7 *Let $B \in \mathcal{L}(Y,Z)$ be a surjective operator, $K \subset Z$ be a closed convex subset and $x \in X$ be different from 0. Let us supply the product space $\mathcal{L}(Y,Z) \times Y$ with the norm $\nu_{\alpha,\beta}$ defined by*

$$\nu_{\alpha,\beta}(W,v)^2 := \alpha(\lambda_\star \otimes \mu)(W)^2 + \beta\mu(w)^2$$

Then the pair $(\overline{W}, \overline{w})$ defined by

$$\begin{cases} \overline{W} = U - \dfrac{\beta Lx}{\alpha + \beta\lambda(x)^2} \otimes B^+((\mathbf{1} - \pi_K^B)(BUx - v) \\ \\ \overline{w} = \dfrac{\alpha}{\alpha + \beta\lambda(x)^2} B^+((\mathbf{1} - \pi_K^B)(BUx - v) \end{cases}$$

is the best approximation of (U,u) in the subset of pairs (W,w) satisfying the constraints

$$BWx - Bw \in K + v$$

Proof — We have to compute the orthogonal right inverse of the linear operator $A \in \mathcal{L}(\mathcal{L}(X,Y) \times Y, Z)$ defined by

$$A(W,w) := BWx - Bw$$

the transpose of which is defined by

$$A^\star q = (x \otimes B^\star q, -B^\star q)$$

Since the duality map is equal to $J_{\alpha,\beta} = \alpha(L^{-1} \otimes M) \times \beta M$, we infer that

$$AJ_{\alpha,\beta}^{-1}A^\star q = \left(\frac{\langle Lx, x\rangle}{\alpha} + \frac{1}{\beta}\right) BM^{-1}B^\star q$$

and that

$$(AJ_{\alpha,\beta}^{-1}A^\star)^{-1}z = \frac{\alpha\beta}{\alpha + \beta\lambda(x)^2}(BM^{-1}B^\star)^{-1}z$$

Therefore, the orthogonal right-inverse is equal to

$$A^+z = J_{\alpha,\beta}^{-1}A^\star(AJ_{\alpha,\beta}^{-1}A^\star)^{-1}z = \left(\frac{\beta Lx}{\alpha+\beta\lambda(x)^2}\otimes B^+z, -\frac{\alpha}{\alpha+\beta\lambda(x)^2}B^+z\right)$$

Now, we supply Z with the final norm $\nu_{\alpha,\beta}^A$ equal to

$$\nu_{\alpha,\beta}^A(z)^2 = \nu_{\alpha,\beta}(A^+z) = \left(\alpha(\lambda_\star\otimes\mu)^2+\beta\mu^2\right)(A^+z)$$

In other words, the final norm

$$\nu_{\alpha,\beta}^A(z) = \sqrt{\frac{\alpha\beta}{\alpha+\beta\lambda(x)^2}}\mu(B^+z)$$

is proportional to the final norm $\mu^B(z) := \mu(B^+z)$ associated with the surjective operator $B\in\mathcal{L}(Y,Z)$.

Consequently, by Proposition 10.1.4, we infer that the solution to

$$\alpha(\lambda_\star\otimes\mu)(\overline{W}-U)^2+\beta\mu(\overline{w}-u)^2) = \inf_{BWx-Bw\in K+v}(\alpha(\lambda_\star\otimes\mu)(W-U)^2+\beta\mu(w-u)^2)$$

is equal to

$$(\overline{W}_{\alpha,\beta},\overline{w}_{\alpha,\beta}) = (U,u)-A^+(\mathbf{1}-\pi_K^B)(BUx-v)$$

This yields that

$$\begin{cases} \overline{W}_{\alpha,\beta} = U-\dfrac{\beta Lx}{\alpha+\beta\lambda(x)^2}\otimes B^+(\mathbf{1}-\pi_K^B)(BUx-v) \\ \overline{w}_{\alpha,\beta} = u+\dfrac{\alpha}{\alpha+\beta\lambda(x)^2}B^+(\mathbf{1}-\pi_K^B)(BUx-v) \end{cases}$$

We also observe that

$$\overline{W}_{\alpha,\beta}x-\overline{w}_{\alpha,\beta} = Ux-u-B^+((\mathbf{1}-\pi_K^B)(BUx-Bu))$$

does not depend of α and β. □

Chapter 11

Set-Valued Analysis

Introduction

There is no need to justify here the introduction of set-valued maps in mathematical economics after the celebrated Gérard Debreu's THEORY OF VALUE [151, Debreu]. However, a differential calculus of these set-valued maps has been designed and efficiently exploited in the 80's — actually, for the very purpose of differentiating the regulation map in order to give a meaning to heavy solutions — and this is what is summarized in this chapter. More details can be found in the monograph SET-VALUED ANALYSIS, [54, Aubin & Frankowska], for instance, and is the forthcoming book [299, Rockafellar & Wets].

We shall present in Chapter 12 another approach to a differential calculus of set-valued maps, called the **mutational calculus**, which is also appropriate to define differential equations (called mutational equations) governing the evolution of subsets we mentioned in Chapter 3, for which we refer to the forthcoming book [35, Aubin].

1. Set-valued analysis is based on the weakest notion of limits of sets which has been introduced by Painlevé in 1902 and popularized by Kuratowski in his famous book **Topologie**, so that they are often called **Kuratowski limits** of sequences of sets.

2. With this concept of limit, one can define and study contingent cones to subsets introduced in the 30's by Georges Bouligand. It happens that this very first concept introduced before many other definitions of tangency remains the one for which the most important theorems in viability theory, in optimization and in nonlinear analysis are valid. However, in order to obtain the convexity of the contingent cone and for having a rich enough calculus, we are often led to posit a regularity assumption that is called sleekness: K is sleek at some $x_0 \in K$ if the set-valued map $x \rightsquigarrow T_K(x)$ is lower semicontinuous at x_0. We also need this property together with lower semicontinuity criteria for proving for

instance that the regulation map is lower semicontinuous, so that we can use the selection theorems of Chapter 4.

3. We still use the concept of upper limits of sets for defining graphical limits of sequences of set-valued maps. This is quite naturally since many properties of set-valued maps are actually properties of their graph (this point of view is known under the name of "graphical approach"). Graphical limits enjoy useful properties different from the classical pointwise limits of single-valued maps which are advantageous for solving many problems in viability theory. Taking graphical limits of difference quotients of set-valued maps, we obtain the concept of contingent derivative of a set-valued map $F : X \rightsquigarrow Y$ at a point (x, y) of its graph. The basic property

$$\text{Graph}(DF)(x,y) = T_{\text{Graph}(F)}(x,y)$$

explains the natural role played by contingent derivatives of set-valued maps in viability theory.

4. Real-valued (extended) functions play a supplementary role because the order relation of $\mathbf{R}$, involved both in optimization and in the study of Lyapunov functions as in Chapter 8. For such properties, the relevant object is the epigraph of the function instead of its mapping character, as it has been recognized since the pioneer work by R.T. Rockafellar. Hence it is natural to singularize the upper epigraphical convergence of a sequence of functions defined from the upper limit of their epigraphs. Taking the upper epigraphical limit of difference quotients yield contingent epiderivative $D_{\uparrow}V(x)$ of a function $V : X \mapsto \mathbf{R} \cup \{+\infty\}$ at x, which satisfies the fundamental property

$$\mathcal{E}p(D_{\uparrow}V)(x) = T_{\mathcal{E}p(V)}(x, V(x))$$

which justifies the natural role played by contingent epiderivatives of extended functions in viability theory, as we did in Chapter 8.

11.1 Upper Limits of Sets

Introduction

In this chapter, X, Y, Z denote finite dimensional vector-spaces. The unit ball is denoted by B (or B_X if the space must be mentioned). Let $K \subset X$, we denote by

$$d_K(x) := d(x,K) := \inf_{y\in K} \|x-y\|$$

the distance from x to K, where we set $d(x,\emptyset) := +\infty$.

Definition 11.1.1 *Let $(K_n)_{n \in \mathbf{N}}$ be a sequence of subsets of X. We say that the subset*

$$\operatorname{Limsup}_{n\to\infty} K_n \;:=\; \left\{x \in X \;\mid\; \liminf_{n\to\infty} d(x, K_n) = 0\right\}$$

which is the set of cluster points of sequences $x_n \in K_n$, *i.e., of limits of subsequences $x_{n'} \in K_{n'}$, is the* upper limit *of the sequence K_n. The subset*

$$\operatorname{Liminf}_{n\to\infty} K_n \;:=\; \{x \in X \;\mid\; \lim_{n\to\infty} d(x, K_n) = 0\}$$

which is the set of limits of sequences $x_n \in K_n$, *is its* lower limit. *A subset K is said to be the* limit *or the* set limit *of the sequence K_n if*

$$K \;=\; \operatorname{Liminf}_{n\to\infty} K_n \;=\; \operatorname{Limsup}_{n\to\infty} K_n \;=:\; \operatorname{Lim}_{n\to\infty} K_n$$

Lower and upper limits are obviously *closed.* We also see at once that

$$\operatorname{Liminf}_{n\to\infty} K_n \;\subset\; \operatorname{Limsup}_{n\to\infty} K_n$$

and that the upper limits and lower limits of the subsets K_n and of their closures $\overline{K_n}$ do coincide, since $d(x, K_n) = d(x, \overline{K_n})$.

Any decreasing sequence of subsets K_n has a limit, which is the intersection of their closures:

$$\text{if } K_n \subset K_m \text{ when } n \geq m, \text{ then } \operatorname{Lim}_{n\to\infty} K_n \;=\; \bigcap_{n\geq 0} \overline{K_n}$$

11.2 Contingent Cones

11.2.1 Definition

The definition of upper limits provide right a way a characterization of contingent cones:

Lemma 11.2.1 *Let $K \subset X$ be a subset of a finite dimensional vector space X and $x \in K$. The* **contingent cone** *$T_K(x)$ is the upper limit of the subsets $(K - x)/h$:*

$$T_K(x) \;:=\; \operatorname{Limsup}_{h\to 0+} \frac{K - x}{h}$$

so that $T_K(x)$ is always a closed cone of "tangent directions" (which is convex when K is convex or, more generally, when the contingent cone is lower semicontinuous, a vector space when K is a smooth manifold).

In order to built a calculus of contingent cones, we need some regularity assumptions more general than differentiability of a manifold or convexity. The concept we need is the one of **sleekness**, which encompasses differentiability and convexity. It means that the contingent cone map is lower semicontinuous.

For that purpose, we need to recall some results on lower semicontinuous maps.

11.2.2 Lower Semicontinuous Maps

Definition 11.2.2 *A set-valued map $F : X \rightsquigarrow Y$ is called* lower semicontinuous *at $x \in \mathrm{Dom}(F)$ if and only if for any $y \in F(x)$ and for any sequence of elements $x_n \in \mathrm{Dom}(F)$ converging to x, there exists a sequence of elements $y_n \in F(x_n)$ converging to y.*

It is said to be lower semicontinuous *if it is lower semicontinuous at every point $x \in \mathrm{Dom}(F)$.*

We refer to SET-VALUED ANALYSIS, [54, Aubin & Frankowska] for more details, although we provide some of the proofs for the convenience of the reader.

Example: Parametrized Set-Valued Maps

Let us consider three finite dimensional vector-space X, Y and Z, a set-valued map

$$P : X \rightsquigarrow Z$$

and a single-valued map

$$c : \mathrm{Graph}(P) \mapsto Y$$

We associate with these data the set-valued map $F : X \rightsquigarrow Y$ defined by

$$\forall\, x \in X, \quad F(x) := (c(x,p))_{p \in P(x)}$$

Proposition 11.2.3 *Assume that c is* continuous *from* Graph(P) *to Y. If P is lower semicontinuous, so is F.*

Proof — Let us consider a sequence $x_n \in \mathrm{Dom}(F)$ converging to $x \in \mathrm{Dom}(F)$ and take $y := c(x,p)$ belonging to $F(x)$, where $p \in P(x)$. Since P is lower semicontinuous, there exists a sequence $p_n \in P(x_n)$ converging to p. Then the sequence $q_n = c(x_n, p_n)$, where q_n belongs to $F(x_n)$, converges to y because c is continuous. Hence F is lower semicontinuous. □

Definition 11.2.4 (Marginal Functions) *Let us consider a set-valued map $F : X \rightsquigarrow Y$ and a function $c : \mathrm{Graph}(F) \mapsto \mathbf{R}$. We associate with them the* marginal function *$g : X \mapsto \mathbf{R} \cup \{\pm\infty\}$ defined by*

$$g(x) := \sup_{y \in F(x)} c(x,y)$$

Theorem 11.2.5 (Maximum Theorem) *Let a set-valued map $F : X \rightsquigarrow Y$ and a function $c : \mathrm{Graph}(F) \mapsto \mathbf{R}$ be given. If c and F are* lower semicontinuous, *so is the marginal function.*

Proof — Let us consider a sequence x_n converging to x, fix $\lambda < g(x)$ and choose $y \in F(x)$ such that $\lambda \leq c(x,y)$. Then there exist elements $q_n \in F(x_n)$ converging to y (because F is lower semicontinuous) and we know that $c(x_n, q_n) \leq g(x_n)$. Since c is lower semicontinuous, we infer that

$$\lambda \leq c(x,y) \leq \liminf_{n\to\infty} c(x_n, q_n) \leq \liminf_{n\to\infty} g(x_n)$$

By letting λ converge to $g(x)$, the claim ensues. □

11.2.3 Sleek Subsets

The main example of lower semicontinuous set-valued maps is provided by the map $x \rightsquigarrow T_K(x)$ associating with any x belonging to a closed convex set K its tangent cone $T_K(x)$ at x.

This "regularity property" of a set K at a point $x \in K$ is not always true for any subset. It is convenient to introduce the following definition:

Definition 11.2.6 *We shall say that a subset K of X is* sleek *at $x \in K$ if the set-valued map*

$$K \ni x' \rightsquigarrow T_K(x') \text{ is lower semicontinuous at } x$$

which is equivalent to say that

$$K \ni x' \rightsquigarrow N_K(x') \text{ is closed at } x$$

We say that K is sleek *if and only if it is sleek at every point of K.*

We define the (Clarke) tangent cone *(or* circatangent cone*) $C_K(x)$ by*

$$C_K(x) := \{v \mid \lim_{h\to 0+, K\ni x'\to x} \frac{d_K(x'+hv)}{h} = 0\}$$

Therefore, with this definition, we obtain:

Theorem 11.2.7 *Any closed convex subset of a finite dimensional vector-space X is sleek.*

We refer to Theorem 4.2.2 of SET-VALUED ANALYSIS for the proof of this Theorem. □

Tangent cones to sleek subsets enjoys almost the same calculus than the tangent cones to closed convex subsets. The results are proved in [54, Aubin & Frankowska], chapter 4.

We see at once that $C_K(x) \subset T_K(x)$ and that if x belongs to Int(K), then $C_K(x) = X$.

It is very convenient to observe that when x belongs to $\overline{K}$,

$$\begin{cases} v \in C_K(x) \text{ if and only if } \forall\ h_n \to 0+,\ \ \forall\, K \ni x_n \to x, \\ \exists\ v_n \to v \text{ such that } \forall\, n,\ \ x_n + h_n v_n \in K \end{cases}$$

The charm of the tangent cone C_K at x is that it is always convex[1]. Unfortunately, the price to pay for enjoying this convexity property of the Clarke tangent cones is that they may often be reduced to the trivial cone $\{0\}$.

For convex subsets K, the Clarke tangent cone and the contingent cone coincide with the closed cone spanned by $K - x$, the tangent cone of convex analysis.

However, we shall show that the Clarke tangent cone and the contingent cone do coincide at those points x where K is sleek:

Theorem 11.2.8 *Let K be a closed subset of a finite dimensional vector-space X. Consider a set-valued map $F : K \rightsquigarrow X$ satisfying*

$$\begin{cases} i) & F \text{ is lower semicontinuous at } x \\ ii) & \exists\, \delta > 0 \quad \text{such that } \forall\, z \in B_K(x,\delta), \;\; F(z) \subset T_K(z) \end{cases}$$

Then $F(x) \subset C_K(x)$.

In particular, if K is sleek at $x \in K$, then $T_K(x) = C_K(x)$ is a closed convex cone.

Before proving this theorem, we need to establish Lemma 11.2.9 We introduce the following notation[2]:

$$D_{\uparrow} d_K(x)(v) \;:=\; \liminf_{h \to 0+} \frac{d_K(x+hv) - d_K(x)}{h}$$

We observe that when $x \in K$, a direction v is contingent to K at x if and only if $D_{\uparrow} d_K(x)(v) \leq 0$.

Lemma 11.2.9 *Let K be a closed subset of a finite dimensional vector-space and $\Pi_K(y)$ be the set of projections of y onto K, i.e., the subset of $z \in K$ such that $\|y - z\| = d_K(y)$. Then the following inequalities:*

$$D_{\uparrow} d_K(y)(v) \;\leq\; d(v, \overline{\text{co}}(T_K(\Pi_K(y))))$$

hold true.

---

[1]Let v_1 and v_2 belong to $C_K(x)$. To prove that $v_1 + v_2$ belongs to this cone, let us choose any sequence $h_n > 0$ converging to 0 and any sequence of elements $x_n \in K$ converging to x. There exists a sequence of elements v_{1n} converging to v_1 such that the elements $x_{1n} := x_n + h_n v_{1n}$ do belong to K for all n. But since x_{1n} does also converge to x in K, there exists a sequence of elements v_{2n} converging to v_2 such that

$$\forall\, n, \;\; x_{1n} + h_n v_{2n} \;=\; x_n + h_n(v_{1n} + v_{2n}) \;\in\; K$$

This implies that $v_1 + v_2$ belongs to $C_K(x)$ because the sequence of elements $v_{1n} + v_{2n}$ converges to $v_1 + v_2$.

[2]this is the contingent epiderivative of the distance functions d_K. See SET-VALUED ANALYSIS, [54, Aubin & Frankowska]) for further details.

Proof — Choose $z \in \Pi_K(y)$. Then

$$\frac{d_K(y+hv)-d_K(y)}{h} \leq \frac{\|y-z+hw\|-\|y-z\|}{h}$$

from which we deduce, letting h go to 0,

$$D_\uparrow d_K(y)(v) \leq \left\langle \frac{y-z}{\|y-z\|}, v \right\rangle$$

Taking $w \in T_K(z)$ and recalling that $\langle y-z, w\rangle \leq 0$, we infer that

$$\forall\, w \in T_K(z),\ \ D_\uparrow d_K(y)(v) \leq \left\langle \frac{y-z}{\|y-z\|}, v-w \right\rangle$$

This inequality is also true for any $w \in \overline{\text{co}}(T_K(z))$, so that

$$D_\uparrow d_K(y)(v) \ \leq\ d(v, \overline{\text{co}}(T_K(z))) \ \square$$

Proof of Theorem 11.2.8

Let us take $x \in K$ and $v \in F(x)$, assumed to be different from 0. Since F is lower semicontinuous at x, we can associate with any $\varepsilon > 0$ a number $\eta \in]0, \delta[$ such that $d(v, F(z)) \leq d(v, F(x)) + \varepsilon = \varepsilon$ for any $z \in B_K(x, \eta)$ (because $d(v, F(x)) = 0$). Therefore, for any $y \in B(x, \eta/4)$ and $\tau \leq \eta/4\|v\|$, the inequality

$$\forall\, z \in \Pi_K(y+\tau v),\ \|z-x\| \leq 2\|y+\tau v - x\| \leq 2\|x-y\| + 2\tau\|v\| \leq \eta$$

implies that

$$\begin{cases} d(v, T_K(\Pi_K(y+\tau v))) \ \leq\ d(v, F(\Pi_K(y+\tau v))) \\ \qquad\qquad\qquad\qquad\quad\ \leq\ d(v, F(x)) + \varepsilon \ = \ \varepsilon \end{cases}$$

We set $g(\tau) := d_K(y+\tau v)$. By Lemma 11.2.9, we obtain

$$\begin{cases} \liminf_{h\to 0+} \left(g(\tau+h) - g(\tau)\right)/h \ = \ D_\uparrow d_K(y+\tau v)(v) \\ \leq\ d(v, T_K(\Pi_K(y+\tau v))) \ \leq\ \varepsilon \end{cases}$$

The function g being Lipschitz, it is almost everywhere differentiable, so that $g'(t) \leq \varepsilon$ for almost all t small enough. Integrating this inequality from 0 to h, we obtain

$$d_K(y+hv) \ = \ g(h) \ = \ g(h) - g(0) \ \leq\ \varepsilon h$$

for any $y \in B(x, \eta/4)$ and $\tau \leq \eta/4\|v\|$. This shows that v belongs to $C_K(x)$.

By taking $F(x) = T_K(x)$, we deduce that $T_K(x) \subset C_K(x)$ whenever K is sleek at $x \in K$, and thus, that they coincide. $\square$

11.2.4 Lower Semicontinuity Criteria

In order to prove that the regulation map Π_K is lower semicontinuous, we can always assume that the pricing map P is lower semicontinuous and we know that the set-valued map $T_K(\cdot)$ is lower semicontinuous whenever the viability set is sleek (and, in particular, smooth or convex).

We thus need **lower semicontinuity criteria** to derive that the regulation map $\Pi_K(\cdot)$ is lower semicontinuous.

Proposition 11.2.10 *Consider two set-valued maps T and P from X to Y and Z respectively and a (single-valued) map c from $X \times Z$ to Y satisfying the following assumptions:*

$$\left\{\begin{array}{ll} i) & T \text{ and } P \text{ are lower semicontinuous with convex values} \\ ii) & c \text{ is continuous} \\ iii) & \forall\, x,\ \ p \mapsto c(x,p) \text{ is affine} \end{array}\right.$$

We posit the following condition:
$\forall\, x \in X,\ \exists\, \gamma > 0,\ \delta > 0,\ c > 0,\ r > 0$ *such that* $\forall\, x' \in B(x,\delta)$ *we have*

$$\gamma B_Y \ \subset\ c(x', P(x') \cap rB_Z) - T(x') \tag{11.2.1}$$

Then the set-valued map $\Pi : X \rightsquigarrow Z$ *defined by*

$$\Pi(x) \ :=\ \{p \in P(x) \ \mid\ c(x,p) \in T(x)\} \tag{11.2.2}$$

is **lower semicontinuous with nonempty convex values.**

Proof — Let us fix $p \in \Pi(x)$ and a sequence x_n converging to x. Since P and T are lower semicontinuous and c is continuous, there exist sequences $p_n \in P(x_n)$ and $q_n \in T(x_n)$ converging to p and $c(x,p)$ respectively. Let us set $\varepsilon_n := \|c(x_n,p_n) - q_n\|$ and $\theta_n := \frac{\gamma}{\gamma+\varepsilon_n} \in]0,1[$. Then ε_n converges to 0. Since

$$\theta_n \varepsilon_n = (1-\theta_n)\gamma$$

we deduce that

$$\left\{\begin{array}{lll} \theta_n(c(x_n,p_n) - q_n) & \in & \theta_n\varepsilon_n B \\ & = & (1-\theta_n)\gamma B \\ & \subset & (1-\theta_n)(c(x_n, P(x_n) \cap rB_Z)) - T(x_n)) \end{array}\right.$$

Therefore, there exist $\widehat{p}_n \in P(x_n) \cap rB_Z$ and $\widehat{q}_n \in T(x_n)$ such that

$$c(x_n, \theta_n p_n + (1-\theta_n)\widehat{p}_n) = \theta_n q_n + (1-\theta_n)\widehat{q}_n$$

This implies that

$$q_n := \theta_n p_n + (1-\theta_n)\widehat{p}_n$$

belongs to $P(x_n)$ and that

$$p_n - q_n = (1-\theta_n)(p_n - \widehat{p}_n) \in (1-\theta_n)(r + \|u\| + 1)B$$

because $\|\widehat{p}_n\| \leq r$, $\|p_n\|$ is bounded, $P(x_n)$ and $T(x_n)$ are convex and c is affine with respect to p. Hence the elements $q_n \in \Pi(x_n)$ converge to p given in $\Pi(x)$. □

We state now another condition which is less symmetric.

Proposition 11.2.11 *Consider a metric space X, two finite dimensional vector spaces Y and Z, two set-valued maps T and P from X to Y and Z respectively and a (single-valued) map c from $X \times Z$ to Y such that*

$$\left\{\begin{array}{ll} i) & P \text{ is lower semicontinuous with convex values} \\ ii) & f \text{ is continuous} \\ iii) & \forall\, x,\ \ p \mapsto c(x,p) \text{ is affine} \\ iv) & \forall\, x,\ \ T(x) \text{ is convex and its interior is nonempty} \\ v) & \text{the graph of the map } x \rightsquigarrow \operatorname{Int}(T(x)) \text{ is open} \end{array}\right.$$

We posit the following condition:

$$\forall\, x \in X,\ \exists\, p \in P(x) \text{ such that } c(x,p) \in \operatorname{Int}(T(x)) \tag{11.2.3}$$

Then the set-valued map Π defined by (11.2.2) is lower semicontinuous with convex values.

Proof

1. — We introduce the set-valued map $S : X \rightsquigarrow Z$ defined by

$$S(x) \ := \ \{p \in P(x) \mid c(x,p) \in \operatorname{Int}(T(x))\} \subset \Pi(x)$$

Assumption (11.2.3) implies that $S(x)$ is not empty. **We claim that S is lower semicontinuous.** Indeed, if $x_n \to x$ and if p belongs to $S(x) \subset P(x)$, there exists $p_n \in P(x_n)$ which converges to p because P is lower semicontinuous. Since

$$(x_n, c(x_n, p_n)) \text{ converges to } (x, c(x,p)) \in \operatorname{Graph}(\operatorname{Int}(T(\cdot)))$$

by continuity of c and since the graph of $\operatorname{Int}(T(\cdot))$ is open, the elements $c(x_n, p_n)$ belong to $\operatorname{Int}(T(x_n))$ for n large enough and thus, the elements p_n belong to $S(x_n)$ and converge to p.

2. — Convexity of $P(x)$ and $T(x)$ implies that $\overline{S(x)} = \Pi(x)$. Indeed, let us fix $p \in \Pi(x)$ and $p_0 \in S(x)$. Then $q_\theta := \theta p_0 + (1-\theta)p$ belongs to $S(x)$ when $\theta \in]0,1[$, because $T(x)$ is convex and $c(x,p_0)$ belongs to the interior of $T(x)$, so that for every $\theta \in]0,1[$,

$$c(x,p) + \theta(c(x,p_0) - c(x,p)) = c(x, u + \theta p_0 - \theta p) = c(x, q_\theta)$$

belongs to the interior of $T(x)$. Then p is the limit of q_θ when $\theta > 0$ converges to 0.

3. — The theorem ensues because the closure of any lower semicontinuous set-valued map is still lower semicontinuous. □

We state the useful **Duality Theorem**

Theorem 11.2.12 *Let $(K_n)_{n\in\mathbf{N}}$ be a sequence of closed convex cones of a finite dimensional vector space X. Then*

$$\operatorname{Liminf}_{n\to\infty} K_n \ = \ \left(\operatorname{Limsup}_{n\to\infty} K_n^-\right)^-$$

(1)	▷	If $K \subset L$ and $x \in \overline{K}$, then $T_K(x) \subset T_L(x)$
(2)	▷	If $K_i \subset X$, $(i = 1, \cdots, n)$ and $x \in \overline{\bigcup_i K_i}$, then $T_{\bigcup_{i=1}^n K_i}(x) = \bigcup_{i \in I(x)} T_{K_i}(x)$ where $I(x) := \{i \mid x \in \overline{K_i}\}$
(3)	▷	If $K_i \subset X_i$, $(i = 1, \cdots, n)$ and $x_i \in \overline{K_i}$, then $T_{\prod_{i=1}^n K_i}(x_1, \ldots, x_n) \subset \prod_{i=1}^n T_{K_i}(x_i)$
(4)	▷	If $g \in \mathcal{C}^1(X, Y)$, if $K \subset X$, $x \in \overline{K}$ and $M \subset Y$, then $g'(x)(T_K(x)) \subset T_{g(K)}(g(x))$ $T_{g^{-1}(M)}(x) \subset g'(x)^{-1} T_M(g(x))$
(5)	▷	If $L \subset X$ and $M \subset Y$ are **closed sleek** subsets, $f \in \mathcal{C}^1(X, Y)$ is a continuously differentiable map and $x \in L \cap f^{-1}(M)$ satisfies the **transversality condition** $f'(x)T_L(x) - T_M(f(x)) = Y$, then $T_{L \cap f^{-1}(M)}(x) = T_L(x) \cap f'(x)^{-1} T_M(f(x))$
(5)a)	▷	If $M \subset Y$ is a *closed sleek* subset, $f \in \mathcal{C}^1(X, Y)$ is a continuously differentiable map and $x \in f^{-1}(M)$ satisfies $\mathrm{Im}(f'(x)) - T_M(f(x)) = Y$, then $T_{f^{-1}(M)}(x) = f'(x)^{-1} T_M(f(x))$
(5)b)	▷	If K_1 and K_2 are **closed sleek** subsets contained in X and $x \in K_1 \cap K_2$ satisfies $T_{K_1}(x) - T_{K_2}(x) = X$, then $T_{K_1 \cap K_2}(x) = T_{K_1}(x) \cap T_{K_2}(x)$
(5)c)	▷	If $K_i \subset X$, $(i = 1, \ldots, n)$, are **closed sleek** and $x \in \bigcap_i K_i$ satisfies $\forall v_i = 1, \ldots, n$, $\bigcap_{i=1}^n (T_{K_i}(x) - v_i) \neq \emptyset$ then, $T_{\bigcap_{i=1}^n K_i}(x) = \bigcap_{i=1}^n T_{K_i}(x)$

Table 11.1: **Properties of Contingent Cones to Sleek Subsets.**

11.2.5 Contingent Cones to Closed Sleek Subsets

We summarize in Table 11.1 the calculus of contingent cones. Formulas (1) to (4) are straightforward. The other properties are valid when K is sleek, and are a consequence of the Constrained Inverse Function Theorem, which we shall prove later[3].

[3]We mention that transversality condition of formula (5) implies the constraint qualification assumption $0 \in \mathrm{Int}(f(L) - M)$ and that the stronger transversality condition

$$\exists\, c > 0 \mid \forall\, x \in K, \;\; B_Y \subset f'(x)(T_L(x) \cap cB_X) - T_M(Ax)$$

implies that if L and M are sleek and f is continuously differentiable, then K is also sleek.

Proposition 11.2.13 *Assume that the resource set is closed and sleek and the consumption sets satisfy*

$$\forall\, x \in K,\ \sum_{i=1}^{n} T_{L_i}(x_i) - T_M(\sum_{i=1}^{n} x_i) \;=\; Y$$

Then

$$T_K(x) \;:=\; \left\{ v \;:=\; (v_1,\ldots,v_n) \in \prod_{i=1}^{n} T_{L_i}(x) \;\middle|\; \sum_{i=1}^{n} v_i \;\in\; T_M\left(\sum_{i=1}^{n} x_i\right) \right\} \tag{11.2.4}$$

11.2.6 Case of Inequality Constraints

We also state the following example of the contingent cone to a set defined by equality and inequality constraints[4]:

Theorem 11.2.14 *Let us consider a closed subset L of a finite dimensional vector-space X and two continuously differentiable maps $g := (g_1,\ldots,g_p) : X \mapsto \mathbf{R}^p$ and $h := (h_1,\ldots,h_q) : X \mapsto \mathbf{R}^q$ defined on an open neighborhood of L.*

Let K be the subset of L defined by the constraints

$$K \;:=\; \{x \in L \mid g_i(x) \geq 0,\ i = 1,\ldots,p,\ \ \&\ h_j(x) = 0,\ j = 1,\ldots,q\}$$

We denote by $I(x) := \{i = 1,\ldots,p \mid g_i(x) = 0\}$ the subset of active constraints.

We posit the following transversality condition at a given $x \in K$:

$$\begin{cases} i) & h'(x)C_L(x) \;=\; \mathbf{R}^q \\ ii) & \exists\, v_0 \in C_L(x) \ \ \text{such that}\ \ h'(x)v_0 = 0 \\ & \text{and}\ \forall\, i \in I(x),\ \ < g_i'(x), v_0 >> 0 \end{cases}$$

Then u belongs to the contingent cone to K at x if and only if u belongs to the contingent cone to L at x and satisfies the constraints

$$\forall\, i \in I(x),\ \ < g_i'(x), u > \geq\ 0\ \ \&\ \forall\, j = 1,\ldots,q,\ \ h_j'(x)u = 0$$

Unfortunately, the graph of $T_K(\cdot)$ is not necessarily closed. However, there exists a closed set-valued map $T_K^\circ(\cdot)$ contained in $T_K(\cdot)$ introduced by N. Maderner. Set

$$\gamma_K(x) \;:=\; \min_{i \notin I(x)} \frac{g_i(x)}{\|g_i'(x)\|} \;\in\;]0, +\infty] \tag{11.2.5}$$

[4]See Proposition 4.3.6 of SET-VALUED ANALYSIS, [54, Aubin & Frankowska].

We observe that γ_K is upper semicontinuous whenever the functions g_i are continuously differentiable. Indeed, let $x_n \in K$ converge to x_0 and $a_n \leq \gamma_K(x_n)$ converge to a_0. Since $g_i(x_0) > 0$ whenever $i \notin I(x_0)$, we infer that $i \notin I(x_n)$ for n large enough. Hence inequalities $a_n\|g_i'(x_n)\| \leq g_i(x_n)$ hold true for any $i \notin I(x_0)$ and imply at the limit that $a_0 \leq \gamma_K(x_0)$.

The growth of the function γ_K is linear whenever we assume that there exists a constant $c > 0$ such that

$$\forall\, i = 1, \ldots, p, \quad \|g_i'(x)\| \;\geq\; c\,\frac{g_i(x)}{\|x\|+1}$$

Theorem 11.2.15 *We posit the assumptions of Theorem 11.2.14. Then the set-valued map $T_K^\diamond(\cdot) : K \rightsquigarrow X$ defined by:*
$u \in T_K^\diamond(x)$ if and only if $u \in T_L(x)$ and

$$\begin{cases} \forall\, i = 1, \ldots, p, & g_i(x) + < g_i'(x), u > \;\geq\; 0 \\ \forall\, j = 1, \ldots, q, & h_j'(x)u \;=\; 0 \end{cases}$$

is contained in the contingent cone $T_K(x)$ and satisfy

$$T_K(x) \cap \gamma_K(x)B \;\subset\; T_K^\diamond(x)$$

Its graph is closed whenever the graph of $x \rightsquigarrow T_L(x)$ is closed.

Proof — Let u belong to $T_K^\diamond(x)$. Then, when $i \in I(x)$, we see that $\langle g_i'(x), u\rangle = g_i(x) + \langle g_i'(x), u\rangle \geq 0$, so that $u \in T_K(x)$.

Conversely, let us choose u in $T_K(x)$ satisfying $\|u\| \leq \gamma_K(x)$. Then either $i \in I(x)$ and $g_i(x) + \langle g_i'(x), u\rangle = \langle g_i'(x), u\rangle \geq 0$ or $g_i(x) > 0$ so that

$$i \notin I(x) \;\;\&\;\; g_i(x) + \langle g_i'(x), u\rangle \;\geq\; g_i(x) \;-\; \|g_i'(x)\|\|u\| \;\geq\; 0$$

because $\|u\| \leq \gamma_K(x) \leq g_i(x)/\|g_i'(x)\|$. Thus, in both cases, $g_i(x)+\langle g_i'(x), u\rangle \geq 0$, so that u belongs to $T_K^\diamond(x)$. □

11.3 Contingent Derivatives of Maps

11.3.1 Graphical Convergence of Maps

Let us consider a sequence of set-valued maps $F_n : X \rightsquigarrow Y$. The set-valued map $F^\sharp := \mathrm{Lim}^\sharp{}_{n\to\infty} F_n$ from X to Y defined by

$$\mathrm{Graph}(\mathrm{Lim}^\sharp{}_{n\to\infty} F_n) := \mathrm{Limsup}_{n\to\infty}\mathrm{Graph}(F_n)$$

is called the *(graphical) upper limit* of the set-valued maps F_n. Even for single-valued maps, this is a weaker convergence than the pointwise convergence: if $f_n : X \mapsto Y$ converges pointwise to f, then, for every $x \in X$, $f(x) \in f^\sharp(x)$. If the sequence is equicontinuous, then $f^\sharp(x) = \{f(x)\}$.

If the sequence of set-valued maps F_n is decreasing, then it has a limit equal to the intersection of the set-valued maps.

The following result justifies the introduction of this concept of convergence:

Theorem 11.3.1 (Convergence Theorem) *Let F_n be a sequence of nontrivial set-valued maps from $K \subset X$ to Y with uniform linear growth: there exists $c > 0$ such that, for any $n \geq 0$,*

$$\forall\, x \in K, \ \ \|F_n(x)\| \ := \ \sup_{y \in F_n(x)} \|y\| \ \leq \ c(\|x\| + 1)$$

Let us consider measurable *functions x_m and y_m from Ω to X and Y respectively, satisfying $y_m(\omega) \in F_m(x_m(\omega))$ for almost all $\omega \in \Omega$.*

If

$$\begin{cases} i) & x_m(\cdot) \text{ converges almost everywhere to a function } x(\cdot) \\ ii) & y_m(\cdot) \ \in \ L^1(\Omega; Y) \text{ and converges } \textit{weakly} \text{ in } L^1(\Omega; Y) \\ & \textit{to a function} \ \ y(\cdot) \in L^1(\Omega; Y) \end{cases}$$

then for almost all $\omega \in \Omega, \ y(\omega) \in \overline{co}F^{\sharp}(x(\omega))$.

The proof is analogous to the proof of the Convergence Theorem 12.1.4 and can also be found in SET-VALUED ANALYSIS, Theorem 7.2.1.

11.3.2 Contingent Derivatives

Definition 11.3.2 *Let $F : X \rightsquigarrow Y$ be a set-valued map from a finite dimensional vector space X to another finite dimensional vector space Y and $y \in F(x)$.*

We introduce the differential quotients

$$u \rightsquigarrow \nabla_h F(x, y)(u) \ := \ \frac{F(x + hu) - y}{h}$$

of a set-valued map $F : X \rightsquigarrow Y$ at $(x, y) \in \text{Graph}(F)$.

The contingent derivative $DF(x, y)$ of F at $(x, y) \in \text{Graph}(F)$ is the graphical upper limit of differential quotients:

$$DF(x, y) \ = \ \text{Lim}^{\sharp}{}_{h \to 0+} \nabla_h F(x, y)$$

By coming back to the original point of view proposed by Fermat, we are able to geometrically characterize the derivatives of set-valued maps through contingent cones to their graph:

Proposition 11.3.3 *Let $F : X \rightsquigarrow Y$ be a set-valued map from a finite dimensional vector space X to another finite dimensional vector space Y and $y \in F(x)$. Then*

$$\mathrm{Graph}(DF(x,y)) = T_{\mathrm{Graph}(F)}(x,y)$$

Proof — Indeed, we know that the contingent cone

$$T_{\mathrm{Graph}(F)}(x,y) = \mathrm{Limsup}_{h\to 0+}\frac{\mathrm{Graph}(F)-(x,y)}{h}$$

is the upper limit of the differential quotients $\frac{\mathrm{Graph}(F)-(x,y)}{h}$ when $h \to 0+$. It is enough to observe that

$$\mathrm{Graph}(\nabla_h F(x,y)) = \frac{\mathrm{Graph}(F)-(x,y)}{h}$$

and to take the upper limit to conclude. □

When $F := f$ is single-valued, we set $Df(x) := Df(x, f(x))$.

We can easily compute the derivative of the inverse of a set-valued map F (or even of a noninjective single-valued map): The contingent derivative of the inverse of a set-valued map F is the inverse of the contingent derivative:

$$D(F^{-1})(y,x) = DF(x,y)^{-1}$$

If K is a subset of X and f is a single-valued map which is Fréchet differentiable around a point $x \in K$, then the contingent derivative of the restriction of f to K is the restriction of the derivative to the contingent cone:

$$D(f|_K)(x) = D(f|_K)(x, f(x)) = f'(x)|_{T_K(x)}$$

Proposition 11.3.4 *If $x \in \mathrm{Int}(\mathrm{Dom}(F))$ and F is Lipschitz around x, then*

$$v \in DF(x,y)(u) \text{ if and only if } \liminf_{h\to 0+} d\left(v, \frac{F(x+hu)-y}{h}\right) = 0$$

If moreover the dimension of Y is finite, then

$$\mathrm{Dom}(DF(x,y)) = X \text{ and } DF(x,y) \text{ is Lipschitz}$$

Proof — The first two statements being obvious, let us check the last one. Let u belong to X and l denote the Lipschitz constant of F on a neighborhood of x. Then, for all $h > 0$ small enough and $y \in F(x)$,

$$y \in F(x) \subset F(x+hu) + lh\|u\|B$$

Hence there exists $y_h \in F(x+hu)$ such that $v_h := (y_h - y)/h$ belongs to $l\|u\|B$, which is compact. Therefore the sequence v_h has a cluster point v, which belongs to $DF(x,y)(u)$. □

11.3.3 Codifferentials

A set-valued map whose graph is a closed cone is called a *closed process*. It is a *closed convex process* if its graph is furthermore convex. Closed convex processes enjoy most of the properties of continuous linear operators, as it is shown in Chapter 2 of SET-VALUED ANALYSIS, [54, Aubin & Frankowska]. The transpose of a closed process $A : X \rightsquigarrow Y$ is the closed convex process $A^\star : Y^\star \rightsquigarrow X^\star$ defined by

$$p \in A^\star(q) \text{ if and only if } \forall\,(x,y) \in \text{Graph}(A),\ \langle p,x\rangle \leq \langle q,y\rangle$$

We define in a symmetric way the *bitranspose* $A^{\star\star} : X \rightsquigarrow Y$ of A, the graph of which is the closed convex cone spanned by the graph of A:

$$\text{Graph}(A) = (\text{Graph}(A))^{--}$$

Definition 11.3.5 *Let $F : X \rightsquigarrow Y$ be a set-valued map and (x,y) belong to its graph. We shall say that the transpose $DF(x,y)^\star : Y^\star \rightsquigarrow X^\star$ of the contingent derivative $DF(x,y)$ is the* codifferential of F at (x,y).

We the observe that

$$p \in DF(x,y)^\star(q) \text{ if and only if } (p,-q) \in N_{\text{Graph}(F)}(x,y)$$

Definition 11.3.6 *We shall say that a set-valued map $F : X \rightsquigarrow Y$ is* sleek *at $(x,y) \in \text{Graph}(F)$ is its graph is sleek at this point and that F is* lower semicontinuously differentiable *if*

$$(x,y,u) \in \text{Graph}(F) \times X \rightsquigarrow DF(x,y)(u)$$

is lower semicontinuous.

Lower semicontinuously differentiable set-valued maps are obviously episleek, the converse being true under boundedness conditions.

To say that F is episleek amounts to saying that the set-valued map $(x,y) \rightsquigarrow \text{Graph}(DF(x,y))$ is lower semicontinuous or that the graph of $(x,y) \rightsquigarrow \text{Graph}(DF(x,y)^\star)$ is closed.

Lower semicontinuously differentiable set-valued maps play the role of continuously differentiable (single-valued) maps. One prove that when the map is lower semicontinuously differentiable, the contingent derivative $DF(x,y)$ is a closed convex process[5].

[5]A closed convex process is a map the graph of which is a closed convex cone. These maps are the set-valued analogues of continuous linear operators and enjoy most of their properties. Actually, the graphs of the lower semicontinuously differentiable maps are sleek, but the converse is not necessarily true. We can derive for instance the following criterion: *Assume that F is sleek on some neighborhood $\mathcal{N}$ of $(x_0,y_0) \in \text{Graph}(F)$. If the boundedness*

11.3.4 A Chain Rule for Contingent Derivatives

Among the formulas given in Chapter 5 of SET-VALUED ANALYSIS, [54, Aubin & Frankowska], we provide one which is used for the computation of contingent derivatives of tubes.

Theorem 11.3.7 *Let X, Y and Z be three finite dimensional vector spaces, $g : Y \mapsto Z$ be a continuously differentiable operator and $F : X \leadsto Z$ be a set-valued map. We associate with it the set-valued map $g^{-1}F : X \leadsto Y$. Let y belong to $(g^{-1}F)(x)$.*

Assume that F is sleek at $(x, g(y))$ and that the "transversality condition"

$$T_{F(x)}(g(y)) + \mathrm{Im}(g'(y)) = Z \tag{11.3.1}$$

holds true. Then

$$D(g^{-1}F)(x,y) = g'(y)^{-1}DF(x,g(y)) \;\&\; D(g^{-1}F)(x,y)^\star = g'(y)^\star DF(x,g(y))^\star$$

Proof — To say that $v \in D(g^{-1}F)(x,y)(u)$ means that there exist sequences $h_n > 0$, u_n and v_n converging to 0, u and v respectively such that

$$g(y + h_n v_n) = g(y) + h_n(g'(y)v_n + e_n) \in F(x + h_n u_n)$$

so that $g'(y)v \in DF(x, g(y))(u)$.

Conversely, assume that the graph of F is sleek. Since the graph of $g^{-1}F$ is the inverse image by $\mathbf{1} \times g$ of the graph of F, we deduce from formula (5) of Table 11.1 that the transversality condition

$$\mathrm{Im}(\mathbf{1} \times g'(y)) - T_{\mathrm{Graph}(F)}(x, g(y)) = X \times Z$$

implies that

$$(\mathbf{1} \times g'(y))^{-1} \left(T_{\mathrm{Graph}(F)}(x, g(y))\right) \subset T_{(\mathbf{1}\times g)^{-1}(\mathrm{Graph}(F))}(x,y)$$

This formula can be translated to

$$\begin{cases} \mathrm{Graph}(g^{-1}DF(x,g(y))) = (\mathbf{1} \times g'(y))^{-1}(\mathrm{Graph}(DF(x,g(y)))) \\ \subset \mathrm{Graph}(D(g^{-1}F)(x,y)) \end{cases}$$

---

property

$$\forall\, u \in X, \quad \sup_{(x,y)\in\mathcal{N}\cap\mathrm{Graph}(F)} \inf_{v\in DF(x,y)(u)} \|v\| < +\infty$$

holds true, then the set-valued map

$$(x, y, u) \in \mathrm{Graph}(F) \times X \leadsto DF(x,y)(u)$$

is lower semicontinuous on $(\mathcal{N} \cap \mathrm{Graph}(F)) \times X$.

It remains to show that transversality condition (11.3.1) implies the general one, i.e., that for any $(\xi,\eta) \in X \times Z$, there exists $(u,v) \in X \times Y$ such that

$$(\xi - u, \eta - g'(y)v) \in T_{\mathrm{Graph}(F)}(x, g(y)) = \mathrm{Graph}(DF(x, g(y)))$$

We can always take $u = \xi$. Since $T_{F(x)}(g(y))$ is contained in $DF(x,g(y))(0)$ and since $\mathrm{Im}(g'(y)) + T_{F(x)}(g(y)) = Z$, we can indeed find v such that $\eta \in g'(y)v + DF(x,g(y))(0)$. □

Corollary 11.3.8 *Let X, Y and Z be three finite dimensional vector spaces, $A \in \mathcal{L}(Y,Z)$ be a linear operator and $F : X \rightsquigarrow Z$ be a set-valued map. We associate with it the set-valued map $A^{-1}F : X \rightsquigarrow Y$. Let y belong to $(A^{-1}F)(x)$.*

Assume that F is sleek at (x, Ay) and that the "transversality condition"

$$T_{F(x)}(Ay) + \mathrm{Im}(A) = Z \tag{11.3.2}$$

holds true. Then

$$D(A^{-1}F)(x,y) = A^{-1}DF(x,Ay) \ \&\ D(A^{-1}F)(x,y)^\star = A^\star DF(x,Ay)^\star$$

11.4 Contingent Epiderivatives of Functions

11.4.1 Epilimits

For reasons motivated both by optimization theory and Lyapunov stability, we involve the order relation on $\mathbf{R}$ by characterizing extended functions $V : X \mapsto \mathbf{R} \cup \{\pm\infty\}$ by their **epigraphs** instead of their graphs.

The epigraph of the lower epilimit

$$\mathrm{lim}^\sharp_{\uparrow n\to\infty} V_n$$

of a sequence of extended functions $V_n : X \mapsto \mathbf{R} \cup \{+\infty\}$ is the upper limit of the epigraphs:

$$\mathcal{E}p(\mathrm{lim}^\sharp_{\uparrow n\to\infty} V_n) := \mathrm{Limsup}_{n\to\infty}\mathcal{E}p(V_n)$$

One can check that

$$\mathrm{lim}^\sharp_{\uparrow n\to\infty} V_n(x_0) = \liminf_{n\to\infty, x\to x_0} V_n(x) \tag{11.4.1}$$

and that if the sequence is increasing, that

$$\mathrm{lim}^\sharp_{\uparrow n\to\infty} V_n(x_0) = \sup_{n\geq 0} V_n(x_0)$$

We refer to Chapter 7 of SET-VALUED ANALYSIS, [54, Aubin & Frankowska] for further details on *epigraphical convergence*.

11.4.2 Contingent Epiderivatives

Definition 11.4.1 *Let $V : X \mapsto \mathbf{R}\cup\{\pm\infty\}$ be a nontrivial extended function and x belong to its domain.*

We associate with it the differential quotients

$$u \rightsquigarrow \nabla_h V(x)(u) \ := \ \frac{V(x+hu)-V(x)}{h}$$

The contingent epiderivative $D_\uparrow V(x)$ of V at $x \in \mathrm{Dom}(V)$ is the lower epilimit of its differential quotients:

$$D_\uparrow V(x) \ = \ \lim{}^\sharp_{\uparrow h\to 0+} \nabla_h V(x)$$

Proposition 11.4.2 *Let $V : X \mapsto \mathbf{R}\cup\{\pm\infty\}$ be a nontrivial extended function and x belong to its domain. Then the contingent epiderivative $D_\uparrow V(x)$ satisfies*

$$\forall\, u \in X, \ \ D_\uparrow V(x)(u) \ = \ \liminf_{h\to 0+,u'\to u} (V(x+hu')-V(x))/h$$

and the epigraph of the contingent epiderivative $D_\uparrow V(\cdot)$ is equal to the contingent cone to the epigraph of V at $(x, V(x))$ is

$$\mathcal{E}p(D_\uparrow V(x)) \ = \ T_{\mathcal{E}p(V)}(x, V(x))$$

Proof — The first statement follows from (11.4.1). For proving the second one, we recall that the contingent cone

$$T_{\mathcal{E}p(V)}(x, V(x)) \ = \ \mathrm{Limsup}_{h\to 0+} \frac{\mathcal{E}p(V)-(x,V(x))}{h}$$

is the upper limit of the differential quotients $\frac{\mathcal{E}p(V)-(x,V(x))}{h}$ when $h \to 0+$. It is enough to observe that

$$\mathcal{E}p(D_\uparrow V(x)) := T_{\mathcal{E}p(V)}(x,y) \ \ \& \ \ \mathcal{E}p(\nabla_h F(x,y)) = \frac{\mathcal{E}p(V)-(x,V(x))}{h}$$

to conclude. □

Consequently, *the epigraph of the contingent epiderivative at x is a closed cone. It is then lower semicontinuous and positively homogeneous whenever V is contingently epidifferentiable at x.*

Definition 11.4.3 *Let $V : X \mapsto \mathbf{R}\cup\{\pm\infty\}$ be an extended function and $x \in$* Dom(V). *We shall say that the function V is* contingently epidifferentiable *at x if for any $u \in X$, $D_\uparrow V(x)(u) > -\infty$ (or, equivalently, if $D_\uparrow V(x)(0) = 0$) and that it is* episleek *at $x \in \mathrm{Dom}(V)$ is its epigraph is sleek at $(x, V(x))$.*

We refer to Chapter 6 of SET-VALUED ANALYSIS, [54, Aubin & Frankowska] for further properties of contingent epiderivatives.

The set-valued map $x \rightsquigarrow \Pi_W(x)$ defined by:

$$\Pi_W(x) \;:=\; \{\, v \in X \mid D_\uparrow V(x)(v) + W(x,v) \leq 0 \,\}$$

plays an important role in the study of evolution along which a function V decreases. We often need to know when it is lower semicontinuous.

Proposition 11.4.4 *Let us assume that V is episleek, that the restriction of V to its domain is continuous, that $W(\cdot,\cdot)$ is continuous and convex with respect to the second argument and that F is lower semicontinuous with closed convex values. If for any x, there exists $\bar{v} \in F(x)$ such that*

$$D_\uparrow V(x)(\bar{v}) + W(x,\bar{v}) \;<\; 0$$

then $x \rightsquigarrow \Pi_W(x)$ is lower semicontinuous at x.

Proof — Let v belong to $\Pi_W(x)$ be chosen and a sequence $x_n \in \mathrm{Dom}(D_\uparrow(V))$ converge to x. Since the set-valued map $\mathcal{E}p(D_\uparrow V(\cdot))$ is lower semicontinuous, and since $(v, -W(x,v))$ belongs to $\mathcal{E}p(D_\uparrow V(x))$, there exist a subsequence (again denoted x_n), a sequence v_n converging to v and a sequence $\varepsilon_n \geq 0$ converging to 0 such that

$$(v_n, -W(x_n,v_n) + \varepsilon_n) \in \mathcal{E}p(D_\uparrow V(x_n))$$

Let us set $a_0 := -W(x,\bar{v}) - D_\uparrow V(x)(\bar{v}) > 0$. Since by assumption the pair $(\bar{v}, -W(x,\bar{v}) - a_0)$ belongs also to $\mathcal{E}p(D_\uparrow V(x))$, we deduce that there exist sequences $\bar{v}_n$ converging to $\bar{v}$ and $a_n > 0$ converging to a_0 such that

$$(\bar{v}_n, -W(x_n,\bar{v}_n) - a_n) \in \mathcal{E}p(D_\uparrow V(x_n))$$

We introduce now $\theta_n := \frac{\varepsilon_n}{2(\varepsilon_n + a_n)} \in [0,1]$ converging to 0, $u_n := (1-\theta_n)v_n + \theta_n \bar{v}_n$ converging to v and $\alpha_n := (1-\theta_n)W_n, v_n) + \theta_n W(x_n,\bar{v}_n) - W(x_n,u_n) \geq 0$ (thanks to the convexity of $W(x_n,\cdot)$). The lower semicontinuity of the contingent cone to the epigraph of V, which is the epigraph of $D_\uparrow V(\cdot)$, implies that these cones are convex. Hence

$$\begin{cases} (u_n, -W(x_n,u_n) - \varepsilon_n/2 - \alpha_n) \\ = (1-\theta_n)(v_n, -W(x_n,v_n) + \varepsilon_n) + \theta_n(\bar{v}_n, -W(x_n,\bar{v}_n) - a_n) \end{cases}$$

belongs to $\mathcal{E}p(D_\uparrow V(x_n))$. This can be written

$$D_\uparrow V(x_n)(u_n) \leq -W(x_n,u_n) - \varepsilon_n - \alpha_n/2 < -W(x_n,u_n)$$

Hence u_n belongs to $\Pi_W(x_n)$ and converges to v. □

We refer to **Set-Valued Analysis** ([54]) for further details on these concepts and their properties.

Chapter 12

The Viability and Invariance Theorems

Introduction

1. We recall the statement of the Viability Theorem of which we provide a simpler[1] proof for the sake of completeness.

   It can be omitted by the non professional mathematician who is more interested to its applications. Like the Brouwer Fixed Point Theorem, the proof of the Viability Theorem is quite involved, and uses most of the theorems of functional analysis.

2. We state next, without proof, the Filippov Theorem, which provides and "localizes" solutions to differential inclusions for Lipschitz Maps.

3. We then adapt the Viability Theorem to nonstationary environments (viability tubes) and to differential inclusions and constraints with memory.

4. Next, we extend these results to the case when both the dynamics of a differential inclusion and the constraints depend upon the history of the evolution, allowing us to take into account delays and cumulated consequences of the past in our mathematical metaphors.

5. Finally, a very short introduction to mutational equations concludes this chapter. We refer to MUTATIONAL AND MORPHOLOGICAL ANALYSIS: TOOLS FOR SHAPE REGULATION AND OPTIMIZATION, [35, Aubin] for many more details on this topic which I think will be useful in economic theory.

More details on these topics can be found in VIABILITY THEORY, [28, Aubin] and DIFFERENTIAL INCLUSIONS, [39, Aubin & Cellina] or [154, Deimling].

[1]due to Hélène Frankowska (personal communication).

12.1 The Viability Theorem

Let us consider the differential inclusion

$$x'(t) \in F(x(t)) \tag{12.1.1}$$

Theorem 12.1.1 (Viability Theorem) *Consider a* Marchaud map $F : X \rightsquigarrow X$ *and a* closed *subset* $K \subset \mathrm{Dom}(F)$ *of a finite dimensional vector space* X.

Then for any initial state $x_0 \in K$, *there exists a* viable *solution on* $[0, \infty[$ *to differential inclusion (6.1.2) if and only if* K *is a* viability domain *of* F.

Furthermore, if we set

$$c_K := \sup_{x \in K} \frac{\|F(x)\|}{\|x\| + 1}$$

then every solution $x(\cdot)$ *starting at* x_0 *satisfies the estimates*

$$\begin{cases} \forall\, t \geq 0, \ \ \|x(t)\| \ \leq \ (\|x_0\| + 1)e^{c_K t} \\ \text{and} \\ \text{for almost all } t \geq 0, \ \ \|x'(t)\| \ \leq \ c_K(\|x_0\| + 1)e^{c_K t} \end{cases}$$

Finally, let $\mathcal{S}_F : X \rightsquigarrow \mathcal{C}(0, \infty, X)$ *denote the* solution map *associating with any initial state* x *the set* $\mathcal{S}_F(x)$ *of solutions to the differential inclusion (6.1.2) starting at* x. *Then the graph of* $\mathcal{S}_F$ *is closed (and the graph of its restriction to any compact subset is compact).*

Since viable absolutely continuous functions $x(\cdot) : [0, T] \mapsto K$ satisfy $x'(t) \in T_K(x(t))$ for almost all $t \in [0, T]$, we could be tempted to derive viability theorems from existence theorems of solutions to differential inclusion $x'(t) \in R_K(x(t))$ where we set $R_K(x) := F(x) \cap T_K(x)$. Unfortunately, this is not possible because $T_K(\cdot)$ may be neither upper semicontinuous nor lower semicontinuous[2]. For instance, it is not upper semicontinuous as soon as inequality constraints are involved: take for example $K := [-1, +1]$. The graph of $T_K(\cdot)$, equal to

$$\{-1\} \times \mathbf{R}_+ \cup] - 1, +1[\times \mathbf{R} \cup \{+1\} \times \mathbf{R}_-$$

is not closed, and not even locally compact.

So we have to devise a specific proof of Theorem 6.1.9.

As it is the case for proving many existence theorems of a solution to a problem, we proceed in three steps:

1. Construct approximate solutions

[2] See section 4.1., p. 178 of DIFFERENTIAL INCLUSIONS for an example of subset K such that $T_K(\cdot)$ is neither upper semicontinuous nor lower semicontinuous.

2. Prove that these approximate solutions converge to some limit

3. Check that this limit is a solution to the problem

We shall not use Euler's method to build approximate solutions, but use instead Zorn's Lemma (i.e., the axiom of choice) to prove the existence of approximate viable solutions in a given time interval.

Then Ascoli's and Alaoglu's Theorem will be used to make the approximate solutions and their derivatives converge. Unfortunately, contrary to differential equations, the convergence of the derivatives is not obtained from the convergence of solutions. We have to use a priori estimates to infer that the derivatives of the approximate solutions converge weakly. But this convergence is too weak to check easily that the limit of the approximate solutions is a solution. To answer this question, specific to differential inclusions, we use the Convergence Theorem, based on Mazur's Theorem, permitting to pass from weak convergence to strong convergence of "convex combinations of tails of the sequences", and from that, an almost everywhere convergence of the derivatives.

The proof of the Viability Theorem shows at least that Functional Analysis is useful !

12.1.1 Necessary Condition

The tangential condition is necessary for K to be viable under F.

Proposition 12.1.2 (Necessary Condition) *Let us assume that*

$$\begin{cases} i) & F : X \rightsquigarrow X \text{ is upper hemicontinuous} \\ ii) & \text{the images of } F \text{ are convex and compact} \end{cases}$$

If K is viable under F, then it is a viability domain.

Proof— Let $x(\cdot)$ be a solution to the differential inclusion starting at x_0. Actually, it is enough to assume that there exists a sequence $t_n \to 0+$ such that $x(t_n) \in K$.

Since F is upper hemicontinuous at x_0, we can associate with any $p \in X^\star$ and $\varepsilon > 0$ an $\eta_p > 0$ such that

$$\forall\, \tau \in [0, \eta_p],\ \langle p, x'(\tau)\rangle \ \leq\ \sigma(F(x(\tau)), p) \ \leq\ \sigma(F(x_0), p) + \varepsilon\|p\|_\star$$

By integrating this inequality from 0 to t_n, setting $v_n := \dfrac{x(t_n) - x_0}{t_n}$ and dividing by $t_n > 0$, we obtain for n larger than some N_p

$$\forall\, p \in X^\star,\ \forall\, n \geq N_p,\ \langle p, v_n\rangle \ \leq\ \sigma(F(x_0), p) + \varepsilon\|p\|_\star$$

Therefore, v_n lies in a bounded subset of a finite dimensional vector space, so that a subsequence (again denoted) v_n converges to some $v \in X$ satisfying

$$\forall\, p \in X^\star,\ \langle p, v\rangle \ \leq\ \sigma(F(x_0), p) + \varepsilon\|p\|_\star$$

By letting ε converge to 0, we deduce that v belongs to the closed convex hull of $F(x_0)$.

On the other hand, since for any n, $x(t_n) = x_0 + t_n v_n$ belongs to K, we infer that v belongs to the contingent cone $T_K(x_0)$. The intersection $F(x_0) \cap T_K(x_0)$ is then nonempty, so that the necessary condition ensues. □

12.1.2 Sufficient Conditions

Construction of Approximate Solutions

We begin by proving that there exist approximate viable solutions to the differential inclusion.

Lemma 12.1.3 *Assume that $K \subset X$ is a viability domain of $F : X \rightsquigarrow X$. Then, for any $\varepsilon > 0$, the set $S_\varepsilon(x_0)$ of a continuous functions $x(\cdot) \in \mathcal{C}(0,1;X)$ satisfying*

$$\left\{ \begin{array}{ll} i) & x(0) = x_0 \\ ii) & \forall\, t \in [0,1], \;\; d(x(t), K) \;\leq\; \varepsilon \\ iii) & \forall\, t \in [0,1], \;\; d((x(t), x'(t)), \mathrm{Graph}(F) \;\leq\; \varepsilon \end{array} \right. \tag{12.1.2}$$

is not empty.

Proof— We denote by $\mathcal{A}_\varepsilon(x_0)$ the set of pairs $(T_x, x(\cdot))$ where $T_x \in [0,1]$ and $x(\cdot) \in \mathcal{C}(0, T_x; X)$ is a continuous functions satisfying

$$\left\{ \begin{array}{ll} i) & x(0) = x_0 \\ ii) & d(x(T_x), K) \;\leq\; \varepsilon T_x \\ iii) & \forall\, t \in [0, T_x], \;\; d(x(t), K) \;\leq\; \varepsilon \\ iv) & \forall\, t \in [0, T_x], \;\; d((x(t), x'(t)), \mathrm{Graph}(F) \;\leq\; \varepsilon \end{array} \right.$$

The set $\mathcal{A}_\varepsilon(x)$ is not empty: take $T_x = 0$ and $x(0) \equiv x_0$.

It is an inductive set for the order relation

$$(T_{x_1}, x_1(\cdot)) \;\preceq\; (T_{x_2}, x_2(\cdot))$$

if and only if

$$T_{x_1} \;\leq\; T_{x_2} \;\;\&\;\; x_2(\cdot)|_{[0, T_{x_1}]} \;=\; x_1(\cdot)$$

Zorn's Lemma implies that there exists a maximal element $(T_x, x(\cdot)) \in \mathcal{A}_\varepsilon(x_0)$. The Lemma follows from the claim that for such a maximal element, we have $T_x = 1$.

If not, we shall extend $x(\cdot)$ by a solution $\widehat{x}(\cdot)$ on an interval $[T_x, S_x]$ where $S_x > T_x$, contradicting the maximal character of $(T_x, x(\cdot))$.

Let us take $\widehat{x} \in K$ achieving the distance between $x(T_x)$ and K:

$$d(x(T_x), \widehat{x}) \;=\; d(x(T_x), K)$$

We then choose a direction $\widehat{u} \in F(\widehat{x}) \cap T_K(\widehat{x})$, which exists by assumption.

We set

$$\forall\, t \in [T_x, 1],\quad \widehat{x}(t) \;:=\; x(T_x) + (t - T_x)\widehat{u}$$

and

$$\alpha \;:=\; \min\left(\varepsilon, \frac{\varepsilon(1 - T_x)}{\|\widehat{u}\|}\right) \;>\; 0$$

By the definition of the contingent cone, there exists $h_x \in]0, \alpha]$ such that

$$d(\widehat{x} + h_x\widehat{u}, K) \;\leq\; \varepsilon h_x$$

We then set $S_x := T_x + h_x > T_x$.

We obtain

$$\begin{cases} d(\widehat{x}(S_x), K) \;=\; d(x(T_x) + h_x\widehat{u}, K) \\ \leq\; d(\widehat{x} + h_x\widehat{u}, K) + d(x(T_x), \widehat{x}) \\ \leq\; \varepsilon h_x + d(x(T_x), K) \;\leq\; \varepsilon(S_x - T_x) + \varepsilon T_x \;=\; \varepsilon S_x \end{cases}$$

We observe that for any $t \in [T_x, S_x]$,

$$\begin{cases} d(\widehat{x}(t), K) \;\leq\; d(\widehat{x}(t), \widehat{x}) \;\leq\; d(\widehat{x}(t), x(T_x)) + d(x(T_x), \widehat{x}) \\ \leq\; (t - T_x)\|\widehat{u}\| + \varepsilon T_x \;\leq\; \alpha\|\widehat{u}\| + \varepsilon T_x \;\leq\; \varepsilon \end{cases}$$

from the very choice of α.

Finally, we note that for any $t \in [T_x, S_x[$, $\widehat{x}'(t) = \widehat{u}$. Therefore, for all $t \in [T_x, S_x]$,

$$d((\widehat{x}(t), \widehat{x}'(t)), \mathrm{Graph}(F) \;\leq\; d((\widehat{x}(t), \widehat{x}'(t)), (\widehat{x}, \widehat{u})) \;\leq\; d(\widehat{x}(t), \widehat{x}) \;\leq\; \varepsilon$$

Therefore, we have extended the maximal solution $(T_x, x(\cdot))$ on the interval $[0, S_x]$ and obtained the desired contradiction. □

Convergence of Approximate Solutions

Consider now a sequence of ε-approximate solutions $x_\varepsilon(\cdot)$, which exist thanks to Lemma 12.1.3. The Convergence Theorem 12.1.4 below (with $\varepsilon = \frac{1}{m}$) and properties (12.1.2)iii) imply that

$$\text{for almost all } t \in [0, T],\quad x'(t) \;\in\; F(x(t))$$

i.e., that $x(\cdot)$ is a solution to differential inclusion (12.1.1). Condition (12.1.2)ii) implies that

$$\forall\, t \in [0, T],\quad x(t) \;\in\; K$$

i.e., that $x(\cdot)$ is viable.

Upper Semicontinuity of the Solution Map

In particular, the graph of the restriction $\mathcal{S}|_L$ of the solution map $\mathcal{S}$ to a compact subset $L \subset \mathrm{Dom}(F)$ (assumed to be nontrivial) is compact.

Let us choose a sequence of elements $(x_{0_n}, x_n(\cdot))$ of the graph of the solution map $\mathcal{S}$. They satisfy:

$$x_n'(t) \in F(x_n(t)) \ \& \ x_n(0) = x_{0_n} \in L$$

A subsequence (again denoted) x_{0_n} converges to some $x_0 \in L$ because L is compact. Thanks to the Convergence Theorem 12.1.4, we deduce that a subsequence of elements $(x_{0_n}, x_n(\cdot))$ of the graph of $\mathcal{S}|_L$ converges to an element $(x_0, x(\cdot))$ of this graph. This shows that it is compact, and thus, that the solution map $\mathcal{S}$ is upper hemicontinuous with compact images.

12.1.3 The Convergence Theorem

Theorem 12.1.4 *Let $F : X \rightsquigarrow X$ be a Marchaud map. Consider a sequence of m-approximate solutions $x_m(\cdot)$ in the sense that*

$$\forall\, t \in [0,T], \ \ d((x_m(t), x_m'(t)), \mathrm{Graph}(F) \ \le \ \frac{1}{m} \tag{12.1.3}$$

such that $x_m(0)$ converges to x_0. Then a subsequence (again denoted by) $x_m(\cdot)$ converges uniformly on compact intervals to a solution $x(\cdot)$ to the differential inclusion $x' \in F(x)$.

Proof— Consider a sequence of m-approximate solutions $x_m(\cdot)$. They satisfy the following a priori estimates:

$$\|x_m(t)\| \le \left(\|x_0\| + 1 + \frac{c+1}{mc}\right) e^{ct} \ \& \ \|x_m'(t)\| \le c\left(\|x_0\| + 1 + \frac{c+1}{mc}\right) e^{ct} \tag{12.1.4}$$

Indeed, the function $t \to \|x_m(t)\|$ being locally Lipschitz, it is almost everywhere differentiable. Therefore, for any t where $x_m(t)$ is different from 0 and differentiable, we have

$$\frac{d}{dt}\|x_m(t)\| \ = \ \left\langle \frac{x_m(t)}{\|x_m(t)\|}, x_m'(t)\right\rangle \ \le \ \|x_m'(t)\|$$

Since there exist elements $u_t \in \frac{1}{m}B_X$ and $v_t \in \frac{1}{m}B_X$ such that

$$x_m'(t) \ \in \ F(x_m(t) + u_t) + v_t$$

we obtain

$$\|x_m'(t)\| \ \le \ c(\|x_m(t)\| + 1 + \frac{1}{m}) + \frac{1}{m}$$

Setting $\varphi(t) := \|x_m(t)\| + 1 + \frac{1}{m}\frac{c+1}{c}$, we infer that $\varphi'(t) \le c\varphi(t)$, and thus

$$\varphi(t) \ \le \ \varphi(0)e^{ct}$$

from which we deduce the estimates (12.1.4).

Estimates (12.1.4) imply that for all $t \in [0,T]$, the sequence $x_m(t)$ remains in a bounded set and that the sequence $x_m(\cdot)$ is **equicontinuous**, because the derivatives $x'_m(\cdot)$ are bounded. We then deduce from Ascoli's Theorem that it remains in a compact subset of the Banach space $C(0,T;X)$, and thus, that a subsequence (again denoted) $x_m(\cdot)$ converges uniformly to some function $x(\cdot)$.

Furthermore, the sequence $x'_m(\cdot)$ being bounded in the dual of the Banach space $L^1(0,T;X)$, which is equal to $L^\infty(0,T;X)$, it is weakly relatively compact thanks to Alaoglu's Theorem[3]. The Banach space $L^\infty(0,T;X)$ is contained in $L^1(0,T;X)$ with a stronger topology[4]. The identity map being continuous for the norm topologies, is still continuous for the weak topologies. Hence the sequence $x'_m(\cdot)$ is weakly relatively compact in $L^1(0,T;X)$ and a subsequence (again denoted) $x'_m(\cdot)$ converges weakly to some function $v(\cdot)$ belonging to $L^1(0,T;X)$. Equations

$$x_m(t) - x_m(s) = \int_s^t x'_m(\tau)d\tau$$

imply that this limit $v(\cdot)$ is actually the weak derivative $x'(\cdot)$ of the limit $x(\cdot)$.

In summary, we have proved that

$$\begin{cases} i) & x_m(\cdot) \text{ converges uniformly to } x(\cdot) \\ ii) & x'_m(t) \text{ converges weakly to } x'(\cdot) \text{ in } L^1(0,T;X) \end{cases}$$

Let us recall that in a Banach space, the closure (for the normed topology) of a set coincides with its weak closure (for the weakened topology[5]).

We apply this result: for every m, the function $y(\cdot)$ belongs to the weak closure of the convex hull $\text{co}(\{y_p(\cdot)\}_{p\geq m}.)$ It coincides with the (strong) closure of $\text{co}(\{y_p(\cdot)\}_{p\geq m})$. Hence we can choose functions

$$v_m(\cdot) := \sum_{p=m}^{\infty} a_m^p y_p(\cdot) \in \text{co}(\{y_p(\cdot)\}_{p\geq m})$$

[3]Alaoglu's Theorem states that any **bounded subset of the dual of a Banach space is weakly compact.**

[4]Since the Lebesgue measure on $[0,T]$ is finite, we know that

$$L^\infty(0,T;X) \subset L^1(0,T;X)$$

with a stronger topology. The weak topology $\sigma(L^\infty(0,T;X),\, L^1(0,T;X))$ (weak-star topology) is stronger than the weakened topology $\sigma(L^1(0,T;X),\, L^\infty(0,T;X))$ since the canonical injection is continuous. Indeed, we observe that the seminorms of the weakened topology on $L^1(0,T;X)$, defined by finite sets of functions of $L^\infty(0,T;X)$, are seminorms for the weak-star topology on $L^\infty(0,T;X)$), since they are defined by finite sets of functions of $L^1(0,T;X)$.

[5]By definition of the weakened topology, the continuous linear functionals and the weakly continuous linear functionals coincide. Therefore, the closed half-spaces and weakly closed half-spaces are the same. The Hahn-Banach Separation Theorem, which holds true in Hausdorff locally convex topological vector spaces, states that closed convex subsets are the intersection of the closed half-spaces containing them. Since the weakened topology is locally convex, we then deduce that closed convex subsets and weakly closed convex subsets do coincide. This result is known as **Mazur's theorem**.

(where the coefficients a_m^p are positive or equal to 0 but for a finite number of them, and where $\sum_{p=m}^{\infty} a_m^p = 1$) which converge strongly to $y(\cdot)$ in $L^1(0,T;X)$. This implies that the sequence $v_m(\cdot)$ converges strongly to the function $y(\cdot)$ in $L^1(0,T;X)$.

Thus, there exists another subsequence (again denoted by) $v_m(\cdot)$ such that[6]

$$\text{for almost all } t \in [0,T], \; v_m(t) \text{ converges to } y(t)$$

— Let $t \in [0,T]$ such that $x_m(t)$ converges to $x(t)$ in X and $v_m(t)$ converges to $y(t)$ in X. Let $p \in X^\star$ be such that $\sigma(F(x(t)),p) < +\infty$ and let us choose $\lambda > \sigma(F(x(t)),p)$. Since F is upper hemicontinuous, there exists a neighborhood $\mathcal{V}$ of 0 in X such that

$$\forall\, u \in x(t) + \mathcal{V}, \quad \text{then } \sigma(F(u),p) \leq \lambda \tag{12.1.5}$$

Let N_1 be an integer such that

$$\forall\, q \geq N_1, \quad x_q \in x(t) + \frac{1}{2}\mathcal{V}$$

Let $\eta > 0$ be given. Assumption (12.1.3) of the theorem implies the existence of N_2 and of elements (u_q, v_q) of the graph of F such that

$$\forall\, q \geq N_2, \;\; u_q \in x_q(t) + \frac{1}{2}\mathcal{V}, \;\; \|y_q(t) - v_q\| \leq \eta$$

Therefore u_q belongs to $x(t) + \mathcal{V}$ and we deduce from (12.1.5) that

$$\begin{cases} \langle p, y_q(t)\rangle & \leq \; \langle p, v_q\rangle + \eta\|p\|_\star \\ & \leq \; \sigma(F(u_q),p) + \eta\|p\|_\star \\ & \leq \; \lambda + \eta\|p\|_\star \end{cases}$$

Let us fix $N \geq \max(N_1, N_2)$, multiply the above inequalities by the nonnegative a_m^q and add them up from $q = 1$ to ∞. We obtain :

$$\langle p, v_m(t)\rangle \leq \lambda + \eta\|p\|_\star$$

[6]Strong convergence of a sequence in Lebesgue spaces L^p implies that some subsequence converges almost everywhere. Let us consider indeed a sequence of functions f_n converging strongly to a function f in L^p. We can associate with it a subsequence f_{n_k} satisfying

$$\|f_{n_k} - f\|_{L^p} \leq 2^{-k}; \quad \cdots < n_k < n_{k+1} < \cdots$$

Therefore, the series of integrals

$$\sum_{k=1}^{\infty} \int \|f_{n_k}(t) - f(t)\|_X^p dt < +\infty$$

is convergent. The Monotone Convergence Theorem implies that the series

$$\sum_{k=1}^{\infty} \|f_{n_k}(t) - f(t)\|_X^p$$

converges almost everywhere. For every t where this series converges, we infer that the general term converges to 0.

By letting m go to infinity, it follows that

$$\langle p, y(t)\rangle \le \lambda + \eta\|p\|_\star$$

Letting now λ converge to $\sigma(F(x(t)),p)$ and η to 0, we obtain:

$$\langle p, y(t)\rangle \le \sigma(F(x(t)),p)$$

Since this inequality is automatically satisfied for those p such that

$$\sigma(F(x(t)),p) = +\infty$$

it thus holds true for every $p \in X^\star$. Hence, the images $F(x)$ being closed and convex, the Separation Theorem implies that $y(t)$ belongs to $F(x(t))$. The Convergence Theorem ensues. □

12.1.4 The Stability Theorem

Let us consider now a sequence of closed subsets K_n viable under a set-valued map F. *Is the upper limit of these closed subsets still viable under F?* The answer is positive.

Theorem 12.1.5 *Let us consider a sequence of* set-valued maps $F_n : X \rightsquigarrow X$ *having the same linear growth. Assume that the upper limit of the graphs of F_n is contained in the graph of F. Then the upper limit of a sequence of closed subsets K_n viable under F_n is viable under F.*

In particular, the intersection of a decreasing family of closed viability domains is a closed viability domain.

Proof — We shall prove that the upper limit $K^\sharp$ of a sequence of subsets K_n viable under F is viable under F.

Let x_0 belong to $K^\sharp$. It is the limit of a subsequence $x_{0n} \in K_n$. Since the subsets K_n are viable under F, there exist solutions $x_n(\cdot)$ to differential inclusion $x' \in F_n(x)$ starting at x_{0n} and viable in K_n. Since the linear growth of the maps F_n is bounded by a constant c, we deduce that

$$\forall\, n \ge 0,\ \|x_n(t)\| \le (\|x_{0_n}\|+1)e^{ct} \ \&\ \|x_n'(t)\| \le c(\|x_{0_n}\|+1)e^{ct}$$

We conclude that a subsequence of these solutions converge uniformly on compact intervals to a solution of the differential inclusion $x' \in F(x)$ thanks to the Convergence Theorem 12.1.4. Since $x_n(t)$ belongs to K_n for all n, we deduce that $x(t)$ does belong to $K^\sharp$ for all $t > 0$.

When the sequence K_n is decreasing, we know that its upper limit is equal to the intersection of the K_n. □

12.2 The Filippov Theorem

Actually, Filippov's Theorem is much more than a mere existence theorem. It also provides an estimate of the distance between a function $y(\cdot)$ and the set $\mathcal{S}_F(x_0)$ of solutions starting at some initial state x_0.

Theorem 12.2.1 (Filippov) *Assume that $F : X \rightsquigarrow X$ is λ-Lipschitz with nonempty closed values on the interior of its domain. Let $y(\cdot)$ be a given absolutely continuous function such that $t \to d(y'(t), F(y(t)))$ is integrable (for the measure $e^{-\lambda s}ds$).*

Then there exists a solution $x(\cdot)$ to differential inclusion (12.1.1) such that, for all $t \geq 0$,

$$\|x(t) - y(t)\| \leq e^{\lambda t}\left(\|x_0 - y(0)\| + \int_0^t d(y'(s), F(y(s)))e^{-\lambda s}ds\right) \quad (12.2.1)$$

and for almost all $t \in [0, T[$,

$$\begin{cases} \|x'(t) \; - \; y'(t)\| \; \leq \; d(y'(t), F(y(t))) \\ +\lambda e^{\lambda t}\left(\|x_0 - y(0)\| \; + \; \int_0^t d(y'(s), F(y(s)))e^{-\lambda s}ds\right) \end{cases}$$

It implies the existence of a solution starting from any initial state with any initial velocity:

Corollary 12.2.2 *Assume that F is Lipschitz on the interior of its domain. Then, for any $x_0 \in \mathrm{Int}(\mathrm{Dom}(F))$ and $v_0 \in F(x_0)$, there exist $T > 0$ and a solution $x(\cdot)$ to differential inclusion (12.1.1) defined on $[0, T]$ and satisfying $x(0) = x_0$ and $x'(0) = v_0$.*

Proof— We apply Filippov's Theorem with $y(t) := x_0 + tv_0$ and $x_0 := y(0)$. Then $d(y'(t), F(y(t))) \leq \lambda t\|v_0\|$ and

$$\begin{cases} \int_0^t d(y'(s), F(y(s)))e^{-\lambda s}ds \\ \leq \; e^{\lambda t}\int_0^t \lambda\tau\|v_0\|e^{-\lambda\tau}d\tau \; \leq \; \frac{\|v_0\|}{\lambda}\left(e^{\lambda t} - 1 - \lambda t\right) \end{cases}$$

Filippov's Theorem implies the existence of a solution $x(\cdot)$ to differential inclusion (12.1.1) starting at x_0 and satisfying

$$\|x(t) - x_0 - tv_0\| \; \leq \; \frac{\|v_0\|}{\lambda}\left(e^{\lambda t} - 1 - \lambda t\right)$$

Dividing by $t > 0$ and letting t converge to $0+$, we infer $x'(0) = v_0$. □

Filippov's Theorem also implies the Lipschitz dependence of the solution map on the initial condition:

Corollary 12.2.3 *Let $y(\cdot) \in \mathcal{S}_F(y_0)$ and assume that F, $y(\cdot)$ satisfy the assumptions of Filippov's Theorem 12.2.1. Then*

$$d(y(t), \mathcal{S}_F(x_0)(t)) \leq \|x_0 - y_0\| e^{\lambda t}$$

so that the solution map $\mathcal{S}_F$ is lower semicontinuous.

Theorem 12.2.4 *Let us assume that $F : X \rightsquigarrow X$ is Lipschitz. Then any solution $x(\cdot)$ to the differential inclusion $x' \in F(x)$ satisfies inequalities*

$$\|x(t) - x(0)\| \leq \|F(x(0))\| \frac{e^{\lambda t} - 1}{\lambda}$$

and

$$x(t) - x(s) \in (t-s)F(x(s)) + \|F(x(0))\| e^{\lambda s} \frac{e^{\lambda(t-s)} - \lambda(t-s) - 1}{\lambda}$$

Proof — Take any solution $x(\cdot) \in \mathcal{S}_F(x_0)$ to the differential inclusion $x' \in F(x)$, which satisfies $\forall\, t \geq s \geq 0$,

$$x(t) - x(s) \in \int_s^t F(x(\tau))d\tau \subset (t-s)F(x(s)) + \lambda \int_s^t \|x(\tau) - x(s)\| d\tau B$$

Then we deduce that

$$\|x(t) - x(s)\| \leq (t-s)\|F(x(s))\| + \lambda \int_s^t \|x(\tau) - x(s)\| d\tau \tag{12.2.2}$$

The Gronwall Lemma implies inequality

$$\|x(t) - x(s)\| \leq \|F(x(s))\| \frac{e^{\lambda(t-s)} - 1}{\lambda} \tag{12.2.3}$$

which, used again in (12.2.2), provides: $\forall\, t \geq s \geq 0$,

$$x(t) - x(s) \in (t-s)F(x(s)) + \|F(x(s))\| \frac{e^{\lambda(t-s)} - \lambda(t-s) - 1}{\lambda} B$$

In particular, taking $s = 0$, we deduce from (12.2.3) that

$$\|x(t) - x(0)\| \leq \|F(x(0))\| \frac{e^{\lambda t} - 1}{\lambda}$$

Therefore, F being Lipschitz, we obtain

$$F(x(t)) \subset (\|F(x(0))\| + \lambda \|x(t) - x(0)\|) B \subset \|F(x(0))\| e^{\lambda t} B$$

Putting these inequalities together, we infer that $\forall\, t \geq s \geq 0$,

$$x(t) - x(s) \in (t-s)F(x(s)) + \|F(x(0))\| e^{\lambda s} \frac{e^{\lambda(t-s)} - \lambda(t-s) - 1}{\lambda}$$

12.3 Time Dependent Differential Inclusions

We consider a set-valued map$F : [0,\infty[\times X \rightsquigarrow X$ and the *time dependent differential inclusion*

$$\text{for almost all } t \in [0,T], \; x'(t) \in F(t,x(t)) \tag{12.3.1}$$

We shall denote by $\mathcal{S}_F(t_0,x_0)$ the set of solutions $x(\cdot)$ to the differential inclusion (12.3.1) starting from x_0 at time $t_0 \geq 0$ and say that $\mathcal{S}_F$ is the solution map.

The **reachable tube** $\vartheta_F(\cdot,K)$ defined by

$$\vartheta_F(t,K) := \{x(t)\}_{x(\cdot)\in\mathcal{S}_F(0,K)}$$

satisfies obviously the semigroup property

$$\vartheta_F(t+h,K) := \vartheta_F(h,\vartheta_F(t,K))$$

We shall characterize in this section the reachable tubes $\vartheta_F(\cdot,K_0)$ of $t \mapsto F(t)$ starting at some compact subset $K_0 \in \mathcal{K}(X)$ as the unique "**closed invariant and backward viable tube**" starting at K_0:

Definition 12.3.1 *A tube K is* viable *under F (or enjoys the* viability property*) if and only if, for all $t_0 \geq 0$ and $x_0 \in K(t_0)$, there exists at least one solution $x(\cdot)$ to the differential inclusion (12.3.1) starting at x_0 at time t_0 which is viable in the tube K.*

It is said to be backward viable *under F if for every $t_0 \in [0,+\infty[$, $x_0 \in K(t_0)$, there exists at least one solution $x(\cdot)$ to the differential inclusion (12.3.1) on the interval $[0,t_0]$ starting at $K(0)$, viable in the tube K on $[0,t_0]$ and such that $x(t_0) = x_0$. It is said to be* bilaterally viable *under F if it is both viable and backward viable, i.e., if and only if $\forall\, t_0 \in [0,+\infty[, \forall\, x_0 \in K(t_0)$, there exists at least one viable solution starting at $K(0)$ and passing through x_0 at time t_0 (in the sense that $x(t_0) = x_0$).*

A tube $K : [0,\infty[\rightsquigarrow X$ is called a viability tube *of a set-valued map $F : [0,\infty[\times X \rightsquigarrow X$ if*

$$\forall\, t \in [0,\infty[,\; \forall\, x \in K(t),\;\; F(t,x) \cap DK(t,x)(1) \neq \emptyset$$

a backward viability tube *if*

$$\forall\, t \in [0,\infty[,\; \forall\, x \in K(t),\;\; F(t,x) \cap -DK(t,x)(-1) \neq \emptyset$$

and a bilateral viability tube *if it is both viable and backward viable. We say that a tube K* is invariant under *F (or enjoys the* invariance property*) if and only if for all t_0 and $x_0 \in K(t_0)$,* all the solutions *to differential inclusion (12.3.1) are viable in the tube K.*

It is called an invariance tube *if*

$$\forall\, t \in [0,\infty[,\; \forall\, x \in K(t),\;\; F(t,x) \subset DK(t,x)(1)$$

12.3.1 Viable Tubes

Remark that a subset K is viable (repspectively invariant) under a time independent set-valued map $F : X \rightsquigarrow X$ if and only if the "stationary tube" $t \rightsquigarrow K(t) := K$ is viable (repspectively invariant) under F.

Our first task is to characterize tubes enjoying the viability property thanks to its "contingent derivative".

We observe that it is enough to know this contingent derivative only in the directions $1, 0$ and -1 because $\tau \rightsquigarrow DK(t,x)(\tau)$ is positively homogeneous. In particular, we note that

$$\begin{cases} DK(t,x)(1) & = \left\{v \in X \mid \liminf_{h\to 0+} d\left(v, \frac{K(t+h)-x}{h}\right) = 0\right\} \\ DK(t,x)(0) & = T_{K(t)}(x) \text{ when } K \text{ is Lipschitz} \end{cases} \tag{12.3.2}$$

Viability Theorem 6.1.9 for autonomous (i.e., time independent) systems can be easily translated in the time dependent case and yield the following summary:

Theorem 12.3.2 *Assume that $F : [0,\infty[\times X \rightsquigarrow X$ is a Marchaud map. Then*

1. *A necessary and sufficient condition for a closed tube to be viable under F is that it is a viability tube, or, equivalently, that*

$$\forall\, x \in K(t),\;\; F(t,x) \cap \overline{co} DK(t,x)(1) \;\neq\; \emptyset$$

2. *For any given closed tube K, there exists a largest closed viability tube $K_\propto$ contained in K, called the* viability kernel *of the tube K.*

   If an initial state $x_s \in K(s)$ at time s does not belong to $K_\propto(s)$, then any solution $x(\cdot)$ to the differential inclusion starting at x_s satisfies

$$\begin{cases} i) & \forall\, t \geq s,\;\; x(t) \;\notin\; K_\propto(t) \text{ as long as } x(t) \;\in\; K(x(t)) \\ ii) & \exists\, T \geq s \;\; \text{such that } x(T) \notin K(T) \quad \square \end{cases}$$

3. *If K_n is a sequence of closed viability tubes, then the graphical upper limit K, which is the tube defined by the*

$$\text{Graph}(K) \;:=\; \text{Limsup}_{n\to\infty}\text{Graph}(K_n)$$

   is a also a closed viability tube.

Furthermore, the solution map $\mathcal{S}_F$ is upper semicontinuous with compact images from Graph(K_F) *to the space $\mathcal{C}(0,\infty;X)$ supplied with the compact convergence topology.*

Proof — We introduce the set-valued map G from Graph(K) to $\mathbf{R}_+ \times \mathbf{R}^n$ defined by

$$G(s,x) := \{1\} \times F(s,x)$$

We observe that $(s(\cdot), x(\cdot))$ is a solution to the differential inclusion

$$\text{for almost all } t, \ (s'(t), y(t)) \in G(s(t), y(t))$$

starting at $(s(0), x(0)) = (t_0, x_0)$ if and only if the function $x(\cdot)$ defined by $x(t) := y(t - t_0)$ is a solution to differential inclusion (12.3.1) starting at x_0 at time t_0. We also note that the tube K is viable under F if and only if its graph is viable under G and that K is a viability tube if and only if its graph is a viability domain of G. It thus remains to translate the time independent results. □

We derive the following characterization of backward viability:

Proposition 12.3.3 *Assume that $F : [0, \infty[\times X \rightsquigarrow X$ is a Marchaud map. Then a closed tube K is backward viable under F if and only if it is a backward viability tube. As a consequence, P is bilaterally viable under F if and only it is a bilateral viability tube.*

Proof — Let $K(\cdot)$ be a backward viability tube starting at K and x_0 belong to $K(t_0)$.

Let us consider the tube $\check{K}_{t_0}(s) := K(t_0 - s)$ defined by

$$\check{K}_{t_0}(s) := \begin{cases} K(t_0 - s) & \text{if } s \in [0, t_0] \\ K & \text{if } s \geq t_0 \end{cases}$$

We observe that

$$D\check{K}_{t_0}(s,x)(\lambda) = DK(t_0 - s, x)(-\lambda)$$

because one can check easily that

$$(\lambda, u) \in T_{\text{Graph}(\check{K}_{t_0})}(s,x)$$

if and only if

$$(-\lambda, u) \in T_{\text{Graph}(K)}(t_0 - s, x)$$

Second, we consider the set-valued map $\check{G}_{t_0}$ defined by

$$\check{G}_{t_0}(s,x) := \begin{cases} \{-1\} \times -F(t_0 - s, x) & \text{if } s \in]0, t_0] \\ [-1, 0] \times \overline{\text{co}}(\{0\} \cup -F(0,x)) & \text{if } s = t_0 \\ \{0\} \times \overline{\text{co}}(\{0\} \cup -F(0,x)) & \text{if } s > t_0 \end{cases}$$

It is a Marchaud map since F is assumed to be a Marchaud map. Then, we observe that K is a backward viability tube if and only if the graph of $\check{K}_{t_0}$ is a viability domain of $\check{G}_{t_0}$.

Therefore, the Viability Theorem implies that this is equivalent to say that the graph of $\check{K}_{t_0}$ is viable under $\check{G}_{t_0}$.

This implies that for every $t_0 \in [0, +\infty[$, $x_0 \in K(t_0)$, there exists a solution $z(\cdot)$ to the backward differential inclusion $z'(t) \in -F(t_0 - t, z(t))$ starting at x_0 at time 0 and viable in the tube $t \rightsquigarrow K(t_0 - t)$ for all $t \in [0, t_0]$. By setting $x(t) := z(t_0 - t)$ when $t \in [0, t_0]$, we infer that $x(\cdot)$ is a solution to the differential inclusion $x' \in F(t, x)$ starting at $x(0) = z(t_0) \in K(0)$ and satisfying $x(t_0) = x_0$. □

12.3.2 Invariant Tubes

Theorem 12.3.4 *Let us assume that* $\forall\, t \geq 0$, $F(t)$ *is Lipschitz and that*

$$\lambda := \sup_{t \geq 0} \|F(t)\|_{\Lambda} < +\infty$$

Let $K(\cdot) : \mathbf{R}_+ \rightsquigarrow X$ *be a Lipschitz tube. It is invariant under* F *if and only if*

$$\forall\, (t, x) \in \mathrm{Graph}(K), \quad F(t, x) \subset \overline{co}(DK(t, x)(1)) \tag{12.3.3}$$

and actually, if and only if K *is an invariance tube under* F.

Proof — The necessity follows from the Filippov Theorem since we know that for any $u \in F(t_0, x_0)$, there exists a solution to the differential inclusion $x'(t) \in F(t, x(t))$ starting at x_0 at time t_0 with the initial velocity u. Since the tube $K(\cdot)$ is assumed to be invariant under F, we deduce that u belongs to $DK(t, x)(1)$, which is contained in $\overline{co}(DK(t, x)(1))$.

Conversely, let $x(\cdot)$ be any solution to differential inclusion $x'(t) \in F(t, x)$ starting at x_0 at time 0. Let us set $g(t) := d(x(t), K(t))$, which is Lipschitz, and thus, almost everywhere differentiable.

Let t be a point such that both $x'(t)$ and $g'(t)$ exist and $x'(t)$ belongs to $F(t, x(t))$. Then there exists $\varepsilon(h)$ converging to 0 with h such that

$$x(t + h) = x(t) + hx'(t) + h\varepsilon(h)$$

and

$$g'(t) = \lim_{h \to 0+} \frac{d(x(t) + hx'(t) + h\varepsilon(h), K(t + h)) - d(x(t), K(t))}{h}$$

Lemma 12.3.5 below implies that

$$g'(t) \leq d(x'(t), \overline{co}(DK(t, \Pi_{K(t)}(x(t))(1)))$$

Choose any y in $\Pi_K(x(t))$. We deduce that:

$$\begin{aligned} &d(x'(t), \overline{co}(DK(t, \Pi_{K(t)}(x(t))(1))) \leq d(x'(t), \overline{co}(DK(t,y)(1))) \\ &\leq d(x'(t), F(t,y)) \\ &\text{(by assumption (12.3.3))} \\ &\leq d(x'(t), F(t,x(t))) + \lambda\|y - x(t)\| \text{ (since } F(t) \text{ is Lipschitz)} \\ &= 0 + \lambda d(x(t), K(t)) = \lambda g(t) \end{aligned}$$

Then g is a solution to

$$\text{for almost all } t \in [0,T], \quad g'(t) \leq \lambda g(t) \quad \& \quad g(0) = 0$$

We deduce that $g(t) = 0$ for all $t \in [0,T]$, and therefore, that $x(t) \in K(t)$ for every $t \in [0,T]$. □

Lemma 12.3.5 *Let $K(\cdot)$ be a closed tube of a finite dimensional vector-space. Then the following inequalities:*

$$\liminf_{h \to 0+} \frac{d(y + hv, K(h)) - d(y, K(0))}{h} \leq d(v, \overline{co}(DK(0, \Pi_{K(0)}(y))(1))))$$

hold true.

Proof — Choose $z \in \Pi_{K(0)}(y)$ and $w \in DK(0,z)(1)$, so that there exist a sequence $h_n > 0$ and w_n converging to 0 and w satisfying $z + h_n w_n \in K(h_n)$. Then

$$\frac{d(y + h_n v, K(h_n)) - d(y, K(0))}{h_n} \leq \frac{\|y - z + h_n(v - w_n)\| - \|y - z\|}{h_n}$$

Passing to the limit, we infer that

$$\liminf_{h \to 0+} \frac{d(y + hv, K(h)) - d(y, K(0))}{h} \leq \left\langle \frac{y - z}{\|y - z\|}, v - w \right\rangle$$

Since this inequality holds true for every $w \in DK(0,z)(1)$, it also holds true for every $w \in \overline{co}(DK(0,z)(1))$. Hence we infer that $\forall\, z \in \Pi_{K(0)}(y)$, and $\forall\, w \in \overline{co}(DK(0,z)(1))$,

$$\liminf_{h \to 0+} \frac{d(y + hv, K(h)) - d(y, K(0))}{h} \leq \|v - w\| \quad \square$$

12.3.3 Characterization of Reachable Tubes

The **reachable tube** $\vartheta_F(\cdot, K)$ is obviously a closed tube whenever F is Marchaud.

Theorem 12.3.6 *The* **reachable tube** $\vartheta_F(\cdot, K)$ *is invariant under* F *and "minimal" in the sense that there is no other tube* K *invariant under* F *starting at* K *and strictly contained in* K.

It is also backward viable under F *and is the largest closed bilateral viability tube starting at* K.

Therefore, the reachable tube $\vartheta_F(\cdot, K)$ *is the only closed invariant and backward viable tube starting at* K.

Proof — The **reachable tube** $\vartheta_F(\cdot, K)$ is invariant under F: Indeed, if $x_0 \in \vartheta_F(t_0, K)$, there exists by definition a solution $x(\cdot)$ to the differential inclusion $x'(t) \in F(t, x(t))$ starting from K at time 0 and passing through x_0 at t_0. Furthermore, all solutions $y(\cdot)$ to the differential inclusion starting at x_0 at time t_0, concatenated to $x(\cdot)$ restricted to the interval $[0, t_0]$ being a solution to the differential inclusion starting at K, so that $\vartheta_F(\cdot, K)$ is invariant.

Let us consider a closed tube $K(\cdot) \subset \vartheta_F(\cdot, K)$ invariant under F starting at K. We claim that it is equal to the reachable tube. Otherwise, there would exist $x_s \in \vartheta_F(s, K)$ such that $x_s \notin K(s)$. Since the reachable tube is backward viable, there exists a solution $x(\cdot)$ to the differential inclusion (12.3.1) starting from $x(0) \in K$ such that $x(s) = x_s$. But starting from $x(0)$, the solution is viable in the tube K since it is invariant under F and satisfies $K(0) = K$. Therefore $x(s)$ belongs to $K(s)$, a contradiction.

The **reachable tube** $\vartheta_F(\cdot, K)$ is obviously backward viable under F.

Let now K be any forward and backward viable tube starting from K at time 0 and let us check that it is contained in the reachable tube. For that purpose, take $x_0 \in K(t_0)$. By Proposition 12.3.3, we know that there exists at least one viable solution starting at $K(0) = K$ and passing through x_0 at time t_0. Hence $x_0 \in \vartheta_F(t_0, K)$.

The **reachable tube** $\vartheta_F(\cdot, K)$ is obviously invariant and backward viable under F, and it is the only one such a tube starting at K. Indeed, if $K(\cdot)$ is invariant under F and starts at K, then we saw that for all $t \geq 0$, $\vartheta_F(t, K) \subset K(t)$. Since it is also bilaterally viable, then for all $t \geq 0$, $K(t) \subset \vartheta_F(t, K))$. □

12.4 Differential Inclusions with Memory

12.4.1 History and Future Spaces

We set

$$\mathcal{H}(X) := \mathcal{C}(-\infty,0;X)$$

the history (or memory) space and

$$\mathcal{A}(X) := \{x(\cdot) \in \mathcal{C}(0,+\infty;X) \text{ such that } x(0) = 0\}$$

the future space. They are supplied with the compact convergence topology. We denote by $\mathcal{H}_\lambda(X)$ and $\mathcal{A}_\lambda(X)$ the subsets of Lipschitz functions with Lipschitz constant λ.

If $\mathcal{K} \subset \mathcal{H}(X)$, we set

$$\mathcal{K}(\tau) := \{\varphi(\tau)\}_{\varphi\in\mathcal{K}}$$

Observe that Ascoli's Theorem states that a *closed subset* $\mathcal{K} \subset \mathcal{H}_\lambda(X)$ *is compact if and only if* $\mathcal{K}(0) := \{\varphi(0)\}_{\varphi\in\mathcal{K}}$ *is bounded*, since it is closed and equicontinuous (by assumption) and pointwise bounded because, for all $\psi \in \mathcal{K}$ and $\tau \leq 0$,

$$\|\psi(\tau)\| \leq \|\psi(\tau) - \psi(0)\| + \|\psi(0)\| \leq \lambda|\tau| + \|\mathcal{K}(0)\|$$

We embed the state space X into $\mathcal{A}(X)$ by identifying a vector x with the function $\psi_x(t) := tx$. The image of the ball λB_X of radius λ under this embedding is contained in $\mathcal{A}_\lambda(X)$.

A first example of closed subset of histories is given by

$$\mathcal{K} := \{\, \varphi \in \mathcal{H}(X) \mid \varphi(0) \in K \,\}$$

where the constraints bear only on the present. Another class is given by Volterra sets defined by

$$\mathcal{K} := \left\{\varphi \mid \varphi(0) \in M\left(\int_{-\infty}^{0} C(-s)\varphi(s)d\mu(s)\right)\right\}$$

where the constraints involve cumulated consequences of the history.

In the discrete case,

$$\mathcal{K} := \left\{\varphi \mid \varphi(0) \in M\left(\sum_{j=-\infty}^{0} C(-i)\varphi(\beta_j)\right)\right\}$$

involves discrete cumulated consequences of the history (delays).

12.4.2 Concatenations

Definition 12.4.1 *Let $h > 0$ be given. The* h-concatenation *(or* concatenation *when there are no ambiguities) $\varphi\Diamond_h\psi$ is the bilinear form from $\mathcal{H}(X) \times \mathcal{A}(X) \mapsto \mathcal{H}(X)$ defined by*

$$(\varphi\Diamond_h\psi)(\tau) := \begin{cases} \varphi(\tau+h) & \text{if } \tau \in]-\infty, h] \\ \varphi(0) + \psi(\tau+h) & \text{if } \tau \in [-h, 0] \end{cases}$$

As an example, the concatenation of $\varphi \in \mathcal{H}(X)$ and $x \in X$ is defined by

$$(\varphi\Diamond_h x)(\tau) = \begin{cases} \varphi(\tau+h) & \text{if } \tau \in]-\infty, h] \\ \varphi(0) + (\tau+h)x & \text{if } \tau \in [-h, 0] \end{cases}$$

We attach to a function $\varphi \in \mathcal{H}(X)$ and a sequence $v_1, \ldots, v_n$ the Euler concatenation defined by

$$\varphi\Diamond_h v_1\Diamond_h v_2\Diamond_h \cdots \Diamond_h v_n$$

which is piecewise linear function on the interval $[-nh, 0]$.

Lemma 12.4.2 *For any Lipschitz constant $\lambda > 0$, the h-concatenation maps $\mathcal{H}_\lambda(X) \times \mathcal{A}_\lambda(X)$ to $\mathcal{H}_\lambda(X)$*

We also observe that

$$\varphi\Diamond_h\psi = \varphi$$

if and only if

$$\begin{cases} i) & \forall\, \tau \in]-\infty, 0],\ \varphi(\tau) = \varphi(\tau - h) \\ & \text{(i.e., } \varphi \text{ is periodic with period } h) \\ ii) & \forall\, t \in [0, h],\ \psi(t) = \varphi(t-h) - \varphi(0) \end{cases}$$

As a consequence, $\varphi\Diamond_h x = \varphi$ if $\varphi \equiv cst$ and $x = 0$.

12.4.3 Contingent Histories

We associate with any continuous function $x(\cdot)$ its *history* $T(t)x$ *up to time* t defined by:

$$\forall\, \tau \in]-\infty, 0],\ T(t)x(\tau) := x(t+\tau)$$

Then $T(t)$ maps $\mathcal{A}(X)$ to $\mathcal{H}(X)$ and satisfies the semi-group property

$$T(t+s)s = T(s)T(t)x$$

In this continuous framework, we define the constraints of the history of the evolution through a closed subset $\mathcal{K} \subset \mathcal{H}(X)$. Viable evolutions with memory are the ones which satisfy

$$\forall\, t \geq 0,\ T(t)x \in \mathcal{K} \tag{12.4.1}$$

Definition 12.4.3 *Let $\mathcal{K} \subset \mathcal{H}(X)$ be a subset of histories and $\varphi \in \mathcal{H}$. We denote by $\mathcal{D}_{\mathcal{K}}(\varphi)$ the set of vectors $v \in X$ such that there exist a sequence $h_n > 0$ converging to 0 and a sequence $v_n \in X$ converging to v satisfying*

$$\forall\, n \geq 0,\ \ \varphi \diamondsuit_{h_n} v_n \ \in\ \mathcal{K}$$

Theorem 12.4.4 *We assume that the function $\gamma : \mathcal{H}(X) \times Z \mapsto X$ is continuous with linear growth and affine with respect to p, that $\mathcal{P} : \mathcal{H}(X) \rightsquigarrow X$ is Marchaud. Assume that $\mathcal{K}$ is contained in $\mathcal{H}_\lambda(X)$ for some $\lambda > 0$.*

Then from any initial history $\varphi \in \mathcal{K}$ starts one solution to

$$\left\{\begin{array}{lll} i) & x'(t) & = \gamma(T(t)x, p(t)) \\ & & \\ ii) & p(t) & \in \mathcal{P}(T(t)x) \end{array}\right. \tag{12.4.2}$$

starting from an initial condition

$$T(0)x \;=\; \varphi, \text{ where } \varphi \text{ is given} \tag{12.4.3}$$

viable in the sense that $T(t)x \in \mathcal{K}$ for all $t \geq 0$ if and only if $\mathcal{K}$ is a viability domain of $(\mathcal{P}, \gamma)$. Let us set

$$\Pi_{\mathcal{K}}(\varphi) \;:=\; \{p \in \mathcal{P}(\varphi) \mid \gamma(\varphi, p) \in \mathcal{D}_{\mathcal{K}}(\varphi)\}$$

Then, if $\mathcal{K}$ is a viability domain of $(\mathcal{P}, \gamma)$, viable solutions are governed by the learning law[7]

$$\text{for almost all } t \geq 0,\ \ p(t) \ \in\ \Pi_{\mathcal{K}}(T(t)x)$$

We denote by $\Lambda : \mathcal{H}(X) \rightsquigarrow \mathcal{H}(X)$ the set-valued map associating with any $\varphi \in \mathcal{H}(X)$ the subset $\Lambda\varphi$ of functions $\nu \in \mathcal{H}(X)$ such that if for every sequence $h_n > 0$ converging to 0, there exists a sequence ν_n converging to ν in $\mathcal{H}(X)$ such that

$$\forall\, n \geq 0,\ \ \varphi \diamondsuit_{h_n} \nu_n(0) \;=\; \varphi + h_n \nu_n$$

Therefore, if $\nu \in \Lambda\varphi$, then

$$\forall\, \tau < 0,\ \ \nu(\tau) := D^\flat_\uparrow \varphi(\tau)(1)$$

is the adjacent derivative of φ at 0 in the forward direction +1.

We consider the case when

$$\mathcal{K} \;:=\; \{\varphi \in \mathcal{H}(X) \ \text{ such that } \ \varphi(0) \ \in\ M(U(\varphi))\} \tag{12.4.4}$$

where $U \in \mathcal{L}(\mathcal{H}(X), Y)$ is a continuous linear operator and where $M : Y \rightsquigarrow X$ is a closed set-valued map.

---

[7]Since the history of the evolution is involved in the regulation map, the regulation law is more than a regulation law: It can be called a learning law.

Theorem 12.4.5 *Assume that M is sleek. We posit the following "surjectivity condition" on U: there exists a constant $\delta > 0$ such that for all $h > 0$,*

$$\begin{cases} \forall\,(v,u) \in Y \times X, \ \exists\, \psi_h \in \mathcal{A}(X) \ \ \textit{such that} \\ \psi_h(h) = u, \ \ U(0\diamond_h\psi_h) = v \\ \text{and satisfying} \\ \|\psi_h\|_{C(0,1;X)} \ \leq \ \delta(\|u\| + \|v\|) \end{cases}$$

Therefore

$$DM(U\varphi,\varphi(0))(U(\Lambda\varphi)) \ \subset \ \mathcal{D}_{\mathcal{K}}(\varphi)$$

The next statement trades surjectivity condition on U with restrictions on the size of the norm of U and the norm $\|DM(y,x)\|$ defined by

$$\|DM(y,x)\| \ := \ \sup_{v\neq 0} \ \inf_{u\in DM(y,x)(v)} \frac{\|u\|}{\|v\|}$$

Theorem 12.4.6 *Assume that for any $(y,x) \in \text{Graph}(M)$, the domain*

$$\text{Dom}(DM(y,x)) \ = \ Y$$

and that

$$\forall\,(y,x) \in \text{Graph}(M), \ \ \|DM(y,x)\| \ \leq \ \beta \ < \ +\infty$$

and that there exists $\gamma > 0$ such that

$$\|U(0\diamond_h\psi)\| \ \leq \ \delta h\|\psi\|$$

Therefore

$$DM(U\varphi,\varphi(0))(U(\Lambda\varphi)) \ \subset \ \mathcal{D}_{\mathcal{K}}(\varphi)$$

These results follow from a more general sufficient condition, which is exposed in chapter 12 of [28, Aubin].

12.4.4 Viability Constraints with Delays

Here, we take $Y := X^p$ and $U(\varphi) := (\varphi(-\theta_1),\ldots,\varphi(-\theta_p))$. Then the surjectivity assumption of Theorem 12.4.5 is satisfied, so that we obtain the following consequence:

Corollary 12.4.7 *Let us consider p positive delays $\theta_1,\ldots,\theta_p$. Assume that*

$$\mathcal{K} \ := \ \{\varphi \in \mathcal{C} \ \ \textit{such that} \ \varphi(0) \ \in \ M(\varphi(-\theta_1),\ldots,\varphi(-\theta_p))\}$$

Then

$$DM(\varphi(-\theta_1),\ldots,\varphi(-\theta_p),\varphi(0))((D_\uparrow^\flat\varphi(-\theta_1)(1),\ldots,(D_\uparrow^\flat\varphi)(-\theta_p)(1)) \ \subset \ \mathcal{D}_{\mathcal{K}}(\varphi)$$

12.4.5 Volterra Viability Constraints

Let us consider a finite dimensional space Y, a set-valued map $M : Y \rightsquigarrow X$ and $A \in L^1(0,\infty;\mathcal{L}(X,Y))$. We consider the case when U is defined by

$$U\varphi := \int_{-\infty}^{0} A(-s)\varphi(s)ds$$

Corollary 12.4.8 *Let us consider* $A \in L^1(0,\infty;\mathcal{L}(X,Y))$ *satisfying*

$$\sup_{t\in[0,1]} \|A(t)\|_{\mathcal{L}(Y,X)} \leq \delta < +\infty$$

Assume that for all $(y,x) \in \mathrm{Graph}(M)$, $\mathrm{Dom}(DM(y,x)) = Y$ *and that*

$$\sup_{(y,x)\in\mathrm{Graph}(M)} \|DM(y,x)\| \leq \beta < +\infty$$

Let $\mathcal{K}$ *be the subset defined by*

$$\mathcal{K} := \left\{\varphi \in \mathcal{C} \ \text{ such that } \varphi(0) \in M\left(\int_{-\infty}^{0} A(-s)\varphi(s)ds\right)\right\}$$

Then

$$DM\left(\int_{-\infty}^{0} A(-s)\varphi(s)ds, \varphi(0)\right)\left(\int_{-\infty}^{0} A(-s)\varphi'(s)ds\right) \subset \mathcal{D}_{\mathcal{K}}(\varphi)$$

Proof — It follows from Theorem 12.4.6. □

12.5 Mutational Equations

We have proposed in Chapter 3 nonempty compact convex subsets $\Phi \subset X$ as candidates for mutations of tubes $t \rightsquigarrow K(t) \subset X$. This provides only one class of mutations, called the **morphological mutations.**

However, it happens that this class may be too small: The larger the class of mutations, the more mutationable tubes.

This is the reason why mutations will be chosen in the space $\mathrm{LIP}(X,X)$ of Lipschitz set-valued maps $\Phi : X \mapsto X$ with compact convex values. This space of mutations contains in particular

1. Lipschitz single-valued maps $\varphi : X \mapsto X$, used in shape optimization,

2. compact convex subsets B, used in mathematical morphology.

When $\Phi \in \mathrm{LIP}(X,X)$, differential inclusion

$$x'(t) \in \Phi(x(t))$$

does have a solution for any initial state x_0, thanks to the Filippov Theorem.

We then regard the reachable map

$$\vartheta_\Phi(h,K) := \{\vartheta_\Phi(h,x)\}_{x\in K}$$

as analogues to the half-lines $x+hv$ for defining "velocities" of tubes. We recall that $\boldsymbol{d}$ denotes the Hausdorff distance between compact sets is defined by

$$\begin{cases} i) & h^\sharp(K,L) = \sup_{x\in K} d(x,L) = \sup_{x\in K}\inf_{y\in L} d(x,y) \\ ii) & \boldsymbol{d}(K,L) := \max(h^\sharp(K,L), h^\sharp(L,K)) \end{cases} \quad (12.5.1)$$

Definition 12.5.1 *Let X be a finite dimensional vector space and $K:\mathbf{R}\rightsquigarrow X$ be tube. A mutation $\Phi \in \overset{\circ}{K}(t)$ of the tube K at t in the forward direction 1 is a set-valued maps Φ in* Lip(X,X) *satisfying*

$$\Phi \in \overset{\circ}{K}(t) \text{ if and only if } \lim_{h\to 0+} \frac{\boldsymbol{d}(K(t+h),\vartheta_\Phi(h,K(t)))}{h} = 0$$

We denote by $\overset{\circ}{K}(t) \subset$ Lip(X,X) *the set of mutations of the tube K at t in the forward direction* 1.

This equivalent to say that

$$\vartheta_\Phi(h,K(t)) \subset B(K(t+h)),\beta(h)h) \;\&\; K(t+h) \subset B(\vartheta_\Phi(h,K(t)),\beta(h)h)$$

where $\beta(h)$ converges to 0 with h.

Theorem 12.5.2 *Assume that K is a set-valued map from $\mathbf{R}$ to X with closed values. Let $\Phi \in \overset{\circ}{K}(t)$ be a mutation of K at t in the forward direction 1. Then:*

$$\forall x \in K(t), \quad DK(t,x)(1) = \Phi(x) + T_{K(t)}(x)$$

Proof

1. We shall prove that if Φ is Lipschitz with closed values, property

$$\vartheta_\Phi(h,K(t)) \subset F(x+h) + h\varepsilon(h)B \quad (12.5.2)$$

implies inclusion

$$\forall x \in K(t), \quad \Phi(x) + T_{K(t)}(x) \subset DK(t,x)(1) \quad (12.5.3)$$

Indeed, let $\varphi \in \Phi(x)$ and $w \in T_{K(t)}(x)$ be given. Then there exist $h_n \to 0^+$, $w_n \to w$ such that $x + h_n w_n \in K(t)$ for all n. Set $z_n(\tau) := x + h_n w_n + \tau\varphi$. Thanks to the Filippov Theorem 12.2.1, there exists a solution $y_n(\cdot)$ to the differential inclusion $x' \in \Phi(x)$ starting at $x+h_n w_n$ at $\tau = 0$ and satisfying $\forall \tau \in [0, h_n]$,

$$\|y_n(\tau) - x - h_n w_n - \tau\varphi\| \leq h_n \mu_n$$

where

$$\mu_n := \frac{1}{h_n}\int_0^\tau e^{\lambda(\tau-s)} d(\varphi, \Phi(x + h_n w_n + s\varphi))ds \to 0$$

Then, taking $\tau := h_n$, we infer that there exists e_n with $\|e_n\| \leq \mu_n$ such that

$$x + h_n w_n + h_n\varphi - h_n e_n = y_n(h_n) \in \vartheta_\Phi(h_n, K(t))$$

By (12.5.2), there exists f_n with $\|f_n\| \leq \varepsilon(h_n)$ such that

$$y_n(h_n) - h_n f_n \in K(t + h_n)$$

This implies that

$$x + h_n(w_n + \varphi - e_n - f_n) \in K(t + h_n)$$

Since $w_n \to w$, $e_n \to 0$ and $f_n \to 0$, we deduce that $w+\varphi \in DK(t,x)(1)$.

2. We shall prove that if Φ is Marchaud, property

$$K(t+h) \subset \vartheta_\Phi(h, K(t)) + h\varepsilon(h)B \tag{12.5.4}$$

implies that

$$\forall x \in K(t), \quad DK(t,x)(1) \ \subset \ \Phi(x) + T_{K(t)}(x) \tag{12.5.5}$$

Let v be fixed in $DK(t,x)(1)$. There exist sequences $h_n \to 0^+$ and $v_n \to v$ such that $x + h_n v_n \in K(t+h_n)$ for all $n \geq 0$. By (12.5.4), there exist $y_n \in K(t)$ and f_n such that $\|f_n\| \leq \varepsilon(h_n)$ and

$$x + h_n v_n - h_n f_n \in \vartheta_\Phi(h_n, y_n) \tag{12.5.6}$$

Let $y_n(\cdot)$ be defined by

$$y_n(\cdot) \ \in \ S_\Phi(y_n) \ \ \& \ \ y_n(h_n) \ = \ y + h_n v_n - h_n f_n$$

Since $y_n \in \vartheta_{-\Phi}(h_n, x + h_n v_n - h_n f_n)$ and since $x + h_n v_n - h_n f_n$ remains in a compact set, so does y_n. The Convergence Theorem 12.1.4 implies that a subsequence (again denoted by) $y_n(\cdot)$ of solutions to the differential inclusion $x' \in \Phi(x)$ starting at y_n when $\tau = 0$ converges uniformly on $[0,1]$ to a solution $x(\cdot)$ starting at x. By letting $n \to \infty$ in (12.5.6), we infer that $y(0) = y$. Furthermore, we can write

$$x + h_n v_n = y_n + \frac{h_n}{h_n}\int_0^{h_n} y_n'(\tau)d\tau + h_n f_n$$

We observe that since Φ is upper semi-continuous, a subsequence of

$$\varphi_n := \frac{1}{h_n}\int_0^{h_n} y_n'(\tau)d\tau \in \frac{1}{h_n}\int_0^{h_n} \Phi(y_n(\tau))d\tau$$

converges to some $\varphi \in \overline{co}(\Phi(x)) = \Phi(x)$: Indeed, for every $\varepsilon > 0$, there exist N such that, for $n \geq N$, $\forall \tau \in [0, h_n]$, $\Phi(y_n(\tau)) \subset \Phi(x) + \varepsilon B$, so that, $\varphi_n \in \overline{co}(\Phi(x) + \varepsilon B)$ converges to some $\varphi \in \overline{co}(\Phi(x))$. Hence $v_n - \varphi_n - f_n = \frac{y_n - y}{h_n}$ converges to $v - \varphi$. Since $y_n \in K(t)$, we infer that $v - \varphi \in T_{K(t)}(x)$. □

We need to adapt the concept of contingent cone for stating the Nagumo Theorem for Tubes:

Definition 12.5.3 *We shall say that a Lipschitz map $\Phi \in \mathrm{LIP}(X, X)$ is contingent to $\mathcal{M}$ at $K \in \mathcal{M}$ if and only if the associated transition ϑ_Φ is contingent to $\mathcal{M}$ at K, i.e.,*

$$\liminf_{h \to 0+} \frac{\boldsymbol{d}_{\mathcal{M}}(\vartheta_\Phi(h, K))}{h} = 0$$

or again, if and only if there exist sequences h_n and ε_n converging to 0 and a sequence of subsets $K_n \in \mathcal{M}$ such that

$$\vartheta_\Phi(h_n, K) \subset K_n + \varepsilon_n h_n B \ \ \& \ \ K_n \subset \vartheta_\Phi(h_n, K) + \varepsilon_n h_n B$$

This contingent cone has been introduced and studied by Luc Doyen under the name of velocity cone.

Theorem 12.5.4 *Let $g : \mathcal{K}(X) \mapsto \mathrm{LIP}(X, X)$ be a continuous map satisfying*

$$\begin{cases} i) & M := \sup_{K \in \mathcal{K}(M)} \|g(K)\|_\Lambda < +\infty \\ ii) & c := \sup_{K \in \mathcal{K}(M)} \|g(K)\|_\infty < +\infty \end{cases} \quad (12.5.7)$$

Then $\mathcal{M} \subset \mathcal{K}(M)$ is a viability domain *of g in the sense that*

$$\forall K \in \mathcal{M}, \ \ g(K) \in T_{\mathcal{M}}(K)$$

if and only if from any $K_0 \in \mathcal{M}$ starts a tube $t \mapsto K(t)$, solution to the mutational equation

$$\forall t \geq 0, \ \ \overset{\circ}{K}(t) \ni g(K(t); \cdot)$$

which is viable in $\mathcal{M}$ *in the sense that*

$$\forall t \geq 0, \ \ K(t) \in \mathcal{M}$$

Proof — We refer to [30,35, Aubin] for a proof of this theorem which goes beyond the limits set to this book. □

We consider a Marchaud map $F : X \rightsquigarrow X$ and the differential inclusion

$$x'(t) \in F(x) \tag{12.5.8}$$

We assume now that the evolution of the tube $K(\cdot) : \mathbf{R}_+ \rightsquigarrow X$ is governed by the mutational equation

$$\overset{\circ}{K}(t) \ni g(K(t))(\cdot) \tag{12.5.9}$$

We would like to characterize the dynamics F and g governing a viable evolution of both the state and the tube in the sense that

$$\forall\, t \geq 0,\ \ x(t) \in K(t)$$

Theorem 12.5.5 *Assume that the map $F : X \rightsquigarrow X$ is Marchaud and that $g : \mathcal{K}(X) \mapsto \text{LIP}(X, X)$ is a continuous map satisfying (12.5.7). The two conditions are equivalent:*

1. *For every compact set K and every $x \in K$,*

$$0 \in g(K)(x) + T_K(x) - F(x) \tag{12.5.10}$$

2. *For any K_0 and for any $x_0 \in K_0$, there exists a solution to the differential inclusion $x' \in F(x)$ starting at x_0 which is viable in the sense that*

$$\forall\, t \geq 0,\ \ x(t) \in K(t)$$

where $K(t)$ is a solution to the mutational equation $\overset{\circ}{K} \ni g(K)$ starting at K_0.

Proof — This is a consequence of Theorems 12.3.2,12.5.4 and12.5.2. □

Chapter 13

Optima

Introduction

There are now many books[1] on convex analysis and convex optimization, so that the purpose of this chapter is mainly to fix the notations and to summarize the main results which we hall use.

1. We begin naturally with the Fermat Rule, which remains the main strategy for obtaining necessary conditions of optimality, from mathematical programming to calculus of variations to optimal control, known in these case the Euler-Lagrange equations and the Pontriagin principle respectively. For convex functions, this requires a generalization of the gradient of an usual function, the **subdifferential**, introduced by Jean-Jacques Moreau and Terry Rockafellar at the dawn of the sixties.

2. The study of subdifferentials of convex functions is dramatically simplified thanks to the introduction of the **Fenchel conjugate** $V^\star : X^\star \mapsto \mathbf{R} \cup \{+\infty\}$ of an extended function $V : X \mapsto \mathbf{R} \cup \{+\infty\}$. The Fenchel transform adapts to the convex case the virtues of the classical Legendre transform: the inverse of the map $x \rightsquigarrow \partial V(x)$ is the map $p \rightsquigarrow \partial V^\star(p)$. Not only a duality theory doubles the number of properties, but inverting the subdifferential in this way allows us to enrich and illuminate many results.

3. The wonderful duality properties of convex lower semicontinuous functions are summarized in a short presentation of duality theory in convex optimization. Solving a convex optimization problems amounts actually to solving two optimization problems, the original one — called the **primal problem** — and the dual one, called the **dual problem**. Conditions

[1] We kept the same notations and the same statements than the ones presented in OPTIMA AND EQUILIBRIA, [19, Aubin] and APPLIED NONLINEAR ANALYSIS, [50, Aubin & Ekeland] for which we refer for more details.

on the dual problem — called under the generic name of **constraint qualification assumptions** guarantee the existence of a solution to the primal problem and conditions on the primal problem warrants the existence of a solution to the dual problem, also known under the name of **Lagrange multiplier**, which enjoys "marginal properties" in the sense that it belongs to the subdifferential of the marginal function.

4. We then study the optimal allocation of scarce resources among consumers in the convex case. Indeed, convex analysis goes much beyond providing the mere existence of an optimal allocation. Duality theory exhibits **prices** that emerge from the problem, which solve an associated **dual optimization problem**. Duality theory of convex analysis reveals **demand and supply maps** on one hand, **change and pricing maps** on the other, which are concealed in this simple optimization problem.

   It shows that **such a price clears the market**, in the sense that the optimal allocation is made of consumptions which belong to the demand maps of each consumer and that the total consumption is in the supply map.

   It demonstrates in a dual way that for such a price, consumers never change their consumptions.

   It exhibits also the **marginal property** of such a price, which measures the marginal variations of the collective utilities when the set of scarce resources is perturbed.

   Last but not least, it conceals the three dynamical algorithms which are the prototypes of a tâtonnement model (which is not viable), a nontâtonnement model and a viable dynamical process exhibiting slow solutions respectively.

   The tâtonnement process is nothing else than the (continuous) gradient method applied to the "dual minimization problem". It states that the variations of the prices are in the excess demand (demand minus supply) maps: It increases whenever demand increases. One can prove that it converges to a price associated with an optimal allocation. But, as it was already mentioned, the associated consumptions do not constitute an allocation whenever the price is not an equilibrium.

   The nontâtonnement process is actually the Hamiltonian system associated with an Hamiltonian which shall be constructed.

   But thanks to the concept of subdifferentiability of nondifferentiable convex functions, one can show that the (continuous) gradient method applied to the initial optimization problem provides the evolution of consumptions which form at each instant allocations and which converges to an optimal allocation.

13.1 Subdifferential and the Fermat Rule

Pierre de Fermat was one of the most important innovators in the history of mathematics. Newton himself recognized explicitly that he got the hint of the differential calculus from Fermat's method of building tangents devised half a century earlier.

Fermat[2] was the one who discovered that the derivative of a (polynomial) function vanishes when it reaches an extremum. This is Fermat's Rule, which has been adapted to more and more general problems by extending the concept of derivatives — and their dual concepts, the gradients — to more and more general classes of "non differentiable functions. For this purpose, the choice of such generalizations should provide convenient necessary conditions for an element to achieve the minimum of a function.

For convex functions, this was done by Jean-Jacques Moreau and Terry Rockafellar who introduced the subdifferential of convex functions in the early 60's:

Definition 13.1.1 *Consider a nontrivial function $V : X \mapsto \mathbf{R} \cup \{+\infty\}$ and $x \in \mathrm{Dom}(V)$. The closed convex subset $\partial V(x)$ defined by*

$$\partial V(x) = \{p \in X^\star \mid \forall\, y \in X,\ \langle p, y - x\rangle \leq V(y) - V(x)\}$$

(which may be empty) is called the **subdifferential** *of V at x. We say that V is* **subdifferentiable at** *x if $\partial V(x) \neq \emptyset$.*

From the definition, we see that the Fermat Rule follows immediately:

Theorem 13.1.2 *Let $V : X \mapsto \mathbf{R} \cup \{+\infty\}$ be a nontrivial extended function. Then the following conditions are equivalent:*

$$\begin{cases} i) & 0 \in \partial V(\overline{x}) \text{ (the Fermat Rule)} \\ ii) & \overline{x} \text{ minimizes } V \end{cases}$$

Therefore, in order to apply Fermat Rule, we need a **Subdifferentiable Calculus**, for which we need the concept of **conjugate functions**. This is how prices will emerge when we shall apply the Fermat Rule to the problem (13.4.1) below of finding an optimal allocation of scarce resources. This is relatively easy in the case of convex extended functions, as we shall report in this chapter.

[2] Fermat also was the first to discover the "principle of least time" in optics, the prototype of the variational principles governing so many physical and mechanical laws. He shared independently with Descartes the invention of analytic geometry and with Pascal the creation of the mathematical theory of probability. His achievements in number theory overshadowed his other contributions, and the Last Fermat Theorem remains a challenge. He was on top of that a poet, a linguist, ... and made his living as a lawyer!

But we have to mention right now that the Fermat Rule replaces the minimization problem by an **equilibrium problem**: Indeed, the inclusion

$$0 \in \partial V(\overline{x})$$

shows that the constant function $x(t) \equiv \overline{x}$ is a solution to the **differential inclusion**

$$\text{for almost all } t \geq 0, \quad x'(t) \in -\partial V(x)$$

(**continuous descent method or subdifferential algorithm**).

Actually, such differential inclusions do have unique solutions: we state here a consequence of the Crandall-Pazy Theorem:

Theorem 13.1.3 *Assume that* $V : X \mapsto \mathbf{R} \cup \{+\infty\}$ *is nontrivial, convex, lower semicontinuous and bounded below. Then, for any initial state* $x_0 \in \mathrm{Dom}(V)$*, there exists a* **unique solution** *to the differential inclusion*

$$\text{for almost all } t \geq 0, \quad x'(t) \in -\partial V(x(t))$$

starting at x_0.

Let $V_0'(x))$ *denote the element of* $\partial V(x)$ *with the smallest norm. Then the solution* $x(\cdot)$ *is* **slow** *in the sense that for almost any* t*, the norm of the velocity* $x'(t)$ *is the smallest one:*

$$\text{for almost all } t \geq 0, \quad x'(t) = -V_0'(x(t))$$

Furthermore, if V *is inf-compact*[3]*, then* $x(t)$ *converges when* $t \to \infty$ *to a limit* $x_\star$ *which achieves the minimum of* V*:*

$$\lim_{t \to \infty} V(x(t)) = \inf_{x \in X} V(x) = V(x_\star)$$

See for instance [39, Aubin & Cellina] for a proof and more details on this topic.

This being underlined, we begin by observing that the concept of subdifferential generalizes the concept of gradient in the following sense:

Proposition 13.1.4 *If* $V : X \mapsto \mathbf{R} \cup \{+\infty\}$ *is convex and differentiable at a point* $x \in \mathrm{Int}(\mathrm{Dom}(V))$*, then*

$$\partial V(x) = \{V'(x)\}$$

[3]this means that the level sets of V are compact.

Proof — First, the gradient $V'(x)$ belongs to $\partial V(x)$, since, V being convex, inequalities

$$\frac{V(x+h(y-x))-V(x)}{h} = \frac{V((1-h)x+hy)-V(x)}{h} \leq V(y)-V(x)$$

imply by letting h converge to 0 that

$$\forall\, y \in X,\ \langle V'(x), y-x\rangle \leq V(y)-V(x)$$

Conversely, if $p \in \partial V(x)$, we obtain, by taking $y = x+hu$ that

$$\langle p,u\rangle \leq \frac{V(x+hu)-V(x)}{h}$$

By letting h converge to 0, we infer that for every $u \in X$, $\langle p,u\rangle \leq \langle V'(x),u\rangle$, so that $p = V'(x)$. □

Actually, this proof and Proposition 11.4.2 of the Tool-Box show that when the convex function V is no longer differentiable, we can still state the same result by using contingent epiderivatives:

Proposition 13.1.5 *If $V : X \mapsto \mathbf{R} \cup \{+\infty\}$ is convex, then*

$$\partial V(x) = \{p \in X^\star \mid \forall\, v \in X,\ \langle p,v\rangle \leq D_\uparrow V(x)(v)\}$$

In other words, the contingent epiderivative of a convex function i the support function of its subdifferential:

$$\forall\, u \in X,\quad \sigma(\partial V(x),u) = D_\uparrow V(x)(u)$$

which is the duality formula generalizing the formula

$$\langle V'(x),u\rangle = DV(x)(u)$$

in the smooth case.

Proposition 13.1.6 *The subdifferential map $x \rightsquigarrow \partial V(x)$ is* monotone *in the sense that*

$$\forall\, x,y \in X,\ \ \forall\, p \in \partial V(x),\ q \in \partial V(y),\ \ \langle p-q, x-y\rangle \geq 0 \tag{13.1.1}$$

Proof — Indeed, since

$$\begin{cases} \forall\, p \in \partial V(x),\ \langle p, x-y\rangle \geq V(x)-V(y) \\ \forall\, q \in \partial V(y),\ \langle q, y-x\rangle \geq V(y)-V(x) \end{cases}$$

we deduce the monotonicity of the subdifferential by adding those two inequalities. □

Monotone maps, and above all, maximal monotone maps, enjoy many of the properties of positive linear operators. We refer to Chapter 3 of [39, Aubin & Cellina] for more details on monotone maps. □

We recall the following important property of convex functions defined on finite dimensional vector-spaces:

Theorem 13.1.7 *An extended convex function V defined on a finite dimensional vector-space is locally Lipschitz and subdifferentiable on the interior of its domain.*

When V is lower semicontinuous, the domain of the subdifferential $x \rightsquigarrow \partial V(x)$ *is dense in the domain of V.*

(See for instance [50, Aubin & Ekeland] for a proof.)

In order to proceed with a subdifferential calculus allowing us to implement the Fermat Rule in the framework of convex functions, we need to relate the concept of subdifferential with the concept of Fenchel's conjugate functions, which is at the root of duality theory of convex analysis.

13.2 Conjugate Functions

There is more to that: lower semicontinuous convex functions enjoy duality properties. In the same way that we associated with cones their polar cones, with closed convex processes their transposes, we can, following Fenchel, associate with lower semicontinuous convex functions **conjugate functions** for the same reasons, and with the same success.

Since the epigraph of a lower semicontinuous convex function is a closed convex subset, it is tempting to compute its support function, and in particular, to observe that

$$\sigma_{\mathcal{E}p(V)}(p,-1) \;=\; \sup_{x\in X,\,\lambda\geq V(x)} (\langle p,x\rangle - \lambda) \;=\; \sup_{x\in X}(\langle p,x\rangle - V(x))$$

Definition 13.2.1 *Let* $V : X \to \mathbf{R}\cup\{+\infty\}$ *be any nontrivial extended function defined on a finite dimensional vector-space X. We associate with it its conjugate function* $V^\star : X^\star \to \mathbf{R}\cup\{+\infty\}$ *defined on the dual of X by*

$$\forall\, p \in X^\star,\;\; V^\star(p) \;:=\; \sup_{x\in X}(\langle p,x\rangle - V(x))$$

Its biconjugate $V^{\star\star} : X \mapsto \mathbf{R}\cup\{\pm\infty\}$ *is defined by*

$$V^{\star\star}(x) \;:=\; \sup_{p\,\in\, X^\star} (\langle p,x\rangle - V^\star(p))$$

We see at once that **the conjugate function of the indicator ψ_K of a subset K is the support function σ_K.**

We deduce from the definition the following convenient inequality

$$\forall x \in X, \ p \in X^\star, \ \langle p, x \rangle \leq V(x) + V^\star(p)$$

known as **Fenchel's Inequality.**

We deduce at once the characterization of the subdifferential in terms of conjugate functions:

Proposition 13.2.2 *Let $V : X \to \mathbf{R} \cup \{+\infty\}$ be a nontrivial extended convex function defined on a finite dimensional vector-space X. Then*

$$p \in \partial V(x) \iff \langle p, x \rangle = V(x) + V^\star(p)$$

This result allows us to derive a subdifferential calculus from the calculus of conjugate functions we shall soon present.

The epigraphs of the conjugate and biconjugate functions being closed convex subsets, the conjugate function is lower semicontinuous and convex and so is its biconjugate when it never takes the value $-\infty$. We observe that

$$\forall x \in X, \ V^{\star\star}(x) \leq V(x)$$

If equality holds, then V is convex and lower semicontinuous. The converse statement, a consequence of the Hahn-Banach Separation Theorem, is the first basic theorem of convex analysis:

Theorem 13.2.3 *A nontrivial extended function $V : X \to \mathbf{R} \cup \{+\infty\}$ is convex and lower semicontinuous if and only if it coincides with its biconjugate.*

In this case, the conjugate function $V^\star$ is nontrivial and **the inverse of the subdifferential $\partial V(\cdot)$ is the subdifferential $\partial V^\star(\cdot)$ of the conjugate function:**

$$p \in \partial V(x) \iff x \in \partial V^\star(p)$$

So, the Fenchel correspondence associating with any function V its conjugate $V^\star$ is a one to one correspondence between the sets of nontrivial lower semicontinuous convex functions defined on X and its dual $X^\star$. Therefore, it plays in the framework of convex analysis the role of the Legendre transform in classical analysis, and provides the same type of consequences, allowing us for instance to pass from Lagrangian formulations to Hamiltonian ones, as we shall mention later.

Taking into account that $-V^\star(0) = \inf_{x \in X} V(x)$, the Fermat Rule yields at once the following characterization of the subdifferential, which embodies the "marginal interpretation" of optimal solutions to convex optimization problems:

Theorem 13.2.4 *Let $V : X \mapsto \mathbf{R} \cup \{+\infty\}$ be a nontrivial lower semicontinuous convex extended function defined on a finite dimensional space X. Then $\partial V^\star(0)$ is the set of minimizers of V.*

As an example, we obtain

Corollary 13.2.5 *Let $K \subset X$ be a closed convex subset. Then*

$$\left\{\begin{array}{ll} i) & \partial\psi_K(x) \;=\; \{p \in X^\star \;\; \textit{such that} \; \langle p, x\rangle \;=\; \sup_{y\in K}\langle p, y\rangle\} \\ & \text{is the } \textbf{normal cone} \text{ to } K \text{ at } x \\ & \\ ii) & \partial\sigma_K(p) \;=\; \{x \in K \;\; \textit{such that} \; \langle p, x\rangle \;=\; \sup_{y\in K}\langle p, y\rangle\} \\ & \text{is the } \textbf{support zone} \text{ of } K \text{ at } p \end{array}\right.$$

The problem of finding an optimal allocation

$$\widehat{x} \in K \;\&\; \sum_{i=1}^{n} V_i(\widehat{x_i}) = \inf_{x\in K}\sum_{i=1}^{n} V_i(x_i)$$

where

$$K \;:=\; \left\{x \;:=\; (x_1,\ldots,x_n) \in \prod_{i=1}^{n} L_i \;\middle|\; \sum_{i=1}^{n} x_i \;\in\; M\right\}$$

can be embedded in the problem of the form

$$v := \inf_{x\in X}[V(x) + W(Ax)].$$

where $X := Y^n$,

$$V(x) \;:=\; \sum_{i=1}^{n} V_i|_{L_i}(x), \;\; W(y) \;:=\; \psi_M(y) \;\&\; Ax \;:=\; \sum_{i=1}^{n} x_i$$

which we shall now study in this simpler and more general framework.

13.3 Duality Theory in Convex Optimization

Suppose we have two finite dimensional vector-spaces X and Y, together with

$$\left\{\begin{array}{ll} i) & \text{a continuous, linear operator } A \in L(X,Y) \\ ii) & \text{two nontrivial, convex, lower semi-continuous functions} \\ & V : X \to \mathbf{R}\cup\{+\infty\} \;\text{ and }\; W : Y \to \mathbf{R}\cup\{+\infty\} \end{array}\right.$$

We shall study the minimization problem

$$v := \inf_{x\in X}[V(x) + W(Ax)] \tag{13.3.1}$$

Note that the function $V + W \circ A$ which we propose to minimize is only nontrivial if its domain $\mathrm{Dom} V \cap A^{-1}(\mathrm{Dom}\, W) \neq \emptyset$, that is to say, if

$$0 \in A(\mathrm{Dom}\, V) - \mathrm{Dom}\, W \tag{13.3.2}$$

In this case, we have $v < +\infty$.

Now we introduce the **dual minimization problem**

$$v^\star := \inf_{q \in Y^\star} [V^\star(-A^\star q) + W^\star(q)] \tag{13.3.3}$$

where $A^\star \in L(Y^\star, X^\star)$ is the transpose of A, $V^\star : X^\star \to \mathbf{R} \cup \{+\infty\}$ is the conjugate of V and $W^\star : Y^\star \to \mathbf{R} \cup \{+\infty\}$ is the conjugate of W. This only makes sense if we assume that

$$0 \in A^\star \,\mathrm{Dom}\, W^\star + \mathrm{Dom} V^\star \tag{13.3.4}$$

and in this case, $v^\star < +\infty$.

Note that we still have the inequality

$$v + v^\star \geq 0 \tag{13.3.5}$$

since, by virtue of Fenchel's inequality,

$$V(x) + W(Ax) + V^\star(-Aq) + W^\star(q) \geq \langle -A^\star q, x\rangle + \langle q, Ax\rangle = 0$$

Consequently, conditions (13.3.2) and (13.3.4) imply that v and $v^\star$ **are both finite.**

13.3.1 The Fenchel Theorem

We shall obtain the existence of an optimal solutions by strengthening condition $0 \in A(\mathrm{Dom}\, V) - \mathrm{Dom}\, W$. We shall call **constrained qualification property** the condition

$$0 \in \mathrm{Int}\,(A\,\mathrm{Dom}\, V - \mathrm{Dom}\, W)$$

The **Slater** condition

$$\exists\, x_0 \in \mathrm{Dom}(V) \quad \text{such that} \quad Ax_0 \in \mathrm{Int}(\mathrm{Dom}(W))$$

which requires that the interior of $\mathrm{Dom}(W)$ is not empty, implies obviously the above constrained qualification property.

When both $\mathrm{Dom}(V)$ and $\mathrm{Dom}(W)$ are cones, this constrained qualification property boils down to

$$A\,\mathrm{Dom}\, V - \mathrm{Dom}\, W \;=\; Y$$

because 0 belongs to the interior of a cone if and only if this cone is the whole space.

It is satisfied in particular when $\mathrm{Dom}(W) = X$ and when A is surjective.

Theorem 13.3.1 *Suppose that $A \in L(X,Y)$ is a linear operator from X to Y and that $V : X \to \mathbf{R} \cup \{+\infty\}$ and $W : Y \to \mathbf{R} \cup \{+\infty\}$ are nontrivial, convex, lower semi-continuous functions. We consider the case in which $0 \in A(\mathrm{Dom}\, V) - \mathrm{Dom}\, W$ and $0 \in A^\star(\mathrm{Dom}\, W^\star) + \mathrm{Dom}\, V^\star$ (which is equivalent to the assumption that v and $v^\star$ are finite).*

If we suppose that

$$0 \in \mathrm{Int}\,(A^\star\, \mathrm{Dom}\, W^\star + \mathrm{Dom}\, V^\star) \tag{13.3.6}$$

then

$$\begin{array}{ll} i) & v + v^\star = 0 \\ ii) & \exists \bar{x} \in X \quad \text{such that} \quad V(\bar{x}) + W(A\bar{x}) = v \end{array} \tag{13.3.7}$$

If we suppose that

$$0 \in \mathrm{Int}\,(A\, \mathrm{Dom}\, V - \mathrm{Dom}\, W) \tag{13.3.8}$$

then

$$\begin{array}{ll} i) & v + v^\star = 0 \\ ii) & \exists \bar{q} \in Y^\star \ \text{such that} \ V^\star(-A^\star\bar{q}) + W^\star(\bar{q}) = v^\star \end{array} \tag{13.3.9}$$

13.3.2 Calculus of Conjugate Functions

Firstly, we note the following elementary propositions.

Proposition 13.3.2 *a) If $V \le W$, then $W^\star \le V^\star$.*
b) If $A \in L(X,X)$ is an isomorphism, then

$$(V \circ A)^\star = V^\star \circ A^{\star-1}.$$

c) If $W(x) := V(x - x_0) + \langle p_0, x\rangle + a$, then

$$W^\star(p) = V^\star(p - p_0) + \langle p, x_0\rangle - (a + \langle p_0, x_0\rangle)$$

d) If $W(x) := V(\lambda x)$, then $W^\star(p) = V^\star\left(\frac{p}{\lambda}\right)$ and if $U(x) := \lambda V(x)$, then $U^\star(p) = \lambda V^\star\left(\frac{p}{\lambda}\right)$

Next, we mention the following result:

Proposition 13.3.3 *Suppose that X and Y are two finite dimensional vector-spaces and that V is a nontrivial, convex function from $X \times Y$ to $\mathbf{R} \cup \{+\infty\}$. Set $W(y) := \inf_{x\in X} V(x,y)$. Then*

$$W^\star(q) = V^\star(0,q) \tag{13.3.10}$$

The Fenchel Theorem 13.3.1 implies the following useful formulas:

Proposition 13.3.4 *Suppose that $A \in L(X,Y)$ is a linear operator and that $V : X \to \mathbf{R} \cup \{+\infty\}$ and $W : Y \to \mathbf{R} \cup \{+\infty\}$ are two nontrivial, lower semi-continuous functions. Suppose further that*

$$0 \in \operatorname{Int}(A \operatorname{Dom} V - \operatorname{Dom} W). \tag{13.3.11}$$

Then, for all $p \in A^\star \operatorname{Dom} W^\star + \operatorname{Dom} V^\star$, there exists $\bar{q} \in Y^\star$ such that

$$\begin{aligned}(V + W \circ A)^\star(p) &= V^\star(p - A^\star\bar{q}) + W^\star(\bar{q}) \\ &= \inf_{q \in Y^\star}(V^\star(p - A^\star q) + W^\star(q))\end{aligned} \tag{13.3.12}$$

13.3.3 Subdifferential Calculus

We can deduce easily from the calculus of conjugate functions a subdifferential calculus.

For instance, Proposition 13.3.4 implies

Theorem 13.3.5 *We consider a linear operator $A \in L(X,Y)$ and two nontrivial, convex, lower semi-continuous functions $V : X \to \mathbf{R} \cup \{+\infty\}$ and $W : Y \to \mathbf{R} \cup \{+\infty\}$.*

We assume further

$$0 \in \operatorname{Int}(A \operatorname{Dom} V - \operatorname{Dom} W). \tag{13.3.13}$$

Then,

$$\partial(V + W \circ A)(x) = \partial V(x) + A^\star \partial W(Ax) \tag{13.3.14}$$

Proof — It is easy to check that $\partial V(x) + A^\star \partial W(Ax)$ is always contained in $\partial(V + W \circ A)(x)$. The inverse inclusion follows from Proposition 3.4. We take $p \in (\partial(V + W \circ A)(x)$. There exists $\bar{q} \in Y^\star$ such that $(V + W \circ A)^\star(p) = V^\star(p - A^\star\bar{q}) + W^\star(\bar{q})$. Thus, from equation (13.3.12),

$$\begin{aligned}\langle p, x\rangle &= V(x) + W(Ax) + (V + W \circ A)^\star(p) \\ &= (V(x) + V^\star(p - A^\star\bar{q})) + (W(Ax) + W^\star(\bar{q})).\end{aligned}$$

Consequently,

$$0 = (\langle p - A^\star\bar{q}, x\rangle - V(x) - V^\star(p - A^\star\bar{q})) + (\langle \bar{q}, Ax\rangle - W(Ax) - W^\star(\bar{q})).$$

Since each of these two expressions is negative or zero, it follows that they are both zero, whence that $\bar{q} \in \partial W(Ax)$ and $p - A^\star\bar{q} \in \partial V(x)$. Thus, we have shown that $p = p - A^\star\bar{q} + A^\star\bar{q} \in \partial V(x) + A^\star \partial W(Ax)$. □

Proposition 13.3.6 *Let W be a nontrivial, convex function from $X \times Y$ to $\mathbf{R} \cup \{+\infty\}$. Consider the function $U : Y \to \mathbf{R} \cup \{+\infty\}$ defined by*

$$U(y) := \inf_{x \in X} W(x, y). \tag{13.3.15}$$

If $\bar{x} \in X$ satisfies $U(y) = W(\bar{x}, y)$, then the following conditions are equivalent:

$$\begin{array}{ll} (a) & q \in \partial U(y) \\ (b) & (0, q) \in \partial W(\bar{x}, y) \end{array} \tag{13.3.16}$$

Proof — Since $U^\star(q) = W^\star(0, q)$, following Proposition 13.3.3, we deduce that q belongs to $\partial U(y)$ if and only if $\langle q, y \rangle = U(y) + h^\star(q) = W(\bar{x}, y) + W^\star(0, q)$, that is, if and only if $(0, q) \in \partial W(\bar{x}, y)$. □

Proposition 13.3.7 *We consider a family of convex functions $x \to V(x, p)$ indexed by a parameter p running over a set P. We assume that*

$$\left\{ \begin{array}{ll} i) & P \text{ is compact} \\ ii) & \text{There exists a neighborhood } \mathcal{N} \text{ of } x \text{ such that,} \\ & \text{for all } y \text{ in } \mathcal{N},\ p \to V(y, p) \text{ is upper semi-continuous.} \\ iii) & \forall p \in P, y \to V(y, p) \text{ is continuous at } x. \end{array} \right. \tag{13.3.17}$$

Consider the upper envelope U of the functions $V(\cdot, p)$, defined by $U(y) = \sup_{p \in P} V(y, p)$. Set

$$P(x) := \{p \in P \mid U(x) = V(x, p)\}. \tag{13.3.18}$$

Then

$$DU(x)(v) = \sup_{p \in P(x)} DV(x, p)(v) \tag{13.3.19}$$

and

$$\partial U(x) = \overline{\text{co}} \left(\bigcup_{p \in P(x)} \partial V(x, p) \right) \tag{13.3.20}$$

Corollary 13.3.8 *Consider n convex functions V_i continuous at a point x. Then*

$$\partial \left(\sup_{i=1,\ldots,n} V_i \right)(x) = \overline{\text{co}} \left(\bigcup_{i \in I(x)} \partial V_i(x) \right) \tag{13.3.21}$$

where $I(x) = \{i = 1, \ldots, n | V_i(x) = \sup_{j=1,\ldots,n} V_j(x)\}$.

13.3.4 Calculus of Tangent Cones to Convex Sets

For the convenience of the reader, we list in the Table 13.1 some useful formulas of the calculus of tangent cones to convex subsets which will be proved later, in which the subsets K, K_i, L, M, ... are assumed to be convex.)

We shall need the following characterization of the normal cone to a convex cone:

Lemma 13.3.9 *Let $K \subset X$ be a convex cone of a normed space X and $x \in K$. Then*

$$p \in N_K(x) \Longleftrightarrow x \in K,\ p \in K^- \ \&\ \langle p, x\rangle = 0 \Longleftrightarrow x \in N_{K^-}(p)$$

where $N_{K^-}(p) := \{x \in K \mid \forall\, q \in K^-,\ \langle q - p, x\rangle \leq 0\}$.

Proof — To say that $p \in N_K(x)$ means that $\langle p, x\rangle = \sigma_K(p)$, which is equal to 0 if and only if $p \in K^-$, and the first statement of the lemma follows. □

Corollary 13.3.10 *Let $L \subset X$ and $M \subset Y$ to closed convex subsets and $A \in \mathcal{L}(X, Y)$ a linear operator linked by the* constraint qualification condition

$$0 \in \mathrm{Int}(AL - M)$$

Then the normal cone to $L \cap A^{-1}(M)$

$$N_{L\cap A^{-1}(M)}(x) = N_L(x) + A^\star N_M(Ax)$$

and the tangent cone by

$$T_{L\cap A^{-1}(M)}(x) = T_L(x) \cap A^{-1}T_M(Ax)$$

Proof — Since $\psi_{L\cap A^{-1}(M)}(x) = \psi_L(x) + \psi_M(Ax)$ and since

$$N_{L\cap A^{-1}(M)}(x) = \partial\psi_{L\cap A^{-1}(M)}(x)$$

we deduce the formula for the normal cones. The one for tangent cones is obtained by polarity and transposition. □

13.3.5 The Duality Theorem

We use the above subdifferential calculus for implementing the Fermat Rules and duality theory for the following general class of convex minimization problems: We consider

1. two finite dimensional spaces X and Y;

Table 13.1: **Properties of Tangent Cones to Convex Sets.**

(1)	▷	If $x \in K \subset L \subset X$, then $T_K(x) \subset T_L(x)$ & $N_L(x) \subset N_K(x)$
(3)	▷	If $x_i \in K_i \subset X_i$, $(i = 1, \cdots, n)$, then $T_{\prod_{i=1}^n K_i}(x_1, \ldots, x_n) = \prod_{i=1}^n T_{K_i}(x_i)$ $N_{\prod_{i=1}^n K_i}(x_1, \ldots, x_n) = \prod_{i=1}^n N_{K_i}(x_i)$
(4)a)	▷	If $A \in \mathcal{L}(X, Y)$ and $x \in K \subset X$, then $T_{A(K)}(Ax) = \overline{A(T_K(x))}$ $N_{A(K)}(Ax) = A^{\star -1} N_K(x)$
(4)b)	▷	If K_1, $K_2 \subset X$, $x_i \in K_i$, $i = 1, 2$, then $T_{K_1+K_2}(x_1 + x_2) = \overline{T_{K_1}(x_1) + T_{K_2}(x_2)}$ $N_{K_1+K_2}(x_1 + x_2) = N_{K_1}(x_1) \cap N_{K_2}(x_2)$ In particular, if $x_1 \in K$ and x_2 belongs to a closed subspace P of X, then $T_{K+P}(x_1 + x_2) = \overline{T_{K_1}(x_1) + P}$ $N_{K+P}(x_1 + x_2) = N_K(x_1) \cap P^\perp$
(5)	▷	If $L \subset X$ and $M \subset Y$ are **closed convex** subsets and $A \in \mathcal{L}(X, Y)$ satisfies the **constraint qualification assumption** $0 \in \mathrm{Int}(M - A(L))$, then, for every $x \in L \cap A^{-1}(M)$, $T_{L \cap A^{-1}(M)} = T_L(x) \cap A^{-1} T_M(Ax)$ $N_{L \cap A^{-1}(M)} = N_L(x) + A^\star N_M(Ax)$
(5)a)	▷	If $M \subset Y$ is *closed convex* and if $A \in \mathcal{L}(X, Y)$ satisfies $0 \in \mathrm{Int}(\mathrm{Im}(A) - M)$, then, for any $x \in A^{-1}(M)$, $T_{A^{-1}(M)}(x) = A^{-1} T_M(Ax)$ $N_{A^{-1}(M)}(x) = A^\star N_M(Ax)$
(5)b)	▷	If K_1, $K_2 \subset X$ are closed convex and satisfy $0 \in \mathrm{Int}(K_1 - K_2)$, then, for any $x \in K_1 \cap K_2$ $T_{K_1 \cap K_2}(x) = T_{K_1}(x) \cap T_{K_2}(x)$ $N_{K_1 \cap K_2}(x) = N_{K_1}(x) + N_{K_2}(x)$
(5)c)	▷	If $K_i \subset X$, $(i = 1, \ldots, n)$, are closed and convex, $x \in \bigcap_{i=1}^n K_i$ and if there exists $\gamma > 0$ satisfying $\forall x_i$ such that $\|x_i\| \leq \gamma$, $\bigcap_{i=1}^n (K_i - x_i) \neq \emptyset$, then $T_{\bigcap_{i=1}^n K_i}(x) = \bigcap_{i=1}^n T_{K_i}(x)$ $N_{\bigcap_{i=1}^n K_i}(x) = \sum_{i=1}^n N_{K_i}(x)$

2. two nontrivial, convex, lower semi-continuous functions

$$\begin{cases} i) & V : X \to \mathbf{R} \cup \{+\infty\} \\ ii) & W : Y \to \mathbf{R} \cup \{+\infty\} \\ iii) & \text{a continuous, linear operator } A \in L(X, Y) \end{cases} \tag{13.3.22}$$

We shall choose elements $y \in Y$ and $p \in X^\star$ as parameters of the optimization problems

$$v(y) := \inf_{x \in X} (V(x) - \langle p, x \rangle + W(Ax + y)) \tag{13.3.23}$$

and

$$w(p) := \inf_{q \in Y^\star} (V^\star(p - A^\star q) + W^\star(q) - \langle q, y \rangle) \tag{13.3.24}$$

which we shall solve at the same time.

We shall say the minimization problems $v(y)$ and $w(p)$ are dual.

Theorem 13.3.11 *a) If*

$$y \in \operatorname{Int}(\operatorname{Dom} W - A \operatorname{Dom} V) \;\&\; p \in \operatorname{Int}(\operatorname{Dom} V^\star + A^\star \operatorname{Dom} W^\star), \tag{13.3.25}$$

then

$$v(y) + w(p) = 0 \tag{13.3.26}$$

and there exist a solution $\overline{x}$ of the problem $v(y)$ and a solution $\overline{q}$ of the dual problem $w(p)$

b) Furthermore, the following conditions are equivalent

$$\begin{cases} i) & \overline{x} \text{ is a solution of the "primal" problem } v \\ ii) & \overline{x} \text{ is a solution of the inclusion } p \in \partial V(\overline{x}) + A^\star \partial W(A\overline{x} + y) \\ iii) & \overline{x} \in \partial w(p) \end{cases} \tag{13.3.27}$$

c) Similarly, the following conditions are equivalent:

$$\begin{cases} i) & \overline{q} \text{ is a solution of the problem } w(p) \\ ii) & \overline{q} \text{ is a solution of the inclusion } y \in \partial W^\star(\overline{q}) - A\partial V^\star(p - A^\star\overline{q}) \\ iii) & \overline{q} \in \partial v(y) \end{cases}$$

d) The solutions $\overline{x}$ and $\overline{q}$ of the problems $v(y)$ and $w(p)$ are solutions of the system of inclusions

$$\begin{array}{ll} i) & p \in \partial V(\overline{x}) + A^\star(\overline{q}) \\ ii) & y \in -A\overline{x} + \partial W^\star(\overline{q}). \end{array} \tag{13.3.28}$$

Remark — An optimal solution of the dual minimization problem $w(p)$ is usually called a ***Lagrange (or Kuhn–Tucker) multiplier***, inclusion (13.3.27)iii) is usually called the ***Euler–Lagrange inclusion*** and inclusion (13.3.11)iii) is the ***Euler–Lagrange dual inclusion.*** The system of inclusions (13.3.28) is usually called the ***Hamiltonian system.*** □

The set-valued map $(x,q) \to (\partial V(x) + A^\star q) \times (-Ax + \partial W^\star(q))$ from $X \times Y^\star$ to its dual $X^\star \times Y$ may be written symbolically in matrix form by

$$\begin{pmatrix} \partial V & A^\star \\ -A & \partial W^\star \end{pmatrix} \tag{13.3.29}$$

The set of solutions $(\overline{x}, \overline{q})$ of the minimization problems v and $w(p)$ may then be written in the suggestive form

$$\begin{pmatrix} \partial V & A^\star \\ -A & \partial W^\star \end{pmatrix}^{-1} \begin{pmatrix} p \\ y \end{pmatrix}$$

This notation highlights the variation of the set of solutions as a function of the parameters $p \in X^\star$ and $y \in Y$.

Remark — When assumption (13.3.25) of Theorem 13.3.11 are satisfied, solution of the problem v is equivalent to solution of the inclusion (set-valued equation)

$$p \in \partial V(\overline{x}) + A^\star \partial W(A\overline{x} + y). \tag{13.3.30}$$

Theorem 13.3.11 indicates another way of solving this problem. This involves first solving the inclusion

$$y \in \partial W^\star(\overline{q}) - A\partial V^\star(p - A^\star\overline{q}) \tag{13.3.31}$$

and then choosing $\overline{x}$ in the set

$$\partial V^\star(p - A^\star\overline{q}) \cap A^{-1}(\partial W^\star(\overline{q}) - y). \tag{13.3.32}$$

This procedure is only sensible if the second inclusion is easier to solve than the first. This clearly depends on the functions V and W. If W is differentiable, it may be better to solve the inclusion (16). If, instead, $V^\star$ is differentiable, it may be easier to solve the inclusion (17), which in this case may be written as

$$A\nabla V^\star(p - A^\star\overline{q}) + y \in \partial W^\star(\overline{q}) \tag{13.3.33}$$

or as

$$\forall q \in Y, \ \langle -A\nabla V^\star(p - A^\star \overline{q}) - y, \overline{q} - q\rangle + W^\star(\overline{q}) - W^\star(q) \leq 0. \quad (13.3.34)$$

Remark: Gradient Methods — For solving the minimization problem v (or $v_\star$), we can use the gradient method to either problem v or $v_\star$.

We deduce from Theorem 13.1.3 the following result:

Theorem 13.3.12 *We posit the assumptions of Theorem 13.3.11. Then, for any initial state $x_0 \in \mathrm{Dom}(V)$, there exists a* unique solution *to the differential inclusion*

$$-x'(t) \ \in \ \partial V(x(t)) + A^\star \partial W(Ax(t))$$

starting at x_0, converging to an optimal solution $\overline{x}$ and satisfying

$$\lim_{t\to\infty}(V(x(t)) + W(Ax(t))) \ = \ \inf_{x\in X}(V(x) + W(Ax))$$

For any initial state $q_0 \in \mathrm{Dom}(V)$, there exists a unique solution *to the differential inclusion*

$$-q'(t) \ \in \ \partial W^\star(q(t)) - A^\star \partial V^\star(-A^\star q(t))$$

starting at q_0, converging to a solution $\overline{q}$ of the dual problem and satisfying

$$\lim_{t\to\infty}(W^\star(q(t)) + V^\star(-A^\star q(t))) \ = \ \inf_{q\in Y^\star}(W^\star(q) + V^\star(-A^\star q))$$

13.4 Three Dynamical Processes for Optimal Allocations

The first type of selection mechanism of an allocation which comes up to the mind is an optimization mechanism permitting to select an allocation. We shall prove that this simple approach conceals the three dynamical processes leading to Pareto Equilibria and/or the associated prices.

13.4.1 Utility Functions

For doing so, we need to introduce a collective utility function

$$U : x := (x_1, \ldots, x_n) \ \in \ X := Y^n = \mathbf{R}^{ln} \ \mapsto \ U(x_1, \ldots, x_n)$$

and to look for an allocation

$$\hat{x} \ \in \ K \ := \ \left\{ x \ := \ (x_1, \ldots, x_n) \ \in \prod_{i=1}^n L_i \ \middle| \ \sum_{i=1}^n x_i \ \in \ M \right\}$$

maximizing the utility function U:

$$\hat{x} \in K \ \& \ U(\hat{x}) = \sup_{x \in K} U(x) \tag{13.4.1}$$

This concept of utility function played (and is still playing) a crucial role in economic theory, and has been at the origin and the them of many heated debates.

Among the first question which arises is the following: who will choose this collective utility function ?, the public interest (who knows it ?), a dictator ?, a planning bureau ?

Since the n consumers are composing the "collectivity" whose behavior is described by the collective utility function U, one generally acquiesces to build U from the utility functions U_i of the consumers. For instance, the collective utility function is a weighted sum of the individual utility functions:

$$U(x) := \sum_{i=1}^{n} \lambda_i U_i(x_i)$$

In this case, the problem is shifted to the one of choosing the weights λ_i attributed to each consumer. This is typically a game-theoretical issue. But even if we assume that this problem of allotting weights among consumers is solved, the question remains to know whether utility functions are the right metaphors for the behavior of consumers. Indeed, the concept of utility function has raised and still raises many issues.

The **cardinality versus ordinality** dispute is by now settled. Many economists did challenge the possibility for any economic agent to associate with any commodity a cardinal number measuring her utility or satisfaction and stressed the fact that utility functions played uniquely an "ordinal" role for comparing two commodities. What matters is the **preference preorder** $\preceq$ defined by

$$x \preceq y \text{ if and only if } U(x) \geq U(y)$$

Recall that a preorder is a reflexive and transitive binary relation, **complete** if any two elements are comparable, **partial** in the opposite case.

We can associate with any preorder the equivalence relation $\sim$ defined by

$$x \sim y \text{ if and only if } x \preceq y \ \& \ y \preceq x$$

The "projection" of the preorder to the factor space $X/\sim$ is then an **order relation**.

If $\varphi : \mathbf{R} \mapsto \mathbf{R}$ is an increasing function, then the utility functions U and $\varphi \circ U$ generate the same preference preorder.

Hence the problem of representing any given preference preorder by a utility function was a real issue, because, in particular, the lexicographic order in $\mathbf{R}^n$ cannot be associated with a continuous utility function. Debreu ended this debate by giving reasonable sufficient conditions for a preorder on a finite dimensional vector-space to be represented by a continuous utility function.

Utility functions do not provide the more judicious representation of a consumer in a dynamical framework. In this case, concepts of change or transformations in a given direction v are better embodied in the various concepts of **directional derivatives.** Starting from a commodity $x \in Y$ in a direction v, the **satisfaction caused by this move** can be described by the infinitesimal utility increment measured by adequate limits of the differential quotients

$$\frac{U(x+hv)-U(x)}{h}$$

when $h \mapsto 0+$ (the usual derivative is no longer sufficient, because in many cases, utility functions being built not only through standard algebraic operations, but also by supremum or infimum, are no longer differentiable in the classical sense. We have to use contingent epiderivatives or other concepts of nonsmooth analysis.)

Nowadays, the concept of **rationality** became synonymous of the narrow notion of making optimal decisions. An individual, regarded as a decision-maker, is then reduced to an utility function postulated to summarize her behavior. Even the broader conceit of the ability of making transitive inferences is more a dream than a reality, as cognitive psychology acknowledges nowadays.

We have seen that in dynamical framework, we can discard utility functions and replace them by change functions in the description of the behavior of consumers, even though, as we shall see, they provide instances of change functions.

However, because of the historical importance of this point of view on one hand, and the importance of optimization theory on the other hand, we shall apply the main results of optimization theory in this framework.

Convexity playing a major role in this study, utility functions are most often assumed to be concave. In order to avoid using both adjectives, convex and concave, we shall avoid maximizing concave functions and we simply shall minimize instead (convex) loss functions. From them, we shall build **demand and supply maps** on one hand, **change and pricing maps** on the other, which are concealed in this simple optimization problem. Duality theory exhibits **prices** that emerge from the problem by solving the **dual optimization problem**. It shows that **such a price clears the market**, in the sense that the optimal allocation is made of consumptions which belong to the demand maps of each consumer and that the total consumption is in the supply map.

It demonstrates in a dual way that for such a price, consumers never change their consumptions.

It exhibits also the **marginal property** of such a price, which measures the marginal variations of the collective utilities when the set of scarce resources is perturbed.

Last but not least, it conceals the three dynamical algorithms which are the prototypes of both a tâtonnement model (which is not viable), the nontâtonnement model and the viable process.

The tâtonnement process is nothing else than the (continuous) gradient method applied to the "dual minimization problem". It states that the variations of the prices are in the excess demand (demand minus supply) maps: It increases whenever demand increases. One can prove that it converges to a price associated with an optimal allocation. But, as it was already mentioned, the associated consumptions do not constitute an allocation whenever the price is not an equilibrium.

The nontâtonnement process is actually the Hamiltonian system associated to an Hamiltonian which shall be constructed.

But thanks to the concept of subdifferentiability of nondifferentiable convex functions, one can show that the (continuous) gradient method applied to the initial optimization problem provides the evolution of consumptions which form at each instant allocations and which converges to an optimal allocation. Indeed, not only Convex Analysis reveals demand and supply maps, but also the **change and pricing** maps with which we shall build the general nontâtonnement models of chapter 3 and beyond.

For simplicity, we shall incorporate the weights λ_i and the consumption sets L_i in the loss functions $V_i : Y \mapsto \mathbf{R} \cup \{+\infty\}$ of the consumers $i = 1, \ldots, n$. Hence, an optimal allocation $\overline{x} = (\overline{x}_1, \ldots, \overline{x}_n)$ is a solution to the minimization problem

$$v := \inf_{x \in K} \sum_{i=1}^{n} V_i(x_i) = \sum_{i=1}^{n} V_i(\overline{x_i})$$

where

$$K := \left\{ x := (x_1, \ldots, x_n) \in \prod_{i=1}^{n} \mathrm{Dom}(V_i) \;\middle|\; \sum_{i=1}^{n} x_i \in M \right\}$$

We shall assume that

$$\forall\, q \in \mathbf{R}^l_+, \quad \inf_{x \in L_i} (\langle q, x \rangle + V_i(x)) > -\infty \tag{13.4.2}$$

and that the set of scarce resources M satisfy

$$\begin{cases} i) & M = M - \mathbf{R}^l_+ \text{ is a closed convex subset} \\ ii) & M \subset \underline{y} - \mathbf{R}^l_+ \end{cases} \tag{13.4.3}$$

From the knowledge of the loss functions and the set of scarce resources we shall extract concealed features on the behavior of the consumers.

13.4.2 Demand and Change Maps

We shall denote by

$$\widetilde{D}_i(q) := \left\{x^i \in \mathbf{R}^l \mid \langle q, x_i\rangle + V_i(x_i)) = \inf_{x\in\mathbf{R}^l}(\langle q, x\rangle + V_i(x))\right\} \quad (13.4.4)$$

the Walrasian demand of consumer i,

$$B_i(q,r) := \{x \in \mathrm{Dom}(V_i) \mid \langle q, x\rangle \leq r\}$$

the **budget set** of consumer i and by

$$D_i(q,r) := \left\{x_i \in B_i(q,r) \mid V_i(x_i) = \inf_{x\in B_i(q,r)} V_i(x)\right\}$$

her **demand set**. The **demand map** is the set-valued map $(q,r) \leadsto D_i(q,r)$. We observe at once that

$$\forall\, \overline{x}_i \in \widetilde{D}_i(q),\ \ \widetilde{D}_i(q) = D_i(q, \langle q, \overline{x}_i\rangle)$$

is independent of the choice of $\overline{x}_i \in D_i(q, \langle q, \overline{x}_i\rangle)$[4].

Change maps $C_i : L_i \times S^l \leadsto Y^\star$ which express the "greed" of consumer i, are defined by:

$$C_i(x,q) := \left\{v \in \mathbf{R}^{l\star} \mid V_i(x) + \langle v+q, x\rangle = \inf_{y\in Y}(V_i(y) + \langle v+q, y\rangle)\right\}$$

In other words, $v \in C_i(x,q)$ is the "velocity" for which the commodity x minimizes the sum of the loss $V_i(y) + \langle q, y\rangle$ and the "greed" $\langle v, y\rangle$.

If the loss function V_i is differentiable at x, then C_i is single-valued and can be written

$$C_i(x,q) = -V_i'(x) - q$$

Otherwise, it can be couched in terms of the subdifferential of the loss function:

$$C_i(x,q) = -\partial V_i(x) - q$$

[4]We also observe that

$$\widetilde{D}_i(q) = \partial(V_i^\star)(-q)$$

Indeed, to say that $x_i \in \partial(V_i^\star)(-q)$ amounts to saying that

$$0 \in \partial V_i(x_i) + q$$

or, equivalently, that

$$\forall\, y \in Y,\ \ V_i(x_i) + \langle q, x_i\rangle \leq V_i(x) + \langle q, x\rangle$$

This can be written

$$\forall\, x \in X,\ \ V_i(x_i) - V_i(x) \leq \langle q, x - x_i\rangle$$

We observe that

$$0 \in C_i(x_i, q) \text{ if and only if } x_i \in \widetilde{D}_i(q)$$

In summary, we can associate with any consumer i represented by a loss function V_i a demand map $\widetilde{D}_i$ associating with any price q a set of commodities minimizing her loss under budgetary constraints and a change map C_i associating with any price q the change of consumption.

13.4.3 Supply and Pricing Maps

We associate now with the set $M \subset Y$ of scarce resources the supply map S_M associating with any $q \in Y^\star$ the supply set $S_M(q) \subset M$ defined by

$$S_M(q) := \left\{ \overline{y} \in M \mid \langle q, \overline{y} \rangle = \sup_{y \in M} \langle q, y \rangle \right\}$$

of scarce resources which maximize the income

$$\sigma_M(q) := \sup_{y \in M} \langle q, y \rangle$$

induced by the available resources.

We also introduce the inverse $N_M := S_M^{-1}$ of the supply map S_M:

$$q \in N_M(y) \text{ if and only if } y \in S_M(q)$$

which we regard as the pricing map.

We observe that assumption

$$M = M - \mathbf{R}^l_+$$

implies that

$$\forall\, y \in M, \ N_M(y) \subset \mathbf{R}^l_+$$

because for any price $q \in N_M(y)$, we have

$$\langle q, y \rangle = \sigma_M(q) = \sigma_M(q) + \sigma_{-\mathbf{R}^l_+}(q) = \begin{cases} \sigma_M(q) & \text{if } q \in \mathbf{R}^l_+ \\ +\infty & \text{if } q \notin \mathbf{R}^l_+ \end{cases}$$

so that q can only be nonnegative. Furthermore, if M is bounded above, we infer that

$$\mathrm{Dom}(\sigma_M) \supset \mathbf{R}^l_+$$

because, for any nonnegative price $q \in \mathbf{R}^l_+$,

$$\sigma_M(q) \leq \langle q, \overline{y} \rangle + \sigma_{-\mathbf{R}^l_+}(q) < +\infty$$

Hence assumption (13.4.3):

$$\begin{cases} i) & M = M - \mathbf{R}^l_+ \text{ is a closed convex subset} \\ ii) & M \subset \underline{y} - \mathbf{R}^l_+ \end{cases}$$

implies that

$$\mathrm{Dom}(\sigma_M) \;=\; \mathbf{R}^l_+$$

We also observe that

$$\forall\, y \in \mathrm{Int}(M),\;\; N_M(y) \;=\; \{0\}$$

13.4.4 Optimal Allocations

Theorem 13.4.1 *Let us assume the set M of scarce resources satisfy assumptions (13.4.3), that the consumption set L_i are closed and convex and that the loss functions V_i are nontrivial, convex and lower semicontinuous and satisfy assumptions (13.4.2). Assume furthermore that*

$$0 \in \mathrm{Int}\left(\sum_{i=1}^{n} L_i - M\right) \tag{13.4.5}$$

Then there exists an optimal allocation $\overline{x} \in K$ which is a solution to the optimal allocation problem

$$v \;:=\; \inf_{x\in K} \sum_{i=1}^{n} V_i(\overline{x_i})$$

where

$$K \;:=\; \left\{ x \;:=\; (x_1, \ldots, x_n) \in \prod_{i=1}^{n} \mathrm{Dom}(V_i) \;\middle|\; \sum_{i=1}^{n} x_i \;\in\; M \right\}$$

Furthermore, there exists a price $\overline{q} \in Y^\star$ such that

a) the price $\overline{q}$ and the allocation $(\overline{x}_1, \ldots, \overline{x}_n)$ satisfy

$$\begin{cases} i) & \forall\, i = 1, \ldots, n,\;\; 0 \in C_i(\overline{x}_i, (\overline{q})) \\ & \text{i.e., each } \overline{x}_i \text{ belongs to consumer } i\text{'s demand set} \\ ii) & \sum_{i=1}^{n} \overline{x}_i \in S_M(\overline{q}) \\ & \text{i.e., } \sum_{i=1}^{n} \overline{x}_i \text{ maximizes the available income } \langle \overline{q}, y \rangle \end{cases}$$

b) the optimal price $\overline{q}$ clears the market *in the sense when it is a solution to the inclusion*

$$0 \in S_M(\overline{q}) - \sum_{i=1}^{n} \widetilde{D}_i(\overline{q})$$

stating that the supply $S_M(\overline{q})$ *is balanced by the total demand* $\sum_{i=1}^{n} \widetilde{D}_i(\overline{q})$.

c) the optimal allocation $\overline{x}$ *is an equilibrium of the associated nontâtonnement process in the sense that*

$$\begin{cases} i) & \forall\, i = 1, \dots n, \;\; 0 \in C_i(\overline{x}_i, \overline{q}) \\ ii) & \overline{q} \in N_M\left(\sum_{i=1}^{n} \overline{x}_i\right) \end{cases}$$

Proof — This a consequence of Theorem 13.3.11, because the constraint qualification assumptions are satisfied.

Assumption (13.4.2):

$$\forall\, q \in \mathbf{R}^l_+, \;\; \inf_{x \in L_i} (\langle q, x\rangle + V_i(x)) \; > \; -\infty$$

states that the negative cone $\mathbf{R}^l_-$ is contained in the domain of each $V_i^\star$.

Since the conjugate function of

$$V(x) \;:=\; \sum_{i=1}^{n} V_i(x_i)$$

is the function defined by

$$V^\star(p) \;=\; \sum_{i=1}^{n} V_i^\star(p_i)$$

we see that

$$\mathrm{Dom}(V^\star) \;=\; \prod_{i=1}^{n} \mathrm{Dom}(V_i^\star) \;=\; \mathbf{R}^{ln}_-$$

We take for operator A the sum:

$$Ax \;:=\; \sum_{i=1}^{n} x_i$$

the transpose of which is equal to

$$A^\star q \;=\; (q, \dots, q)$$

Therefore, property

$$0 \;\in\; \mathrm{Int}(\mathrm{Dom}(V^\star) + A^\star \mathrm{Dom}(W^\star))$$

is satisfied because $\mathrm{Dom}(V^\star) + A^\star\mathrm{Dom}(W^\star) = \mathbf{R}^{ln}_- + A^\star\mathbf{R}^{ln}_+ = \mathbf{R}^{ln}$.

We observe that assumption

$$M \;=\; M - \mathbf{R}^l_+$$

implies that

$$\forall\, y \in M, \;\; N_M(y) \;\subset\; \mathbf{R}^l_+$$

because for any price $q \in N_M(y)$, we have

$$\langle q, y\rangle = \sigma_M(q) = \sigma_M(q) + \sigma_{-\mathbf{R}^l_+}(q) = \begin{cases} \sigma_M(q) & \text{if } q \in \mathbf{R}^l_+ \\ +\infty & \text{if } q \notin \mathbf{R}^l_+ \end{cases}$$

so that q can only be nonnegative. Furthermore, if M is bounded above, we infer that

$$\mathrm{Dom}(\sigma_M) \supset \mathbf{R}^l_+$$

because, for any nonnegative price $q \in \mathbf{R}^l_+$,

$$\sigma_M(q) \leq \langle q, \overline{y}\rangle + \sigma_{-\mathbf{R}^l_+}(q) < +\infty$$

Hence assumption (13.4.3):

$$\begin{cases} i) & M = M - \mathbf{R}^l_+ \text{ is a closed convex subset} \\ ii) & M \subset \underline{y} - \mathbf{R}^l_+ \end{cases}$$

implies that

$$\mathrm{Dom}(\sigma_M) = \mathbf{R}^l_+ \quad \square$$

We shall also prove that this price $\overline{q}$ enjoys a **marginal property**. We introduce perturbations on the resources and we define the **marginal function** v associating with any resource y the optimal value

$$v(y) := \inf_{x \in K(y)} \sum_{i=1}^{n} V_i(\overline{x_i})$$

where

$$K(y) := \left\{ x := (x_1, \ldots, x_n) \in \prod_{i=1}^{n} \mathrm{Dom}(V_i) \;\middle|\; \sum_{i=1}^{n} x_i \in M - y \right\}$$

We observe that $v(0) = v$. Naturally, the marginal function v is not necessarily differentiable. But it is convex, and we can extend the concept of differential to a concept of **subdifferential**. With this notion, we prove that $\overline{q}$ **belongs to the subdifferential of the marginal function at** $y := 0$:

Proposition 13.4.2 *We posit the assumptions of Theorem 13.4.1. Then $\overline{q}$ is a solution of the dual problem if and only if it belongs to the subdifferential $\partial v(y)$ of the marginal function v defined by*

$$v(y) := \inf_{x \in K(y)} \sum_{i=1}^{n} V_i(\overline{x_i})$$

where

$$K(y) := \left\{ x := (x_1, \ldots, x_n) \in \prod_{i=1}^{n} \mathrm{Dom}(V_i) \;\middle|\; \sum_{i=1}^{n} x_i \in M - y \right\}$$

Proof — The marginal function v can be written $v(y) = \inf_{U(x,y)}$ where

$$U(x,y) := \sum_{i=1}^{n} V_i(\overline{x_i}) + \psi_M \left(\sum_{i=1}^{n} x_i + y \right)$$

By Proposition 13.3.6, we know that $\overline{p} \in \partial v(0)$ if and only if $(0, \overline{q})$ belongs to $\partial U(\overline{x}, 0)$ since $v(0) = U(\overline{x}, 0)$. The latter inclusion can be written

$$\begin{cases} \forall\, i = 1, \ldots, n, \;\; 0 \in \partial V_i(\overline{x}) + N_M \left(\sum_{i=1}^{n} \overline{x}_i \right) \\ \overline{q} \in N_M \left(\sum_{i=1}^{n} \overline{x}_i \right) \end{cases}$$

Therefore, $\overline{q}$ is the solution to the dual problem. □

13.4.5 A Tâtonnement Process

Since we have associated with any consumer her demand map $\widetilde{D}_i$ and we have defined a supply map S_M, we can define the continuous **tâtonnement** process defined by

$$q'(t) \in \sum_{i=1}^{n} \widetilde{D}_i(q(t)) - S_M(q(t)) \tag{13.4.6}$$

which is a metaphor for the law of supply and demand. Then the following consequence of the Crandall-Pazy Theorem holds true:

Theorem 13.4.3 *We posit the assumptions of Theorem 13.4.1. For any positive initial price q_0, there exists a unique solution $q(\cdot)$ to the tâtonnement process (13.4.6) starting at q_0 with converges to an equilibrium $\overline{q}$ when $t \to +\infty$.*

Proof — Indeed, the Crandall-Pazy Theorem implies that the gradient inclusion

$$-q'(t) \in \partial \left(\sigma_M(q) + \sum_{i=1}^{n} V_i^\star(-q) \right)$$

has a unique solution converging to one solution of the dual optimal allocation problem. Under the assumption of Theorem 13.4.1, this differential inclusion can be written in the form

$$-q'(t) \in \partial\sigma_M(q(t)) - \sum_{i=1}^{n} \partial V_i^\star(-q(t)) \;=\; S_M(q(t)) - \sum_{i=1}^{n} \widetilde{D}_i(q(t)) \quad \square$$

Unfortunately, we have seen that tâtonnement processes are not viable.

13.4.6 A Nontâtonnement Process

The price $\overline{q}$ and the allocation $(\overline{x}_1, \ldots, \overline{x}_n)$ form a saddle point of the Lagrangian (for the terminology in use in optimization theory) defined by

$$H(x_1, \ldots, x_n, p) := \sum_{i=1}^{n} V_i(x_i) + \left\langle p, \sum_{i=1}^{n} x_i \right\rangle - \sigma_M(p)$$

We can also derive from the duality a second system of differential inclusions

$$\begin{cases} i) & -x'(t) \in \partial(\sum_{i=1}^n V_i(x_i(t))) + q(t) \\ ii) & -q'(t) \in \partial\sigma_M(q) - \sum_{i=1}^n x_i(t) \end{cases}$$

This an algorithm of the form of a dynamical system proposed by Arrow-Hahn, Negishi, Smale, Uzawa among many others, which unfortunately shares with the tâtonnement process the flaw of not being necessarily viable. □

The Crandall-Pazy Theorem imply the existence of an unique solution since the right-hand side is a maximal monotone operator.

13.4.7 A Viable Process

However, we can design another dynamical process which is viable, through change maps $C_i : L_i \times S^l \rightsquigarrow Y$ which express the satisfaction of consumer i.

We then define the dynamical behavior of consumer i by the differential inclusion

$$x_i'(t) \in C_i(x_i(t), q(t))$$

which is controlled by the price $q(t)$.

Theorem 13.4.4 *We posit the assumptions of Theorem 13.4.1. From any initial allocation $x_0 = (x_{01}, \ldots, x_{0n}) \in K$ starts a* unique solution $x(\cdot) = (x_1(\cdot), \ldots, x_n(\cdot))$ *to the system of differential inclusions*

$$x_i'(t) \in C_i(x_i(t), q(t))$$

where the price $q(t)$ satisfies

$$\forall t \geq 0, \ q(t) \in N_M\left(\sum_{i=1}^{n} x_i(t)\right)$$

and which are viable in *K in the sense that*

$$\forall t \geq 0, \ x_i(t) \in L_i \ (i = 1, \ldots, n) \ \& \ \sum_{i=1}^{n} x_i(t) \in M$$

Furthermore, the total loss $t \to \sum_{i=1}^{n} V_i(x_i(t))$ decreases and the allocation $x(t)$ converges to an optimal allocation $\overline{x}$ when $t \to +\infty$. If denote by

$$c_i^0(x, p) := \Pi_{-\partial V_i(x)}(p) - p$$

the element of minimal norm of the change map $C_i(x,p)$ *and* $p^0(x) \in N_M\left(\sum_{i=1}^n x_i\right)$ *the element which minimizes* $\|c_i^0(x,p)\|$ *on* $N_M\left(\sum_{i=1}^n x_i\right)$. *Then* $x(\cdot)$ *is actually the solution to the differential equation*

$$x'(t) \;=\; c_i^0(x(t), p^0(x(t)))$$

Proof — We know that the gradient inclusion

$$-x'(t) \in \partial\left(\sum_{i=1}^n V_i(x_i(t)) + \psi_M\left(\sum_{i=1}^n x_i(t)\right)\right)$$

has a unique solution converging to one solution of the optimal allocation problem. Under the assumption of Theorem 13.4.1, it can be written in the form

$$\begin{cases} \forall\, i = 1,\ldots,n, \;\; x_i'(t) \;\in\; -\partial V_i(x_i(t)) - q(t) \;=:\; C_i(x_i(t), q(t)) \\ \text{where} \\ q(t) \;\in\; N_M\left(\sum_{i=1}^n x_i(t)\right) \end{cases}$$

It is clear that $c_i^0(x,p^0(x))$ is the element of $-\partial\left(\sum_{i=1}^n V_i(x_i) + \psi_M\left(\sum_{i=1}^n x_i\right)\right)$ with minimal norm. □

Remark — The unique viable solution is a **slow** solution since it chooses the velocity with the smallest norm in the set

$$\{C_i(x_i,p)\}_{p\in N_M(\sum_{i=1}^n x_i)}$$

For any fixed q, the element $c_i^0(x,q)$ of minimal norm is equal to

$$c_i^0(x,q) \;=\; \Pi_{-\partial V_i(x)}(q) - q$$

so that the velocity with minimal norm is equal to $c_i(x, q_M^0(x))$ where $q_M^0(x) \in N_M\left(\sum_{i=1}^n x_i\right)$ minimizes $\|c_i(x,q)\|$ on $N_M\left(\sum_{i=1}^n x_i\right)$.

Therefore, the viable allocations are actually solutions to the system of differential equations

$$x_i'(t) \;=\; c_i^0(x_i(t), q_M^0(x(t)))$$

One can regard the map q_M^0 as a feedback map — a pricing map — governing the evolution of viable allocations of M improving the total welfare of the consumers. It can be interpreted as a planning process. Furthermore, this feedback map is obtained through a minimization procedure, because $p^0(x) \in N_M\left(\sum_{i=1}^n x_i\right)$ minimizes $\|c_i^0(x,p)\|$ on $N_M\left(\sum_{i=1}^n x_i\right)$. Observe that $q(t) = 0$ **whenever the total consumption** $\sum_{i=1}^n x_i(t)$ **belongs to the interior of the set** M **of scarce resources.** □

Therefore, under convexity assumptions, one can derive from the problem of optimal allocation many more informations than the mere existence of an optimal allocation. First, the concept of price emerges, and we can associate the concepts of demand maps and supply maps. We stated that there exists a price $\overline{q}$ which clears the marker: the total optimal consumption is in the supply set and each optimal consumption belongs to the demand set. This price has marginal properties.

The tâtonnement process can be defined, and, given an initial price, it has a unique solution converging to an equilibrium price.

We can also derive decentralized dynamical processes of each consumer, described by a differential inclusion controlled by prices. Starting from any initial allocation, there exists a unique allocation evolving according to this controlled dynamical process which converges to an optimal allocation when $t \mapsto +\infty$.

Chapter 14

Equilibria

Introduction

This chapter surveys the basic equilibrium theorems, essentially, the Arrow-Debreu Theorem for the existence of a Walrasian equilibrium and the Equilibrium Theorem, as well as the Kakutani Fixed Point theorems and the Inverse-Function Theorem.

1. The proofs of the Equilibrium Theorem and of the Arrow-Debreu Theorem depend upon the Brouwer Fixed Point Theorem, via a more versatile equivalent statement, the Ky Fan Inequality, which is so useful that we do not hesitate to recall it.

2. We then use the Ky Fan inequality for studying the classical Walrasian equilibrium based on demand functions describing the static behavior of the consumers and on an allocation of the income function among the consumers.

3. Next, we deduce from the Ky Fan inequality a version of the equilibrium theorem general enough to cover many applications.

4. Equilibrium Theorems imply readily Fixed-Point Theorems, including the celebrated Kakutani Theorem that John von Neumann "ordered", so to speak, for solving problems in economic and game theories.

5. We used the adaptation of the Constrained Inverse Function Theorem which extends the Graves Theorem when the equilibrium is subjected to constraints in the study of qualitative economics. Its proof is based on Ekeland's Variational Principle which plays a role analogous to the Ky Fan Inequality for those problems set on complete spaces instead of compact spaces. Actually, Inverse Function Theorems are more stability theorems than simply results on the existence of solutions.

14.1 The Ky Fan Inequality

We shall prove that the 1910 Brouwer Fixed Point Theorem[1]:

Theorem 14.1.1 (Brouwer Fixed Point Theorem) *Let K be a compact convex subset of a finite dimensional vector-space and $f : K \mapsto K$ be a continuous (single-valued) map. Then there exists a fixed point $f(\overline{x}) = \overline{x} \in K$ of f in K.*

is equivalent to the Ky Fan inequality:

Theorem 14.1.2 (Ky Fan Inequality) *Let K be a compact convex subset of a finite dimensional vector-space[2] X and $\varphi : X \times X \mapsto \mathbf{R}$ be a function satisfying*

$$\left\{\begin{array}{lll} i) & \forall\, y \in K, & x \mapsto \varphi(x,y) \text{ is lower semicontinuous} \\ ii) & \forall\, x \in K, & y \mapsto \varphi(x,y) \text{ is concave} \\ iii) & \forall\, y \in K, & \varphi(y,y) \leq 0 \end{array}\right. \tag{14.1.1}$$

Then, there exists $\overline{x} \in K$, a solution to

$$\forall\, y \in K, \;\; \varphi(\overline{x},y) \leq 0 \tag{14.1.2}$$

Proof

— **Proof of Ky Fan Inequality** We shall derive a contradiction from the negation of the conclusion:

$$\forall\, x \in K, \; \exists\, y \in K \text{ such that } \varphi(x,y) > 0$$

so that K can be covered by the subsets

$$\mathcal{V}_y := \{x \in K \mid \varphi(x,y) > 0\}$$

which are open by assumption (14.1.1) *i*). Since K is compact, it can be covered by n such open subsets $\mathcal{V}_{y_i}$. Let us consider a **continuous partition of unity**[3] $(\alpha_i)_{i=1,\ldots,n}$ associated with this open covering of K and define the map $f : K \mapsto X$ by

$$\forall\, x \in K, \;\; f(x) := \sum_{i=1}^{n} \alpha_i(x) y_i$$

---

[1] See [12, Appendix B] for a proof of the Brouwer Fixed Point Theorem based on Sperner's Lemma and [50, Chapter II] for a proof based on differential geometry.

[2] Actually, this theorem holds true for any Hausdorff locally convex topological vector space and in particular, for weak topologies of Banach spaces X.

[3] **A continuous partition of unity associated with a covering of K by n open subsets $\mathcal{V}_i$** is a sequence of n continuous maps $\alpha_i : K \mapsto \mathbf{R}$ such that,

$$\forall\, x \in K, \; \sum_{i=1}^{n} \alpha_i(x) = 1, \; \forall\, i = 1,\ldots,n, \; \alpha_i(x) \geq 0 \;\&\; \text{support}(\alpha_i) \subset \mathcal{V}_i$$

Such continuous partitions of unity do exist when K is a compact metric space.

It maps K to itself because K is convex and the elements y_i belong to K. It is also continuous, so that Brouwer's Fixed Point Theorem implies the existence of a fixed point $\overline{y} = f(\overline{y}) \in K$ of f. Assumption (14.1.1) *ii*) imply that

$$\varphi(\overline{y},\overline{y}) = \varphi(\overline{y},\sum_{i=1}^{n}\alpha_i(\overline{y})y_i) \geq \sum_{i=1}^{n}\alpha_i(\overline{y})\varphi(\overline{y},y_i)$$

Let us introduce

$$I(\overline{y}) := \{i = 1,\ldots,n \mid \alpha_i(\overline{y}) > 0\}$$

It is not empty because $\sum_{i=1}^{n}\alpha_i(\overline{y}) = 1$. Furthermore

$$\sum_{i=1}^{n}\alpha_i(\overline{y})\varphi(\overline{y},y_i) = \sum_{i\in I(\overline{y})}\alpha_i(\overline{y})\varphi(\overline{y},y_i) > 0$$

because, whenever i belongs to $I(\overline{y})$, $\alpha_i(\overline{y}) > 0$, so that $\overline{y}$ belongs to $\mathcal{V}_{y_i}$, and thus, by the very definition of this subset, $\varphi(\overline{y},y_i) > 0$. Hence, we have proved that $\varphi(\overline{y},\overline{y})$ is strictly positive, a contradiction of assumption (14.1.1) *iii*).

— **Proof of Brouwer's Theorem** We associate with the continuous map $f : K \mapsto K$ the function φ defined on $K \times K$ by

$$\varphi(x,y) := \langle f(x) - x, y - x\rangle$$

which satisfies obviously the assumptions (14.1.1) of Ky Fan Inequality. Then there exists $\overline{x} \in K$ such that inequality (14.1.2) holds true. By taking $y := f(\overline{x}) \in K$, we infer that $\|f(\overline{x}) - \overline{x}\| \leq 0$, and thus, that $\overline{x}$ is a fixed point of f. □

Remark — Minimax and Lop-sided Minimax Inequalities

The Ky Fan inequality implies readily the von Neumann Minimax Theorem:

Theorem 14.1.3 (Minimax) *Let X and Y be finite dimensional vector spaces*[4]*, $L \subset X$ and $M \subset Y$ be compact convex subsets and $f : L \times M \mapsto \mathbf{R}$ be a real valued function satisfying*

$$\begin{cases} i) & \forall\, y \in M,\ \ x \mapsto f(x,y) \text{ is lower semicontinuous and convex} \\ ii) & \forall\, x \in L,\ \ y \mapsto f(x,y) \text{ is upper semicontinuous and concave} \end{cases}$$

Then there exists a saddle point $(\bar{x},\bar{y}) \in L \times M$ *of* f*:*

$$\forall\,(x,y) \in L \times M,\ \ f(\bar{x},y) \leq f(\bar{x},\bar{y}) \leq f(x,\bar{y})$$

Proof — We apply the Ky Fan Inequality with $K := L \times M$ and φ defined by

$$\varphi((\bar{x},\bar{y}),(x,y)) := f(\bar{x},y) - f(x,\bar{y}) \ \Box$$

Actually, we often need a weaker version of the Minimax Theorem, called the Lop-Sided Minimax Theorem, relying only on the Separation Theorem. We recall its statement:

[4]actually, Hausdorff locally convex topological vector spaces.

Theorem 14.1.4 (Lop-Sided Minimax Theorem) *Let X and Y be finite dimensional vector-space*[5]*, $L \subset X$ be a compact convex subset, $M \subset Y$ be a convex subset and $f : L \times M \mapsto \mathbf{R}$ be a real valued function satisfying*

$$\begin{cases} i) & \forall\, y \in M, \quad x \mapsto f(x,y) \text{ is lower semicontinuous and convex} \\ ii) & \forall\, x \in L, \quad y \mapsto f(x,y) \text{ is and concave} \end{cases}$$

Then there exists $\bar{x} \in L$ satisfying

$$\forall\, y \in M, \quad f(\bar{x},y) \leq \inf_{x\in L}\sup_{y\in M} f(x,y) = \sup_{y\in M}\inf_{x\in L} f(x,y)$$

Proof — We refer to Theorem 6.2.7. of APPLIED NONLINEAR ANALYSIS for an instance of proof using only the Separation Theorem. □

14.2 The Walrasian Model

With the Walras model, we find an equilibrium price, which is an equilibrium for the tâtonnement process. The behavior of consumers is described by **demand functions or demand maps**: Knowing the equilibrium price, the choice of a consumer is given through the demand function or the demand map. The problem is then to find a price which provides an allocation of a scarce resource.

We shall begin with a simple version of the 1954 Arrow-Debreu Theorem, a prototype of many of the theorems stating the existence of an equilibrium. This theorem is easily derived from a very versatile tool, the Ky Fan inequality, which is equivalent to the 1910 Brouwer Fixed Point Theorem, the corner stone of Nonlinear Analysis.

But using the Brouwer Fixed Point Theorem amounts to transforming the problem under investigation to a fixed point problem, for which one can apply this fixed point theorem or its numerous variations or extensions. The "fixed point" format being quite rigid, one encounters the risk of more or less considerable loss of information by doing so. It also often happens that this transformation may require additional assumptions and useless technical difficulties.

This is the reason why so many statements logically equivalent to the Brouwer Fixed Point Theorem — constituting the corpus of nonlinear analysis — have been designed to be readily adapted to classes of specific problems.

Among these equivalent results, Ky Fan Inequality incorporates, in some sense, more **labor-value**[6], and therefore, might be more useful.

[5] or, more generally, a Banach space or even, a Hausdorff locally convex topological vector spaces.

[6] Although two statements $\mathcal{P}$ and $\mathcal{Q}$ may well be equivalent, it is common experience that the proof of one of the implications, say $\mathcal{P} \Longrightarrow \mathcal{Q}$, is more difficult or involves deeper results

We shall prove later that Ky Fan's Inequality is equivalent to the Brouwer Fixed Point Theorem and use it to derive the Arrow-Debreu Theorem for both the simple Walras allocation model and for general excess demand maps.

This Arrow-Debreu theorem asserts that contrary to a primary intuition, a deep mathematical metaphor, in a extremely simplified framework, justifies the functioning of a decentralized mechanism. Meanwhile, it has been improved in many ways, naturally, and adapted to more and more complicated situations. We shall concentrate here only on the basic features.

Yet, this model is quite imperfect, and we have already mentioned some flaws: This a static model, and its dynamical version is not viable.

14.2.1 The Numéraire

The **numéraire** is a specific commodity used to compare arbitrary commodities. The **value** of a commodity can be expressed in amounts of units of numéraire judged equivalent to this commodity.

The choice of a numéraire requires a consensus among the economic agents trading the commodities, the faith or belief that everyone agrees on the common value of the unit of numéraire. This is why the numéraire is called a fiduciary good.

Economic history shows the evolution of numéraires, from specific and useful goods (camels, cows, etc.) to seldom employable goods (like shells, gold, etc.) to paper money and now, to abstract figures concealed in computer memories of some banks. Nowadays, after Richard Nixon abandoned on August 13, 1971 the Gold Exchange Standard by renouncing the Bretton Woods agreement signed in 1944, the numéraire is no longer gold, but is made of an explicit, although abstract, commodity bundle used to make an **index**.

On should note confuse **money** with either **accounting units** or **numéraire**. Money has no definite meaning, conveys many different kinds of concepts and play different roles (storage of values, reward for risk taking, etc.).

Here, **sociopsychology**, in the sense of psychology of masses, plays an important role, since the consensus on the choice of a numéraire must be reached before we can use it in an economic model.

Hence, the choice of a numéraire and its value depend upon the set of economic agents which accept it and evolves with time. The space of values is one dimensional space, the unit of which is the unit of account, the Lira of May 15, 1992 for instance.

---

than the proof of the converse. In this case, one can say that $\mathcal{Q}$ **incorporates more labor value** than $\mathcal{P}$ and thus, expect as a general rule that the statement $\mathcal{Q}$ may be more useful than $\mathcal{P}$.

By the way, the most pleasant or intuitive statements are quite often the ones with the least labor value.

Actually, in complex economies of today, there are many different fiduciary goods, which add to the space of physical commodities (subject to inviolable scarcity constraints) a more and more complex space of fiduciary commodities (subject to psychological constraints, resulting from unknown psychological mechanisms governing the emergence of fashions, etc.). This aspect of things will not be taken into account in this chapter, naturally, but they should be kept in mind in order not to rely too much to the very unassuming and crude mathematical description of our humble economic problem.

In a static world, one can change the scale of prices (or price level) without altering the demand of a consumer (this is called **absence of monetary illusion**).

So, we need a further condition to fix the price level. This is done by fixing the value of a numéraire $\omega \in Y$: the unit of gold, a commodity basket entering the composition of an **index**, etc.:

$$\langle p, \omega \rangle = \mu$$

There are no longer "real" numéraires nowadays, but **commodity indexes**, the value of which is observed and measured (rather than being fixed in an evolving — and not really controllable — world).

This is time for a warning that in evolving models, the value of the numéraire evolves (although it should remain constant to satisfy the expectations (or dreams) of economists and finance ministers).

Here, we shall take for numéraire the commodity

$$\omega := (1, 1, \ldots, 1) \ \& \ \mu := 1$$

We shall agree to take for price set the **price simplex** S^l defined by the normalization rule:

$$S^l := \left\{ p \in \mathbf{R}_+^{l\star} \mid \sum_{h=1}^{l} p^h = 1 \right\}$$

14.2.2 The Arrow-Debreu Theorem

We present a very short description of the Walras model, based on the representation of consumers by n **demand functions**

$$d_i : (p, r) \in S^l \times \mathbf{R} \to d_i(p, r) \in L_i$$

associating to a price p and an income r the ith consumption of Mrs. i.

This a very crude description of the behavior of a consumer. Later on, we shall attenuate the "automatic feature" by accepting set-valued demand functions, that are called **demand maps**. As we have seen, some demand map can be derived from utility or loss functions.

Demand functions are required to obey a **budgetary rule** called the **Walras law**. It states that it is **forbidden to spend more than her income.**

This can be translated mathematically in the following form:

$$\forall\, p \in S^l, \quad \langle p, d_i(p,r)\rangle \;\leq\; r$$

since $\langle p, d_i(r,p)\rangle$ is the amount of monetary units spent to acquire $d_i(p,r)$ when p is the price and r is the income.

When M is the given subset of resources, we regard its support function $\sigma_M(p) := \sup_{y\in M} < p, y >$ as the **total income.**

The second basic ingredient of the Walras model is the **allocation among the consumers of (part of) the total income**

$$\sum_{i=1}^{n} r_i(p) \;\leq\; \sigma_M(p)$$

derived from the set of available resources. (This is an assumption, and not a conclusion: the Walras model associates an equilibrium with any allocation of the income, fair or unfair).

Therefore, when a consumer is described by her demand function d_i and is supplied with an income function r_i, her decision depends upon the knowledge of the price: she associates with a price p the commodity $d_i(p, r_i(p))$.

The role of a decentralized mechanism is then to provide every consumer a common price, i.e., a message summarizing the unknown set of available resources and the behavior of all consumers, such that the resulting consumptions $d_i(p, r_i(p))$ do form an allocation, i.e., satisfy

$$0 \;\in\; \sum_{i=1}^{n} d_i(p, r_i(p)) - M \;\;=:\;\; E(p)$$

where $E(p)$ is the **excess demand.**

Theorem 14.2.1 (Arrow-Debreu) *Assume that the demand functions d_i of the n consumers are continuous and that the set of scarce resources is closed, convex, bounded above and satisfy $M = M - \mathbf{R}^l_+$.*

Assume also that the n income functions $r_i(p)$ are continuous and satisfy $\sum_{i=1}^{n} r_i(p) \leq \sigma_M(p)$.

If the individual Walras law is satisfied, then there exists at least a Walras equilibrium.

This theorem is the prototype of many theorems of general equilibrium theory. This theorem is a simple consequence of the Theorem 14.3.2 below:

Remark — Walras Law

The advantage of the individual Walras law is that it is a decentralized condition. If each consumer conforms to it, then an equilibrium does exist thanks to the above theorem.

But the proof shows that it is enough to assume that the weaker — but not decentralized — (collective) Walras law

$$\sum_{i=1}^{n}\langle p, d_i(p, r_i)\rangle \ \leq\ \sum_{i=1}^{n} r_i$$

is satisfied. This weaker version admits monetary transactions among consumers to modify their initial income functions. □

14.2.3 Walras Equilibria of Excess Demand Maps

We started with the simplest model, but the proof of the Arrow-Debreu Theorem from the Ky Fan Inequality shows that this theorem can be readily extended to the case when

- demand functions $d_i(p,r) : S^l \times \mathbf{R} \mapsto L_i$ are replaced by **demand maps** $D_i(p,r) : S^l \times \mathbf{R} \rightsquigarrow L_i$
- a **supply map** $S_M : S^l \rightsquigarrow M$ is introduced (instead of the constant supply map $S_M(p) \equiv M$)

We have seen examples of such demand and a supply maps in the optimal allocation model.

We recall the definition of **support function** σ_K and **lower support function** $\sigma_K^\flat$ of a subset $K \subset X$:

$$\forall\, p \in X^\star,\ \ \sigma_K(p) \ :=\ \sup_{x\in K}\langle p, x\rangle \ \ \&\ \ \sigma_K^\flat(p) \ :=\ \inf_{x\in K}\langle p, x\rangle \ =\ -\sigma_K(-p)$$

and that the Separation Theorem 3.3.2 can also be written in the form:

$$\overline{co}(K) \ =\ \left\{x \ \in\ X \mid \forall\, p\ \in\ X^\star,\ \ \langle p, x\rangle\ \geq\ \sigma_K^\flat(p)\right\}$$

We associate with the supply map its **income function** r defined by

$$\forall\, q \in S^l,\ \ r(q) \ :=\ \sigma_{S_M(q)}(q) \ \leq\ \sigma_M(q)$$

We shall assume that this income function is allocated among consumers by providing them income functions r_i satisfying

$$\forall\, p \in S^l,\ \ \sum_{i=1}^{n} r_i(p) \ \leq\ r(p)$$

Describing the behavior of economic agents by these supply and demand map, we define the **excess demand map** E by

$$\forall\, p \in S^l,\ \ E(p) \ :=\ \sum_{i=1}^{n} D_i(p, r_i)(p) - S_M(p) - \mathbf{R}_+^l$$

The **tâtonnement process** is described by the following differential inclusion

$$p'(t) \in E(p(t))$$

in the continuous version and

$$p_{n+1} - p_n \in \delta_n E(p_n)$$

in the discrete version to describe the **Law of Supply and Demand**: the price increases when the demand increases.

Hence, the **Walras equilibrium** is a price $\overline{p}$ which clears the market in the sense that

$$0 \in E(\overline{p})$$

Consider now the case when **excess demand map** E is derived from demand and supply maps:

$$\forall\, p \in S^l,\ \ E(p) := \sum_{i=1}^{n} D_i(p, r_i)(p) - S_M(p) - \mathbf{R}^l_+$$

We shall assume that the values $D_i(p,r)$ and $S_M(p)$ of the demand and supply functions are closed and convex. Therefore, they can be characterized by their support functions:

$$D_i(p,r) = \left\{x \mid \forall\, q \in Y^\star,\ \sigma^\flat_{D_i(p,r)}(q) \le \langle q, x\rangle\right\}$$

and

$$S_M(p) = \left\{y \mid \forall\, q \in Y^\star,\ \langle q, y\rangle \le \sigma_{S_M(p)}(q)\right\}$$

We shall assume also that the demand maps satisfy the **Walras law** which can be written in the form

$$\forall\, q \in S^l,\ \ \sigma^\flat_{D_i(q,r)}(q) \le r$$

or the **collective Walras law**

$$\forall\, q \in S^l,\ \ \sum_{i=1}^{n} \sigma^\flat_{D_i(q,r_i)}(q) \le \sum_{i=1}^{n} r_i \qquad (14.2.1)$$

We derive the following consequence:

Theorem 14.2.2 *Let us assume that the demand and supply maps are upper hemicontinuous, that the values $D_i(p,r)$ are convex and compact and that the values $S_M(p)$ of the supply demand are closed, convex, bounded above and satisfy*

$$\forall\, p \in S^l,\ \ S_M(p) = S_M(p) - \mathbf{R}^l_+$$

Assume also that the n income functions $r_i(p)$ are continuous and satisfy $\sum_{i=1}^{n} r_i(p) \le \sigma_{S_M(p)}(p)$.

If the collective Walras law (14.3.1) is satisfied, then there exists at least a Walras equilibrium $\overline{p} \in S^l$, a solution to

$$0 \in \sum_{i=1}^{n} D_i(\overline{p}, r_i)(\overline{p}) - S_M(\overline{p})$$

Proof — Since the images $D_i(p,r)$ are convex and compact and the images $S_M(p)$ are closed and convex, we infer that the images

$$E(p) := \sum_{i=1}^{n} D_i(p, r_i(p)) - S_M(p) = E(p) - \mathbf{R}^l_+$$

are closed, convex and bounded above.

Since

$$\sigma^\flat_{E(p)}(q) := \sum_{i=1}^{n} \sigma^\flat_{D_i(p,r_i(p))}(q) - \sigma_{(S_M(p)}(q)$$

we infer that $E(\cdot)$ is upper hemicontinuous. Finally, it satisfies

$$\sigma^\flat_{E(q)}(q) = \sum_{i=1}^{n} \sigma^\flat_{D_i(p,r_i(p)}(q) - \sigma_{S_M(q)}(q) \le \sum_{i=1}^{n} r_i(q) - r(q) \le 0$$

Hence the assumptions of Theorem 14.3.2 below are satisfied and the conclusion ensues. □

Theorem 14.2.3 *Let us assume that an excess demand map $E : S^l \leadsto Y$ is upper hemicontinuous and that its images are closed, convex, bounded above and satisfy*

$$\forall\, p \in S^l,\ E(p) = E(p) - \mathbf{R}^l_+$$

Assume also that it satisfies the Walras law

$$\forall\, q \in S^l,\ \sigma^\flat_{E(q)}(q) \le 0$$

Then there exists at least a Walras equilibrium $\overline{p} \in S^l$, a solution to

$$0 \in E(\overline{p})$$

Proof — We apply Ky Fan's Inequality (Theorem 14.1.2) to the following function φ defined on $S^l \times S^l$ by

$$\varphi(p,q) := \sum_{i=1}^{n} \sigma^\flat_{E(p)}(q)$$

It is obviously finite (because $E(p)$ is bounded above) and concave with respect to q. Since the excess demand map is assumed to be upper hemicontinous, we infer that φ is lower semicontinuous with respect to p.

Walras law implies that

$$\varphi(q,q) = \sum_{i=1}^{n} \sigma^\flat_{E(q)}(q) \le 0$$

Hence, the assumptions of Ky Fan's Inequality Theorem are satisfied, so that there exists a price $\bar{p} \in S^l$ such that

$$\forall\, q \in S^l,\ \varphi(\bar{p},q) = \sum_{i=1}^{n} \sigma^\flat_{E(\overline{p})}(q) \le 0$$

The above inequality is still true for $q \in \mathbf{R}^l_+$.

We thus deduce that for any

$$\forall\, q \in \mathbf{R}^l, \quad \varphi(\bar{p}, q) = \sum_{i=1}^{n} \sigma^{b}_{E(\bar{p})-\mathbf{R}^l_+}(q) \;\leq\; 0$$

Hence we derive from the Separation Theorem that 0 belongs to the closed convex set $E(\overline{p})$ since this set is closed and convex. □

We did observe that if $p(t)$ is a price supplied by the Walras tâtonnement process and if it is not an equilibrium, it cannot be implemented because the associated total demand $\sum_{i=1}^{n} d_i(p, r_i(p))$ is not necessarily available. Therefore, by forbidding consumers to transact as long as the prices are not equilibria this model is not viable.

But the question which was asked despite this drawback is the convergence of solutions $p(t)$ of the above tâtonnement process to a limit, which then is an equilibrium. This issue, as well as either the Lyapunov stability or the asymptotic stability of a Walrasian equilibrium, require the use of adequate Lyapunov functions which decrease along the price paths $p(t)$ in such a way that one can derive the convergence of $p(t)$ when t goes to infinity. For example, an excess demand satisfying the property

$$\forall\, p \neq \bar{p}, \quad \left\langle p - \bar{p}, \sum_{i=1}^{n} d_i(p, r_i(p)) \right\rangle \;<\; \sigma_M(p - \bar{p})$$

is shown to converge to an equilibrium. Each differentiable Lyapunov function $p \mapsto V(p)$ entails a condition on the excess demand map because the property

$$V(p(t)) \;\leq\; V(p(0))e^{-\alpha t}$$

holds for any price path if and only if

$$\left\langle V'(p), \sum_{i=1}^{n} d_i(p, r_i(p)) \right\rangle + \alpha V(p) \;\leq\; \sigma_M(V'(p))$$

The function $V(p) := \frac{1}{2}\|p - \bar{p}\|^2$ provides the above condition. Another (non-differentiable) Lyapunov function proposed in the economic literature has been $V(p) := \max_{i=1,\dots n} \frac{p_i}{\bar{p}_i}$. (We saw in Chapter 8 that the Lyapunov method can be extended to any lower semicontinuous function.) Observe however that for any desired Lyapunov function V, a modified tâtonnement of the form

$$p'(t) \;\in\; E(V'(p(t)))$$

satisfies $t \mapsto V(p(t))$ does not increase under the usual Walras law.

If one of these functions V are not Lyapunov functions, Theorem 8.2.1 allows us to derive the existence of the smallest lower semicontinuous extended Lyapunov function V_∞ larger than V which decreases along the path $p(\cdot)$ of the Walrasian tâtonnement:

$$V(p(t)) \leq V_\infty(p(t)) \leq V_\infty(p_0)e^{-at}$$

where the domain $\mathrm{Dom}(V_\infty)$ provides the domain of attraction of the equilibrium.

14.3 Equilibrium Theorem on Convex Domains

By modifying slightly the proof of the Equilibrium Theorem 6.1.15, we can prove the existence of zeros of a set-valued map from a finite dimensional vector space X to another space Y:

Theorem 14.3.1 *Let K be a convex compact subset of a finite dimensional vector space X and F be an upper hemicontinuous set-valued map with closed convex values from K to another finite dimensional vector space Y.*

Let us consider also a continuous map $B : K \mapsto \mathcal{L}(X,Y)$. If K, F and B are related by the condition

$$\forall\, x \in K,\quad F(x) \cap \overline{B(x)T_K(x)} \neq \emptyset$$

then

$$\begin{cases} i) \quad \exists\, \bar{x} \in K \quad \text{such that } 0 \in F(\bar{x}) \\ ii) \quad \forall\, y \in K,\, \exists\, \hat{x} \in K \quad \text{such that } B(\hat{x})y \in B(\hat{x})\hat{x} + F(\hat{x}) \end{cases}$$

As an example, we derive the existence of a solution to the equation $f(\bar{x}) = 0$ where the solution $\bar{x}$ must belong to a compact convex subset K:

Theorem 14.3.2 *Let X and Y be finite dimensional vector spaces, $K \subset X$ be a compact convex subset, $\Omega \supset K$ be an open neighborhood of K and $f : \Omega \mapsto Y$ be a continuously differentiable single-valued map. Assume that*

$$\forall\, x \in K,\quad -f(x) \in f'(x)T_K(x)$$

Then there exists a solution $\bar{x} \in K$ to the equation $f(\bar{x}) = 0$. In particular, when $x_0 \in K$ is given, there exists a sequence of elements $x_n \in K$ satisfying

$$\forall\, n \geq 0,\quad f'(x_n)(x_n - x_{n-1}) = -f(x_n)$$

i.e., the implicit version of the Newton algorithm.

Proof — We take $F(x) := \{f(x)\}$ and $B(x) = -f'(x)$ in Theorem 14.4.1. □

Proof of Theorem 14.4.1 — We proceed by contradiction, assuming that the conclusion is false. Hence, for any $x \in K$, 0 does not belong to $F(x)$. Since the images of F are closed and convex, the Hahn-Banach Separation Theorem implies that there exists $p_x \in X^\star$ such that $\sigma(F(x), p_x) < 0$.

By setting

$$\mathcal{V}_p := \{x \in K \mid \sigma(F(x), p) < 0\}$$

the negation of the existence of an equilibrium of F in K implies that K can be covered by the subsets $\mathcal{V}_p$ when p ranges over the dual of X. These subsets are open by the very definition of upper hemicontinuity of F. So K can be covered by n such open subsets $\mathcal{V}_{p_i}$. Let us consider a continuous partition of unity $(\alpha_i)_{i=1,\ldots,n}$ associated with this finite open covering and introduce the function $\varphi : K \times K \to \mathbf{R}$ defined by

$$\varphi(x,y) := \sum_{i=1}^{n} \alpha_i(x)\langle p_i, B(x)(x-y)\rangle$$

Being continuous with respect to x and affine with respect to y, the assumptions of Ky Fan's Inequality (Theorem 14.1.2) are satisfied. Hence there exists $\overline{x} \in K$ such that for $\overline{p} := \sum_{i=1}^{n} \alpha_i(\overline{x})p_i$ we have

$$\forall\, y \in K, \quad \varphi(\overline{x}, y) = \langle \overline{p}, B(\overline{x})(\overline{x}-y) >\rangle \leq 0$$

The above inequality means that $-B(\overline{x})^\star\overline{p}$ belongs to the polar cone $T_K(\overline{x})^-$ of the convex subset K at $\overline{x}$. Therefore, taking a sequence $u_n \in T_K(\overline{x})$ such that $B(\overline{x})u_n$ converges to some

$$v \in F(\overline{x}) \cap \overline{B(\overline{x})T_K(\overline{x})}$$

(which exists by the tangential condition), we infer that

$$\left\{ \begin{array}{l} \sigma(F(\overline{x}), \overline{p}) \\ \geq < \overline{p}, v > = \lim_{n\to\infty} < \overline{p}, B(\overline{x})u_n > = \lim_{n\to\infty} < B(\overline{x})^\star\overline{p}, u_n > \geq 0 \end{array} \right.$$

We shall now contradict this inequality. We set

$$I(\overline{x}) := \{i = 1, \ldots, n \mid \alpha_i(\overline{x}) > 0\}$$

which is not empty. Hence

$$\sigma(F(\overline{x}), \overline{p}) \leq \sum_{i \in I(\overline{x})} \alpha_i(\overline{x})\sigma(F(\overline{x}), p_i) < 0$$

because, for any $i \in I(\overline{x})$, $\alpha_i(\overline{x}) > 0$, and thus, $\overline{x}$ belongs to the subset $\mathcal{V}_{p_i}$, which means precisely that $\sigma(F(\overline{x}), p_i) < 0$. The latter inequality is then a contradiction of the previous one.

Take now $y \in K$ and introduce the set-valued map $G : X \rightsquigarrow Y$ defined by

$$G(x) := F(x) + B(x)(x-y)$$

which also satisfies the assumptions of our theorem. Then there exists a zero $\widehat{x} \in K$ of G, which is a solution to the inclusion $B(\widehat{x})y \in B(\widehat{x})\widehat{x} + F(\widehat{x})$. □

Remark: The Nonconvex Case — For nonconvex compact subsets, the most general Fixed Point Theorem (known to the author) is due to [75, Ben El Merchaiek & Kryszewski]. Convexity subsumes two kinds of properties required for such a Fixed-Point Theorem in nonconvex compact sets. Such a theorem requires

1. a non zero "Euler characteristic" of K. The Euler characteristic of K is a topological invariant : it is the same for all sets homeomorphic to K. The Euler characteristic of convex sets, and more generally, of acyclic sets, is equal to 1. The Euler characteristic of star-shaped or contractible subsets is different from 0.

2. a regularity property: A subset K is said to be a $\mathcal{L}$-retract if there exists a neighborhood $\mathcal{V}$ of K and a retraction $r : X \mapsto K$ and a constant $\lambda > 0$ such that for all

$$x \in \mathcal{V}, \quad \|r(x) - x\| \leq \lambda d(x, K)$$

If the projector Π_K onto a closed subset is single-valued (in which case K is said to have property ρ according to [288, Plaskacz], or to be a "proximal neighborhood retract" according to [218, Górniewicz] or "proximally smooth" in nonsmooth analysis), then K is obviously a $\mathcal{L}$-retract. For such closed subsets, the Clarke tangent cone $C_K(x)$ and the contingent cone $T_K(x)$ do coincide thanks to [288, Plaskacz]. In this case, the above theorem was derived in [289, Plaskacz] through the Viability Theorem.

The Ben El Merchaiek-Kryszewski Theorem states that *if K is a compact $\mathcal{L}$-retract the Euler characteristic of which is non zero, then an upper semicontinuous with closed convex images has an equilibrium whenever K satisfies the stronger tangential condition*

$$\forall\, x \in K, \quad F(x) \cap C_K(x) \neq \emptyset \quad \square$$

14.4 Fixed Point Theorems

We begin by showing that Theorem 6.1.15 implies the Kakutani Fixed Point Theorem[7], which is the set-valued version of the Brouwer Fixed Point Theo-

[7]called Ky Fan's Fixed Point Theorem in infinite dimensional spaces.

The story began in 1910 with the Brouwer Fixed Point Theorem, which was proved later in 1926 via the Three Polish Lemma, the three Poles being Knaster, Kuratowski and Mazurkiewicz. Knaster saw the connection between Sperner's Lemma and the fixed point theorem, Mazurkiewicz provided a proof corrected by Kuratowski. The extension to Banach spaces was proved in 1930 by their colleague Schauder.

Von Neumann did need the set-valued version of this Fixed Point Theorem in game theory, which was proved by Kakutani in 1941.

rem. Therefore, these theorems are all equivalent to the Brouwer Fixed Point Theorem and the Ky Fan Inequality.

Theorem 14.4.1 (Kakutani Fixed Point Theorem) *Let K be a convex compact subset of a finite dimensional vector space X and $G : X \leadsto K$ be an upper hemicontinuous set-valued map with nonempty closed convex values. Then G has a fixed point[8] $\bar{x} \in K \cap G(\bar{x})$.*

Proof — We set $F(x) := G(x) - x$, which is also upper hemicontinuous with convex values. Since K is convex, then $K - x \subset T_K(x)$, and since $G(K) \subset K$, we deduce that K is a viability domain of F because

$$F(x) \subset K - x \subset T_K(x)$$

Hence there exists a viable equilibrium $\bar{x} \in K$ of F, which is a fixed point of G. □

Actually, we do not need to assume that G maps K to itself. It is enough to assume that K is a viability domain of $F := G - \mathbf{1}$, which can be written in the following form

$$\forall\, x \in K, \quad G(x) \cap (x + T_K(x)) \neq \emptyset \tag{14.4.1}$$

This leads to the following

Definition 14.4.2 (Inward & Outward Maps) *A map $G : K \leadsto X$ satisfying property (14.5.1) is said to be* **inward**. *It is called* **outward** *if*

$$\forall\, x \in K, \quad G(x) \cap (x - T_K(x)) \neq \emptyset$$

Since K is a viability domain of $F := G - \mathbf{1}$ when G is inward and of $F_- := \mathbf{1} - G$ when G is outward, and since the equilibria of F and F_- are fixed points of G, we obtain the useful

Theorem 14.4.3 *Let K be a convex compact subset of a finite dimensional vector space X and $G : K \leadsto X$ be an upper hemicontinuous map with nonempty closed convex values. If G is either inward or outward, it has a fixed point*

$$\bar{x} \in K \cap G(\bar{x})$$

Remark — Fixed-Point Theorems are easily derived from Equilibrium Theorems. The derivations of Equilibrium Theorems from Fixed-Point Theorems is more laborious. It goes through the acquirement of theorems on periodic solutions to periodic differential inclusions. These are obtained by applying Fixed-Point Theorems to the reachable map of the differential inclusion. When the differential inclusion is time independent, we exploit the fact that an equilibrium is a periodic solution of period 0, thus obtained by taking limits of periodic solutions when the period converges to 0. □

[8] which can be regarded as an equilibrium for the discrete set-valued dynamical system $x_{n+1} \in G(x_n)$.

14.5 The Inverse Function Theorem

We shall deduce from Ekeland's Theorem a "nonlinear extension" of Graves' Theorem, which allows one to solve (locally) a "constrained problem" of the form

$$\text{find } x \in K \quad \text{such that } f(x) = y$$

for any right-hand side y in a neighborhood of some

$$y_0 := f(x_0) \in f(K)$$

When K is equal to the whole space X, a solution to this unconstrained problem is given by Graves' version of the Inverse-Function Theorem[9].

Theorem 14.5.1 (Constrained Inverse Function Theorem) *Let X and Y be finite dimensional vector spaces, K be a closed sleek subset of X and $f : X \mapsto Y$ be a continuously differentiable at $x_0 \in K$ map. Assume that*

$$f'(x_0)T_K(x_0) = Y$$

Then $f(x_0)$ belongs to the interior of $f(K)$ and there exist constants $l > 0$ and $\eta > 0$ such that for every $y_i \in B(f(x), \eta)$ (i=1,2), there exist $x_i \in K \cap B(x, \eta)$ satisfying

$$\|x_1 - x_2\| \leq l\|y_1 - y_2\|$$

(It is said that the set-valued map

$$y \rightsquigarrow f^{-1}(y) \cap K$$

is pseudo-Lipschitz *around $(f(x_0), x_0)$).*

Definition 14.5.2 *If y is given in $F(x)$, F is said to be* pseudo-Lipschitz *around $(x, y) \in$ Graph(F) if there exist a positive constant λ and neighborhoods $\mathcal{U} \subset$ Dom(F) of x and $\mathcal{V}$ of y such that*

$$\forall\, x_1, x_2 \in \mathcal{U}, \quad F(x_1) \cap \mathcal{V} \subset F(x_2) + \lambda\|x_1 - x_2\| B_Y$$

This property happens to be very useful, and has been extensively studied by Terry Rockafellar (see for instance [304, Rockafellar]). Its role in the convergence of the Newton Method for solving several classes of inclusions by Asen Dontchev (see [156,157,160, Dontchev], [163, Dontchev & Rockafellar]).

Like Ky Fan's Inequality, Ekeland's Theorem provides less elegant but more efficient statements than the Banach-Picard Fixed Point Theorem. It allows an easy proof of the above theorem. Ekeland's Theorem provides an approximate minimizer of a bounded from below lower semicontinuous function in a given neighborhood of a point. This localization property is very useful and explains the importance of this result which has been used extensively since its discovery in 1974.

[9] For recent reviews, see [158, Dontchev] and [162, Dontchev & Hager].

Theorem 14.5.3 (Ekeland's Variational Principle) *Let*

$$V : K \mapsto \mathbf{R} \cup \{+\infty\}$$

be a lower semicontinuous nontrivial extended bounded from below function defined on a complete metric space K. Let $x_0 \in \mathrm{Dom}(V)$ and $\varepsilon > 0$ be fixed. Then there exists $\overline{x} \in K$, a solution to

$$\left\{ \begin{array}{ll} i) & V(\overline{x}) + \varepsilon d(x_0, \overline{x}) \ \leq \ V(x_0) \\ & \\ ii) & \forall\, x \neq \overline{x}, \ \ V(\overline{x}) \ < \ V(x) + \varepsilon d(x, \overline{x}) \end{array} \right. \tag{14.5.1}$$

When K is a vector space X, we can reformulate this theorem in terms of contingent epiderivatives of the function V.

Theorem 14.5.4 *Let X be a Banach space, $V : X \mapsto \mathbf{R}_+ \cup \{+\infty\}$ be a nontrivial lower semicontinuous bounded from below function and $x_0 \in \mathrm{Dom}(V)$ be a given point of its domain. Then, for any $\varepsilon > 0$, there exists a solution $x_\varepsilon \in \mathrm{Dom}(V)$ to:*

$$\left\{ \begin{array}{ll} i) & V(x_\varepsilon) + \varepsilon \|x_\varepsilon - x_0\| \ \leq \ V(x_0) \\ & \\ ii) & \forall\, u \ \in \ X, \ \ 0 \ \leq \ D_\uparrow V(x_\varepsilon)(u) + \varepsilon \|u\| \end{array} \right. \tag{14.5.2}$$

Nut not only are we interested in knowing whether a solution to the constrained problem does exist, but we wish to approximate it by solutions to the approximate problems

$$\text{find } \ x_n \ \in \ K_n \quad \text{such that} \ \ f_n(x_n) = y_n$$

where y_n converges to y_0, f_n converges to f and K_n to K in the following sense:

Definition 14.5.5 *We say that the maps f_n are* consistent *with f on K_n at $x_0 \in K$ if there exist elements $x_{0n} \in K_n$ converging to x_0 such that $f_n(x_{0n})$ converges to $f(x_0)$.*

We have to add to this consistency assumption a stability assumption to be able to approximate the solution of the constrained problem. Hence Theorem 14.6.1 follows from

Theorem 14.5.6 *Consider a sequence of continuous single-valued maps f_n from X to Y, a sequence of closed subsets $K_n \subset X$ and an element x_0 in the lower limit of the subsets K_n such that the maps f_n are* consistent *with f on K_n at x_0.*

We assume that f_n are Fréchet differentiable on a neighborhood of x_0 and verify the following stability assumption*:*

there exist constants $c > 0$, $\alpha \in [0,1[$ *and* $\eta > 0$ *such that*

$$\forall\, x \in K_n \cap B(x_0,\eta),\quad B_Y \subset f_n'(x)(T_{K_n}(x) \cap cB_X) + \alpha B_Y \qquad (14.5.3)$$

Then there exist $l > 0$ *and* $\gamma > 0$ *such that for any* $x_{0n} \in B_{K_n}(x_0,\gamma)$ *satisfying* $f_n(x_{0n}) \in B(f(x_0),\gamma)$, *we have*

$$\forall\, y_n \in B(f(x_0),\gamma),\quad d\left(x_{0n}, f_n^{-1}(y_n) \cap K_n\right) \leq l\|y_n - f_n(x_{0n})\|$$

We refer to Theorem 4.3.1 of SET-VALUED ANALYSIS for a proof.

We conclude this section with a local uniqueness criterion of a solution to a constrained nonlinear problem:

Proposition 14.5.7 *Let* K *be a subset of a finite dimensional vector-space* X *and* $f : X \mapsto Y$ *a single-valued map to a finite dimensional vector space* Y. *Let* $x_0 \in K$ *and assume that* f *is Fréchet differentiable at* x_0.

If

$$T_K(x_0) \cap \ker(f'(x_0)) = \{0\}$$

then x_0 *is the only local solution* $x \in K$ *to the equation* $f(x) = f(x_0)$, *in the sense that there exists a neighborhood* $\mathcal{N}(x_0)$ *of* x_0 *such that*

$$\forall\, x \in \mathcal{N}(x_0) \cap K,\quad x \neq x_0,\quad f(x) \neq f(x_0)$$

Proof — Assume the contrary: for all $n > 0$, there exists

$$x_n \in K \cap B(x_0, 1/n),\quad x_n \neq x_0$$

such that $f(x_n) = f(x_0)$. Let us set

$$h_n := \|x_n - x_0\| > 0$$

which converges to 0 and

$$v_n := \frac{x_n - x_0}{h_n}$$

Since v_n belongs to the unit sphere, which is compact, a subsequence (again denoted) v_n converges to some element v of the unit sphere.

This limit v belongs also to the contingent cone $T_K(x_0)$ because, for all $n > 0$, $x_0 + h_n v_n = x_n$ belongs to K. Furthermore, f being Fréchet differentiable at x_0, we obtain the equation $f'(x_0)v_n = \varepsilon(h_n)$ where $\varepsilon(h_n)$ converges to 0 with h_n. We thus infer that v belongs also to the kernel of $f'(x_0)$. Since $v \neq 0$, we derive a contradiction of the assumption. □

which converges to 0 and

$$v_n := \frac{x_n - x_0}{h_n}$$

Since v_n belongs to the unit sphere, which is compact, a subsequence (again denoted) v_n converges to some element v of the unit sphere.

This limit v belongs also to the contingent cone $T_K(x_0)$ because, for all $n > 0$, $x_0 + h_n v_n = x_n$ belongs to K. Furthermore, f being Fréchet differentiable at x_0, we obtain the equation $f'(x_0)v_n = \varepsilon(h_n)$ where $\varepsilon(h_n)$ converges to 0 with h_n. We thus infer that v belongs also to the kernel of $f'(x_0)$. Since $v \neq 0$, we derive a contradiction of the assumption. □

Chapter 15

Stochastic Viability

Introduction

Although I do not advocate the use of stochastic differential equations to provide metaphors of the kind of uncertainty we meet in theoretical economics, I know that it is widely used and justified in some specific domains, such as finance. Since viability constraints — even of a stochastic nature — appear naturally, Giuseppe da Prato and I adapted some of the basic viability theorems to stochastic differential inclusions.

The first aim of this chapter is then to combine two ways for representing uncertainty through stochastic differential inclusions: a "stochastic uncertainty", driven by a Wiener process, and a "contingent uncertainty", driven by a set-valued map [1]:

$$d\xi \in F(\xi(t))dt + g(\xi(t))dW(t),$$

where F is a set-valued maps with values in X, g is a map from X into $\mathcal{L}(Y,X)$ and W is an Y-valued Wiener process. It describes the contribution of two kinds of randomness: a contingent one, represented by a set-valued drift, and a stochastic one described by a diffusion. Actually, the diffusion can also be assumed set-valued.

The second objective is to study the evolution of stochastic dynamical economies with continuous dynamics and state dependent prices

$$\begin{cases} d\xi = c(\xi(t),p(t))dt + g(\xi(t),p(t))dW(t) \\ \text{for almost all } (\omega,t) \in \Omega \times [0,\infty[, \;\; p_\omega(t) \in P(x_\omega(t)) \end{cases}$$

under scarcity constraints

$$\forall\, t \geq 0, \;\; \text{for almost all } \omega \in \Omega, \;\; \xi_\omega(t) \in K$$

[1] which can even be a "fuzzy set-valued map" as in [47, Aubin & Dordan].

For that purpose, we have to extend to "stochastic differential inclusions" the Viability Theorem for nonstochastic differential inclusions, where both the drift and the diffusion are set-valued.

Both problems fit to the following framework. If H is a set-valued map from X to $X \times \mathcal{L}(Y, X)$, we introduce stochastic differential inclusions of the form

$$d\xi \in H(\xi, dt \oplus dW(t)).$$

We look for "weak solutions" $\xi(\cdot)$ in the sense of [339, Stroock & Varadhan] (also known under the name of "martingale solutions" in the sense of [139, Da Prato & Zabczyk]): there exist a probability space $(\widehat{\Omega}, \widehat{\mathcal{A}}, \widehat{\mathbf{P}})$ with a filtration $\{\widehat{\mathcal{F}}_t\}$ and a Wiener process $\widehat{W}$ such that

$$\widehat{\xi}(t) = \widehat{\xi}(0) + \int_0^t \widehat{u}(s)ds + \int_0^t \widehat{v}(s))d\widehat{W}(s)$$

where for almost all $(\omega, t) \in \widehat{\Omega} \times [0, \infty[$,

$$(\widehat{u}_\omega(t), \widehat{v}_\omega(t)) \in H(\widehat{\xi}_\omega(t)).$$

We also want to prove the **(stochastic) viability property** of K with respect to H: from any initial random variable x in K, starts at least one solution $\xi(\cdot)$ to the stochastic differential inclusion which is **viable in** K, in the sense that

$$\forall\, t \in [0, \infty[, \text{ for almost all } \omega \in \Omega, \;\; \xi_\omega(t) \in K$$

whenever the stochastic dynamics H satisfy adequate tangential conditions.

1. For that purpose, we use the concept of stochastic contingent set $\mathcal{T}_K(x)$ introduced in [42, Aubin & Da Prato].

2. Then we shall prove in essence that the subset K enjoys the viability property with respect to H whenever

   $$H(x) \cap \mathcal{T}_K(x) \neq \emptyset,$$

   for any $x \in K$.

   For instance, this condition means that for every $\mathcal{F}_0$-random variable x viable in K

   - when K is a vector subspace,

     $$\forall x \in H,\; H(x) \cap (K \times K) \neq \emptyset$$

   - When K is the unit sphere, there exists $(u, v) \in H(x)$ such that

     $$vx = 0 \;\&\; \langle x, u\rangle + \frac{1}{2}\|v\|^2 = 0$$

- When K is the unit ball, there exists $(u,v) \in H(x)$ such that

$$vx = 0 \ \& \ \langle x,u\rangle + \frac{1}{2}\|v\|^2 \leq 0$$

Here the norm $\|\cdot\|$ is the Hilbert–Schmidt norm, that is

$$\|v\|^2 = \text{Tr}\,[vv^*]$$

We mention that an elementary calculus of stochastic tangent sets to direct images, inverse images and intersections of closed subsets can be found in [42, Aubin & Da Prato].

Let us conclude this introduction by mentioning that the "dual" concept of invariance, requiring that all solutions remain viable in K follows from [42, Aubin & Da Prato] and [138, Da Prato & Frankowska]: under adequate Lipschitzianity assumptions the invariance follows from the condition

$$H(x) \subset \mathcal{T}_K(x)$$

3. We then extend to the stochastic case the Lyapunov second method for studying stability of solutions to stochastic differential inclusions, by adopting the viability approach used about dynamical inequalities for differential inclusions as in Chapter 8, observing that a function enjoys the Lyapunov property if and only if its epigraph is a viability domain of an auxiliary system.

15.1 Stochastic Tangent Sets

Let us consider a σ–complete probability space $(\Omega, \mathcal{F}, \mathbf{P})$, an increasing family of σ–sub–algebras $\mathcal{F}_t \subset \mathcal{F}$ and two finite dimensional vector-spaces $X := \mathbf{R}^n$, $Y := \mathbf{R}^m$.

The constraints are defined by a closed subset $K \subset X$. We denote by $\mathcal{K}$ the subset

$$\mathcal{K} := \{u \in L^2(\Omega, \mathcal{F}, P) \mid \text{for almost all } \omega \in \Omega,\ \ u_\omega \in K\}$$

Definition 15.1.1 (Stochastic Contingent Set) *Let us consider a $\mathcal{F}_\tau$-measurable random variable $x \in \mathcal{K}$. We define the* stochastic contingent set *$\mathcal{T}_K(x)$ to K at x as the set of pairs $(\gamma, v) \in \mathbf{R}^n \times \mathcal{L}(Y,X)$ satisfying the following property: For any $\alpha, \rho > 0$, there exist $h \in]0,\alpha[$ and $\mathcal{F}_{\tau+t}$-measurable random variables a^h and b^h with values in $\mathbf{R}^n$, such that*

$$\begin{cases} i) & \mathbf{E}(\|a^h\|^2) \leq \rho^2 \\ ii) & \mathbf{E}(\|b^h\|^2) \leq \rho^2 \\ iii) & \mathbf{E}(b^h) = 0 \\ iv) & b^h \text{ is independent of } \mathcal{F}_\tau \end{cases}$$

and satisfying

$$x + h\gamma + v(W(\tau+h) - W(\tau)) + ha^h + \sqrt{h}b^h \in \mathcal{K}$$

Definition 15.1.2 *We shall say that* $H : X \rightsquigarrow X \times \mathcal{L}(Y,X)$ *is a Marchaud set-valued map if its graph is closed, its images are convex and its growth is linear in the sense that*

$$\forall\, x \in X,\ \forall (u,v) \in H(x),\ \ \max\left(\|u\|^2,\ \|v\|^2\right) \le c\left(\|x\|^2 + 1\right) \qquad (15.1.1)$$

We consider the stochastic differential inclusion

$$d\xi \ \in\ H(\xi, dt \oplus dW(t)) \qquad (15.1.2)$$

where H is a Marchaud set-valued map.

Definition 15.1.3 *We say that a stochastic process* $\xi(t)$ *is a weak or* martingale solution *to the stochastic differential inclusion (15.1.2) starting at* ξ_0 *if there exist a probability space* $(\widehat{\Omega}, \widehat{\mathcal{A}}, \widehat{\mathbf{P}})$ *with a filtration* $\{\widehat{\mathcal{F}}_t\}$ *and a* $\{\widehat{\mathcal{F}}_t\}$*-Wiener process* $\widehat{W}$ *such that*

$$\begin{array}{ll} i) & \widehat{\xi}(t) \ =\ \widehat{\xi}(0) + \int_0^t \widehat{u}(s)ds + \int_0^t \widehat{v}(s))d\widehat{W}(s) \\ ii) & \text{for almost all } (\omega,t) \in \Omega \times [0,\infty[,\ \ (\widehat{u}_\omega(t), \widehat{v}_\omega(t)) \in H(\widehat{\xi}_\omega(t)) \end{array} \qquad (15.1.3)$$

where $\widehat{\xi}_0$ *has the same law than* ξ_0*.*

Definition 15.1.4 *We shall say that a stochastic process* $x(\cdot)$ *is* viable in K *if and only if*

$$\forall\, t \ge 0,\ \ x(t)\ \in\ \mathcal{K} \qquad (15.1.4)$$

i.e., if and only if

$$\forall\, t \ge 0,\ \text{ for almost all } \omega \in \Omega,\ \ \xi_\omega(t)\ \in\ K$$

We shall say that K *enjoys the* (stochastic) viability property *with respect to* H *— or, in short, that* K *is viable under* H *— if for any* $\mathcal{F}_0$*-measurable random variable* ξ_0 *in* K *starts one* $\mathcal{F}_0$*-measurable martingale solution* $t \mapsto \xi(t)$ *to the stochastic differential inclusion (15.1.2) starting at* x *which is* viable in K.

15.2 Stochastic Viability

Theorem 15.2.1 (Stochastic Viability) *Let* K *be a closed subset of* X*. We assume that* H *is a Marchaud set-valued map. If*

$$\forall x \in K,\ \ H(x) \cap \mathcal{T}_K(x) \neq \emptyset,$$

then, from any initial $\mathcal{F}_0$*-random variable* $\xi_0 \in \mathcal{K}$ *starts a martingale solution to the stochastic differential inclusion (15.1.2) which is viable in* $\mathcal{K}$*.*

It is useful to translate this theorem in terms of stochastic dynamical economies with state-dependent constraints bearing on the prices where the set $H(x) := \{(c(x,p), g(x,p))\}_{p\in P(x)}$ is parametrized by prices:

Theorem 15.2.2 (Viable Stochastic Dynamical Economy) *Let K be a closed subset of X. We assume that the set-valued map $P : X \mapsto Z$ is Marchaud and that the maps c, g : Graph$(P) \mapsto X$ are continuous and affine with respect to p.*

We introduce the feedback map $x \mapsto \mathcal{P}_K(x)$ associating with every $\mathcal{F}_t$-random variable x in $\mathcal{K}$ the set of $\mathcal{F}_t$-random variable $p \in P(x)$ such that

$$(c(x,p), g(x,p)) \in \mathcal{T}_{\mathcal{K}}(x)$$

If for every $\mathcal{F}_t$-measurable random variable x in $\mathcal{K}$, the set $\mathcal{P}_K(x)$ is not empty, then, from any initial $\mathcal{F}_0$-measurable random variable $\xi_0 \in \mathcal{K}$ starts a martingale solution to the stochastic dynamical economy

$$\begin{cases} d\xi = c(\xi(t), p(t))dt + g(\xi(t), p(t))dW(t) \\ \text{for almost all } (\omega, t) \in \Omega \times [0, \infty[, \; p_\omega(t) \in P(x_\omega(t)) \end{cases} \tag{15.2.1}$$

which is viable in $\mathcal{K}$. The evolution of viable martingale solutions is regulated by

$$\textit{for almost all } (\omega, t) \in \Omega \times [0, \infty[, \; p_\omega(t) \in \mathcal{P}_K(\xi_\omega(t))$$

Proof of Theorem 15.2.1 — It is enough to prove the existence of a solution to (15.1.2) on an arbitrary closed line interval, $[0, 1]$ for instance. We set

$$\|u(\cdot)\|_\infty := \sup_{t\in[0,1]} \|u(t)\| \;\&\; \|\xi(\cdot)\|_\alpha := \sup_{t\neq s} \frac{\|\xi(t) - \xi(s)\|}{|t-s|^\alpha}$$

We shall first provide a priori estimates, second construct approximate solutions to the stochastic differential inclusion and third, prove that they converge to a solution to our problem.

A priori estimates

We shall prove that the linear growth assumption implies that :
$\forall\, T > 0, \; \exists \gamma > 0$ such that

$$\max\left\{\sup_{t\in[0,T]} \mathbf{E}(\|\xi(t)\|^2), \; \sup_{t\in[0,T]} \mathbf{E}(\|u(t)\|^2), \; \sup_{t\in[0,T]} \mathbf{E}(\|v(t)\|^2)\right\} \le \gamma$$

For simplicity, set $T = 1$. Hence

$$\mathbf{E}(\|\xi(t)\|^2) \leq 2\mathbf{E}(\|\xi_0\|^2) + 2\mathbf{E}\left(\left\|\int_0^t u(s)ds\right\|^2\right)$$

$$+2\mathbf{E}\left(\left\|\int_0^t v(s)dW(s)\right\|^2\right) + 4\mathbf{E}\left(\left\langle\int_0^t u(s)ds, \int_0^t v(s)dW(s)\right\rangle\right)$$

$$\leq 2(\mathbf{E}(\|\xi_0\|^2 + t\int_0^t \mathbf{E}(\|u(s)\|^2)ds + \int_0^t \mathbf{E}(\|v(s)\|^2)ds)$$

because

$$\mathbf{E}\left(\left\|\int_0^t \varphi(s)ds\right\|^2\right) \leq t\int_0^t \mathbf{E}(\|\varphi(s)\|^2)ds, \tag{15.2.2}$$

and

$$\mathbf{E}\left(\left\|\int_0^t \varphi(s)dW(s)\right\|^2\right) = \int_0^t \mathbf{E}(\|\varphi(s)\|^2)ds \tag{15.2.3}$$

Using linear growth assumption (15.1.1), and setting $\varphi(s) := \mathbf{E}(\|\xi(s)\|^2)$, we deduce that

$$\varphi(t) \leq (c+1) + 2\varphi(0) + 2(c+1)\int_0^t \varphi(s)ds.$$

Then the claim follows from the Gronwall lemma. □

Construction of Approximate Solutions

We begin by constructing approximate viable solutions to the stochastic differential inclusion.

Let B denote the unit ball of $X \times \mathcal{L}(X,Y)$.

Lemma 15.2.3 *Let K be a closed subset of X. Assume that H is a bounded set-valued map. Then there exists a constant $\gamma > 0$ such that, for any $\varepsilon > 0$, the set $S_\varepsilon(\xi_0)$ of adapted stochastic processes*

$$\xi(t) \;=\; \xi(0) + \int_0^t u(s)ds + \int_0^t v(s)dW(s)$$

on $[0,1]$ satisfying $\xi(0) = \xi_0$,

$$\textit{for almost all } (\omega,t) \in \Omega \times [0,1], \;\; (u_\omega(t), v_\omega(t)) \;\in \gamma B$$

and

$$\begin{cases} i) & \mathbf{E}\left(\sup_{t\in[0,1]} d^2(\xi(t),\mathcal{K})\right) \leq 2\varepsilon^2 \\ ii) & \mathbf{E}\left(\text{ess sup}_{t\in[0,1]} d^2\left((\xi(t),u(t),v(t)),\text{Graph}(H)\right)\right) \;\leq\; 2\varepsilon^2 \\ iii) & \max\left(\mathbf{E}\left(\|\xi(\cdot)\|_\alpha\right),\; \mathbf{E}(\|u(\cdot)\|_\infty),\; \mathbf{E}(\|v(\cdot)\|_\infty)\right) \;\leq\; \gamma, \end{cases} \tag{15.2.4}$$

is not empty.

Proof — Let us fix $\varepsilon > 0$. We denote by $\mathcal{A}_\varepsilon(\xi_0)$ the set of pairs $(T_\xi, \xi(\cdot))$ where $T_\xi \in [0,1]$ and $\xi(\cdot)$ is a stochastic process satisfying $\xi(0) = \xi_0$ and

$$\begin{cases} i) & \forall\, t \in [0, T_\xi],\ \mathbf{E} d^2(\xi(T_\xi), \mathcal{K}) \ \le\ \varepsilon^2 T_\xi \\ ii) & \mathbf{E}\left(\sup_{t\in[0,T_\xi]} d^2(\xi(t), \mathcal{K})\right) \le 2\varepsilon^2 \\ iii) & \mathbf{E}\left(\text{ess sup}_{t\in[0,T_x]} d^2\left((\xi(t), u(t), v(t)), \text{Graph}(H)\right)\right) \ \le\ 2\varepsilon^2 \\ iv) & \max\left(\mathbf{E}\left(\|\xi(\cdot)\|_\alpha\right),\ \mathbf{E}(\|u(\cdot)\|_\infty),\ \mathbf{E}(\|v(\cdot)\|_\infty)\right) \ \le\ \gamma \end{cases} \tag{15.2.5}$$

The set $\mathcal{A}_\varepsilon(\xi)$ is not empty: take $T_\xi = 0$ and $\xi(0) \equiv \xi_0$. It is an inductive set for the order relation

$$(T_{\xi_1}, \xi_1(\cdot)) \ \preceq\ (T_{\xi_2}, \xi_2(\cdot))$$

if and only if

$$T_{\xi_1} \ \le\ T_{\xi_2} \ \ \&\ \ \xi_2(\cdot)|_{[0,T_{\xi_1}]} \ =\ \xi_1(\cdot)$$

Zorn's Lemma implies that there exists a maximal element $(T_\xi, \xi(\cdot)) \in \mathcal{A}_\varepsilon(\xi_0)$. The Lemma follows from the claim that for such a maximal element, we have $T_\xi = 1$. If not, we shall extend $\xi(\cdot)$ by a stochastic process $\widehat{\xi}(\cdot)$ on an interval $[T_\xi, S_\xi]$ where $S_\xi > T_\xi$, contradicting the maximal character of $(T_\xi, \xi(\cdot))$.

Since $\xi_\omega(T_\xi)$ is $\mathcal{F}_{T_\xi}$ measurable, the projection map $\Pi_K(\xi_\omega(T_\xi))$ is also $\mathcal{F}_{T_\xi}$-measurable (see Theorem 8.2.13, p. 317 of [54, Aubin & Frankowska]). Then there exists a $\mathcal{F}_{T_\xi}$-measurable selection $y_\omega \in \Pi_K(\xi_\omega(T_\xi))$, which we call a projection of the random variable $\xi(T_\xi)$ onto K.

For simplicity, *we set $x = \xi(T_\xi)$ and thus choose a projection $y \in \Pi_K(x)$*. We take

$$\rho := \frac{\varepsilon\sqrt{1 - T_\xi}}{2} > 0$$

and we set

$$c^2 \ :=\ \max_{(u,v)\in\gamma B} (\mathbf{E}(\|u\|^2), 4\mathbf{E}(\|v\|^2)) \ <\ +\infty \tag{15.2.6}$$

We then introduce

$$\beta \ :=\ \min\left(\varepsilon, \frac{\varepsilon^2(1 - T_x)}{2c^2}\right) \ >\ 0$$

which is positive whenever $T_\xi < 1$.

We know that there exists $(u, v) \in H(y)$ such that (u, v) belongs to the stochastic contingent set $\mathcal{T}_\mathcal{K}(y)$: There exist $h \in]0, \beta]$ and $\mathcal{F}_{T_x+h}$-random variables a^h and b^h such that

$$\begin{cases} i) & \mathbf{E}(\|a^h\|^2) \le \rho^2 \\ ii) & \mathbf{E}(\|b^h\|^2) \le \rho^2 \\ iii) & \mathbf{E}(b^h) \ =\ 0 \\ iv) & b^h \text{ is independent of } \mathcal{F}_t \end{cases} \tag{15.2.7}$$

and satisfying

$$y + v(W(T_x + h) - W(T_x)) + hu + ha^h + \sqrt{h}b^h \in \mathcal{K} \tag{15.2.8}$$

We then set $S_x := T_x + h > T_x$ and we *define the stochastic process* $\widehat{\xi}(t)$ *on the interval* $[T_x, S_x]$ by

$$\widehat{\xi}(t) \ := \ x + (t - T_x)u + v(W(t) - W(T_x)) \tag{15.2.9}$$

Therefore, setting $h := t - T_x$,

$$d_{\mathcal{K}}^2(\widehat{\xi}(S_x)) - d_{\mathcal{K}}^2(\widehat{\xi}(T_x)) \ \le \left\|x - y - ha^h - \sqrt{h}b^h\right\|^2 - \|x - y\|^2 =$$

$$\|ha^h + \sqrt{h}b^h\|^2 - 2\left\langle x - y, ha^h\right\rangle - 2\left\langle x - y, \sqrt{h}b^h\right\rangle$$

We take the expectation in both sides of this inequality and estimate each term of the right hand-side. First, we use the estimate

$$\mathbf{E}(\|ha^h + \sqrt{h}b^h\|^2) \ \le \ 2h(h\mathbf{E}(\|a^h\|^2) + \mathbf{E}(\|b^h\|^2))$$

thanks to estimates (15.2.2), and (15.2.3).

Next,

$$\mathbf{E}\left(\left\langle x - y, a^h\right\rangle\right) \ \le \ \mathbf{E}\left(\|x - y\|^2\right)^{\frac{1}{2}} \left(\mathbf{E}\left(\left\|a^h\right\|^2\right)\right)^{\frac{1}{2}}$$

and we observe that

$$\mathbf{E}\left\langle x - y, \frac{1}{\sqrt{h}}b^h\right\rangle \ = \ 0$$

since b^h is independent of $x - y$ and $\mathbf{E}(b^h) = 0$. We obtain, by the very choice of ρ,

$$\mathbf{E}(d^2(\widehat{\xi}(S_x), \mathcal{K})) \ = \ \mathbf{E}(d^2(\widehat{\xi}(T_x + h), \mathcal{K}))$$

$$\le \ \mathbf{E}d^2(\widehat{\xi}(T_x), \mathcal{K}) + 2h\mathbf{E}\left(\|x - y\|^2\right)^{\frac{1}{2}} \left(\mathbf{E}\left(\left\|a^h\right\|^2\right)\right)^{\frac{1}{2}}$$

$$+2h(h\mathbf{E}(\|a^h\|^2) + \mathbf{E}(\|b^h\|^2))$$

$$\le \ \mathbf{E}d^2(\widehat{\xi}(T_x), \mathcal{K}) + h\left(\mathbf{E}\left(\|x - y\|^2\right) + 3\mathbf{E}(\|a^h\|^2) + \mathbf{E}(\|b^h\|^2)\right)\Big)$$

$$\le \ \varepsilon^2 T_x + h(\varepsilon^2 T_x + 4\rho^2) \ \le \ \varepsilon^2 T_x + h\varepsilon^2 \ = \ \varepsilon^2 S_x$$

by (15.2.5)i).

Hence $\widehat{\xi}(\cdot)$ satisfies (15.2.5)i) for S_x.

We observe also that for any $t \in [T_x, S_x]$,

$$d_{\mathcal{K}}^2(\widehat{\xi}(t)) \ \le \ \|\widehat{\xi}(t) - y\|^2$$

and that

$$\|\widehat{\xi}(t)-y\|^2 = \|x-y+(t-T_x)u+v(W(t)-W(T_x))\|^2$$

$$= d_K^2(x)+2\langle x-y,(t-T_x)u+v(W(t)-W(T_x))\rangle$$

$$+\|(t-T_x)u+v(W(t)-W(T_x))\|^2$$

$$\leq d_K^2(x)+\|x-y\|+2\left\|(t-T_x)u+v(W(t)-W(T_x))\right\|^2$$

$$\leq 2d_K^2(x)+4(t-T_x)^2\left\|u\right\|^2+4\left\|v(W(t)-W(T_x))\right\|^2$$

Therefore

$$\sup_{t\in[T_x,S_x]}\|\widehat{\xi}(t)-y\|^2$$

$$\leq 2d_K^2(x)+4(S_x-T_x)^2\left\|u\right\|^2+4\sup_{t\in[T_x,S_x]}\left\|v(W(t)-W(T_x))\right\|^2$$

By taking the expectations and by using inequality

$$\mathbf{E}\left(\sup_{t\in[a,b]}\left\|\int_a^b h(s)dW(s)\right\|^2\right) \leq 4\int_a^b \|h(s)\|^2 ds$$

we obtain

$$\mathbf{E}\left(\sup_{t\in[T_x,S_x]}\|\widehat{\xi}(t)-y\|^2\right) \leq 2\varepsilon^2 T_x+4(S_x-T_x)c^2 \leq 2\varepsilon^2 T_x+4\beta c^2$$

since $\max(\mathbf{E}(\|u\|^2),4\mathbf{E}(\|v\|^2))=c^2$ and since $S_x-T_x\leq\beta$. By the choice of $\beta\leq\dfrac{\varepsilon^2(1-T_x)}{2c^2}$, we infer that

$$\mathbf{E}\left(\sup_{t\in[T_x,S_x]}\|\widehat{\xi}(t)-y\|^2\right) \leq 2\varepsilon^2 T_x+4\beta c^2 \leq 2\varepsilon^2 T_x+2\varepsilon^2(1-T_x) = 2\varepsilon^2 \tag{15.2.10}$$

Hence $\widehat{\xi}(\cdot)$ satisfies (15.2.5)ii) for S_x.

We then observe that $\forall\, t\in[S_x,T_x]$,

$$(\widehat{\xi}(t),u,v) = (y,u,v)+(\widehat{\xi}(t)-y,0,0) \in \mathrm{Graph}(H)+\sup_{t\in[T_x,S_x]}\|\widehat{\xi}(t)-y\|B$$

Therefore,

$$\mathbf{E}\left(\text{ess sup}(\sup_{t\in[T_x,S_x]} d^2\left((\widehat{\xi}(t),u,v),\mathrm{Graph}(H)\right)\right)\leq\mathbf{E}\left(\sup_{t\in[T_x,S_x]}\|\widehat{\xi}(t)-y\|^2\right)\leq 2\varepsilon^2$$

We introduce now the concatenation $\widetilde{\xi}(\cdot)$ of $\xi(\cdot)$ defined on the interval $[0,T_x]$ and $\widehat{\xi}(\cdot)$ on $[T_x,S_x]$ defined by

$$\widetilde{\xi}(t) := \begin{cases} \xi(t) & \text{if} \quad t \in [0, T_x] \\ \widehat{\xi}(t) & \text{if} \quad t \in [T_x, S_x] \end{cases}$$

One observe easily that

$$\|\widetilde{\xi}\|_\alpha \leq 2^{\alpha-1}\max(\|\xi\|_\alpha, \|\widehat{\xi}\|_\alpha) \tag{15.2.11}$$

We have to estimate $\mathbf{E}(\|\widehat{u}\|_\infty)$ and $\mathbf{E}(\|\widehat{\xi}\|_\alpha)$. The first estimate follows from

$$\mathbf{E}\left(\|u\|_\infty\right) \leq \sup_{(u,v)\in\gamma B} \mathbf{E}(\|u\|)$$

Writing

$$\widehat{\xi}(t) := \xi_1(t) + \xi_2(t)$$

where

$$\xi_1(t) := x + (t - T_x)u \ \& \ \xi_2(t) := \int_{T_x}^t v dW(s)$$

we infer that

$$\mathbf{E}\left(\|\xi_1\|_\alpha^2\right) \leq \sup_{(u,v)\in\gamma B)} \mathbf{E}(\|u\|) \ \& \ \mathbf{E}\left(\|\xi_2\|_\alpha^2\right) \leq c_{\alpha,p} \sup_{(u,v)\in\gamma B} \mathbf{E}(\|v\|^p)$$

thanks to inequality

$$\mathbf{E}\left(\left\|\int_a^b a(s)dW(s)\right\|_\alpha^2\right) \leq c_{\alpha,p}\int_a^b \|a(s)\|^p ds,$$

(see [139, Da Prato & Zabczyk]).)

Tightness

Definition 15.2.4 *We shall say that probability laws $\mathcal{L}(\xi_n)$ are* tight *if we can associate with any $\varepsilon > 0$ a compact subset K_ε such that*

$$\mathbf{P}(\xi_n \in K_\varepsilon) \geq 1 - \varepsilon$$

Lemma 15.2.5 *if the expectations of the norms*

$$\mathbf{E}(\|u_n(\cdot)\|_\infty) \leq \gamma, \ \mathbf{E}(\|v_n(\cdot)\|_\infty) \leq \gamma, \ \mathbf{E}(\|\xi_n(\cdot)\|_\alpha) \leq \gamma$$

are bounded, the probability laws $\mathcal{L}(\xi_n, u_n, v_n)$ are tight.

Proof — We have to associate with any $\varepsilon > 0$ a compact subset K_ε such that

$$\mathbf{P}((\xi_n, u_n, v_n) \in K_\varepsilon) \geq 1 - \varepsilon$$

Since the expectations of the norms are bounded, we deduce from the Chebichev inequality that

$$\mathbf{P}\left(\|\xi\|_\alpha \geq \frac{\gamma}{\varepsilon}\right) \leq \varepsilon \,,\ \mathbf{P}\left(\|u\|_\infty \geq \frac{\gamma}{\varepsilon}\right) \leq \varepsilon \;\&\; \mathbf{P}\left(\|v\|_\infty \geq \frac{\gamma}{\varepsilon}\right) \leq \varepsilon$$

Therefore, it is enough to take

$$K_\varepsilon := \frac{\gamma}{\varepsilon}(B_\alpha \times B_\infty \times B_\infty)$$

which is compact thanks to Ascoli's Theorem and the weak compactness of the unit ball of L^∞ in L^1. Indeed, the complement of K_ε is the product of the sets $\{\|\xi\|_\alpha \geq \frac{\gamma}{\varepsilon}\}$, $\{\|u\|_\infty \geq \frac{\gamma}{\varepsilon}\}$ and $\{\|v\|_\infty \geq \frac{\gamma}{\varepsilon}\}$, whose measure is larger than or equal to ε. □

Convergence of Martingale Solutions

The existence of martingale solutions follows from Lemma 15.2.3 (with $\varepsilon := \frac{1}{n}$) and the following Martingale Convergence Theorem.

Theorem 15.2.6 (Martingale Convergence Theorem) *Let K be a closed subset of X. We assume that H is a Marchaud set-valued map.*

Consider a sequence of $\mathcal{F}_t$-adapted stochastic processes

$$\xi_n(t) = \xi_n(0) + \int_0^t u_n(s)ds + \int_0^t v_n(s)dW(s)$$

on $[0,1]$ satisfying

$$\textit{for almost all}\ (\omega, t) \in \Omega \times [0,1],\ (u_{n_\omega}(t), v_{n_\omega}(t)) \in \gamma B$$

and

$$\begin{array}{ll} i) & \mathbf{E}\left(\sup_{t\in[0,1]} d^2(\xi_n(t), \mathcal{K})\right) \to 0 \\ ii) & \mathbf{E}\left(\text{ess sup}_{t\in[0,1]} d^2\left((\xi_n(t), u_n(t), v_n(t)), \text{Graph}(H)\right)\right) \to 0 \\ iii) & \max\left(\mathbf{E}\left(\|\xi_n(\cdot)\|_\alpha\right), \mathbf{E}(\|u_n(\cdot)\|_\infty), \mathbf{E}(\|v_n(\cdot)\|_\infty)\right) \leq \gamma. \end{array} \quad (15.2.12)$$

Then there exist a probability space $(\widehat{\Omega}, \widehat{\mathcal{A}}, \widehat{\mathbf{P}})$ with a filtration $\{\widehat{\mathcal{F}}_t\}$ and a $\{\widehat{\mathcal{F}}_t\}$-Wiener process $\widehat{W}$, other stochastic processes $(\widehat{\xi}, \widehat{u}, \widehat{v})$ and $(\widehat{\xi}_n, \widehat{u}_n, \widehat{v}_n)$ such that their laws are equal:

$$\mathcal{L}(\widehat{\xi}_n, \widehat{u}_n, \widehat{v}_n) = \mathcal{L}(\xi_n, u_n, v_n)$$

and such that a subsequence (again denoted by) $\widehat{\xi}_n$ converges almost surely to $\widehat{\xi}$ in $\mathcal{C}([0,1];X)$, $(\widehat{u}_n,\widehat{v}_n)$ converges almost surely to $(\widehat{u},\widehat{v})$ in $L^1_\sigma([0,1],X\times\mathcal{L}(Y,X))$ which satisfy

$$\left\{\begin{array}{ll} i) & \widehat{\xi}(t) \;=\; \widehat{\xi}(0)+\int_0^t \widehat{u}(s)ds+\int_0^t \widehat{v}(s)d\widehat{W}(s) \\ ii) & \textit{for almost all } (\omega,t)\in\Omega\times[0,1],\;\; (\widehat{u}_\omega(t),\widehat{v}_\omega(t)) \;\in\; H(\widehat{\xi}_\omega(t)) \\ iii) & \textit{for almost all}\,\omega\in\widehat{\Omega},\;\; \forall\, t\in[0,1],\;\; \xi_\omega(t)\in K \end{array}\right. \tag{15.2.13}$$

Proof— Let us denote by $\mathcal{B}_\infty$ the unit ball of $L^\infty([0,1],X\times\mathcal{L}(Y,X))$. Then the subset $L^1_\sigma([0,1],\gamma\mathcal{B}_\infty)$ of functions of $L^1([0,1],X\times\mathcal{L}(Y,X))$ with values in the relatively compact subset $\gamma\mathcal{B}_\infty$ of $X\times\mathcal{L}(Y,X)$ is a Polish space (i.e., a separable complete metric space) when L^1 is supplied with the weakened topology. We thus consider the random variable (ξ_n,u_n,v_n) taking its values in the Polish space

$$\Phi := \mathcal{C}([0,1],X)\times L^1_\sigma([0,1],\gamma\mathcal{B}_\infty)$$

Their probability laws $\mathcal{L}(\xi_n,u_n,v_n)$ are defined by

$$\mathcal{L}(\xi_n,u_n,v_n)(A\times B\times C) \;:=\; \mathbf{P}((\xi_n(\cdot),u_n(\cdot),v_n(\cdot))\in A\times B\times C)$$

so that, for any function $\varphi\in\mathcal{C}(\Phi,\mathbf{R})$, we have by definition

$$\begin{array}{l} \int_\Omega \varphi(\xi_n(\cdot),u_n(\cdot),v_n(\cdot))\mathbf{P}(d\omega) \\ = \int_\Phi \varphi(y(\cdot),u(\cdot)v(\cdot))\mathcal{L}(\xi_n,u_n,v_n)(dy,du,dv) \end{array} \tag{15.2.14}$$

Since the expectations of the norms $\mathbf{E}(\|u_n(\cdot)\|_\infty)\le\gamma$, $\mathbf{E}(\|v_n(\cdot)\|_\infty)\le\gamma$ and $\mathbf{E}(\|\xi_n(\cdot)\|_\alpha)\le\gamma$ are bounded, we know by Lemma 15.2.5 below that their probability laws $\mathcal{L}(\xi_n,u_n,v_n)$ defined by (15.2.14) are tight. Therefore, they converge "weakly" to some measure $\mathcal{L}_\star$.

By the Skorohod Theorem (see for instance Theorem 2.4 of [139, Da Prato & Zabczyk])), there exist a probability space $(\widetilde{\Omega},\widetilde{\mathcal{A}},\widetilde{\mathbf{P}})$ with a filtration $\{\widetilde{\mathcal{F}}_t\}$, stochastic processes $\widetilde{\xi}_n$ and $\widetilde{\xi}$ and stochastic processes $(\widetilde{u}_n,\widetilde{v}_n)$ and $(\widetilde{u},\widetilde{v})$ such that

$$\mathcal{L}(\widetilde{\xi}_n,\widetilde{u}_n,\widetilde{v}_n) \;=\; \mathcal{L}(\xi_n,u_n,v_n)\;\&\;\mathcal{L}(\widetilde{\xi},\widetilde{u},\widetilde{v}) \;=\; \mathcal{L}_\star$$

and converging to $(\widetilde{\xi},\widetilde{u},\widetilde{v})$ in the sense that

$$\left\{\begin{array}{ll} i) & \text{almost surely, } \widetilde{\xi}_n \text{ converges to } \widetilde{\xi} \text{ in } \mathcal{C}([0,1];X) \\ ii) & \text{almost surely, } (\widetilde{u}_n,\widetilde{v}_n) \text{ converges weakly to } (\widetilde{u},\widetilde{v}) \text{ in } \mathcal{L}^1_\sigma([0,1];\gamma\mathcal{B}_\infty) \end{array}\right.$$

Therefore

$$\begin{array}{l} \int_\Omega \varphi(\xi(\cdot),u(\cdot),v(\cdot))\mathbf{P}(d\omega) \\ = \int_\Phi \varphi(y(\cdot),u(\cdot),v(\cdot))\mathcal{L}_\star(dy,du,dv) \end{array}$$

Set

$$M_n(t) := \xi_n(t) - \xi(0) - \int_0^t u_n(s)ds \;\&\; \widetilde{M}_n(t) := \tilde{\xi}_n(t) - \tilde{\xi}(0) - \int_0^t \tilde{u}_n(s)ds$$

Then M_n is a martingale with respect to the filtration $\sigma(\xi_n, u_n)$ whose quadratic variation is

$$\langle\langle M_n(t)\rangle\rangle = \int_0^t \|v_n(s)\|^2 ds$$

because, for any bounded and continuous function φ,

$$\begin{cases} \mathbf{E}\left([M_n(t) - M_n(s)]\varphi(\xi_n(s), u_n(s), v_n(s))\right) \\ = \mathbf{E}\left(\left[\int_s^t v_n(s)dW(s)\right]\varphi(\xi_n(s), u_n(s), v_n(s))\right) = 0 \end{cases}$$

Therefore, $\widetilde{M}_n$ is a martingale because for any bounded and continuous function φ,

$$\begin{cases} \mathbf{E}\left([\widetilde{M}_n(t) - \widetilde{M}_n(s)]\varphi(\tilde{\xi}_n(s), \tilde{u}_n(s), \tilde{v}_n(s))\right) \\ = \mathbf{E}\left([M_n(t) - M_n(s)]\varphi(\xi_n(s), u_n(s), v_n(s))\right) = 0 \end{cases}$$

Its quadratic variation is equal to

$$\langle\langle \widetilde{M}_n(t)\rangle\rangle = \int_0^t \|\tilde{v}_n(s)\|^2 ds.$$

Since

$$\sup_{n\geq 0} \mathbf{E}(\|M_n(\cdot)\|^2_{\mathcal{C}([0,1];X)}) < +\infty$$

we deduce[2] that the martingales converge to $\widetilde{M}$ defined by

$$\widetilde{M}(t) := \tilde{\xi}(t) - \tilde{\xi}(0) - \int_0^t \tilde{u}(s)ds$$

which is a martingale because, passing to the limit in (15.2), we obtain

$$\mathbf{E}\left([\widetilde{M}(t) - \widetilde{M}(s)]\varphi(\tilde{\xi}(s), \tilde{u}(s), \tilde{v}(s))\right) = 0$$

for any bounded and continuous function φ.

Therefore, by the Martingale Representation Theorem (see for instance Theorem 8.2 of [139, Da Prato & Zabczyk])), there exist another probability space $(\widehat{\Omega}, \widehat{\mathcal{A}}, \widehat{P})$ with a filtration $\{\widehat{\mathcal{F}}_t\}$, a Wiener process $\widehat{W}$, other stochastic processes $(\hat{\xi}_n, \hat{u}_n, \hat{v}_n)$ and $(\hat{\xi}, \hat{u}, \hat{v})$ such that

$$\mathcal{L}(\hat{\xi}_n, \hat{u}_n, \hat{v}_n) = \mathcal{L}(\tilde{\xi}_n, \tilde{u}_n, \tilde{v}_n) \;\&\; \mathcal{L}(\hat{\xi}, \hat{u}, \hat{v}) = \mathcal{L}(\tilde{\xi}, \tilde{u}, \tilde{v})$$

and a martingale $\widehat{M}$

$$\widehat{M}(t) = \int_0^t \hat{v}(s)d\widehat{W}(s)$$

[2]see for instance criterion II-5-2 of [278, Neveu].

the quadratic variation of which is

$$\langle\langle\widehat{M}(t)\rangle\rangle = \int_0^t \|\widehat{v}(s)\|^2 ds$$

such that

$$\widehat{\xi}(t) = \widehat{\xi}(0) + \int_0^t \widehat{u}(s)ds + \widehat{M}(t) = \widehat{\xi}(0) + \int_0^t \widehat{u}(s)ds + \int_0^t \widehat{v}(s)d\widehat{W}(s)$$

It remains to prove that for almost all $(\omega, t) \in \widehat{\Omega} \times [0,1]$, $(\widehat{u}_\omega(t), \widehat{v}_\omega(t)) \in H(\widehat{\xi}_\omega(t)$.

By definition,

$$\mathbf{E}\left(\text{ess sup}_{t\in[0,1]} d^2((\xi_n(t), u_n(t), v_n(t)), \text{Graph}(H))\right)$$

$$= \int_\Phi \text{ess sup}_{t\in[0,1]} d^2((y(t),, u(t), v(t)), \text{Graph}(H))\mathcal{L}(\xi_n(\cdot), u_n(\cdot), v_n(\cdot))(dy, du, dv)$$

$$= \int_\Phi \text{ess sup}_{t\in[0,1]} d^2((y(t), u(t), v(t)), \text{Graph}(H))\mathcal{L}(\widehat{\xi}_n(\cdot), \widehat{u}_n(\cdot), \widehat{v}_n(\cdot))(dy, du, dv)$$

$$= \mathbf{E}\left(\text{ess sup}_{t\in[0,1]} d^2((\widehat{\xi}_n(t), \widehat{u}_n(t), \widehat{v}_n(t)), \text{Graph}(H))\right)$$

converges to 0 by Lemma 15.2.3.

Now, we know that the $(\widehat{u}_n(\cdot), \widehat{v}_n(\cdot))$ are bounded in $L^\infty(\widehat{\Omega} \times [0,1]; X \times \mathcal{L}(Y,X))$ so that a subsequence (again denoted by) $(\widehat{u}_n, \widehat{v}_n)$ converges weakly to some $(\widehat{u}, \widehat{v})$ in $L^\infty(\widehat{\Omega} \times [0,1]; X \times \mathcal{L}(Y,X))$, and thus, weakly in $L^1(\widehat{\Omega} \times [0,1]; X \times \mathcal{L}(Y,X))$. On the other hand, for almost all $(\omega, t) \in \widehat{\Omega} \times [0,1]$, $\widehat{\xi}_{n,\omega}(t)$ converges to $\widehat{\xi}_\omega(t)$. By the Chebichev inequality, we deduce that

$$\mathbf{P}\left(\text{ess sup}_{t\in[0,1]} d^2\left((\widehat{\xi}_n(t), \widehat{u}_n(t), \widehat{v}_n(t)), \text{Graph}(H)\right) \geq \lambda\right) \leq \frac{2}{\lambda n^2}$$

Taking $\lambda := \frac{1}{n}$, we infer that

$$\mathbf{P}\left(\text{ess sup}_{t\in[0,1]} d^2\left((\widehat{\xi}_n(t), \widehat{u}_n(t), \widehat{v}_n(t)), \text{Graph}(H)\right) \geq \frac{1}{n}\right) \leq \frac{2}{n}$$

Therefore, $\text{ess sup}_{t\in[0,1]} d^2\left((\widehat{\xi}_n(t), \widehat{u}_n(t), \widehat{v}_n(t)), \text{Graph}(H)\right)$ converges to 0 in probability, and thus, there exists a subsequence (again denoted by) $(\widehat{\xi}_n, \widehat{u}_n, \widehat{v}_n)$ such that

$$\text{almost surely, } \text{ess sup}_{t\in[0,1]} d^2\left((\widehat{\xi}_n(t), \widehat{u}_n(t), \widehat{v}_n(t)), \text{Graph}(H)\right) \to 0$$

Hence,

$$\text{almost surely, } (\widehat{\xi}_n(t), \widehat{u}_n(t), \widehat{v}_n(t)) \in B(\text{Graph}(H), \frac{1}{n})$$

Now, by using the Convergence Theorem 12.1.4, we infer from the fact that F is a Marchaud set-valued map, that

$$\text{for almost all } (\omega,t) \in \widehat{\Omega} \times [0,1], \quad (\widehat{\xi}(t), \widehat{u}(t), \widehat{v}(t)) \in \text{Graph}(H)$$

Let us prove now that the stochastic process $\widehat{\xi}(\cdot)$ is viable in K. By definition,

$$\begin{array}{l} \mathbf{E}\left(\sup_{t\in[0,1]} d^2(\xi_n(t),K)\right) \\ = \int_{\Phi} \sup_{t\in[0,1]} d^2(y(t),K)\mathcal{L}(\xi_n(\cdot),u_n(\cdot),v_n(\cdot))(dy,du,dv) \end{array}$$

Since the left-hand side converges to 0, we observe that

$$0 = \int_{\Phi} \sup_{t\in[0,1]} d^2(y(t),K)\mathcal{L}(\widehat{\xi}(\cdot),\widehat{u}(\cdot),\widehat{v}(\cdot))(dy,du,dv)$$

thanks to Lemma 12.1.3. We thus deduce that

$$\int_{\widehat{\Omega}} \sup_{t\in[0,1]} d^2(\widehat{\xi}(t),K)\widehat{\mathbf{P}}(d\omega) = 0$$

thanks to (15.2.14). Hence, for every $t \geq 0$, $\widehat{\xi}(t) \in K$ almost surely. □

15.3 Stochastic Lyapunov Functions

We shall say that an extended function $V : \Omega \times X \mapsto \mathbf{R}_+ \cup \{+\infty\}$ is a "$\mathcal{F}_0$-epirandom function" if its epigraph $\omega \in \Omega \rightsquigarrow \mathcal{E}p(V_\omega) \subset X \times \mathbf{R}$ is $\mathcal{F}_0$-measurable (we refer for instance to chapter 8 of SET-VALUED ANALYSIS, [54, Aubin & Frankowska] for the definition and properties of measurable set-valued maps). An epirandom function is said to be "lower semicontinuous" if the epigraphs $\mathcal{E}p(V_\omega)$ are closed.

15.3.1 Stochastic Contingent Epiderivatives

Let $V : X \mapsto \mathbf{R} \cup \{+\infty\}$ be a nontrivial epirandom function and x belong to its domain. We begin by adapting Lemma 8.1.2 to the stochastic case:

Lemma 15.3.1 *For all $w \geq V(x)$, if $((\gamma,\delta),(u,v))$ belongs to $\mathcal{T}_{\mathcal{E}p(V)}(t,x,w)$, then $((\gamma,\eta),(u,v))$ belongs to $\mathcal{T}_{\mathcal{E}p(V)}(t,x,w')$ either when $w' > w$ or when $w' = w$ and $\eta \geq \delta$.*

Proof — Fix $t = 0$ for simplicity. Assume that $((\gamma,\delta),(u,v))$ belongs to $\mathcal{T}_{\mathcal{E}p(V)}(0,x,w)$ and fix $\eta \geq \delta$. We know that there exist $h^n \to 0$ and stochastic errors e^n, ε^n such that

$$(x + h_n\gamma + W(h_n)u + e^n, w + h_n\delta + W(h_n)v + \varepsilon^n) \in \mathcal{E}p(V)$$

almost everywhere. Therefore,

$$\begin{cases} (x + h_n\gamma + W(h_n)u + e^n, w' + h_n\eta + W(h_n)v + \varepsilon^n \\ = (x + h_n\gamma + W(h_n)u + e^n, w + h_n\delta + W(h_n)v + \varepsilon^n) \\ + (0, w' - w + h(\eta - \delta)) \\ \in \mathcal{E}p(V) + \mathbf{R}_+ = \mathcal{E}p(V) \; \square \end{cases}$$

This implies that the set of pairs (γ, δ) such that $((\gamma, \delta), (u, v))$ belongs to $\mathcal{T}_{\mathcal{E}p(V)}(0, x, V(x))$ is the epigraph of a function:

Definition 15.3.2 *Let V be a $\mathcal{F}_t$-epirandom extended function, x belong to the domain of V. We shall say that the epirandom function $\mathcal{D}_\uparrow V(t, x, u, v)$ whose epigraph is the set of pairs (γ, δ) such that $((\gamma, \delta), (u, v))$ belongs to $\mathcal{T}_{\mathcal{E}p(V)}(0, x, V(x))$ is the* **stochastic contingent epiderivative** *of V at (x, u, v).*

When V is smooth, the Itô formula yields the formula of the stochastic contingent epiderivative:

Proposition 15.3.3 *If $V : X \mapsto \mathbf{R}_+$ is twice continuously differentiable, we have*

$$\mathcal{D}_\uparrow V(t, x, u, \langle V'(x), u\rangle)(\gamma) = \langle V'(x), \gamma\rangle + \frac{1}{2}V''(x)(u, u)$$

Another interesting and useful example is provided by distance functions which are no longer twice continuously differentiable. By adapting the proof of Lemma 1.5.6 of VIABILITY THEORY, [28, Aubin], we obtain:

Proposition 15.3.4 *Let $\mathcal{K}$ be a random set-valued variable, x a $\mathcal{F}_t$-adapted stochastic process and let us choose a $\mathcal{F}_t$-measurable projection $y \in \Pi_K(x))$. Then, for any pair of $\mathcal{F}_t$-random variable (γ, v) in the stochastic contingent set $\mathcal{T}_K(0, y)$, the following estimate*

$$\mathcal{D}_\uparrow d_K^2(t, x, u, 2\langle x - y, u - v\rangle)(\gamma) \leq 2\,\langle x - y, \gamma - \delta\rangle + \|u - v\|^2$$

holds true.

15.3.2 The Characterization Theorem

We consider a stochastic differential inclusion (15.1.2) and a time-dependent function $w(\cdot)$ defined as a solution to the differential equation

$$w'(t) + \varphi(w(t)) = 0 \qquad (15.3.1)$$

where $\varphi : \mathbf{R}_+ \to \mathbf{R}$ is a given Lipschitz function used as parameter in what follows.

Our problem is to characterize nonnegative epirandom extended functions $V : \Omega \times X \to \mathbf{R}_+ \cup \{+\infty\}$ satisfying dynamical inequality

$$\forall\, t \geq 0, \;\; V(x(t)) \leq w(t) \text{ almost everywhere}, \quad w(0) = V(x(0)) \qquad (15.3.2)$$

along the solution $x(\cdot)$ to the stochastic differential inclusion (15.1.2) and the solution $w(\cdot)$ to the differential equation (15.3.1).

Theorem 15.3.5 *Let V be a nonnegative and $H : X \rightsquigarrow X \times \mathcal{L}(Y,X)$ be a Marchaud map. If*

$$\forall\, x \in X, \quad \inf_{(u,v)\in H(x)} \mathcal{D}_{\uparrow}V(t,x,v,0)(u) + \varphi(V(x)) \;\leq\; 0 \tag{15.3.3}$$

then for any initial state $x_0 \in \text{Dom}(V)$, the solutions $x(\cdot)$ to (15.1.2) and $w(\cdot)$ to (15.3.1) satisfy dynamical inequality (15.3.2).

Proof — Obviously, the system (15.1.2), (15.3.1) has a solution satisfying (15.3.2) if and only if the system of stochastic differential inclusions

$$\begin{cases} d\xi \;\in\; H(\xi, dt \oplus dW(t)) \\ \\ dw \;=\; -\varphi(w)dt \end{cases}$$

has a solution starting at $(x_0, V(x_0))$ viable in $\mathcal{K} := \mathcal{E}p(V)$. We thus have to check that $\mathcal{K}$ is a viability domain for this system if and only if V is a Lyapunov function for H with respect to φ.

If $\mathcal{K}$ is a viability domain of this system, by taking $z = (x, V(x))$, we infer that

$$((u, -\varphi(V(x))), (v, 0)) \in \mathcal{T}_{\mathcal{K}}(t, x, V(x))$$

This implies that

$$\mathcal{D}_{\uparrow}V(t,x,v,0)(u) \;\leq\; -\varphi(V(x))$$

so that (15.3.3) is satisfied.

Conversely, (15.3.3) implies that the pair $((u, \varphi(V)(x)), (v, 0))$ belongs to $\mathcal{T}_{\mathcal{K}}(t, x, V(x))$. Setting $t = 0$, $\delta := \mathcal{D}_{\uparrow}V(t,x,v,0)(u)$, we deduce that there exist stochastic errors e^h, ε^h such that

$$(x + vW(h) + hu + e^h, V(x) - h\varphi(V(x)) + \varepsilon^h) \;\in\; \mathcal{K} := \mathcal{E}p(V)$$

so that $((u, -\varphi(V(x)))), (v, 0) \in \mathcal{T}_{\mathcal{K}}(0, x, V(x))$.

If $w > V(x)$, this implies that for small h

$$\begin{cases} (x + hu + vW(h) + e^h, w - h\varphi(w) + \varepsilon^h) \\ = \;(x + hu + vW(h) + e^h, V(x) + h\delta + \varepsilon^h) \\ +(0, w - V(x) - h(\mathcal{D}_{\uparrow}V(0,x,v,0)(u) + \varphi(w))) \in \mathcal{K} + \{0\} \times \mathbf{R}_+ = \mathcal{K} \end{cases}$$

so that $((u, -\varphi(V(x))), (v, 0)) \in \mathcal{T}_{\mathcal{K}}(0, x, w)$. □

15.3.3 Functionally Absorbing Sets

Using norms as Lyapunov functions, we can check that a ball is absorbing:

Definition 15.3.6 *We say that a subset K is* functionally absorbing *if for any initial state x_0, there exist a $T_{\|x_0\|}$ and a solution x to the stochastic differential inclusion (15.1.2) such that*

$$\text{for almost all } \omega, \ \forall\, t \geq T_{\|x_0\|}, \ \ x_\omega(t) \in K$$

When $T_{\|x_0\|}$ depends upon ω, we say that K is "absorbing".

We can recognize absorbing sets by checking whether the norm is a stochastic Lyapunov function for some Lipschitz function φ:

Corollary 15.3.7 *If there exists a Lipschitz function φ such that*

$$\begin{cases} i) & \langle x, g(x)\rangle \ = \ 0 \\ ii) & 2\langle x, f(x)\rangle + \|g(x)\|^2 + \varphi(\|x\|^2) \ \leq \ 0 \\ iii) & \varphi(0) \ \leq \ 0 \ \& \ \liminf_{w\to\infty} \varphi(w) \ > \ 0 \end{cases}$$

then there exist a $T_{\|x_0\|}$ and a solution x to the stochastic differential inclusion (15.1.2) such that

$$\forall\, t \geq T_{\|x_0\|}, \ \ \|x_\omega(t)\|^2 \ \leq \ \rho \text{ for almost all } \omega$$

so that $B(0,\rho)$ is functionally absorbing.

15.3.4 Attracting Sets

Using distance functions as Lyapunov functions, we can study attracting sets:

Definition 15.3.8 *We shall say that a closed random set $\mathcal{K}$ is a* stochastic attracting set *if from any x_0, the solution $x(\cdot)$ to the stochastic differential inclusion (15.1.2) satisfies*

$$\lim_{t\to+\infty} d_{\mathcal{K}}(x(t)) \ = \ 0 \ \text{ almost everywhere}$$

We can recognize attracting sets by checking whether the distance function to $\mathcal{K}$ is a stochastic Lyapunov function for some Lipschitz function φ:

Corollary 15.3.9 *Assume that $\mathcal{K}$ is invariant by H in the sense that that for every $\mathcal{F}_t$-random variable x, there exists a $\mathcal{F}_t$-measurable projection $y \in \Pi_K(x)$ such that*

$$\forall\,(u,v) \in H(x), \ \ (u,v) \ \in \ \mathcal{T}_K(t,y) \qquad (15.3.4)$$

If there exists a Lipschitz function φ such that there exists $(\gamma,\beta) \in \mathcal{T}_K(t,y)$ satisfying

$$\begin{cases} i) & \langle x-y, v-\beta\rangle = 0 \\ ii) & 2\langle x-y, u-\gamma\rangle + \|v-\beta\|^2 + \varphi(d_K^2(x)) \leq 0 \end{cases}$$

then

$$\forall\, t \geq 0,\ d_K^2(x(t)) \leq w(t)$$

so that K is an attracting set when 0 is asymptotically stable for the differential equation $w' + \varphi(w) = 0$.

We shall use these results for the reachable tubes

$$\varpi_{H,\,\omega}(t,K) := \{x_\omega(t)\}_{x \in \mathcal{S}_H(t,K)(x_0),\, x_0 \in K} \subset X$$

and

$$\varpi_H(t,K) := \{x(t)\}_{x \in \mathcal{S}_H(t,K)(x_0),\, x_0 \in K} \subset L^2(\Omega, \mathcal{F}_t, P) \subset L^2(\Omega, \mathcal{F}, P)$$

Therefore, there are among many other concepts of attractors at least the two following ones:

Definition 15.3.10 *We denote by $K \subset X$ any bounded subset. We shall say that the random set-valued variable*

$$\mathrm{Attr}H,\ \omega := \overline{\bigcup_{K \subset X} \mathrm{Limsup}_{t\to\infty} \varpi_{H,\,\omega}(t,K)}$$

is the "stochastic attractor" *(or* "pathwise attractor" *of the stochastic differential inclusion (15.1.2) and that*

$$\mathrm{Attr}_H := \overline{\bigcup_{K \subset X} \sigma - \mathrm{Limsup}_{t\to\infty} \varpi_H(t,K)}$$

is the "functional attractor" *of the stochastic differential inclusion (15.1.2).*

The existence of a functionally absorbing set implies the existence of a functional attractor:

Theorem 15.3.11 *Assume that the stochastic differential inclusion (15.1.2) has a bounded functionally absorbing set B. Then the functional and the stochastic attractors are not empty and satisfy*

$$\begin{cases} \mathrm{Attr}_{H,\,\omega} = \mathrm{Limsup}_{t\to\infty} \varpi_{H,\,\omega}(t,B) \\ \& \\ \mathrm{Attr}_H = \mathrm{Limsup}_{t\to\infty} \varpi_H(t,B) \end{cases}$$

Bibliographical Comments

Chapter 1

As it was mentioned, Chapter 1 was written with Jean Cartelier in [38, Aubin & Cartelier]. We refer to [314, Saari] for a survey about the paradoxes of the Walras process following the papers of [334,335, Sonnenschein], improved by [261, Mantel] and [152, Debreu]. The Global Newton Method has been introduced by [333, Smale] and other price mechanisms in [305, Saari & Simon]. This and other problems about demand functions are revisited in [113, Chiappori & Ekeland].

Chapter 2

The thorough study of the simplest model with one commodity and one agent is taken from [31, Aubin].

These examples use the concept of viability kernel of a closed subset under a differential inclusion which has been introduced in [21, Aubin] in the 1985 meeting in honor of Professor Ky Fan for studying evolution under bounded inflation, as well as heavy solutions introduced in [52,53, Aubin & Frankowska].

Furthermore, the first algorithms allowing us to compute the viability kernels, which run on personal computers for small dimensions appeared in [212, Frankowska & Quincampoix] and [319, Saint-Pierre], and have been developed by their collaborators. They have been adapted to obtain the examples presented in this chapter. For instance, the recent book [357, Zhang] provides many dynamical economic models to which the viability approach should be applied.

The Keynesian model was studied in [111, Cartelier & Müllers].

The section on technology and climate risk is a simplified version of [174, Doyen, Gabay & Hourcade]) whereas the study on management of renewable resources is obtained from [173, Doyen & Gabay]. The viability analysis of an ecological-economic model of the French Guyana shrimp fisheries is taken from the paper [73, Béné & Doyen].

The demographic examples are due to [83,84, Bonneuil] and the predator-prey model is extracted *verbatim* from [86, Bonneuil & Müllers].

The examples on oscillators are taken form [274, Müllers].

The dynamical study of predator-prey systems with two states and two regulee, which are still amenable to explicit analytical formulation, was conducted by Noël Bonneuil and Katharina Müllers in [86, Bonneuil & Müllers]. The mathematical model of Barth's study on Nomads of South Persia is taken from [85, Bonneuil], in which one can find several mathematical models of Fredrik Barth's studies in social sciences and in particular, anthropology.

The few pages devoted to a viability analysis come from personal discussions with Jerzy Zabczyk.

Chapter 3

The first viability theorems were motivated by dynamical economic theory in [14,13, Aubin], [46, Aubin & Day] and later, in [337, Stacchetti]. We refer to [144, Day] for a more general economic interpretation of economic models based on a viability approach.

The concept of contingent derivative was motivated by dynamical economic theory to take into account time-dependent sets of available resources in [15, Aubin].

The concept of mutational equations has been introduced in [29,30, Aubin] [49, Aubin, Doyen & Mattioli] and [170, Doyen]. We refer to [220,221, Gorre] and [35, Aubin] for further informations on this recent topic.

The theorems on viable solutions to differential inclusions with memory were as well motivated by economics, from the papers [225,226, Haddad] which provided the general criteria to article [179, Duluc & Vigneron] in which the characterization for economies with memory appeared.

Chapter 4

Dynamical economies are control systems under viability (or state) constraints, and this chapter translates in an economic language the contents of Chapter 6 of [28, Aubin].

The section on qualitative economies is taken from [164,165, Dordan], [166, Dordan & Suppo], [48, Aubin & Dordan] and also from discussions with Katharina Müllers. See [165, Dordan] for a modern reference of a mathematical treatment of qualitative physics, which is nothing else than the following up of "comparative statics" initiated by economists such as Samuelson and Lancaster.

The study of slow solutions started with [186, Falcone & Saint-Pierre] and [318, Saint-Pierre]. The strategy followed here is based on selection procedures due to Hélène Frankowska which appeared in [52,53, Aubin & Frankowska]. The proof of the Michael and the Steiner selection procedures can be found in [54, Aubin & Frankowska]. The Steiner point was introduced by Steiner for convex polytopes of $\mathbf{R}^2$.

The concepts of connection matrices, of an static index of connectionist complexity, the equivalence theorem between slow evolutions and maximal decentralization and the study of slow evolutions of technological matrices appear here for the first time. They have been motivated by the introduction of tensor algebra in neural networks in [33, Aubin]. These ideas have been used in the context of network organizations in processes of technological choice in [51, Aubin & Foray].

Chapter 5

The design of an economy built from "viability multipliers" appears here for the first time, as well as the concept of frustrations which resulted from discussions with Jean Cartelier.

Fuzzy sets were introduced in static cooperative games in [12,16,17, Aubin] and in [39, Aubin & Cellina] in the dynamical case. The idea of using psychological profiles and social groups is taken from [60, Aubin, Louis-Guerin & Zavalloni] in the static case. The treatment of social mobility results from discussions with Jean Cartelier. Other techniques used for tackling the problem of conventions, herd behavior and mimetic contagion can be found in [91, Boyer & Orlean] and [280,281,282, Orlean] where further bibliographic references on this problem can be found.

Chapter 6

Differential inclusions are now the topic of several books, such as [39, Aubin & Cellina], [28, Aubin], [154, Deimling] among other sources.

The proof of Theorem 6.1.3 stated in [224, Guseinov, Subbotin & Ushakov] given here is due to Frankowska (see [57, Aubin & Frankowska], and the roots of the proof of Theorem 6.1.6 can be found in [40, Aubin & Clarke].

The relaxation of the convexity assumptions in the existence theorems of differential inclusions follows from discussions with Arrigo Cellina. The Shapley-Folkman Theorem, which is sophisticated version of the Carathéodory Theorem, can be found for instance in [8, Arrow-Hahn], Appendix B, [12, Aubin], Theorem 14.2.2, p. 463, and [338, Starr]. Fot a thorough treatment of the cores and equilibria of large economies, we refer to [229, Hildenbrand].

The equilibrium theorems derived from the viability theorem are taken from [28, Aubin].

The concept of *reachable maps*, which extend to the set-valued case the concept of semi-groups of operators, goes back to Barbashin and Roxin in the sixties. The concept of infinitesimal generator of the semi-group of reachable maps (using contingent derivatives) is due to [204, Frankowska]. For an exposition of permanence and its links with viability properties, we refer to [62, Aubin & Sigmund], [232, Hofbauer & Sigmund] and [28, Aubin].

The concept of viability kernel has been introduced in [21, Aubin] for studying heavy evolutions and evolutions under bounded inflation. This concept of viability kernel happened to be a very useful tool for studying other problems, such as the construction of absorbing sets and attractors, reformulating Lyapunov stability, solving the target problem, characterizing and constructing Lyapunov functions, devising the Montagnes Russes Algorithm for finding a global minimum of a lower semicontinuous function, characterizing and constructing minimal time functions, for building invariant manifolds of a system of differential inclusions, constructing feedback maps under bounded inflation, deriving the differential equation governing heavy solutions, constructing cascades in hierarchical dynamical games, etc., without mentioning in details their biological applications which also contributed to the motivation of viability theory in the first place (see the papers by Noël Bonneuil, Pierre Cardaliaguet, Jean Cartelier, Olivier Dordan, Luc Doyen, Hélène Frankowska, Daniel Gabay, Anne Gorre, Katharina Müllers, Laurent Najman, Marc Quincampoix, Patrick Saint-Pierre and Nicolas Seube).

The viability kernel algorithms has been devised by Patrick Saint-Pierre in [319, Saint-Pierre] and adapted by Luc Doyen, Philippe Lacoude, Katharina Müllers, Nicolas Seube and Patrick Saint-Pierre himself to many other examples.

Hélène Frankowska and Marc Quincampoix proved the convergence of a "sharper" in [212, Frankowska & Quincampoix] (see also section 4.5 of chapter 4 of VIABILITY THEORY, [28, Aubin] and its bibliography for more informations on this very important algorithm).

Patrick Saint-Pierre proved in [319, Saint-Pierre] that when F is Marchaud, the upper limit of the discrete viability kernels is contained in the viability kernel and, together with Marc Quincampoix in [298, Quincampoix & Saint-Pierre], that they are equal when F is also Lipschitz and the approximation is slightly modified.

Hitting and exit times were introduced in [28, Aubin] and the study of complements of viability kernels and absorbing properties are taken from [293, Quincampoix]. The **crisis functional** which measures the time spent by the trajectory outside of a set has been introduced by Luc Doyen & Patrick Saint-Pierre in [177, Doyen & Saint-Pierre] and applied in several studies such as [174, Doyen, Gabay & Hourcade]) and [173, Doyen & Gabay].

We refer to the wonderful book [313, Saari] and his other contributions for a thorough study of paradoxes and chaos in decision making as well as other areas of social and mechanical sciences.

Chapter 7

As mentioned above, evolution under bounded inflation, as well as heavy solutions have started with [52,53, Aubin & Frankowska]. It uses the concept of viability kernel of a closed subset under a differential inclusion which has been introduced in [21, Aubin]. The strategy followed here based on selection procedures is due to Hélène Frankowska.

The concept of heavy solutions is close to the concepts related to singular perturbations when one group of variables (here, the ones describing the prices) evolves slowly and when the velocity of the second group of variables (here, the commodities) is multiplied by a scalar λ going to ∞. In the case of heavy solutions, however, we just specify that the prices evolve as slowly as possible, without requiring that the velocities of the commodities become infinite for studying the limiting situation. Basic theorems on singular perturbations of differential equations has been extended by A. Dontchev, M. Quincampoix, A. Kurzhanski and V. Veliov among other authors. We refer to [295,297,296, Quincampoix] for further bibliographical comments.

The lock-in property was mentioned in the framework of urn processes in [6, Arthur, Ermoliev & Kaniowski] based on mathematical techniques used in [230, Hill, Lane & Sudderth]. See [167, Dosi], [141, David & Foray], [168, Dosi & Nelson], [199, Foray & Freeman] and [326, Silverberg, Dosi & Orsenigo] among for further developments in which more bibliography can be found on this important problem.

The concept of hierarchical games and selection by cascades is due to Katharina Müllers and the section dealing with this problem is taken from [273, Müllers].

Chapter 8

The idea of contingent epiderivative goes back to R.T. Rockafellar, to whom I borrowed the "arrow notation" in $D_{\uparrow}V(x)$ recalling the use of the epigraph.

The treatment of Lyapunov functions through viability of their epigraphs goes back to [20, Aubin]. See also the survey [121, Clarke, Ledyaev, Stern & Wolenski] for a "normal" point of view. The existence of smallest Lyapunov functions appeared in [23] and the characterization of its domain as the basin of exponential attraction was obtained in collaboration with Christopher Byrnes. The Montagnes Russes algorithm for global optimization is due to [61, Aubin & Najman] for functions defined on vector spaces and by [219, Gorre] for functions defined on metric spaces using mutational equations.

The study on minimal time functions time is taken from [103, Cardaliaguet, Quincampoix & Saint-Pierre] and the one on crisis functional from [177, Doyen & Saint-Pierre].

The contents of the section on infinite horizon optimal control is taken from [58, Aubin & Frankowska]. Earlier results on this topic can be found in [13,15, Aubin] (in the framework of monotone solutions) and in chapter 6 of [39, Aubin & Cellina], where the precise definition of "almost convergence" and further asymptotic properties of monotone solutions are presented. See also [106, Carlson, Haurie & Leizarowitz]. But we refer above all to the monograph CONTROL OF NONLINEAR SYSTEMS AND DIFFERENTIAL INCLUSIONS, [209, Frankowska], for an exhaustive exposition of generalized solutions (both contingent and viscosity) to Hamilton-Jacobi-Bellman equations as well as other issues in control theory. Daniel Gabay pointing out the relations between the viability kernel algorithm applied for approximating the value function of an infinite horizon optimal control problem and algorithms obtained in [80, Bertsekas] for computing the value function of stochastic optimal control problems. For necessary conditions of infinite-horizon optimal problems in the convex case, we refer to [40, Aubin & Clarke].

The section on intertemporal optimal evolution under bounded inflation is taken in [37, Aubin, Byrnes & Seube].

Chapter 9

The literature on differential games is too large to do it justice. It began with the celebrated book [239, Isaacs]. For a short survey of differential games, we refer to the monograph [79, Bernhard] and to [67, Basar & Bernhard]. See also [242,243, Krasovski-Subbotin] among many other references.

The contents of this chapter is taken from [26,25] and has been extended in [292, Quincampoix] for the time dependent case. The section on discriminating kernels is due to [102,98,100, Cardaliaguet] and [104, Cardaliaguet, Quincampoix & Saint-Pierre].

The numerical approximation of the discriminating kernels can be found in [102,105, Cardaliaguet, Quincampoix & Saint-Pierre].

Chapter 10

The contents of this chapter is taken from [32, Aubin]. See [33, Aubin] for the use of tensor algebra in the study of neural networks algorithms such as the back-propagation algorithm and the intimate connections between tensor products and the Hebbian Rule.

Chapter 11

We refer to the monograph [54, Aubin & Frankowska] for an introductory presentation of set-valued analysis, and to the forthcoming book [299, Rockafellar & Wets] for a more comprehensive and recent exposition. We refer to [71, Beer] for an exhaustive exposition of topologies on the spaces of closed subsets and [155, Delfour & Zolesio] for an up-to-date view on this topic.

Chapter 12

One can find an extensive bibliography on differential inclusions and their applications to control and viability theory in the books [39, Aubin & Cellina] and [28, Aubin] for instance.

In a nutshell, the first independent investigations of differential inclusions began with the papers by Zaremba [354,355] and Marchaud [265,264], where the derivatives of the nondifferentiable solutions where taken in the contingent and paratingent sense respectively, following the definitions introduced at the beginning of the thirties by Bouligand. They were resurrected in the fifties and sixties by the Polish school around Ważewski [345] when he made the link between the Marchaud-Zaremba theory and optimal control problems. In the same period, and independently, Filippov, in [192,193,194,195], and the Russian school around him, developed a series of fundamental theorems on differential inclusions.

The original proof of Filippov's Theorem appeared in [194, Filippov]. It can be also found in [39, Aubin] for set-valued maps continuous with respect to the time.

The proof of the Viability Theorem we use is due to Hélène Frankowska (private communication).

The Convergence Theorem (or Mazur's Theorem) has been used by several authors studying differential inclusions with upper semicontinuous convex-valued right-hand sides with closed convex values.

The section on viability tubes is taken from [59, Aubin & Frankowska]. Actually, the concept of viability tubes goes back to Ważewski under the name of *tuyau*, in the framework of a fundamental method to determine the viability kernel of a viability tube by a topological approach.

The theorems on viable solutions to differential inclusions with memory are taken from the papers [225,226, Haddad] which provided the general criteria to article and to [179, Duluc & Vigneron] in which the characterization for differential inclusions with memory appeared.

The theorems on mutational equations come from [29,30, Aubin], [49, Aubin, Doyen & Mattioli], [170, Doyen] and [35, Aubin], to which we refer for further informations on this recent topic.

Chapter 13

We refer to the books of [302, Rockafellar], [19, Aubin], [231, Hiriart-Urruty J.-B. & Lemarechal] and the forthcoming book [299, Rockafellar & Wets] among many other books for an account of convex analysis. For a short introduction to convex analysis, one can also consult [19, Aubin].

There are many books devoted to the Walrasian tâtonnement methods, among them we only refer to [8, Arrow-Hahn], [151, Debreu] and [228, Hildenbrand & Kirman]. The purpose of this chapter was to link the tâtonnement and nontâtonnement algorithms with the gradient method applied to the constrained original optimal allocation problem, its dual problem (Walrasian tâtonnement) and its Hamiltonian formulation (Smale's nontâtonnement of [333, Smale]) and the derivation of its convergence properties from [136, Crandall & Pazy] and [94, Brézis] in a framework where the famous counterexample of [321, Scarf] is not operative. Monotone maps have been introduced by Zarantonello in 1960. For a comprehensive account of monotone maps, see for instance the book [94, Brézis]. Maximal monotone maps were characterized by Minty in 1965 and the fact that the subdifferential is maximal monotone is due to [301, Rockafellar]. The Theorem on nonlinear semigroups is due to Crandall-Pazy, [136]. See for instance [39, Aubin & Cellina] for further developments and bibliography.

Chapter 14

The literature on general equilibrium and nonlinear analysis is too broad to give any fair account in this short summary.

Ky Fan's Inequality has been proved in [188, Fan] and the proof we gave is derived from Browder in finite dimensional vector-spaces and from [12, Aubin] in the general case, where extensions are provided and further bibliographical comments can be found. The Equilibrium Theorem was derived from Ky Fan's papers [187,189, Fan] in [12, Aubin]. The "viability version" of the equilibrium theorem has a long history and goes back at least to one Halpern's paper. Browder and Ky Fan both played a major role in this history. See also recent contributions of [329, Simons] and [352, Yuan]. One can find in [50, Aubin & Ekeland] further developments on the existence of solutions to inclusions. For the nonconvex case, we refer to [288,289, Plaskacz] and [75, Ben El Merchaiek & Kryszewski] where one can find further references.

Ekeland's variational principle has been proved in [182, Ekeland] and used very often since then. For more recent views, see [50, Aubin & Ekeland], [184, Ekeland] and their bibliography.

The Inverse Function Theorem for surjective single-valued maps in Banach spaces is due to Graves in 1935. Ljusternik proved in 1934 that the tangent space to $f^{-1}(0)$ is the kernel of $f'(x)$ when the Fréchet derivative f' is continuous and surjective. See [158, Dontchev] and [162, Dontchev & Hager] for more recent contributions.

The extension of this theorem to problems with constraints and set-valued maps in finite dimensional spaces appeared in [251, Lebourg] and [50, Aubin & Ekeland]. The use of the localization property of Ekeland's Variational Principle is due to [251, Lebourg]. Actually, inverse-function theorems imply the pseudo-Lipschitzianty of the inverse, a property which happened to be very useful, and which has been extensively studied by Terry Rockafellar (see for instance [304, Rockafellar]). Its role in the convergence of the Newton Method for solving several classes of inclusions by Asen Dontchev (see [156, 157,160, Dontchev], [163, Dontchev & Rockafellar]).[164,165, Dordan]

This theorem has been extended to the case when the surjectivity of the first derivative is lacking, and also, when the set-valued map is defined on metric spaces, thanks to the concept of *high order variations* introduced in [202,203,207,205, Frankowska]. Such results were motivated by control theory, in particular by local controllability of control systems in [204,206, Frankowska]. See [207, Frankowska] and the forthcoming monograph [209,

Frankowska] for an exhaustive presentation of the theory and other applications.

Chapter 15

The results on stochastic viability are taken from [42,43,44,45, Aubin & Da Prato]. For stochastic replicator systems, see for instance [200, Foster & Young]. In [43, Aubin & Da Prato], the Nagumo Theorem was extended to stochastic differential equations with Lipschitz or monotone dynamics. The aim of [45, Aubin & Da Prato] is to combine two ways for representing uncertainty through stochastic differential inclusions: a "stochastic uncertainty", driven by a Wiener process, and a "contingent uncertainty", driven by a set-valued map. It is also devoted to viability of random set-valued variables by those stochastic differential inclusions, characterized in terms of stochastic tangent sets to random set-valued variables.

Bibliography

[1] ALIPRANTIS C., BROWN D. & BURKINSHAW O. (1989) EXITENCE AND OPTIMALITY OF COMPETITIVE EQUILIBRIA, Springer-Verlag

[2] ALLEN P.M. (1994) *Coherence, chaos and evolution in the social context*, Futures, 26, 583-597

[3] ALLEN P.M. (1994) *Evolutionary complex systems: models of technological change*, in CHAOS AND ECONOMIC THEORY, Leydersdoff L. & van den Besselaar, Eds. Pinter

[4] ALLEN P.M. (1996) *Viability and evolutionary complex systems*, In C. R. of Montpeller workshop on "systèmes viables"

[5] ALLEN P. M. & MC GLADE J.M. (1987) *Modelling complex human systems: a fisheries example*, European J. Operational Research, 30, 147-167

[6] ARTHUR W.B., ERMOLIEV Y.M. & KANIOWSKI Y.M. (1983) *The generalized urn problem and its application*, Kibernetika, 1, 49-56

[7] ARTHUR W.B., ERMOLIEV Y.M. & KANIOWSKI Y.M. (1987) *Path dependent processes and the emergence of macro-structure*, European J. Oper. Res., 30, 294-303

[8] ARROW K. & HAHN F. (1971) *General competitive analysis*, (Holden Day, San Francisco, CA)

[9] AUMANN R.J. (1964) *Markets with a continuum of traders*, Econometrica, 32, 39-50

[10] AUMANN R.J. (1965) *Integrals of set-valued maps*, J. Math. Anal. Appl., 12, 1-12

[11] AUMANN R.J. (1966) *Existence of competitive equilibria in markets with a continuum of traders*, Econometrica, 34, 1-17

[12] AUBIN J.-P. (1979) MATHEMATICAL METHODS OF GAME AND ECONOMIC THEORY, North-Holland (Studies in Mathematics and its applications, Vol. 7, 1-619)

[13] AUBIN J.-P. (1980) *Monotone Trajectories of Differential Inclusions : a Darwinian Approach*, Methods of Operations research, 37, 19-40

[14] AUBIN J.-P. (1981) *A dynamical, pure exchange economy with feedback pricing*, J. Economic Behavior and Organizations, 2, 95-127

[15] AUBIN J.-P. (1981) *Contingent derivatives of set-valued maps and existence of solutions to nonlinear inclusions and differential inclusions*, Advances in Mathematics, Supplementary studies, Ed. Nachbin L., 160-232

[16] AUBIN J.-P. (1981) *Cooperative fuzzy games*, Math. Op. Res., 6, 1-13

[17] AUBIN J.-P. (1981) *Locally lipchitz cooperative games*, J. Math. Economics, 8, 241-262

[18] AUBIN J.-P. (1982) MÉTHODES EXPLICITES DE L'OPTIMISATION, Dunod (Traduction Anglaise: Explicit Methods of Optimization, Dunod, 1985)

[19] AUBIN J.-P. (1983) L'ANALYSE NON LINÉAIRE ET SES MOTIVATIONS ÉCONOMIQUES, Masson (English version: OPTIMA AND EQUILIBRIA, (1993), Springer-Verlag)

[20] AUBIN J.-P. (1986) *A viability approach to Lyapunov's second method*, In DYNAMICAL SYSTEMS, Kurzhanski A. and Sigmund K., Lectures Notes in Economics and Math. Systems, Springer-Verlag, 287, 31-38

[21] AUBIN J.-P. (1987) *Smooth and Heavy Solutions to Control Problems*, in NONLINEAR AND CONVEX ANALYSIS, Eds. B-L. Lin & Simons S., Proceedings in honor of Ky Fan, Lecture Notes in pure and applied mathematics, June 24-26, 1985,

[22] AUBIN J.-P. (1988) *L'évolution des systèmes historiques vu à travers la théorie de la viabilité*, Stratégique, 1, 129-142

[23] AUBIN J.-P. (1989) *Smallest Lyapunov Functions of Differential Inclusions*, J. Differential and Integral Equations, 2,

[24] AUBIN J.-P. (1989) *Contingent Isaacs' equations of a differential game*, in DIFFERENTIAL GAMES AND APPLICATIONS, Eds. T. Başar & P. Bernhard, Lecture Notes in Control and Information Sciences, Springer-Verlag, 51-61

[25] AUBIN J.-P. (1989) *Contingent Isaacs' equations of a differential game*, in DIFFERENTIAL GAMES AND APPLICATIONS, Eds. T. Başar & P. Bernhard, Lecture Notes in Control and Information Sciences, Springer-Verlag, 51-61

[26] AUBIN J.-P. (1990) *Differential games: a viability approach*, SIAM J. on Control and Optimization, 28, 1294-1320

[27] AUBIN J.-P. (1990) *Fuzzy differential inclusions*, Problems of Control and Information Theory, 19, 55-67

[28] AUBIN J.-P. (1991) VIABILITY THEORY, Birkhäuser, Boston, Basel

[29] AUBIN J.-P. (1992) *A Note on Differential Calculus in Metric Spaces and its Applications to the Evolution of Tubes*, Bulletin of the Polish Academy of Sciences, 40, 151-162

[30] AUBIN J.-P. (1993) *Mutational Equations in Metric Spaces*, Set-Valued Analysis, 1, 3-46

[31] AUBIN J.-P. (1994) *Une métaphore mathématique de l'évolution économique : La "théorie de la viabilité" à Dauphine*, in PARIS-DAUPHINE, VINGT-CINQ DE SCIENCES D'ORGANISATION, C. Le Pen (Ed), Masson, 326-352

[32] AUBIN J.-P. (1994) INITIATION À L'ANALYSE APPLIQUÉE, Masson

[33] AUBIN J.-P. (1996) NEURAL NETWORKS AND QUALITATIVE PHYSICS: A VIABILITY APPROACH CAMBRIDGE, University Press, Cambridge

[34] AUBIN J.-P. (1996) *Une Métaphore Mathématique du princpe de précaution*, Natures, Sciences, Sociétés, 4, 146-154

[35] AUBIN J.-P. (to appear) MUTATIONAL AND MORPHOLOGICAL ANALYSIS: TOOLS FOR SHAPE REGULATION AND OPTIMIZATION, Birkhäuser, Boston, Basel

[36] AUBIN J.-P. (in preparation) LA MORT DE DEVIN, L'ÉMERGENCE DU DÉMIURGE

[37] AUBIN J.-P., BYRNES C. & SEUBE N. (in preparation) *Optimal solutions of control problems subjected to bounded chattering*

[38] AUBIN J.-P. & CARTELIER J. (1993) *Decentralized ecodynamics: an alternative view*, IIASA WP-93-45

[39] AUBIN J.-P. & CELLINA A. (1984) DIFFERENTIAL INCLUSIONS, Springer-Verlag, Grundlehren der math. Wiss.

[40] AUBIN J.-P. & CLARKE (1977) *Monotone invariant solutions to differential inclusions*, J. London Math. Soc., 16, 357-366

[41] AUBIN J.-P. & CLARKE (1979) *Shadow prices and duality for a class of optimal control problems*, SIAM J. Opt. and Control, 17, 567-586

[42] AUBIN J.-P. & DA PRATO G. (1990) *Stochastic Viability and Invariance*, Annali Scuola Normale di Pisa, 27, 595-694

[43] AUBIN J.-P. & DA PRATO G. (1995) *Stochastic Nagumo's Viability Theorem*, Stochastic Analysis and Applications, 13, 1-11

[44] AUBIN J.-P. & DA PRATO G. (1995) *Stochastic Lyapunov method*, NoDEA 2, 511-525

[45] AUBIN J.-P. & DA PRATO G. (1998) *The viability theorem for stochastic differential inclusions*, Stochastic Analysis and Applications, 16,

[46] AUBIN J.-P. & DAY R. H. (1980) *Homeostatic trajectories for a class of adaptive economic systems*, J. Economic Dyn. Control, 2, 185-203

[47] AUBIN J.-P. & DORDAN O. (1996) *Fuzzy Control Systems and Toll Sets*, In HANDBOOK OF FUZZY SYSTEMS, Hung Nguyen Ed.

[48] AUBIN J.-P. & DORDAN O. (in preparation) *Qualitative economies*

[49] AUBIN J.-P., DOYEN L. & MATTIOLI J. (to appear) *Mutations and contingent derivatives of set-valued maps: How they are related*,

[50] AUBIN J.-P. & EKELAND I. (1984) APPLIED NONLINEAR ANALYSIS, Wiley-Interscience

[51] AUBIN J.-P. & FORAY D. (to appear) *The Emergence of Network Organizations in Processes of Technological Choice: a Viability Approach*, IIASA WP 96

[52] AUBIN J.-P. & FRANKOWSKA H. (1984) *Trajectoires lourdes de systèmes contrôlés*, Comptes-Rendus de l'Académie des Sciences, PARIS,298, 521-524

[53] AUBIN J.-P. & FRANKOWSKA H. (1985) *Heavy viable trajectories of controlled systems*, Annales de l'Institut Heanri-Poincaré, Analyse Non Linéaire, 2, 371-395

[54] AUBIN J.-P. & FRANKOWSKA H. (1990) SET-VALUED ANALYSIS, Birkhäuser, Boston, Basel, Berlin

[55] AUBIN J.-P. & FRANKOWSKA H. (1990) *Inclusions aux dérivées partielles gouvernant des contrôles de rétroaction* , Comptes-Rendus de l'Académie des Sciences, Paris, 311, 851-856

[56] AUBIN J.-P. & FRANKOWSKA H. (1991) *Systèmes hyperboliques d'inclusions aux dérivées partielles*, Comptes-Rendus de l'Académie des Sciences, Paris, 312, 271-276

[57] AUBIN J.-P. & FRANKOWSKA H. (1995) *Partial differential inclusions governing feedback controls*, J. Convex Analysis, 2, 19-40

[58] AUBIN J.-P. & FRANKOWSKA H. (1996) *The Viability Kernel Algorithm for Computing Value Functions of Infinite Horizon Optimal Control Problems*, J.Math. Anal. Appl., 201, 555-576

[59] AUBIN J.-P. & FRANKOWSKA H. (à paraître) *Set-valued Solutions to the Cauchy Problem for Hyperbolic Systems of Partial Differential Inclusions*, NoDEA

[60] AUBIN J.-P., LOUIS-GUERIN C. & ZAVALLONI M. (1979) *Comptabilité entre conduites sociales réelles dans les groupes et les représentations symboliques de ces groupes : un essai de formalisation mathématique*, Math. Sci. Hum., 68, 27-61

[61] AUBIN J.-P. & NAJMAN L. (1994) *L'algorithme des montagnes russes pour l'optimisation globale*, Comptes-Rendus de l'Académie des Sciences, Paris, 319, 631-636

[62] AUBIN J.-P. & SIGMUND K. (1988) *Permanence and Viability*, J. of Computational and Applied Mathematics, 22, 203-209

[63] AXELROD R. (1984) THE ECOLUTION OF COOPERATION, Cambridge University Press

[64] BALASKO Y. (1988) FOUNDATIONS OF THE THEORY OF GENERAL EUILIBRIUM, Academic Press

[65] BARTH F. (1961) NOMADS OF SOUTH PERSIA, Oslo University Press

[66] BARTH F. (1981) PROCESS AND FORM IN SOCIAL LIFE, Routledge & Kegan Paul

[67] BASAR T. & BERNHARD P. (1991) H^∞-OPTIMAL CONTROL AND RELATED MINIMAX DESIGN PROBLEMS. A DYNAMIC GAME APPROACH, Birkhäuser

[68] BAUMOL W. (1970) ECONOMIC DYNAMICS, Macmillan

[69] BAUMOL W. (1970) ECONOMIC DYNAMICS, Macmillan

[70] BEBERNES J. W. & SCHUUR J. D. (1970) *The Ważewski topological method for contingent equations*, Annali di Mat. Pura ed Appli., 87, 271-280

[71] BEER G. (1993) TOPOLOGIES ON CLOSED AND CLOSED CONVEX SETS, Kluwer Academic Publisher

[72] BENASSY J.-P. (1975) *Neo-Keynesian disequilibrium theory in a monetary economy*, The review of Economic Studies, 42, 503-523

[73] BENE C, & DOYEN L. (1996) *Storage and viability of a fishery with resource and market dephases seasonalitiees*, Preprint

[74] BENETTI C. & CARTELIER J. (1980) MARCHANDS, SALARIAT ET CAPITALISTES, Maspero

[75] BEN EL MERCHAIEK H. & KRYSZEWSKI W. (1995) *Equilibres dans les ensembles non convexes* , C.R. Acad. Sci. Paris, 320, 573-576

[76] BENSOUSSAN A. & LIONS J.-L. (1978) APPLICATIONS DES INÉQUATIONS VARIATIONNELLES EN CONTRÔLE STOCHASTIQUE, Dunod.

[77] BENSOUSSAN A. & LIONS J.L (1982) CONTRÔLE IMPULSIONNEL ET INÉQUATIONS QUASI-VARIATIONNELLES, Dunod, Paris

[78] BENSOUSSAN A. (1982) STOCHATIC CONTROL BY FUNCTIONAL ANALYSIS METHODS, North-Holland

[79] BERNHARD P. (1976) COMMANDE OPTIMALE, DÉCENTRALISATION ET JEUX DYNAMIQUES, Dunod

[80] BERTSEKAS D. (1976) DYNAMIC PROGRAMMING AND STOCHASTIC CONTROL, Academic Press

[81] BERTSEKAS D. & TSITSIKLIS J. N. (1989) PARALLEL AND DISTRIBUTED COMPUTATION, Prentice Hall

[82] BONNEUIL N. (1990) *Contextual and structural factors in fertility behaviour*, Popu. Eng. Select, 2

[83] BONNEUIL N. (1994) *Malthus, Boserup and population viability*, Mathematical population studies, 4,

[84] BONNEUIL N. (1994) *Capital accumulation, inertia of consumption and norms of reproduction*, Journal of population economics, 7, 49-62

[85] BONNEUIL N. (to appear) *Viabilité chez Fredrik Barth*, Annales Economie et Société

[86] BONNEUIL N. & MULLERS K. (1997) *Viable populations in a predator-prey system*, J. Mathematical Biology

[87] BONNISSEAU & CORNET B. (1986) *Valuation equilibrium and Pareto optimum in nonconvex economies*, Center for Operations Research & Econometrics - Université Catholique de Louvain

[88] BONNISSEAU & CORNET B. (1986) *Fixed-point theorems and Morse's lemma for lipschitzian functions*, Center for Operations Research & Econometrics Université Catholique de Louvain

[89] BONNISSEAU & CORNET B. (to appear) *Existence of marginal cost pricing equilibria in an economy with several non convex firms*, Preprint Université Paris 1

[90] BOULIGAND G. (1932) INTRODUCTION À LA GÉOMÉTRIE INFINITÉSIMALE DIRECTE, Gauthier-Villars

[91] BOYER R. & ORLEAN A. (1992) *How do conventions evolve*, Journal of Evolutionary Economics, 2, 165-177

[92] BRESSAN A., CELLINA A. & COLOMBO G. (1988) *Upper semicontinuous differential inclusions without convexity*, SISSA

[93] BRESSAN A. (to appear) *Upper and lower semicontinuous differential inclusions: a unified approach*, Controllability & optimal control

[94] BREZIS H. (1973) OPÉRATEURS MAXIMAUX MONOTONES ET SEMI-GROUPES DE CONTRACTION DANS LES ESPACES DE HILBERT, North-Holland

[95] BROCK W.A. & MALLIARIS A.G. (1989) DIFFERENTIAL EQUATION, STABILITY AND CHAOS IN DYNAMIC ECONOMICS, North Holland

[96] BURGER E. (1963) INTRODUCTION TO THE THEORY OF GAMES,

[97] CANARD N.-F. (1802) PRINCIPES D'ÉCONOMIE POLITIQUE

[98] CARDALIAGUET P. (1994) Domaines discriminants en jeux différentiels, Thèse de l' Université de Paris-Dauphine

[99] CARDALIAGUET P. (1995) *A differential game with one target and two players*, Bull. Polish Acad. Sc., 43, 69-74

[100] CARDALIAGUET P. (à paraître) *Semipermeable surfaces, Isaacs equation and appliction to a differentiable game with two players and one target*, Applied Math. Opt.

[101] CARDALIAGUET P. & PLASKACZ S. (1996) *Viability and invariance for differential game with applications to Hamilton-Jacobi-Isaacs equations*, in TOPOLOGY IN NONLINEAR ANALYSIS, Banach Center Publications, Vol. 38, 149-158

[102] CARDALIAGUET P., QUINCAMPOIX M. & SAINT-PIERRE P. (1994) *Some algorithms for differential games with two players and one target*, MAM, 28, 441-461

[103] CARDALIAGUET P., QUINCAMPOIX M. & SAINT-PIERRE P. (1994) *Temps optimaux pour des problèmes avec contraintes et sans contrôlabilité locale* Comptes-Rendus de l'Académie des Sciences, Série 1, Paris, 318, 607-612

[104] CARDALIAGUET P., QUINCAMPOIX M. & SAINT-PIERRE P. (1995) *Contribution à l'étude des jeux différentiels quantitatifs et qualitatifs avec contrainte sur l'état*, Comptes-Rendus de l'Académie des Sciences, 321, 1543-1548

[105] (1995) CARDALIAGUET P., QUINCAMPOIX M. & SAINT-PIERRE P. (1995) *Numerical methods for optimal control and differential games*, Cahiers de Mathématiques de la Décision 9510

[106] CARLSON D.A., HAURIE A.B. & LEIZAROWITZ A. (1987) INFINITE HORIZON OPTIMAL CONTROL, Springer-Verlag

[107] CARMICHAEL H.L. & MACLEOD W.B. (1994) *Gift giving and the evolution of cooperation*,

[108] CARTELIER J. (1987) *Introduzione a un'economia politica eterodossa*, Metamorfosi, 5,

[109] CARTELIER J. (1996) LA MONNAIE, Flammarion

[110] CARTELIER J. & FROESCHLE C. (1989) *A model of a monetary economy: principles and study of stability*, Metroeconomica, 40, 189-209

[111] CARTELIER J. & MULLERS K. (1994) *An elementary Keynesian model: A preliminary approach*, IIASA WP 94-095

[112] CHIAPPORI P.A. & EKELAND I. (1996) *Problèmes d'aggrégation en théorie du consommateur et calcul différentiel extérieur*, C.R. Acad. Sci. Paris, 233, 565-570

[113] CHIAPPORI P.A. & EKELAND I. (to appear) *Exterior differential calculus and aggregation theory: a presentation and some new results*,

[114] CLARKE F.H. (1983) OPTIMIZATION AND NONSMOOTH ANALYSIS, Wiley-Interscience

[115] CHANDLER A.D. (1977) THE VISIBLE HAND: THE MANAGERIAL REVOLUTION IN AMERICAN BUSINESS, Belknap Press

[116] CHANDLER A.D. (1990) SCAPE AND SCOPE, Belknap Press

[117] CHIARAMONTE F. (in preparation) *Dealing with diversity among economic agents: can viability theory help us to understand?*

[118] CHERENE L.J. (1978) SET VALUED DYNAMICAL SYSTEMS AND ECONOMIC FLOW, Springer-Verlag

[119] CLARK C.W. (1976) MATHEMATIC BIO-ECONOMICS: THE OPTIMAL MANAGEMENT OF RENEWABLE RESOURCE, Wiley

[120] CLARKE F.H. (1983) OPTIMIZATION AND NONSMOOTH ANALYSIS, Wiley-Interscience

[121] CLARKE F., LEDYAEV Y., STERN R. & WOLENSKI P. (1995) *Qualitative properties of trajectories of control systems: a survey*, J. Dyanmical Control Sys.,1, 1-48

[122] CLÉMENT-PITIOT H. & DOYEN L. (to appear) *Modèle bio-economique et viabilité: le cas de la ressource halieutique*,

[123] CLÉMENT-PITIOT H. & DOYEN L. (1994) *Le comportement des acteurs sur le marché des produits de la pêche: une approche de viabilité*, Rapport CNRS.

[124] CLÉMENT-PITIOT H. & DOYEN L. (to appear) *Exchange rates dynamics, target zone and viability*

[125] CLOWER R. W. (1984) *A reconsideration of the microfoundations theory*, in MONEY AND MARKETS. Walker D. Ed., Cambridge University Press

[126] CORNET B. & HADDAD G. (1985) *Viability theorems for second order differential inclusions*, Israel J. Math.

[127] CORNET B. & LAROQUE G. (1986) *Lipchitz properties of solutions in mathematical programming*, J. Optim. Theory and Appl.

[128] CORNET B. (1976) *On planning procedures defined by multivalued differential equations*, Système dynamiques et modèles économiques (C.N.R.S)

[129] CORNET B. (1981) *Regular properties of tangent and normal cones*, Cahiers de Maths. de la Décision #8130

[130] CORNET B. (1983) *An existence theorem of slow solutions for a class of differential inclusions*,

[131] CORNET B. (1983) *Monotone dynamic processes in a pure exchange economy*,

[132] CORNET B. (1984) *Regular properties of open tangent cones*, Center for Operations Research & Econometrics - Université Catholique de Louvain

[133] CORNET B. (1984) *Existence of equilibria in economies with increasing returns*, Center for Operations Research & Econometrics - Université Catholique de Louvain

[134] CORNET B. (1986) *The second welfare theorem in nonconvex economies*, Preprint Université Paris 1

[135] CORNET B. (1987) *Topological properties of the set of attainable production plans in a nonconvex production economy*, Center for Operations Research & Econometrics

[136] CRANDALL M.G. & PAZY A. (1972) *Nonlinear evolution equations in Banach spaces*, Israel J. Math., 11, 57-94

[137] DALLE J.-M. (1995) *Dynamiques d'adoption, coordinaiton et diversité: la diffusion des standars technologiques*, Revue Economique, 46.

[138] DA PRATO G. & FRANKOWSKA H. (1994) *A Stochastic Filippov Theorem*, Stochastic Analysis and Applictions, 12(4), 409–426.

[139] DA PRATO G. & ZABCZYK J. (1992) STOCHASTIC EVOLUTION EQUATIONS IN INFINITE DIMENSIONS Cambridge University Press.

[140] D'AUTUME A. (1985) MONNAIE, CROISSANCE ET DÉSÉQUILIBRE, Economica

[141] DAVID P.A. & FORAY D. (1993) *Percolation structures, Markov random fields and the economics of EDI standards diffusion*, in GLOBAL TELECOMMUNICATION STATEGIES AND TECHNOLOGICAL CHANGES, Pogorel G. Ed., Elsevier

[142] DAVID P., FORAY D. & DALLE J.-M. (1996) *Marshallian externalities and the emergence and spatial stability of technological enclaves*, Economics of Innovation and New Technologies,

[143] DAY R.H. & WALTER J-L (1989) *Economic growth in the very long run*, in ECONOMICS COMPLEXITY : CHAOS, SUNSPOTS, BUBBLES AND NONLINEARITY, Cambridge University Press

[144] DAY R.H. (1984) *Disequilibrium economic dynamics: A post-Schumpeterian contribution*, JEBO, 5, 57-76

[145] DAY R.H. (1985) *Dynamical systems theory and complicated economic behavior*, Environment and Planning B: Planning and Design, 12, 55-64

[146] DAY R.H. (1987) *Economic development in the very long run - on the multiple-phase interaction of population, technology, and social infrastructure*, Preprint University of Southern California

[147] DAY R.H. (1994) COMPLEX ECONOMIC DYNAMICS, VOL. I, AN INTRODUCTION TO DYNAMICAL SYSTEMS AND MARKET MECHANIMS, MIT Press

[148] DAY R.H. (1995) *L'existence hors de l'équilibre*, Revue économique

[149] DAY R.H. (to appear) COMPLEX ECONOMIC DYNAMICS, VOL. II, AN INTRODUCTION TO MACROECONOMIC DYNAMICS, MIT Press

[150] DEBREU G. & SCARF H. (1963) *A limit theorem on the core of an economy*, International Economic review, 4, 235-246

[151] DEBREU G. (1959) THEORY OF VALUE, Wiley

[152] DEBREU G. (1974) *Excess demand functions*, J. Math. Econ., 1, 15-23

[153] DEBREU G. (1983) TWENTY PAPERS OF GÉRARD DEBREU, Cambridge University Press

[154] DEIMLING K. (1992) MULTIVALUED DIFFERENTIAL EQUATIONS, De Gruyter, Berlin

[155] DELFOUR M. & ZOLESIO J.-P. (1994) *Shape analysis via oriented distance functions*, Journal of Functional Analysis, 123, 129-201

[156] DONTCHEV A.L. (1994) *Characterizations of Lipschitz stability in optimization*,

[157] DONTCHEV A.L. (1996) *Local convergence of the Newton method for generalized equations*, CRAS, 322, 327-331

[158] DONTCHEV A.L. (1996) *The Graves Theorem revisited*, J. Convex Analysis, 3, 1-9

[159] DONTCHEV A.L. (to appear) *Uniform convergence of the Newton method for Aubin continuous maps*, Serdica Math. Journal

[160] DONTCHEV A.L. (to appear) *Local analysis of a Newton-type method based on partial linearization*, in MATHEMATICS OF NUMERICAL ANALYSIS: REAL NUMBER ALGORITHMS, Renegar J., Shub M. & Smale S. Eds.

[161] DONTCHEV A.L., HAGER W.W., POORE A.B.& YANG B. (1995) *Optimality, stability and convergence in nonlinear control*, Appl. Math. Opt., 31, 297-326

[162] DONTCHEV A.L. & HAGER W.W. (1994) *An inverse function theorem for set-valued maps*, Proc. Amer. Math. Soc., 121, 481-489

[163] DONTCHEV A.L. & ROCKAFELLAR R.T. (to appear) *Characterizations of strong regularity for variational inequalities over polyhedral sets*, SIAM J. Optimization

[164] DORDAN O. (to appear) *Mathematical problems arising in qualitative simulation of a differential equation*, Artificial Intelligence

[165] DORDAN O. (1995) PHYSIQUE QUALITATIVE, Masson

[166] DORDAN O. & SUPPO C. (1996) *Qualitative analysis applied to population dynamics*, In PROCEEDINGS OF HOUSTON CONFERENCE ON MATHEMATICAL POPULATION DYNAMICS

[167] DOSI G. (1988) *Sources, procedures and microeconomics effects of innovation*, J. Econ. Litt., 26, 1120-1171

[168] DOSI G. & NESLON R.R. (1993) *Evolutionary Theories in Economics: Assessment and Prospects*, IIASA WP 93-064

[169] DOYEN L. (1993) *Inverse function theorems and shape optimization*, SIAM Journal on Control and optimization

[170] DOYEN L. (1993) *Filippov and invariance theorems for mutational inclusions for tubes*, Set-Valued Analysis, 1, 289-303

[171] DOYEN L. (1994) *Lagrangian multipliers in Shape Optimization under shape constraints*, Proceedings of IFIP Workshop Boundary Control, Centre de Mathématiques Appliquées, Ecole des Mines, Sophia-Antipolis

[172] DOYEN L. (1994) *Shape Lyapunov functions*, J. Math. Anal. Appl.

[173] DOYEN L. & GABAY D. (to appear) *Economie des ressources renouvelables et viabilité*,

[174] DOYEN L., GABAY D. & HOURCADE J.-C. (to appear) *Risque climatique, technologie et viabilité*,

[175] DOYEN L., NAJMAN L. & MATTIOLI J. (submitted) *Mutational Equations of Morphological Dilation Tubes*, Journal of Mathematical Imaging and Vision

[176] DOYEN L. & QUINCAMPOIX M. (to appear) *Multi-targets control problems*, Cahiers du CEREMADE

[177] DOYEN L. & SAINT-PIERRE P. (1995) *Scale of viability and minimal time of crisis* Preprint 9513 Cahiers du CEREMADE,

[178] DOYEN L. & WEBER J. (1995) *Usage des richesses, théorie, modèle et viabilité*,

[179] DULUC R. & VIGNERON C. (1990) *Linear functional viability constraints*, Proceedings of the 9th International Conference on Analysis and Optimization of Systems, Nice, June 1990, Lecture Notes in Control and Information Sciences, Springer-Verlag,

[180] DUMENIL G. & LEVY D. (1987) *La concurrence classique à la croisée des chemins*, CEPREMAP

[181] EKELAND I & TEMAM R. (1976) CONVEX ANALYSIS AND VARIATIONAL PROBLEMS, North Holland, Amsterdam

[182] EKELAND I. (1974) *On the variational principle*, J. Mathematical Analysis and Applications, 47, 324-353

[183] EKELAND I, (1974) THÉORIE DES JEUX ET APPLICATIONS ÉCONOMIQUES, PUF

[184] EKELAND I. (1979) *Nonconvex minimization problems*, Bull. Amer. Math. Soc., 1, 443-474

[185] EKELAND I, (1979) ELÉMENTS D'ÉCONOMIE MATHÉMATIQUE, Hermann

[186] FALCONE M. & SAINT-PIERRE P. (1987) *Slow and quasi-slow solutions of differential inclusions*, J. Nonlinear Anal.,T.,M.,A., 3, 367-377

[187] FAN KY (1969) *Extension of two fixed-point theorems of F.E. Browder*, Math. Z., 112, 234-240

[188] FAN KY (1972) *A Minimax Inequality and Applications*, Inequalities III, 103-113, Shisha Ed., Academic Press

[189] FAN KY (1979) *Fixed point theorems and related theorems for non-compact convex sets*, In GAME THEORY AND RELATED TOPICS, 151-156, North-Holland

[190] FELDMAN A. M. (1973) *Bilateral trading processes, pairwise optimality and Pareto optimality*, Review of Economic Studies

[191] FENCHEL W. (1949) *On conjugate convex functions*, Mimeographed Lecture Notes, Princeton University

[192] FILIPPOV A.F. (1958) *On some problems of optimal control theory*, Vestnik Moskowskogo Universiteta, Math. no.2, 25-32

[193] FILIPPOV A.F. (1962) *On certain questions in the theory of optimal control*, SIAM J. Control, 1, 76-84

[194] FILIPPOV A.F. (1967) *Classical solutions of differential equations with multivalued right hand side*, SIAM J. on Control, 5, 609-621

[195] FILIPPOV A.F. (1988) DIFFERENTIAL EQUATIONS WITH DISCONTINUOUS RIGHTHAND SIDE, Kluwer Academic Publishers

[196] FISHER F. M. (1983) DISQUILIBRIUM FOUNDATIONS OF EQUILIBRIUM ECONOMICS, Cambridge University Press

[197] FLAM S. D. & ZOWE J. (to appear) *A primal-dual differential method for convex programming*, in OPTIMIZATION AND NONLINEAR ANALYSIS, Ioffe A., Marcus M. & Reich S. Eds. Longman

[198] FLAM S. D. (to appear) *Paths to constrained Nash equilibria*, Applied Math. Opt.

[199] FORAY D. & FREEMAN C. Eds (1992) TECHNOLOGY AND THE WEALTH OF NATIONS, Pinter

[200] FOSTER D. & YOUNG P.H. (1990) *Stochastic evolutionary game dynamics*, Theoretical Populaiton Biology, 38, 219-232

[201] FRANKE R. (1987) PRODUCTION PRICES AND DYNAMICAL PROCESSES OF THE GRAVITATION OF MARKET PRICES, Peter Lang

[202] FRANKOWSKA H. (1986) *Theorème d'application ouverte pour des correspondances*, Comptes-Rendus de l'Académie des Sciences, PARIS, Série 1, 302, 559-562

[203] FRANKOWSKA H. (1987) *Théorèmes d'application ouverte et de fonction inverse*, Comptes Rendus de l'Académie des Sciences, PARIS, Série 1, 305, 773-776

[204] FRANKOWSKA H. (1987) *Local controllability and infinitesimal generators of semigroups of set-valued maps*, SIAM J. on Control and Optimization, 25, 412-432

[205] FRANKOWSKA H. (1989) *High order inverse function theorems*, in ANALYSE NON LINÉAIRE, Attouch, Aubin, Clarke, Ekeland Eds., Gauthier-Villars & C.R.M. Université de Montréal, 283-304

[206] FRANKOWSKA H. (1989) *Local controllability of control systems with feeedbacks*, Journal of Optimization Theory and Applications, 60, 277-296

[207] FRANKOWSKA H. (1990) *Some inverse mapping theorems*, Ann. Inst. Henri Poincaré, Analyse Non Linéaire, 3, 183-234

[208] FRANKOWSKA H. (1993) *Lower semicontinuous solutions of Hamilton-Jacobi-Bellman equation*, SIAM J. on Control and Optimization,

[209] FRANKOWSKA H. (to appear) SET-VALUED ANALYSIS AND CONTROL THEORY, Systems and Control: Birkhäuser, Boston, Basel, Berlin

[210] FRANKOWSKA H. & DA PRATO G. (1994) *A stochastic Filippov theorem*, Stochastic Calculus

[211] FRANKOWSKA H. & QUINCAMPOIX M. (1990) *Un algorithme déterminant les noyaux de viabilité pour des inclusions différentielles avec contraintes*, Comptes-Rendus de l'Académie des Sciences, PARIS, Série 1, 312, 31-36

[212] FRANKOWSKA H. & QUINCAMPOIX M. (1991) *Viability kernels of differential inclusions with constraints*, Mathematics of Systems, Estimation and Control,

[213] FREEMAN C. (1982) THE ECONOMICS OF INDUSTRIAL INNOVATION, Pinter Publishers

[214] FUDENBERG D. & TIROLE J. (1991) GAME THEORY, MIT Press

[215] GABAY D. (1994) *Modeling the articulation between the economy and the environment*, in ENVIRONEMENT, ECONOMICS, AND THEIR MATHEMATICAL MODELS, J.-L. Diaz & J.-L. Lions Eds., 67-86, Masson.

[216] GAUTHIER D. (1986) MORALS BY AGREEMENT, Cambridge University Press

[217] GOMEZ G. L. (1992) DYNAMIC PROBABILISTIC MODELS AND SOCIAL STRUCTURES, Reidel

[218] GORNIEWICZ L. (1994) *Topological approach to differential inclusions*, in TOPOLOGICAL METHODS FOR DIFFERENTIAL EQUATIONS AND INCLUSIONS, Granas A. & Frigon Eds, Kluwer, 129-190

[219] GORRE A. (1996) *The "Montagnes Russes" algorithm in mutationnal spaces*, in PARAMETRIC OPTIMIZATION IV, Springer–Verlag.

[220] GORRE A. (à paraître) *Evolution of operable tubes under constraints* Journal of Mathematical Analysis and Applications

[221] GORRE A. (1996) *Evolution de Tubes Opérables gouvernée par des équations mutationnelles*, Thèse de l' Université de Paris-Dauphine

[222] GRAHAM M. (1935) *Modern theory of exploiting a fishery and applications to North-Sea trawling*, J. Com. Perm. Intern. Exploitation des Mers, 10, 264-274

[223] GUERRIEN B. (1989) CONCURRENCE, FLEXIBILITÉ ET STABILITÉ, Economica

[224] GUSEINOV H. G., SUBBOTIN A. I. & USHAKOV V. N. (1985) *Derivatives for multivalued mappings with applications to game theoretical problems of control*, Problems of Control and Information Theory, Vol.14, 155-167

[225] HADDAD G. (1981) *Monotone trajectories of differential inclusions with memory*, Isr. J. Math., 39, 83-100

[226] HADDAD G. (1981) *Monotone viable trajectories for functional differential inclusions*, J. Diff. Eq., 42, 1-24

[227] HADDAD G. (1981) *Topological properties of the set of solutions for functional differential differential inclusions*, Nonlinear Anal. Theory, Meth. Appl., 5, 1349-1366

[228] HILDENBRAND W. & KIRMAN A.P. (1988) EQUILIBRIUM ANALYSIS, North-Holland

[229] HILDENBRAND W. (1972) CORE AND EQUILIBRIA OF LARGE ECONOMIES, Princeton University Press

[230] HILL B.M., LANE D. & SUDDERTH W. (1980) *A strong law for some generalized urn processes*, The Annals of Probability, 8, 214-226

[231] HIRIART-URRUTY J.-B. & LEMARECHAL C. (1994) CONVEX ANALYSIS AND MINIMIZATION ALGORITHMS, Springer-Verlag (2 volumes)

[232] HOFBAUER J. & SIGMUND K. (1988) THE THEORY OF EVOLUTION AND DYNAMICAL SYSTEMS, Cambridge University Press, London Math. Soc. # 7

[233] HOURCADE J.-C. (1991) *Calcul économique et construction sociale des irréversibilités. Leçons de l'histoire énergétique récente.*, in FIGURES DE L'IRRÉVERSIBILITÉ EN ÉCONOMIE, Editions de l'EHESS

[234] HOURCADE J.C. (1993) *Analyse économique et gestion des risques climatiques*, Sciences Nature et Sociétés, Vol 2, DUNOD.

[235] HOURCADE J.C. & CHAPUIS T. (1995) *No regret potentiall: a viability approach to integrated assesment of climate policy*, Energetic politic, vol 23, n4-5

[236] IANNELLI M. (1995) MATHEMATICAL THEORY OF AGE-STRUCTURED POPULATION DYNAMICS, Giardini, Pisa

[237] INGRAO B. & ISRAEL G. (1987) LA MANO INVISIBILE, Laterza

[238] KALECKI (1971) SELECTED ESSAYS AND THE DYNAMICS OF THE CAPITALISTIC ECONOMY, Cambridge University Press

[239] ISAACS R. (1965) DIFFERENTIAL GAMES, Wiley

[240] KEYNES J. M. (1936) THE GENERAL THEORY OF EMPLOYMENT, INTEREST AND MONEY, in THE COLLECTED WRITINGS OF JOHN MAYNARD KEYNES, Vol. VII (MacMillan 1973)

[241] KIRMAN A (1991) *Epidemics of opinions and speculative bubbles in financial markets*, in MONEY AND FINANCIAL MARKETS, Ed. Taylor, Macmillan

[242] KRASOVSKI N. N. & SUBBOTIN A. I. (1974) POSITIONAL DIFFERENTIAL GAMES, Nauka, Moscow

[243] KRASOVSKI N. N. & SUBBOTIN A. I. (1988) GAME-THEORETICAL CONTROL PROBLEMS, Springer-Verlag

[244] KRASOVSKI N. N. (1986) THE CONTROL OF A DYNAMIC SYSTEM, Nauka, Moscow

[245] KRASOVSKII N.N. (1963) STABILITY OF MOTION, Stanford : Stanford University press

[246] KRIVAN V. (1995) *Individual behavior and population dynamics*, in LECTURE NOTES IN BIOMATHEMTICS AND BIOINFORMATICS'95, Candev M. Ed. DATECS Pub., 7-11

[247] KRIVAN V. (1996) *Demographic models for subdivided populations: The renewal equation approach*, Theoretical Population Biology, 70, 291-313

[248] KURATOWSKI K. (1958) TOPOLOGIE, VOLS. 1 AND 2 4TH. ED. CORRECTED, Panstowowe Wyd Nauk, Warszawa. (Academic Press, New York, 1966)

[249] KURZHANSKI A.B. (1986) *On the analytical properties of viability tubes of trajectories of differential systems*, Doklady Acad. Nauk SSSR, 287, 1047-1050

[250] LASOTA A. & MACKEY M.C. (1985) PROBABILISTIC PROPERTIES OF DETERMINISTIC SYSTEMS, Cambridge University Press

[251] LEBOURG G. (personal communication)

[252] LEIPONEN A. & MULLERS K. (1995) *Competences, bounded learning and innovation in a customer flow model*, IIASA WP-95

[253] LU HU (1993) *Analysis of a dynamic decentralized productive economy with monopolistic price-setting firms*, Cahiers de Mathématiques de la décision

[254] LYAPUNOV A.M. (1907) *Problème general de la stabilite du mouvement*, Ann. Fac. Toulouse, vol. 9, 203-474 ; also in : Annals of mathematics study, # 17; Princeton University Press, 1949.

[255] MACLEOD W.B. &MACOLMSON J. (1989) *Implicit contracts, incentive compatibility and involontary unemployment*, Econometrica, 57, 312-322

[256] MADERNER N. (1992) *Regulation of control systems under inequality constraints*, J. Math. Anal. Appl.,170, 591-599

[257] MADERNER N. (to appear) *Continuously regulated control systems*, J. Nonlinear Analysis, T.M.A.

[258] MADDEN P. *Why the Edgeworth process assumption isn't that bad*, Review of Economic Studies, 45, 279-283

[259] MAGILL M. & QUINZI M. (to appear) THEORY OF INCOMPLETE MARKETS, MIT Press

[260] MANKIW, N. G. & ROMER, D. (ed.) (1991) NEW KEYNESIAN ECONOMICS, MIT Press

[261] MANTEL (1976) *Homothetic preferences and community excess demand functions*, Journal of Economic Theory, 12, 197-201

[262] MAS-COLLEL A. (1985) THE THEORY OF GENERAL ECONOMIC EQUILIBRIUM: A DIFFERENTIAL APPROACH, Cambridge University Press

[263] MARSHALL A. (1920) PRINCIPLES OF ECONOMICS, Macmillan

[264] MARCHAUD H. (1934) *Sur les champs de demi-cônes et les équations différentielles du premier ordre*, Bull. Sc. Math., 62, 1-38

[265] MARCHAUD H. (1936) *Sur les champs continus de demi-cônes convexes et leurs intégrales*, Compos. Math. 1, 89-127

[266] MÉGIE G. (1992) *Changements globaux: incertitudes scientifiques*, séminaire du PRISTE-CNRS et du programme ECLAT du ministère de l'environnement.

[267] MICHAEL E. (1956) *Continuous selections I*, Annals of Math., 63, 361-381

[268] MICHAEL E. (1956) *Continuous selections II*, Annals of Math., 64, 562-580

[269] MICHAEL E. (1957) *Continuous selections III*, Annals of Math., 65, 375-390

[270] MONTORO L.M. & HERNANDEZ J.A.M. (to appear) *Higher order differential inclusions and viability*,

[271] MOULIN H. (1980) THÉORIE DES JEUX POUR L'ÉCONOMIE ET LA POLITIQUE, Hermann

[272] MOULIN H. (to appear)*Axiomatic cost sharing: Shapley-Shubik, Aumann-Shapley and beyond*,

[273] MULLERS K. (1994) *Cascades for dynamical games*, J. Math. Systems, Estimation and Control

[274] MULLERS K. (1994) *Regulation of control systems with oscillatory control by viability constraints*, Cahiers de Mathématiques de la Décision

[275] MURRAY J. (1990) BIOMATHEMATICS, Springer-Verlag

[276] NELSON R. & WINTER S. (1982) AN EVOLUTIONARY THEORY OF ECONOMIC CHANGE, The Belknap Press of Harvard University Press

[277] NEUMANN (von) J. & MORGENSTERN O. (1944) THEORY OF GAMES AND ECONOMIC BEHAVIOR,

[278] NEVEU J. (1964) BASES MATHÉMATIQUES DU CACUL DES PROBABILITÉES, Masson.

[279] NORDHAUS W.D. (1992) *How fast shall we graze the global commons?*, American Economic Review 72(2), pp. 242-246.

[280] ORLEAN A. (1992) *Contagion des opinions et fonctionnement des marchés financiers*, Revue Economique, 43, 685-697

[281] ORLEAN A. (1994) *Bayesian interactions and collective dynamics of opinion : herd behavior and mimetic contagion*, JEBO

[282] ORLEAN A. (1994) *Apprentissage collectif versus apprentissage individuel :un modèle de décentralisation par les règles*,

[283] PAVEL N. H. (1984) DIFFERENTIAL EQUATIONS, FLOW INVARIANCE ANS APLLICATIONS, Pitman Research Notes in Mathematics # 113

[284] PAZY A. (1983) *Semi-groups of linear operators and applications*, Springer Verlag, Berlin

[285] PEARCE D. (1990)*Evaluating the socio-economic impact of climate change: an introduction*, in Climate Change- Evaluating the Socio-Economic Impact, OECD, Paris.

[286] PEARL R.(1930) THE BIOLOGY OF POPULATION GROWTH, Knopf

[287] PIANKA E. R. (1978) EVOLUTIONARY ECOLOGY. Harper & Row Publishers

[288] PLASKACZ S. (1990) *Periodic solutions of differential inclusions on compact sets of* $\mathbf{R}^n$, J. Math. Anal. Appl., 148, 202-212

[289] PLASKACZ S. (1992) *On the solution set for differential inclusions*, Boll. U. M. I., 6 A, 387-384

[290] POLANYI K. (1968) THE GREAT TRANSFORMATION, Beacon Press

[291] QUINCAMPOIX M. (1990) *Frontières de domaines d'invariance et de viabilité pour des inclusions différentielles avec contraintes*, Comptes-Rendus de l'Académie des Sciences, Paris, 311, 411-416

[292] QUINCAMPOIX M. (1990) *Playable differentiable games*, J. Math. Anal. Appl., 194-211

[293] QUINCAMPOIX M. (1992) *Differential inclusions and target problems*, SIAM J. Control and Optimization, 30, 324-335

[294] QUINCAMPOIX M. (1992) *Enveloppes d'invariance pour des inclusions différentielles Lipschitziennes : applications aux problèmes de cibles*, Comptes-Rendus de l'Académie des Sciences, Paris, 314, 343-347

[295] QUINCAMPOIX M. (1993) *Contribution à l'étude des perturbations singulières pour les systèmes contrôlés et les inclusions différentielles*, Comptes-Rendus de l'Académie des Sciences, Paris, 316, 133-138

[296] QUINCAMPOIX M. (1996) *Contributions à l'étude de quelques problèmes en théorie du contrôle et de jeux différentiels* Habilitation à Diriger des Recherches de l' Université de Paris-Dauphine

[297] QUINCAMPOIX M. (à paraître) *Singular perturbations for control systems and for differential inclusions*,

[298] QUINCAMPOIX M. & SAINT-PIERRE (1995) *An algorithm for viability kernels in Hölderian case: Approximation by discrete viability kernels*, J. Math. Syst. Est. and Control, Summary : 115-120

[299] ROCKAFELLAR R. T. & WETS R. (to appear) VARIATIONAL ANALYSIS,

[300] ROCKAFELLAR R.T. (1967) MONOTONE PROCESSES OF CONVEX AND CONCAVE TYPE, Mem. of AMS # 77

[301] ROCKAFELLAR R.T. (1970) *On the maximal monotonicity of subdifferential mappings*, Pacific J. Math., 33, 209-216

[302] ROCKAFELLAR R.T. (1970) CONVEX ANALYSIS, Princeton University Press

[303] ROCKAFELLAR R.T. (1981) THE THEORY OF SUBGRADIENTS AND ITS APPLICATIONS TO PROBLEMS OF OPTIMIZATION CONVEX AND NONCONVEX FUNCTIONS , Helderman Verlag, W. Berlin 1981

[304] ROCKAFELLAR R.T. (1985) *Lipschitzian properties of multifunctions*, J. Nonlinear Anal. T.M.A., 9, 867-885

[305] SAARI D.G. & SIMON C. (1978) *Effective prIce mechanisms*, Econometrica, 46, 1097-1125

[306] SAARI D. & WILLIAMS S. (1986) *On the local convergence of economic mechanisms*, Journal of Economic Theory, 40, 152-167

[307] SAARI D.G. (1984) *The ultimate of chaos resulting from weighted voting systems*, Advances in Applied Mathematics, 5, 286-308

[308] SAARI D. (1985) *Iterative price mechanisms*, Econometrica, 53, 1117-1131

[309] SAARI D. G. (1986) *The source of some paradoxes from social choice and probability*, IIASA CP-8625

[310] SAARI D.G. (1987) *Some informational requirements for convergence*, J. on Complexity, 3, 302-311

[311] SAARI D. (1991) *Erratic behavior in economic models*, JEBO, 16, 3-35

[312] SAARI D.G. (1992) *The aggregate excess demand function and other aggregation procedures*, Economic Theory, 2, 359-388

[313] SAARI D.G. (1994) GEOMETRY OF VOTING, Springer-Verlag

[314] SAARI D.G. (1995) *Mathematical complexity of simple economics*, Notices of AMS

[315] SAARI D.G. (1995) *A chaotic exploration of aggregation paradoxes*, SIAM Review, 37, 37-52

[316] SAARI D. (to appear) *Connecting and resolving Sen's and Arrow's theorems*,

[317] SAARI D.G. (to appear) THE WAVERING INVISIBLE HAND, MIT Press

[318] SAINT-PIERRE P. (1990) *Approximation of slow solutions to differential inclusions*, Applied Mathematics & Optimisation

[319] SAINT-PIERRE P. (1994) *Approximation of the viability kernel*, Applied Mathematics & Optimisation, 29, 187-209

[320] SAINT-PIERRE P. (to appear) *Newton's method for set-valued maps*, Set-Valued Analysis

[321] SCARF H.E (1960) *Some examples of global instability of the competitive equilibrium*, International Economic Review, 1, 157-172

[322] SCARF H.E (1967) *The core of a N-person game*, Econometrica, 35, 50-69

[323] SCARF H.E (1973) *The computation of economic equilibria*, Yale University Press

[324] SHUMPETER J. (1934) THEORY OF ECONOMIC DEVELOPMENT, Harvard University Press

[325] SEUBE N. (to appear) *Robust stabilization of uncertain systems*, J. Math. Anal. Appl.

[326] SILVERBERG G., DOSI G. & ORSENIGO L. (1988) *Innovation, diversity and diffusion: a self-organization model*, The economic Journal, 98, 1032-1054

[327] SIMON H. (1976) MODELS OF MAN, Wikey

[328] SIMONS S. (1986) *Cyclical coincidences of multivalued maps*, J. Math. Soc. Japan, 38, 516-525

[329] SIMONS S. (1995) *Minimax Theorems and their proofs*, in MINIMAX AND APPLICATIONS, Du D.Z. & Pardalos P.M. Eds. Kluwe, 1-23

[330] SCHAEFFER M.B. (1954) *Some aspects of the dynamics of Populations*, Bull. Inter. Amer. Trop Tuna Comm. 1, 26-56.

[331] SHAPLEY L.S. (1972) *On balanced games without side-payments*, in MATHEMATICAL PROGRAMMING, T.C. HU and S. ROBINSON Eds., Academic Press, 261-290

[332] SIGMUND K. (1993) GAMES OF LIFE, Oxford University Press

[333] SMALE S. (1976) *Exchange processes with price adjustement*, Journal Mathematical Economics 3, 211-226

[334] SONNENSCHEIN H. (1972) *Market excess demand functions*, Econometrica, 40, 549-563

[335] SONNENSCHEIN H. (1973) *Do Walras' Identity and continuous characterize the class of community excess demand functions?*, Journal of Economic Theory, 6, 345-354

[336] SONNENSCHEIN H. (1973) *The utility hypothesis and market demand theory*, Western Economic Journal, 2, 404-410

[337] STACCHETTI (1985) *Analysis of a dynamic, decentralized exchange economy*, Journal of Mathematical Economics, 14, 241-259

[338] STARR R. (1969) *Quasi Equilibrium in markets with nonconvex preferences*, Econometrica, 37, 25-38

[339] STROOCK D.W. & VARADHAN S.R.S. (1979) *Multidimensional diffusion processes*, Springer-Verlag.

[340] TARAFDAR E. & YUAN XIAN-ZHI (to appear) *The set-valued dynamical system and its applications to Pareto optima*, Acta Applicandae Mathematicae

[341] VELIOV V. M. (to appear) *On the time-discretization of control systems*, SIAM. J. Control

[342] VELIOV V. M. (to appear) *Stability-like properties of differential inclusions*, preprint

[343] VERHULST P.F. (1838) *Notice sur la loi que la population suit dans son accroissement*, Corr. Math. et Phys., 10, 113-121

[344] WILEN J.E. *Bio-economics of renewable resource use*, in HANDBOOK OF NATURAL RESOURCE ET ENERGY ECONOMICS, vol I, Kneese et Sweeney.

[345] WAŻEWSKI T. (1990) SELECTED PAPERS, PWN (Polish Scientific Publishers)

[346] WEINTRAUB E. (1979) MICROFOUNDATION: THE COMPATIBILITY OF MICROECONOMICS AND MACROECONOMICS , Cambridge University Press

[347] YORKE J.A. (1967) *Invariance for ordinary differential équations*, Math. Systems Theory, 1, 353-372

[348] YORKE J.A. (1969) *Invariance for contingent equations*, Lecture Notes in Operations Research and Mathematical Economics, 12, Springer Verlag, Berlin, 379-381

[349] YORKE J.A. (1970) *Differential inequalities and non-lipschitz scalar functions*, Math. Systems Theory, 4, 140-153

[350] YORKE J.A. (1971) *Another proof of the Liapounov convexity theorem*, SIAM J. Control Optim. 9, 351-353

[351] YORKE J.A. (1972) *The maximum principle and controllability of nonlinear differential equations*, SIAM J. of Control and Optimization, 10, 334-338

[352] YUAN XIAN-ZHI G. (to appear) NONLINEAR ANALYSIS AND APPLICATIONS ABSTRACT ECONOMIES, VARIATIONAL INEQUALITIES AND GAME THEORY, Marcel Dekker

[353] ZADEH L.A. (1965) *Fuzzy sets*, Information and Control, 8, 338-353

[354] ZAREMBA S.C. (1934) *Sur une extension de la notion d'équation diférentielle*, Comptes Rendus Acad. Sciences 199, 545-548

[355] ZAREMBA S.C. (1936) *Sur les équations au paratingent*, Bull. Sc. Math., 60, 139-160

[356] ZHAO J. (1992) *The hybrid solutions of an N-person game*, Game and Economic behavior, 4, 145-160

[357] ZHANG WEI-BIN (1996) *Knowledge and value: economic structures with time and space*, Umea University Press

Index

MIX
Papier aus verantwortungsvollen Quellen
Paper from responsible sources
FSC® C105338

If you have any concerns about our products,
you can contact us on
ProductSafety@springernature.com

In case Publisher is established outside the EU,
the EU authorized representative is:
Springer Nature Customer Service Center GmbH
Europaplatz 3, 69115 Heidelberg, Germany

Printed by Libri Plureos GmbH
in Hamburg, Germany